*U*NDERSTANDING *O*UR *E*NVIRONMENT:

An Introduction

UNDERSTANDING OUR ENVIRONMENT:

An Introduction

William P. Cunningham
University of Minnesota

WCB Wm. C. Brown Publishers
Dubuque, Iowa • Melbourne, Australia • Oxford, England

Book Team

Editor *Margaret J. Kemp*
Developmental Editor *Robin Steffek*
Production Editor *Rachel K. Daack*
Designer *LuAnn Schrandt*
Art Editor *Joseph P. O'Connell*
Photo Editor *Rose Deluhery*
Permissions Coordinator *Gail I. Wheatley*

Wm. C. Brown Publishers
A Division of Wm. C. Brown Communications, Inc.

Vice President and General Manager *Beverly Kolz*
Vice President, Publisher *Kevin Kane*
Vice President, Director of Sales and Marketing *Virginia S. Moffat*
National Sales Manager *Douglas J. DiNardo*
Marketing Manager *Patrick Reidy*
Advertising Manager *Janelle Keeffer*
Director of Production *Colleen A. Yonda*
Publishing Services Manager *Karen J. Slaght*
Permissions/Records Manager *Connie Allendorf*

Wm. C. Brown Communications, Inc.

President and Chief Executive Officer *G. Franklin Lewis*
Corporate Senior Vice President, President of WCB Manufacturing *Roger Meyer*
Corporate Senior Vice President and Chief Financial Officer *Robert Chesterman*

Cover © Martha Cooper/Peter Arnold, Inc.

Copyedited by Julie Bach

The credits section for this book begins on page 372 and is considered an extension of the copyright page.

For Mary with love and gratitude

BRIEF CONTENTS

Preface xiii

Part I:
Humans and Nature: An Overview 1

Chapter 1:
An Introduction to Environmental Science 2

Chapter 2:
Matter, Energy and Life 21

Chapter 3:
Biological Communities and Biomes 42

Chapter 4:
Human Populations 64

Part II:
Basic Concepts 90

Chapter 5:
Environmental Resource Economics 91

Chapter 6:
Environmental Health and Toxicology 112

Part III:
Earth and Biological Resources 133

Chapter 7:
Food and Agriculture 134

Chapter 8:
Forests, Rangelands, Parks, and Wilderness 160

Chapter 9:
Biological Resources 183

Chapter 10:
Air Resources 205

Chapter 11:
Water Resources 227

Chapter 12:
Energy Resources 248

Part IV:
Toward a Sustainable Future 275

Chapter 13:
Solid, Toxic, and Hazardous Waste 276

Chapter 14:
Urbanization and Sustainable Cities 298

Chapter 15:
Toward a Sustainable Future 319

Appendix A: World Map 346
Appendix B: Units of Measurement Metric/English Conversion 348
Appendix C: Environment Opportunities for Students 349
Appendix D: Environment Organizations 352
Glossary 355
Credits 372
Index 375

DETAILED TABLE OF CONTENTS

Preface *xiii*
Acknowledgments *xvi*

Part 1:
Humans and Nature: An Overview 1

Chapter 1:
An Introduction to Environmental Science 2
Objectives 2
Introduction 2
A Marvelous Planet 2
  Environmental Science 3
  Environmental Dilemmas 4
A Brief History of Resource Use, Technology, and Development 6
  Tool Making Revolution 6
  Agricultural Revolution 6
  Industrial Revolution 7
Attitudes toward Nature 9
  Stewardship 9
  Domination 9
  Consumerism 10
A Brief History of Conservation and Environmentalism 10
  Conservation or Preservation 11
  Environmentalism 11
  Global Citizenship 12
North versus South: A Divided World 12
  Rich and Poor Countries 14
  Political Economies 15
  Development 16
Environmental Futures 16
  Pessimism 16
  Optimism 16
  Communal Cooperation and Social Justice 17
  Is There Hope? 17
Summary 18
Review Questions 19
Exercises in Critical or Reflective Thinking 19
Key Terms 20
Suggested Readings 20

Chapter 2:
Matter, Energy, and Life 21
Objectives 21
Introduction 21

From Atoms to Cells 22
  Atoms, Molecules, and Compounds 22
  Organic Chemicals 23
  Cells, the Fundamental Units of Life 23
Energy and Matter 24
  Energy Types and Quality 24
  Conservation of Matter 24
  Thermodynamics and Energy Transfers 25
Energy for Life 25
  Solar Energy: Warmth and Light 25
  How Does Photosynthesis Capture Energy? 26
From Species to Ecosystem 27
  Populations, Communities, and Ecosystems 27
  Food Chains, Food Webs, and Trophic Levels 29
  Ecological Pyramids 31
Material Cycles and Life Processes 31
  The Water Cycle 32
  The Carbon Cycle 35
  The Nitrogen Cycle 36
  Phosphorus Cycles 38
Summary 39
Review Questions 40
Questions for Critical or Reflective Thinking 40
Key Terms 41
Suggested Readings 41

Chapter 3:
Biological Communities and Biomes 42
Objectives 42
Introduction 42
Some Properties of Communities 42
  Productivity 42
  Diversity 43
  Complexity 43
  Structure 45
Who Lives Where, and Why? 48
  Limiting Factors and Tolerance Limits 48
  Natural Selection and Adaptation 48
  The Ecological Niche 49
Species Interactions and Community Dynamics 51
  Predation 51
  Competition 52
  Symbiosis 53
Communities in Transition 55
  Ecological Succession 55
  Introduced Species and Community Change 57
Biomes 57
  Deserts 57
  Grasslands 58
  Forests 59
  Aquatic Biomes 60

Summary 60
Review Questions 61
Questions for Critical or Reflective
  Thinking 62
Key Terms 62
Suggested Readings 63

Chapter 4:
Human Populations 64
Objectives 64
Introduction 64
Population Biology 65
  Biotic Potential 65
  Catastrophic Declines and Population
    Oscillations 65
  Growth to a Stable Population 65
  Mortality and Survivorship 66
  Density-dependent and Density-independent
    Factors 66
Human Population History 66
Limits to Growth: Some Opposing
  Views 69
  Malthusian Checks on Population 69
  Malthus and Marx Today 70
  Can Technology Make the World More
    Habitable? 70
  Can More People Be Beneficial? 71
Human Demography 71
  How Many of Us Are There? 71
  Fertility and Birth Rates 71
  Mortality and Death Rates 72
  Population Growth Rates 73
  Life Span and Life Expectancy 73
  Living Longer: Demographic Implications 74
  Emigration and Immigration 76
Population Growth: Opposing Factors 77
  Pronatalist Pressures 77
  Birth Reduction Pressures 78
Birth Dearth? 79
Demographic Transition 80
  Development and Population 80
  An Optimistic View 80
  A Pessimistic View 81
  A Social Justice View 81
  Ecojustice 81
  Infant and Child Mortality 81
Family Planning and Fertility Control 82
  Traditional Fertility Control 82
  Current Birth Control Methods 82
  New Developments in Birth Control 83
The Future of Human Populations 86
Summary 86
Review Questions 87
Questions for Critical or Reflective
  Thinking 87
Key Terms 88
Suggested Readings 88

Part II:
Basic Concepts 90
Chapter 5:
Environmental Resource Economics 91
Objectives 91
Introduction 91
Economic Context 91
Resources and Reserves 93
  Defining Resources 93
  Economic Categories 94
Economic Development and Resource
  Use 95
  Frontier Economy 95
  Industrial Economy 95
  Postindustrial Economy 96
Population, Technology, and Resource
  Scarcity 96
  Supply, Price, and Demand Relationships 96
  Market Efficiencies and Technological
    Development 97
  Population Effects 97
  Factors that Mitigate Scarcity 98
  Increasing Environmental Carrying Capacity 98
Limits to Growth 99
  Computer Models of Resource Use 99
  Why Not Conserve Resources? 100
Resource Economics 101
  Internal and External Costs 101
  Intergenerational Justice and Discount
    Rates 101
  Cost/Benefit Ratios 103
  Distributing Intangible Assets 104
  International Development 104
  International Trade 106
  Jobs and the Environment 106
Sustainable Development:
  The Challenge 107
Summary 108
Review Questions 109
Questions for Critical or Reflective
  Thinking 109
Key Terms 109
Suggested Readings 110
Chapter 6:
Environmental Health and Toxicology 112
Objectives 112
Introduction 112
Types of Environmental Health
  Hazards 112
  Infectious Organisms 113
  Chemicals 116
  Natural and Synthetic Toxins 119
  Physical Agents, Trauma, and Stress 119
  Diet 120

Movement, Distribution, and Fate of
    Toxins 122
    Solubility 122
    Bioaccumulation and Biomagnification 123
    Persistence 123
Mechanisms for Minimizing Toxic
    Effects 123
    Metabolic Degradation and Excretion 124
    Repair Mechanisms 124
Measuring Toxicity 124
    Animal Testing 124
    Toxicity Ratings 125
    Acute versus Chronic Doses and Effects 126
    Detection Limits 126
Risk Assessment and Acceptance 127
    Assessing Risk 127
    Accepting Risk 127
Establishing Public Policy 128
Summary 130
Review Questions 131
Questions for Critical or Reflective
    Thinking 131
Key Terms 131
Suggested Readings 132

Part III:

**The Earth and Biological
Resources   133**

Chapter 7:
Food and Agriculture   134
Objectives 134
Introduction 134
World Hunger 135
    Famines 135
    Chronic Food Shortages 136
    Nutritional Requirements 137
    Energy Needs 137
    Nutritional Needs 138
World Food Resources 139
    Major Crops 139
    Meat and Milk 139
    Fish and Seafood 140
    Increases in World Food Production 140
    International Food Trade 141
Soil Resources 142
    Soil, A Renewable Resource 142
    Soil Organisms 143
    Soil Profiles 143
    Erosion and Land Degradation 144
    Erosion in the United States 145
    Erosion in Other Countries 146
Other Agricultural Resources 147
    Water 147
    Fertilizer 147
    Energy 148
How Many People Can the World
    Feed? 148
    Bringing New Land into Cultivation 149
    The Green Revolution 150

Toward a Just and Sustainable
    Agriculture   150
    Soil Conservation 152
    Using Reduced Tillage Systems 153
    Integrated Pest Management 153
Summary 157
Review Questions 157
Questions for Critical or Reflective
    Thinking 158
Key Terms 158
Suggested Readings 158

Chapter 8:
*Forests, Rangelands, Parks,
    and Wilderness   160*
Objectives 160
Introduction 160
World Land Area Characteristics 161
World Forests 161
    Forest Distribution 161
    Forest Products 162
    Forest Management 163
Tropical Forests 164
    The Diminishing Forests 164
    A Cycle of Destruction 164
    Forest Protection 167
    Debt-for-Nature Swaps 168
Northern Forests 168
    Ancient Forests 168
    Wilderness and Wildlife Protection 169
    Harvest Methods 169
Rangelands 170
    Range Management 170
    Overgrazing and Desertification 170
    Forage Conversion by Domestic Animals 171
    Harvesting Wild Animals 171
Rangelands in the United States 173
    State of the Range 173
National Parks 174
    The Origin of American Parks 174
    The National Park System Today 174
    Current Problems 174
Wilderness Areas 175
World Parks and Preserves 176
    Conservation and Economic Development 177
Land Ownership and Land Reform 178
    Who Owns How Much? 178
    Land Reform 178
    Resettlement Projects 179
Summary 180
Review Questions 180
Questions for Critical or Reflective
    Thinking 180
Key Terms 181
Suggested Readings 181

Chapter 9:
Biological Resources   183
Objectives 183
Introduction 183
Biological Resources 184
    How Many Species Are There? 184

Biological Abundance 184
Benefits from Biological Resources 185
  Food 185
  Industrial and Commercial Products 185
  Medicine 185
  Ecological Benefits 186
  Aesthetic and Cultural Benefits 186
Destruction of Biological Resources 188
  Kinds of Losses 188
  Natural Causes of Extinction 189
  Mass Extinctions 189
  Human-Caused Extinction 189
  Current Extinction Rates 189
Ways Humans Cause Biological
  Losses 190
  Direct Impacts 190
  Indirect Damage 193
Biological Resources Management 197
  Hunting and Fishing Laws 197
  The Endangered Species Act 197
  Recovery Programs 198
Convention on International Trade in
  Endangered Species 198
  Habitat Protection 199
  Ecosystem Management 199
  Size and Location of Nature
    Preserves 199
  Social and Economic Factors 201
Summary 202
Review Questions 202
Questions for Critical or Reflective
  Thinking 202
Key Terms 203
Suggested Readings 203

Chapter 10:
Air Resources 205
Objectives 205
Introduction 205
Atmosphere and Climate 205
  Solar Energy Warms the Earth 205
Types and Sources of Air Pollution 208
  Primary, Secondary, and Fugitive
    Emissions 208
  Conventional or "Criteria" Pollutants 208
  Unconventional or "Noncriteria" Pollutants 213
  Indoor Air Pollution 213
Effects of Air Pollution 215
  Human Health 215
  Plant Pathology 216
  Acid Deposition 217
  Forest Damage 218
  Damage to Buildings and
    Monuments 219
Air Pollution Control 219
  Particulate Removal Techniques 220
  Sulfur Removal 220
  Nitrogen Oxide Control 221
  Hydrocarbon Emission Controls 221
Current Conditions and Prospects for the
  Future 222
Summary 223

Review Questions 224
Questions for Critical or Reflective
  Thinking 224
Key Terms 225
Suggested Readings 225
Chapter 11:
Water Resources 227
Objectives 227
Introduction 227
Water Resources 227
  Major Water Compartments 228
  Replenishing Freshwater Supplies 228
Water Availability and Use 229
  Regional Water Supplies 229
  Seasonal Variability 230
  Types of Water Use 230
  Use by Sector 230
Fresh Water: A Scarce Resource 232
  Water Shortages 232
  Depleting Groundwater Supplies 232
  Subsidence and Saltwater Infiltration 234
Increasing Supplies: Proposals and
  Projects 234
  Desalination 234
  Water Transfer Projects 234
  Dam Difficulties 236
Water Conservation 237
  Industrial and Agricultural Conservation 238
  Price Mechanisms 238
Water Pollution 238
  What Is Water Pollution? 238
  Ocean Pollution 239
  Waterborne Diseases 239
  Oxygen-demanding Wastes 239
  Toxic, Inorganic Water Pollutants 240
  Salts, Acids, and Other Nonmetallic
    Pollutants 240
  Toxic Organic Chemicals 241
Current Water Quality Conditions 242
  Areas of Progress 242
  Remaining Problems 242
  Surface Waters in Other Countries 242
  Groundwater 244
Water Legislation 245
Summary 245
Review Questions 246
Questions for Critical or Reflective
  Thinking 246
Key Terms 247
Suggested Readings 247
Chapter 12:
Energy Resources 248
Objectives 248
Introduction 248
  Major Energy Sources 248
  Per Capita Consumption 249
  How Energy Is Used 250
Coal 251
  Coal Resources and Reserves 251
  Mining 251

Oil 252
  Oil Resources and Reserves 252
  Unconventional Oil 253
Natural Gas 253
  Natural Gas Reserves 253
  Unconventional Gas Sources 253
Nuclear Power 255
  Reactor Fuel 255
  Kinds of Reactors in Use 256
  Alternative Reactor Designs 256
  Breeder Reactors 257
Radioactive Waste Management 260
  Citizen Protests Against Nuclear Power 261
Energy Conservation 262
  Saving Energy 262
  Utility Conservation Programs 263
Solar Energy 263
  Passive Solar Heat Collectors 263
  Active Solar Heat Systems 263
  High-Temperature Solar Energy Collection 264
  Photovoltaic Solar Energy Conversion 265
  Storing Electrical Energy 267
Biomass 268
  Methane from Biomass 268
  Alcohol from Biomass 269
Hydropower 269
Wind Energy 270
Summary 272
Review Questions 272
Questions for Critical or Reflective Thinking 273
Key Terms 273
Suggested Readings 273

Part IV:
Toward a Sustainable Future 275

Chapter 13:
Solid, Toxic, and Hazardous Waste 276
Objectives 276
Introduction 276
Solid Waste 276
  The Waste Stream 277
  Waste Disposal Methods 278
  Exporting Waste 280
  Incineration and Resource Recovery 280
  Recycling 282
  Composting 284
  Energy from Waste 284
  Shrinking the Waste Stream 285
  Producing Less Waste 286
Hazardous and Toxic Wastes 286
  What Is Hazardous Waste? 287
  Hazardous Waste Disposal 288
  Waste Lagoons and Injection Wells 289
  Warehousing and Illegal Dumping 289
  Options for Hazardous Waste Management 290
Summary 295
Review Questions 295

Questions for Critical or Reflective Thinking 296
Key Terms 296
Suggested Readings 296
Chapter 14:
Urbanization and Sustainable Cities 298
Objectives 298
Introduction 298
Urbanization 298
  What Is a City? 299
  World Urbanization 299
Causes of Urban Growth 302
  Immigration Push Factors 302
  Immigration Pull Factors 302
  Government Policies 302
Current Urban Problems 303
  The Developing World 303
  The Developed World 307
Transportation and City Growth 307
City Planning 310
  Garden Cities and New Towns 311
  Cities of the Future 311
  Suburban Redesign 313
Sustainable Development in the Third World 315
Summary 316
Review Questions 316
Questions for Critical or Reflective Thinking 317
Key Terms 317
Suggested Readings 317
Chapter 15:
Toward a Sustainable Future 319
Objectives 319
Introduction 319
Environmental Ethics and Philosophies 319
  Rights, Duties, and Obligations 320
  Stewardship and Relationships 320
  Modernism and Technology 320
  Deep, Shallow, Social, or Progressive? 322
Green Consumerism 322
  How Much Is Enough? 322
  Shopping for "Green" Products 323
  Blue Angels and Green Seals 326
  Limits of Green Consumerism 326
  Paying Attention to What's Important 327
Collective Actions 327
  Student Environmental Groups 328
  Mainline Environmental Organizations 331
  Broadening the Environmental Agenda 331
  Radical Environmental Groups 331
  Anti-environmental Backlash 333
Global Issues 333
  Sustainable Development 334
  Achieving Our Goals 334
  International Nongovernmental Organizations 335
Green Government and Politics 336
  Green Politics 336
  National Legislation 336

The Courts 337
The Executive Branch 339
Environmental Impact Statements 339
International Environmental Treaties and
    Conventions 339
Summary 341
Review Questions 342
Questions for Critical or Reflective
    Thinking 342
Key Terms 343
Suggested Readings 343

Appendix A: World Map 346
Appendix B: Units of Measurement Metric/English
    Conversion 348
Appendix C: Environmental Opportunities for
    Students 349
Appendix D: Environmental Organizations 352
Glossary 355
Credits 372
Index 375

LIST OF BOXES

▼

Box 1.1
Acting Locally: Thinking About Thinking   4
Box 1.2
Understanding Principles and Issues: Science as a Way of Knowing   7
Box 1.3
Thinking Globally: The Earth Summit   13
Box 1.4
Acting Locally: What's Your Environmental Perspective?   18

Box 2.1
Thinking Globally: Is the Earth a Superorganism?   28
Box 2.2
Understanding Principles and Issues: A "Water Planet"   33

Box 3.1
Thinking Globally: Tropical Rainforests: Life in Layers   46
Box 3.2
Understanding Principles and Issues: How the Camel Got Its Hump   50
Box 3.3
Understanding Principles and Issues: Partners for Life   54

Box 4.1
Understanding Principles and Issues: Wolves and Moose on Isle Royale: A Case Study   67
Box 4.2
Thinking Globally: "Now Because of My Large Family, I Am a Rich Man"   73
Box 4.3
Thinking Globally: China's One-Child Family Program   76
Box 4.4
Acting Locally: Preparing a Personal Fertility Plan   84

Box 5.1
Understanding Principles and Issues: Market-based Incentives for Environmental Protection   101
Box 5.2
Thinking Globally: Microlending at the Grammeen Bank   105

Box 6.1
Thinking Globally: The Child Survival Revolution   113
Box 6.2
Acting Locally: Electromagnetic Fields and Your Health   119

Box 7.1
Thinking Globally: Feeding a Fifth of the World   136
Box 7.2
Understanding Principles and Issues: Sustainable Agriculture in America   151
Box 7.3
Understanding Principles and Issues: Focus on Pesticides   154
Box 7.4
Acting Locally: What Can We Do Personally?   156

Box 8.1
Thinking Globally: Murder in the Rainforest   165
Box 8.2
Thinking Globally: Restoring a Dry Tropical Forest   167
Box 8.3
Understanding Principles and Issues: Buffalo Commons   172

Box 9.1
Acting Locally: Ecotourism   187
Box 9.2
Understanding Principles and Issues: Alien Invaders in the Great Lakes   195

Box 10.1
*Thinking Globally: The Greenhouse Effect
and Global Climate Change* 207
Box 10.2
*Thinking Globally: A Hole in the Ozone
Shield* 211
Box 10.3
Acting Locally: Radon in Indoor Air 214

Box 11.1
*Understanding Principles and Issues: The
Ogallala Aquifer: An Endangered
Resource* 233
Box 11.2
*Thinking Globally: The Aral Sea Is
Dying* 235
Box 11.3
*Acting Locally: You Can Make
a Difference* 237
Box 11.4
Thinking Globally: Minamata Disease 241

Box 12.1
*Thinking Globally: Chernobyl: The Worst
Possible Accident?* 258
Box 12.2
*Acting Locally: Personal Energy
Efficiency: What Can You Do?* 264

Box 13.1
*Acting Locally: Building a Home
Compost Pile* 284

Box 13.2
*Understanding Principles and Issues:
The Forgotten Wastes of Love Canal* 290
Box 13.3
*Acting Locally: What to Do with
Household Hazardous Wastes* 292
Box 13.4
*Thinking Globally: Toxic Waste
Management in Denmark* 294

Box 14.1
*Understanding Principles and Issues:
Crowding, Stress, and Crime in Cities* 304
Box 14.2
Acting Locally: Noise 308
Box 14.3
*Thinking Globally: Tapiola and
Farsta* 312

Box 15.1
*Thinking Globally: Restoration of the
Bermuda Cahow* 321
Box 15.2
*Thinking Globally: Curitiba, An
Environmental Showcase* 324
Box 15.3
*Acting Locally: Green Business and
Environmental Jobs* 325
Box 15.4
*Understanding Principles and Issues:
Cleaning Up the Nashua River* 329

PREFACE

▼

Environmental science is a dynamic, rapidly changing field of vital importance to each of us. The decisions we make now about resource use, waste disposal, population growth, and pollution control will have tremendous impacts on our lives and those of future generations. We live in an increasingly interconnected global village with worldwide communications and transportation systems that make us aware of day-to-day conditions in places our parents never heard of. What happens on the other side of the planet has profound effects on us; we, in turn, have impacts on people of whom we may be totally unaware. Problems such as global climate change, destruction of stratospheric ozone, ocean pollution, loss of biodiversity, and famine and war in developing countries are discussed every day in newspapers and on radio and television. Although many problems clamor for our attention, I hope that readers also will come to appreciate the beauty, complexity, and value of both the natural world and the cultural and technological environments that humans have built.

Among the central environmental concerns addressed in this book are (1) How does the natural world work? (2) What have we done and what are we doing now to our environment both for good and for ill? and (3) What can we do to ensure a sustainable future for ourselves, other species, and future generations? As citizens of a global community, we depend on a foundation of scientific principles as well as a knowledge of human history and social systems to answer these questions, help protect our environment, and build a better world.

My purpose in writing this book is to provide a concise, informative, and interesting survey of the key concepts of environmental science that will help readers become informed environmental citizens. I have tried to present information honestly and factually, without biases or prejudgment so as to encourage readers to form their own opinions about what should be done. Most of all, I hope that this book will inspire readers to become involved—individually and with others—to work toward solutions to our pressing environmental problems.

AUDIENCE

This book is intended for use in a one-semester or one-quarter course in environmental science or human ecology. It is written in a casual, conversational style. The vocabulary and level of discussion are straightforward and self-evident enough to be accessible to beginning students with little or no science or math background. I have avoided excessive scientific or technical detail, concentrating instead on key concepts. At the same time, enough data and scientific theory are included so that the book will stand alone without additional reference material. I hope that this book will remain valuable as a reference source and that students will add it to their personal libraries when their course is finished.

DESIGN

Environmental science texts have tended to grow larger and larger in recent years, some reaching the size and cost of small encyclopedias. These books contain far more material than can be covered in a one-semester or one-quarter course. The unused portions are a waste of resources both in terms of the paper and ink used to print them and in the student funds to buy them. This book is condensed to include only the key concepts that students need to understand the principles of environmental science. Illustrations have been carefully selected to explain essential principles, and the type is designed to be visually attractive and readable while reducing wasteful white space on the pages. The paper stock is recycled from 70 percent postconsumer and 30 percent preconsumer waste (on average). Printing is done with soy oil-based inks.

ORGANIZATION

This book is divided into fifteen chapters (one per week for a semester course) that are grouped into four parts. The parts and chapters follow a logical progression from basic principles through a survey of environmental problems to a discussion of what we can do to build a sustainable future. Because some readers may choose to omit certain chapters or study them in another sequence, however, they are written so they can be combined in any order.

- Part one presents an overview of the major themes of population, resource consumption, technological impact, and the relation between poverty and environmental degradation. It then surveys the principles of ecology and population dynamics.

- Part two adds key concepts of resource economics and environmental health that help us understand how we value environmental services and why we are concerned about environmental diseases and contaminants.
- Part three examines physical and biological resource issues. It shows how we are depleting or degrading vital assets on which we and other species depend, as well as ways we can protect, replenish, or restore our ecological capital.
- Part four looks at some specifically human issues such as urbanization and production of toxic or hazardous wastes. This section concludes by examining what we can do individually and collectively to build a sustainable future.

THEMES

Global Concern

Geographic literacy is a vital tool for citizens of an increasingly interconnected world. By studying key examples from around the earth, students are exposed to cultural diversity and global concerns. Understanding our environment implies wisdom as well as knowledge, care, sympathy, and affection for the earth and all its inhabitants that go beyond a mere mastery of facts.

Balanced Approach

This book achieves a balance between an unwarranted technological optimism on one hand and the paralyzing "doom and gloom" common in many environmental texts on the other hand. By adopting a cautious optimism tempered by healthy skepticism, I hope to demonstrate the seriousness of our environmental problems and yet generate confidence that we can find reasonable, progressive solutions for them.

Critical Thinking

With new issues appearing and new information available every day, students need skills in evaluating conflicting theories and claims. Learning how to think rationally about issues is more important than simply memorizing facts. Critical thinking based on clear, logical, systematic analysis is a special emphasis of this book. Throughout the text, material is presented in a way that fosters self-awareness, understanding of diverse views, and recognition that there may be more than one possible solution to our problems. Critical thinking questions at the end of each chapter help students plan how to study and analyze information. Somewhat less analytical and more contemplative reflective thinking questions ask students to

pause for a moment to consider underlying causes and deeper meanings in the facts they learn. Both types of questions appear under the heading Questions for Critical or Reflective Thinking.

Stewardship and Ethics

Human intelligence and enterprise give us tools for managing and transforming our environment, but they also create obligations and responsibilities. Technology provides means to restore damaged ecosystems and to make the world a better place as well as gives us the power to do great harm. Our ability to change and shape our environment gives us a responsibility to be caretakers, not only for future generations but also for other creatures. Many difficult ethical dilemmas arise where we have conflicting desires for environmental resources. Throughout this book, I point out ethical concerns and relate the choices we face to moral values.

Sustainable Development and Ecojustice

Poor people are often both the victims of environmental degradation and part of its cause. Those who struggle for daily survival have few options for sustainable resource use and environmental protection. A central thesis of this book is that a better standard of living and more options for the poorest of our neighbors are essential if we are to preserve and enhance our environment. This concern for the plight of the poor is not merely a pragmatic issue of saving our common environment, however; it also is a matter of social justice. Everyone has a right to food, shelter, decent work at a fair wage, and a chance to express their human potential. The environment in which we live is a crucial part of that set of rights.

Personal and Collective Actions

Active learning requires asking questions such as How does this affect me? and What can I do about this situation? I hope that students will be inspired by what they learn in this book to see ways that we each can work to improve our world. Nearly every chapter includes suggestions for individual and collective actions for environmental protection.

LEARNING AIDS

This book is designed to be useful as a self-education tool for students. To clarify important points and to help create a strategy for learning, each chapter begins with a brief **outline** and a set of **objectives.** Headings within the chapter act as signposts to show relationships and to keep track of the orderly flow of information and concepts. **Key terms,** indicated by boldface type, are defined in context where they are first used and are listed in the glossary for quick reference. The glossary also contains additional useful vocabulary terms.

At the end of each chapter a **summary** reviews the material just covered and gives the student a quick overview of important themes. Two sets of **questions** are appended to each chapter: Objective Questions help students review facts and key concepts while Questions for Critical and Reflective Thinking help them analyze what they have learned and what it all means. These open-ended questions serve to stimulate discussion and further learning. They invite the reader to dig beneath facts to understand how science works, to evaluate the reliability of information presented, and to think deeply about the significance of what they have read.

Boxed essays and **case studies** in each chapter demonstrate applications of the principles discussed in the text. These boxes are divided into three types: some emphasize thinking globally to expand the student's geographical perspective, others invite students to learn about local solutions, while still others cover important issues and situations. **Suggested Readings** are drawn from a wide variety of scientific and popular literature to suggest additional resources to students. These articles will help guide students as a starting point for further research, term papers, and group discussions.

Finally, the writing is itself designed to facilitate learning and promote scientific literacy. It is personal and accessible, engaging the reader in the substance of the text. Presentation of alternative points of view encourages critical reading and thinking and promotes a problem-solving mindset. While this book is condensed to essential concepts and examples, it is not watered down or simplistic. Students are challenged by up-to-date information and a thoughtful presentation. Explicit attention is given to representing science as a dynamic, human activity, rather than a static, dogmatic authority.

SUPPLEMENTARY MATERIALS

Instructor's Manual

The Instructor's Manual and Test Item File prepared by Mary Ann Cunningham is designed to assist instructors as they plan and prepare for classes using *Understanding Our Environment*. Additional learning resources such as relevant films, videotapes, and computer programs are suggested. The Test File provides approximately fifty questions per chapter (including multiple-choice, fill-in-the-blank, and essay questions) to assist the instructor in preparing tests and quizzes (ISBN 0-697-20479-0).

Computerized Testing Program and Service

WCB provides a computerized test generator for use with this text. It allows you to quickly create tests based on questions provided by **WCB** and requires no programming experience to use. The questions are provided on diskette in a test item file. **WCB** also provides support services, via mail or phone, to assist in the use of the test generator software, as well as in the creation and printing of tests.

A computerized grade management system is also available for instructors. This allows you to track student performance on exams and assignments. Reports based on this information can be generated for your review.

Software to generate quizzes can also be provided. These quizzes can be used to allow students to prepare for exams on their own.

Student Study Guide

Professor Darby Nelson (Anoka-Ramsey Community College, Coon Rapids, Minnesota) has prepared a study guide to accompany *Understanding Our Environment* that will help students formulate a strategy for efficient learning as well as prepare for tests. Each chapter is divided into three parts: BUILDING THE BASE reviews the text material, IN YOUR OWN WORDS invites students to express what they have learned, and THINKING THINGS THROUGH provides practice in critical thinking. BE ALERT FOR BOXES interspersed throughout the study guide point out issues of special concern while TIME OUT SECTIONS provide breaks in the study to inject interesting perspectives and insights that aid in the understanding of difficult material. (ISBN 0-697-20480-4)

Transparencies

There are 25 color overhead *Transparencies* available free to adopters. These acetates feature key illustrations that will enhance your lecture or course outlines. (ISBN 0-697-22598-4).

Transparency Masters

Transparency Masters are also available. These can be used to prepare course handouts or to project as overhead transparencies, depending on the instructor's needs. (00-697-21989-5).

ACKNOWLEDGMENTS

Many people have contributed to the accomplishment of this book. The whole Cunningham family has been an invaluable source of assistance and support throughout the process. Mary is initial editor as well as my chief critic and confidant. Mary Ann and John did background research and wrote drafts of parts of the manuscript. Captain Peg and First Mate Mark navigated all over the Northeast Kingdom in search of the quintessential New England village. In addition, Mary Ann wrote the Instructor's Manual and Test Item File. Bill Lynn helped with research and design of the book, and he, Paul Roebuck, and Gene Wahl each read drafts of one or more chapters. My colleague, Dennis O'Melia, of Inver Hills Community College allowed me to preview some of the material included here in his class. He and his students had many useful suggestions for clarifying and improving the text.

Much of the contents of this text are based on previously published material from *Environmental Science: A Global Concern*. I acknowledge the contributions of my co-author Barbara W. Saigo to that book and to the development of this one. I thank the entire WCB book team for their support, patience, and encouragement. The editorial team of Kevin Kane, Meg Johnson, and Robin Steffek have been enormously helpful. With their encouragement and assistance, this has been a smooth and enjoyable writing project. To everyone at Wm. C. Brown Communications who has worked on this book, I express my appreciation.

Obviously, little of the contents of a book as comprehensive as this can be the original research or personal experience of the author. I owe a debt of gratitude to the entire scientific community through whose work we know about the natural world and our effects on it. I also owe an equal debt to the environmental community for its dedication to informing the public about the issues discussed in this book as well as its efforts to preserve nature. If errors persist in the text in spite of the able assistance of many individuals who have helped with research and proofreading, I accept responsibility for the final product. If you, the reader, find errors of fact or interpretation, I would appreciate your assistance in correcting them. I gratefully acknowledge the helpful comments of users of previous editions of my textbooks whose suggestions have helped me immeasurably in preparing this book.

And finally, I am especially indebted to the following reviewers for their invaluable comments on the draft manuscript chapters during the development of this book.

Thomas B. Cobb
Bowling Green State University

Loretta M. Bates
Concordia College

Malcolm P. Frisbie
Eastern Kentucky University

Patrick K. Gladu
Lindsey Wilson College

Andrew S. Hopkins
Alverno College

Edward P. Laine
Bowdoin College

Ann Lopez
San Jose City College

Daniel J. Steffek
Clarke College

Charles Wick
Vermillion Community College

PART I: HUMANS AND NATURE: AN OVERVIEW

▼

This, as citizens, we all inherit. This is ours, to love and live upon, and use wisely down all the generations of the future.

Ansel Adams and Nancy Newhall, *This Is the American Earth*

1
AN INTRODUCTION TO ENVIRONMENTAL SCIENCE

We travel together, passengers on a little spaceship, dependent upon its vulnerable reserve of air and soil; all committed for our safety to its security and peace; preserved from annihilation only by the care, the work, and I will say, the love that we give to our fragile craft. We cannot maintain it half fortunate, half miserable; half confident, half despairing; half slave to the ancient enemies of man, half free in a liberation of resources. No craft, no crew can travel safely with such vast contradictions. On their resolution, then, depends the survival of us all.

Adlai E. Stevenson II,
U.S. Ambassador to the United Nations,
Farewell address, 1965

Objectives

After studying this chapter you should be able to:

- Define environment and environmental science and identify some important environmental issues that we face today.

- Discuss patterns of resource use and environmental impact in historical stages of human development as well as the attitudes toward nature that characterize cultures and societies.

- Explain how scientific methods contribute to our knowledge of the world.

- Outline conservation or environmental history and distinguish between utilitarian conservation and biocentric preservation.

- Identify the attitudes and skills necessary for critical thinking.

- Understand differences between the First, Second, and Third Worlds as well as differences between more-and less-developed countries.

- Give some reasons for both pessimism and optimism about our environmental future.

INTRODUCTION

Welcome to environmental science, a subject I hope you will find interesting and useful. This chapter is an overview of some of the important questions, issues, history, and philosophy of this timely and important discipline. First, however, let's consider the extraordinary planet that is the subject of our study.

A MARVELOUS PLANET

Imagine that you were an astronaut returning to Earth after a long trip to the moon or Mars. What a relief it would be after the hostile, desolate environment of space to come back to this beautiful blue-green planet (fig. 1.1). We live in a wonderfully bountiful and hospitable world. Compared to outer space, temperatures here are mild and constant. Plentiful supplies of clean air, fresh water, and fertile soil are regenerated spontaneously.

Perhaps most amazing is the diversity of life on this planet. Millions of remarkable species coexist in complex, interrelated communities. Towering trees and huge animals live together with and depend upon tiny

life-forms such as viruses, bacteria, and fungi. Together they create delightfully diverse and complex communities including dense, moist forests, vast, sunny savannas, and richly colorful coral reefs. From time to time we should stop to remember that, in spite of the challenges and complications of life on Earth, we are incredibly lucky to be here.

Figure 1.1 We live in a bountiful and beautiful world. Ours is a unique and irreplaceable planet.

Environmental Science

How are the landscapes of Earth and the life they support created and sustained? How do the communities of organisms coexist and interact? And what are our roles in this complex, magnificent world? These are among the important questions of **environmental science**—the subject of this book. But our environment consists of more than the natural world. We humans have a unique ability to modify our surroundings and to create more comfortable conditions than we find in nature. **Environment** (from the Old French *environ,* to encircle or surround) can be defined as (1) the circumstances or conditions that surround an organism or group of organisms, or (2) the complex of social or cultural conditions that affect an individual or community (fig. 1.2). In this book, we will consider both the natural environment and the "built" or technological environment as well as some aspects of the cultural and social environments in which humans dwell.

Figure 1.2 Our environment includes both natural and built or technological components. Cultural and social environments are also important parts of our lives.

Among the central questions of environmental science with which we will be concerned throughout this book are

- How does nature work?
- What have we done and what are we doing to our environment?
- Why do environmental problems occur and what can we do about them?
- What can we do to ensure a sustainable future for ourselves and future generations?

Keep these questions in mind as you go through the following chapters. You will find that they don't have simple, straight-forward answers. Environmental science—like most of life—is filled with ambiguity and uncertainty. You will need **critical thinking** skills (box 1.1) to help you through the maze of conflicting data and interpretation about these important environmental questions.

Environmental Dilemmas

In January 1989, *Time* magazine declared the "Endangered Earth" to be Planet of the Year. The cover photo showed a battered and wilted globe trussed up in cord and dumped on a trash heap. The introductory essay said that "the earth's future is clouded by man's reckless ways: overpopulation, pollution, waste of resources, and wanton destruction of natural habitat." The entire issue was devoted to "the looming ecological crisis and proposals for urgent action." Environmental problems have assumed a high level of importance on our national agenda. More than 80 percent of the people polled in a New York Times/CBS opinion survey agreed that "protecting the environment is so important that requirements and standards cannot be too high, and continuing environmental improvements must be made regardless of cost" (fig. 1.3).

Acting Locally: Box 1.1

Thinking about Thinking

How can we know what to believe when the facts are confusing and experts disagree? As you learn about environmental science—in this book and elsewhere—you will find many issues about which the data are indecisive, leading reasonable people to disagree on how they should be interpreted. How can we choose between competing claims? Is it simply a matter of what feels good at any particular moment, or are there objective ways to evaluate arguments? Critical thinking skills can help us form a rational basis for deciding what to believe and do. These skills foster reflective and systematic analysis to help us bring order out of chaos, discover hidden ideas and meanings, develop strategies for evaluating reasons and conclusions in arguments, and avoid jumping to conclusions. Developing rational analytic skills is an important part of your education and will give you useful tools for life.

Certain attitudes, tendencies, and dispositions are essential for critical or reflective thinking. Among these are

- *Skepticism and independence.* Question authority. Don't believe everything you hear or read, including this book. Even the experts can be wrong.
- *Open-mindedness and flexibility.* Be willing to consider differing points of view and entertain alternative explanations.
- *Accuracy and orderliness.* Strive for as much precision as the subject permits or warrants. Deal systematically with parts of a complex whole.
- *Persistence and relevance.* Stick to the main point and avoid allowing diversions or personal biases to lead you astray.

- *Contextual sensitivity and empathy.* Consider the total situation, relevant context, feelings, level of knowledge, and sophistication of others as you study situations. Try to put yourself in another person's place to understand his or her position.
- *Decisiveness and courage.* Draw conclusions and take a stand when the evidence warrants doing so.
- *Humility.* Realize that you may be wrong and that you may have to reconsider in the future.

Critical thinking is sometimes called metacognition or "thinking about thinking." It is not critical in the sense of finding fault but rather is an attempt to rationally plan how to think about a problem. It requires a self-conscious monitoring of the process while you are doing it and an evaluation of how your strategy worked and what you learned when you have finished. Assembling, understanding, and evaluating data are important steps, but critical thinking looks beyond simple facts to ask what reasons underlie an argument as well as what implications flow from a set of claims.

These are some steps in critical thinking.

1. *Identify and evaluate premises and conclusions in an argument.* What is the basis for the claims made? What evidence is presented to support these claims, and what conclusions are drawn from this evidence? If the premises and evidence are correct, does it follow that the conclusions are necessarily true?
2. *Acknowledge and clarify uncertainties, vagueness, equivocation, and contradictions.* Do the terms used have more than one meaning? If so, are all participants in the argument using

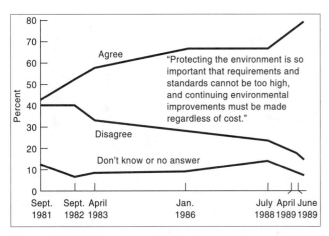

Figure 1.3 Public support for environmental protection is at an all-time high.

Source: Data from *New York Times*/CBS news polls taken since 1981.

Clearly many environmental issues merit our immediate attention. Warnings about impending global climate change, destruction of stratospheric ozone, proliferation of toxic waste dumps, accumulation of poisonous chemicals in fish, birds, and other wildlife, and an unprecedented loss of natural variety are only a few of the issues that worry us. What we are doing to our world and what that may mean for our future and the future of our children is of paramount concern as we enter the twenty-first century. I trust that's why you're reading this book.

There is hope, however. You will see in subsequent chapters that in many areas we have made progress toward controlling pollution and reducing wasteful resource uses. Still, there is much more we can do both individually and collectively to restore and protect our environment. Young people today may be in a unique position to address these issues because for the first time in history our society now has the resources, motivation, and knowledge to do something about our environmental problems. If we don't act quickly, however, this may be our last opportunity to do so.

the same meanings? Are ambiguity or equivocation deliberate? Can all the claims be true simultaneously?

3. *Distinguish between fact and values.* Can the claims be tested? (If so, these are statements of fact and should be verifiable by gathered evidence.) Are claims or appeals being made about what we *ought* to do? (If so, these are value statements and probably cannot be verified objectively.) For example, claims of what we *ought* to do to be moral or righteous or to respect nature are generally value statements.

4. *Recognize and interpret assumptions.* Given the backgrounds and views of the protagonists in this argument, what underlying reasons might there be for the premises, evidence, or conclusions presented? Does anyone have an ax to grind or a personal agenda concerning this issue? What do they think I know, need, want, believe? Is a subtext based on race, gender, ethnicity, economics, or some belief system distorting this discussion?

5. *Determine the reliability or unreliability of a source.* What makes the experts qualified in this issue? What special knowledge or information do they have? What evidence do they present? How can we determine whether the information offered is accurate, true, or even plausible?

6. *Recognize and understand conceptual frameworks.* What are the basic beliefs, attitudes, and values that this person, group, or society holds? What dominating philosophy or

ethics control their outlook and actions? How do these beliefs and values affect the way people view themselves and the world around them? If there are conflicting or contradictory beliefs and values, how can these differences be resolved?

In logic, an argument is made up of one or more introductory statements, called the premises, and a conclusion that supposedly follows from the premises. It is useful to distinguish between these kinds of statements. Premises usually claim to be based on facts; conclusions are usually opinions and values drawn from or used to interpret those facts. Words that often introduce a premise include *as, because, assume that, given that, since, whereas, and we all know that.* Words that often indicate a conclusion or statement of opinion or values include *and so, thus, therefore, it follows that, consequently, the evidence shows, we can conclude that.* Remember, even if the facts in a premise are correct, the conclusions drawn from them may not be.

As you go through this book, you will have many opportunities to practice these critical thinking skills. Try to distinguish between statements of fact and opinion. Ask yourself if the premises support the conclusions drawn from them. Although I will try to present controversies fairly and evenhandedly, I, like everyone, have biases and values—some that I may not even recognize—that affect how I present arguments. Watch for areas in which you must think for yourself and use your critical thinking skills.

Modified with permission from Karen J. Warren, Associate Professor of Philosophy, Macalester College (St. Paul, Minnesota).

A BRIEF HISTORY OF RESOURCE USE, TECHNOLOGY, AND DEVELOPMENT

Inevitably, all living organisms use resources and alter their environment. This ability is a unique and essential characteristic of life. We humans, however, are now making changes on an unprecedented and dangerous scale. In this section, we will review how culture, science, and technology have contributed to our ability to modify our environment for better or worse.

Tool-making Revolution

Our hominid (humanlike) ancestors probably separated from other primates about 4 million years ago. *Homo sapiens* (intelligent humans) first appeared around 200,000 years ago, probably in Africa, but perhaps on other continents as well. These early people lived by hunting and gathering. Compared to other species, they were slow and weak, yet their ability to think, plan, communicate, and use tools set them apart from other animals and allowed them to prosper.

Our early ancestors' first applications of technology were probably using wood or stone tools for defense or to obtain food and building fires for warmth, protection, or cooking. This **tool-making revolution** profoundly affected life on earth by transforming what would have been a relatively weak and puny species into the dominant members of many communities. For more than 95 percent of our history, most humans have been nomadic hunters and gatherers. These patterns probably still affect our customs and attitudes.

Many hunting and gathering cultures developed a rich understanding of nature through thousands of years of close observation and experience (fig. 1.4). Cultural and religious traditions created a respect for nature and generally maintained a balance between population and resources. Those of us who spend most of our lives in the artificial environment of cities can hardly imagine the depth of these peoples' knowledge and their sensitivity to subtle environmental clues. It is a great tragedy that much of this wisdom is being lost as traditional cultures disappear.

On the other hand, we shouldn't assume that all such cultures lived in perfect harmony with nature and had only beneficent effects on their environment. We know that some hunters killed entire herds of animals by driving them off cliffs. Others used fire to maintain grasslands or to open forests to improve hunting, sometimes with disastrous results. The disappearance of numerous species of large grazing animals in North America and elsewhere at the end of the last Ice Age may have been caused by super efficient predation by Stone Age hunters as well as climate change. Bone caches from

Figure 1.4 Indigenous and tribal cultures have environmental wisdom that may help us understand our proper relationship with nature.

Hawaii, New Zealand, and other islands suggest that dozens of bird species in Polynesia disappeared after humans arrived. Similarly, fourteen species of large lemurs—some the size of gorillas—and twelve flightless birds—including the 500-kg (1,110-lb) elephant bird—became extinct on Madagascar shortly after people from Southeast Asia and Africa settled there. Perhaps early humans were no more or less rapacious than we are, only less numerous, less powerful, and less able to bring about permanent changes in their environments.

Agricultural Revolution

About ten thousand years ago, the **agricultural revolution** began as people living in what are now Syria and Turkey began domesticating animals and cultivating crop plants. We generally assume that, once discovered, agriculture spread rapidly because it provided a larger and more stable food supply than did hunting and gathering. Some archaeologists argue, however, that people gave up hunting and gathering only because the wild animals and plants on which they depended were disappearing at the end of the last Ice Age due to overhunting and changing climate. Studies suggest that people knew about farming for a long time but continued hunting as long as game was available. Farming was probably considered much less exciting and much more work than hunting but became imperative as wild food supplies were exhausted.

Because farming can support ten to a hundred times more people per unit area than can hunting, populations increased rapidly as farming techniques spread across the Middle East between eight thousand and four thousand years ago. Agriculture enabled or perhaps required people to settle in permanent villages rather than continue the nomadic life they had previously followed. This settled life and the need to work together to drain fields, protect themselves from floods, and keep records of what was planted, where, and by whom may well have led to important developments in writing, politics, religion, technology, and civilization.

Agricultural and pastoral life often gives people a strong sense of stewardship or responsibility for the livestock, crops, and land on which they live. The farmer has a caring, nurturing relationship with the plants and animals she or he works with on a daily basis (fig. 1.5). Living permanently in a particular place may create a greater familiarity and intimacy with the land than does nomadic life. In spite of this sense of commitment and permanence, however, treatment of the land by farming communities has often been less than ideal. A long history of soil erosion, forest destruction, desertification, and water pollution associated with agriculture will be discussed later in this book.

Industrial Revolution

Although people have done useful work with externally powered machines for thousands of years, a dramatic change has occurred in the past two or three centuries as machines have become increasingly important in our lives. This **industrial revolution**

Figure 1.5 Many farmers have a strong sense of stewardship and an intimate knowledge of the land. In caring for their crops, they also care for the environment.

was made possible by advances in science and technology that have given us tremendous power to understand and change our world. Science has become a pervasive force in our lives, one that everyone should understand at least at a basic level (box 1.2).

Understanding Principles and Issues: Box 1.2

Science as a Way of Knowing

What comes to mind when you hear the word science? Do you imagine weird people wearing lab coats surrounded by gurgling glassware and complex, sinister-looking machines doing unspeakable things to cute little animals? Do you expect science courses to be difficult, disagreeable, and full of obscure minutiae? For many people today science seems to be a Faustian bargain (in a German legend, Dr. Faustus sells his soul to the devil in exchange for power and wealth). As French sociologist Alfred Sauvy said, "Science has indeed succeeded in making men live longer and worse."

Should there be limits to what scientists are allowed to study? Would we have been better off if we had never started science investigations in the first place? Although

there have been misapplications of science, I hope to persuade you that science also has many benefits and that the methods of science can be useful in your everyday life.

In the best sense, science is a methodical, precise, objective way to study the natural world. It is often an exciting and satisfying enterprise that requires creativity, skill, and insight. Science takes many different forms and is done in assorted ways by widely diverse people. When done right, science is neutral and unbiased. In the wrong hands it can be perverted and misused, but this is usually a human fault, not necessarily the fault of science itself.

Most likely you are not a scientist, but you probably use the scientific technique often without being aware of it. Suppose, for example, your flashlight doesn't work. The problem could be in the batteries, the bulb, or the switch—or all three could be faulty. How can you determine which component (or components) is the cause? A series of methodical steps to test each component is in order. First, you might try new batteries. If the flashlight still doesn't light, you might replace the bulb with one you know works. If neither of these steps helps, perhaps your flashlight has a faulty switch. You might try the original battery and bulb in a different flashlight to test the switch. By testing the variables one at a time, you should be able to identify the problem.

What you have been following is the scientific method of inquiry. Science is a way of knowing about the physical world based on an ordered cycle of observations, identification, description, experimental investigation, and explanation of results. The general flow of a scientific study is shown in box figure 1.2.1.

We always start with an observation: In this case, my flashlight doesn't work. From this observation we formulate a hypothesis or a provisional explanation: The batteries must be dead. To be useful, our hypothesis must enable us to make predictions that we can test. Tests of our hypothesis must be controlled in the sense that only one variable is changed at a time. If we change both the bulb and the batteries simultaneously, we won't know which component is at fault. Testing one component with two others that we know work should show whether the first component works. Finally, we carry out the experiment, collect data—in this case, look for light—and draw conclusions.

Often the result of our first experiment gives us information that leads to further hypotheses and additional experiments. For example, if the batteries are okay, then we suspect the bulb is burned out and we design a way to test that hypothesis. In every case, prior knowledge and experience help us design experiments and interpret results. Eventually, with evidence from a group of related investigations, we create a theory to explain a set of general principles.

It probably seems obvious that an orderly series of observations and experiments are the proper way to discover facts such as why a flashlight doesn't work. Medieval scholars, however, rejected direct observation of nature, depending instead on explanations based on religious dogma and deduced rationally by Aristotelian logics from first principles. In their view, there was room for interpretation of accepted truth but little reason for experimentation and no possibility for new explanations. Data that didn't fit preconceived notions were ignored.

The introduction of modern scientific methods in the sixteenth and seventeenth centuries by Francis Bacon, René Descartes, Galileo Galilei, Isaac Newton, and others powered one of the most significant revolutions in human history. Rather than accept earlier doctrines of medieval scholasticism, these scientists relied on empirical evidence gathered from observation and experimentation. Supernatural and mystical explanations of phenomena were discarded

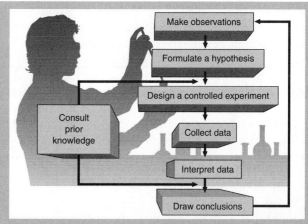

Figure 1.2.1 Science is a methodical, precise, organized way to study the natural world.

in favor of pragmatic ones. Instead of accepting a synthesis based on first principles, these scientists built up a body of knowledge piece by piece like a puzzle. The methods they developed are the roots of all modern science and technology.

Not all questions can be investigated by reductive, experimental approaches. Often ethics, practicality, or aesthetics require us to take other approaches. We don't consider it acceptable, for instance, to study the effects of toxins on human health by deliberately poisoning people under controlled conditions, no matter how useful the information might be. In this case, we might resort to a historical retrospective study of people poisoned accidentally. Or we use laboratory animals as surrogates. Social sciences usually have quite different methods of testing hypotheses than do physical sciences.

A common misconception is that science proves theories. You probably have heard claims that something has been proven "scientifically." In fact, scientific interpretations are always conditional. The evidence from experiments may support a particular hypothesis or provide contrary evidence to it, but we can never be absolutely sure that we have a final proof. A possibility always exists that some new evidence will require a modification of our conclusions. We may even have to discard them and formulate an entirely new set of theories.

Because of this uncertainty, you can often find distinguished scientists who draw diametrically opposed conclusions from the same data. This may seem confusing and demoralizing. How can we know what to believe when the experts disagree? Critical thinking skills help us evaluate how information is gathered and how interpretations are drawn, by whom, and for what purposes. An essential part of these skills is to keep an open mind and be alert for new data and new interpretations.

Figure 1.6 Science and technology give us power to improve our lives, but they also make it possible to make bigger mistakes faster than ever before, as is the case in these industrial smokestacks shown in the 1890s.

Think of the differences between your life and the lives of your ancestors a century ago. For many people in the nineteenth century, life was little different in terms of personal comfort or longevity than it had been for the Romans nearly two thousand years earlier. Most houses lacked central heat and running water. Candles and lanterns were the only sources of light at night. Both transportation and communication were slow and difficult. Most people never traveled more than a few miles from home in their entire lifetime. Medicine was primitive and dangerous; little could be done for most diseases, and commonly used treatments were often worse than the conditions they were intended to cure.

Contrast this picture with the comfort and convenience of your own life. Most of us would find it difficult to give up the benefits of modern society, yet many of the luxuries we enjoy have serious environmental and human costs. Industries producing materials on which we depend consume resources and create pollution at a prodigious rate (fig. 1.6). Industrialization has also tragically separated individuals, classes, and nations into haves and have-nots. Whether the root causes of these problems are contained in technology or in human nature, technology clearly gives us the power to make mistakes faster and on a larger scale than ever before. On the other hand, technology may also create options to avoid environmental damage in the future or to repair mistakes made in the past.

ATTITUDES TOWARD NATURE

What moral philosophies characterize the historical stages we have just examined? How does our treatment of the environment reveal our attitudes about ourselves and the world around us? Although many different views of our place in creation have existed through history, we can identify three broad sets of beliefs about the proper relationship between humans and nature.

Stewardship

Many tribal or indigenous peoples, both hunters and gatherers and people in traditional agricultural societies, have a strong sense of responsibility or **stewardship** toward a particular part of nature. They see themselves as custodians of natural resources, working together with human and nonhuman forces to sustain life. Humility and reverence are essential in this worldview, where humans are seen as partners in natural processes rather than as masters over them; not outside of nature but part of it.

This attitude is advocated by many people today. The sense of cooperation and humility it encourages may be an essential part of sustainable societies of the future. A sense of stewardship need not reject science or technology, however. If we are part of nature, then our intelligence and discoveries are part of nature too. As stewards of our environment, we can use the power of science and technology to maintain and improve rather than destroy or degrade the world.

Domination

While Western views of cosmology (the origin and structure of the universe) and teleology (the purpose or meaning of creation) have contributed to scientific and technological progress, our views about ourselves and our place in creation often have been used to justify domination and exploitation of nature. In 1967, historian Lynn White Jr., in an influential and widely quoted paper entitled *The Historic Roots of Our Ecological Crisis,* traced this tradition of domination to the Biblical injunction to "be fruitful, multiply, fill the earth and conquer it. Be masters of the fish of the sea, the birds of heaven, and all living animals on the earth" (Genesis 1:28). In antiquity, White claimed, "every tree, every spring had its guardian spirit. Before one cut a tree, dammed a brook, or killed an animal, it was important to placate the spirit in charge of that particular situation. By destroying pagan animism, Christianity made it possible to exploit nature in a mood of indifference to feelings of natural objects."

Although many people agree with White's analysis, others argue that this biblical passage has been translated and interpreted inaccurately. The original, they claim, meant something more like stewardship than

Figure 1.7 Our view of our place in nature determines how we behave. Do we have a right to use resources in any way we please regardless of our impact on other species and future generations?

conquest and domination. White himself pointed out that Judeo-Christian culture also has a long tradition of care for nature. He recommended St. Francis of Assisi as an inspiration and a patron saint for all environmentalists. Still, it is clear that many of us behave as if we have the right to use resources and damage nature in any way that we see fit. Many people seem to have little regard for the rights of other species or future generations (fig. 1.7).

Consumerism

Today it appears that many of us give little thought to whether our actions are ethical or not. For many people,

acquiring and consuming things seems to be the reason for existence; success in our society is often based on how fast we extract, use, and discard resources rather than on the permanence, beauty, or even utility of what we make or buy. We live on the throughput or the cash flow of our system. The more we use, the richer and more satisfied we think we are (fig. 1.8).

As a result, we consume an inordinate amount of the world's resources (table 1.1). In one day the average American consumes about 450 kg (nearly 1,000 lbs) of raw materials, including 18 kg (40 lbs) of fossil fuels, 13 kg (29 lbs) of other minerals, 12 kg (26 lbs) of farm products, 10 kg (22 lbs) of wood and paper, and 450 liters (119 gallons) of water. Every year we throw away some 160 million tons of garbage, including 50 million tons of paper, 67 billion cans and bottles, 25 billion styrofoam cups, 18 billion disposable diapers, and 2 billion razors.

This profligate consumption of resources and disposal of wastes now threaten the life-support systems on which we all depend. Unless we find ways to curb our desires and produce the things we truly need in less destructive ways, the sustainability of human life on our planet is questionable.

A BRIEF HISTORY OF CONSERVATION AND ENVIRONMENTALISM

Although traditional practices and restraints of earlier societies are clearly examples of resource conservation, most of us consider modern environmental consciousness to have developed in the past 150 years. There are at least three distinct stages in this process: conservation or preservation, environmentalism, and global citizenship. Each is concerned with a different set of problems and each results in a unique group of solutions. These waves or stages are not necessarily mutually exclusive; parts of each persist today in the environmental movement.

Figure 1.8 Is consumption our highest purpose?
© 1990 Bruce vonAlten. Used with permission.

TABLE 1.1 *U.S. consumption and waste*	
The United States, with 5 percent of the world's population,	
Consumes	**Produces**
26 percent of all oil	50 percent of toxic wastes
24 percent of aluminum	26 percent of nitrogen oxides
20 percent of copper	25 percent of sulfur oxides
19 percent of nickel	22 percent of chlorofluorocarbons
13 percent of steel	22 percent of carbon dioxide

Conservation or Preservation

Many historians consider the 1864 publication of *Man and Nature* by geographer George Perkins Marsh to be the beginning of the first wave of nature protection in the United States. Marsh, who was a lawyer, politician, diplomat, and educator, traveled widely around the Mediterranean as part of his diplomatic duties in Turkey and Italy. He saw the damage done by cutting trees on steep hillsides and by excessive grazing of goats and sheep. Appalled by the wanton destruction and profligate squandering of resources on the American frontier, he warned of its consequences. Largely as a result of his book, national forest reserves were established in 1873 to protect dwindling timber supplies and endangered watersheds.

Among those influenced by Marsh's warnings were President Theodore Roosevelt and his chief conservation adviser, Gifford Pinchot (fig. 1.9). Together with naturalists John Muir and William Brewster, publisher George Bird Grinnell, and others, Roosevelt and Pinchot established the framework of our national forest, park, and wildlife refuge system, passed game protection laws, and tried to stop some of the most flagrant abuses of the public domain. This first wave of environmental concern in America was focused primarily on saving resources from wasteful consumption.

An important philosophical split occurred between Pinchot and Muir about the purposes and methods of conservation. Pinchot, who was the first native-born professional forester in the United States, promoted pragmatic, professional, scientific resource management. This might be called **utilitarian conservation.** He argued that the purpose of saving forests was "not because they are beautiful or because they shelter wild creatures of the wilderness, but only to provide homes and jobs for people." Resources should be used "for the greatest good, for the greatest number, for the longest time."

By contrast, John Muir, first president of the Sierra Club, argued that nature deserves to exist for its own sake, regardless of its usefulness to humans. His philosophy might be called **biocentric preservation**

Figure 1.9 President Theodore Roosevelt and naturalist John Muir spent several days in 1906 camping in Yosemite National Park and discussing resource conservation.

because it emphasizes habitat protection as a fundamental right of other species. Muir wrote, "The world, we are told, was made for man. A presumption that is totally unsupported by the facts. . . . Nature's object in making animals and plants might possibly be first of all the happiness of each one of them. . . . Why ought man to value himself as more than an infinitely small unit of the one great unit of creation?"

Environmentalism

The tremendous expansion of industry during and after the Second World War brought a new set of concerns to conservationists. The birth of the second wave of nature protection might also be traced to another powerful book. *Silent Spring,* written by Rachel Carson (fig. 1.10) and published in 1962, awakened the public to the threats of pollution and toxic chemicals to humans as well as other species. The movement she engendered might be called **environmentalism** because it is concerned with the entire environment, built as well as natural. Among the leaders of this new movement are activist David Brower and scientist Barry Commoner. Brower, as executive director successively of the Sierra Club, Friends of the Earth, and Earth Island Institute, introduced many of the techniques of modern environmentalism, including using mass media for publicity campaigns, litigation, book and calendar publishing, and intervention in regulatory hearings. Commoner, who was trained as a molecular biologist, has been a leader in analyzing the links between science, technology, and society.

Figure 1.10 Rachel Carson's book *Silent Spring* was a landmark in modern environmental history. She alerted readers to the dangers of indiscriminate pesticide use.

Under the leadership of a number of other brilliant and dedicated activists and scientists, the environmental agenda was expanded in the 1960s and 1970s to include issues such as human population growth, atomic weapons testing and atomic power, fossil fuel extraction and use, recycling, air and water pollution, wilderness protection, and a host of other pressing problems. Environmentalism has become well established on the public agenda since the first national Earth Day in 1970. A majority of Americans now consider themselves environmentalists, although there is considerable variation in what that means.

The largest environmental groups have become big businesses. Their total annual income in 1990 was estimated to be $2.3 billion, probably a one hundred-fold increase since the publication of *Silent Spring.* Among the largest and most influential groups in the United States are Greenpeace, the National Wildlife Federation, the World Wildlife Fund, Audubon, the Sierra Club, the Nature Conservancy, Ducks Unlimited, the Wilderness Society, the Natural Resources Defense Council, and the Environmental Defense Fund. These organizations probably total somewhere around 10 million members. Most of them have offices in Washington, D.C., where large professional staffs lobby Congress and follow national issues.

These large national groups are often criticized by grass-roots activists who claim that mainstream organizations are too eager to compromise, curry favor with policymakers, and preserve their inside contacts. Dissatisfaction with the conservative policies of the major groups has led to the formation of a plethora of local, countercultural, special interest groups. The policies and actions of the mainstream organiza-tions have been criticized by philosophers Bill Duvall and George Sessions as examples of **shallow ecology.** They call for radical reforms and alternative lifestyles of **deep ecology.** We will discuss the implications of these terms in chapter 15.

Global Citizenship

Photographs of the earth from space (see fig. 1.1) provide a powerful icon for the third wave of environmental concern. They remind us how small, fragile, beautiful, and rare our home planet is. As our attention shifts from questions of preserving particular pieces of wilderness or preventing pollution of a specific watershed or airshed, we begin to worry about the life-support systems of the whole planet. We now see that we are changing global weather systems and atmospheric chemistry, reducing the natural variety of organisms, and degrading ecosystems in ways that could have devastating effects both on humans and on all other life-forms. Protecting our environment has become an international cause, and it will take international cooperation to bring about the necessary changes.

Among the trendsetters of this worldwide environmental movement are British economist Barbara Ward, French-American scientist René Dubos, Norwegian prime minister Gro Harlem Brundtland, and Canadian diplomat Maurice Strong. All were leaders of either the 1972 United Nations Environmental Conference in Stockholm or the 1992 U.N. Earth Summit in Rio de Janiero (box 1.3). Once again, new issues have become part of the agenda as our field of vision widens. Questions are being raised about the linkages among poverty, injustice, oppression and exploitation of humans and nature, economic development, and the environment. A key concept of this third wave is **sustainable development,** a term introduced in *Our Common Future,* published in 1987 by the Brundtland Commission on Environment and Development.

Many definitions of sustainable development have been offered, but the one in the Brundtland report is probably the best: "to meet the needs of the present without compromising the ability of future generations to meet their own needs." The report continues: "A world in which poverty is endemic will always be prone to ecological and other catastrophes. . . . Meeting essential needs requires not only a new era of economic growth for nations in which the majority are poor, but an assurance that those poor get their fair share of the resources required to sustain that growth."

NORTH VERSUS SOUTH: A DIVIDED WORLD

We live in a world of haves and have-nots; some of us live in luxury while others lack the basic necessities for a decent, healthy, productive life. The World Bank

The Earth Summit

*F*or twelve days in June 1992, more than 35,000 environmental activists, politicians, and business representatives, along with 9,000 journalists, 25,000 troops, and uncounted vendors, taxi drivers, and assorted others converged on Rio de Janeiro, Brazil, for the United Nations Conference on Environment and Development. Known as the Earth Summit, this was the largest environmental conference in history; in fact, it was probably the largest nonreligious meeting ever held. Like a three-ring environmental circus, this conference brought together everyone from pin-striped diplomats to activists in blue jeans to indigenous Amazonian people in full ceremonial regalia (box fig. 1.3.1). We can't say yet whether the conventions and treaties discussed at the Earth Summit will be effective, but they have the potential to make it the most important environmental meeting ever held.

The first United Nations Environmental Conference met in Stockholm in 1972, exactly twenty years before the Rio meeting. Called by the industrialized nations of Western Europe primarily to discuss their worries about transboundary air pollution, the Stockholm conference had little input from less-developed countries and almost no representation from nongovernmental organizations (NGOs). Some major accomplishments came out of the conference, however, including the United Nations Environment Program, the Global Environmental Monitoring System, the Convention on International Trade in Endangered Species (CITES), and the World Heritage Biosphere Reserve Program, which identifies particularly valuable areas of biological diversity. A humane and compassionate companion book to the Stockholm Conference entitled *Only One Earth* was written by René Dubos and Barbara Ward.

In 1983, the United Nations established an independent commission to address the issues raised at the Stockholm conference and to propose new strategies for global environmental protection. Chaired by Norwegian prime minister Gro Harlem Brundtland, the commission spent four years in hearings and deliberations. The developing world gained a significantly greater voice as it became apparent that environmental problems affected the poor more than the rich. The commission's final report, published in 1987 as the book *Our Common Future*, is notable for coining the term "sustainable development" and for linking environmental problems to social and economic systems.

In 1990, preparations for the Earth Summit began. Maurice Strong, the Canadian environmentalist who chaired the Stockholm conference, was chosen to lead once again. The U.N. scheduled a series of four working meetings called PrepComs to work out detailed agendas and agreements to be ratified in Rio. The first PrepCom was held in Nairobi in August 1990. The second and third meetings were held in Geneva in March and August

Figure 1.3.1 Then Senator Al Gore meets with indigenous people at the United Nations Earth Summit in Rio de Janiero, Brazil in 1992 to discuss protection of the tropical rain forest.

1991. The final PrepCom was held in New York in March 1992. Twenty-one issues were negotiated at these conferences, including biodiversity, climate change, deforestation, environmental health, marine resources, ozone, poverty, toxic wastes, and urban environments. Notably, population crisis was barely mentioned in the documents because of opposition by religious groups.

Intense lobbying and jockeying for power marked the two-year PrepCom process. As the date for the Rio conference neared, it appeared that several significant treaties would be ratified in time for presentation to the world community. Among these were treaties on climate change, biological diversity, and forests and a general Earth Charter that would be an environmental bill of rights for all people. A comprehensive 400-page document called *Agenda 21* presented a practical action plan spelling out policies, laws, institutional arrangements, and financing to carry out the provisions of these and other treaties and conventions. Chairman Strong estimated that it would take $125 billion per year in aid to help the poorer nations of the world protect their environment.

In the end, however, the United States refused to accept much of the PrepCom work. During PrepCom IV in New York City, for instance, 139 nations voted for mandatory stabilization of greenhouse gases at 1990 levels by the year 2000, laying the groundwork for what promised to be the showcase treaty of the Earth Summit. Only the U.S. delegation opposed it. After behind-the-scenes arm-twisting and deal making, the targets and compulsory aspects of the treaty were stripped away, leaving only a weak shell to take to Rio. Many environmentalists felt betrayed. Similarly, the United States, alone among the industrialized world, refused to sign the biodiversity treaty, the forest protection convention, or the promise to donate 0.7 percent of its gross domestic product to less-developed countries for environmental protection. The United States' excuse was that these treaties were too restrictive for American businesses and might damage the U.S. economy.

Many environmentalists went to Rio intending to denounce U.S. intransigence. Even Environmental Protection Agency chief William Reilly, head of the U.S. delegation, wrote a critical memo to his staff saying that the Bush administration was slow to engage crucial issues, late in assembling a delegation, and unwilling to devote suffcent resources to the meeting. *Newsweek* magazine entitled one article about the summit "The Grinch of Rio," saying that to much of the world, the Bush administration represented the major obstacle to environmental protection.

Not all was lost at the Rio meeting, however. Delegates from many countries made important contacts and began direct negotiations. They made great strides in connecting poverty to environmental destruction. They placed issues such as sustainable development and justice prominently on the negotiating table for the first time. Furthermore, the meeting provided a unique forum for discussing the disparity between the rich industrialized northern nations and the poor, underdeveloped southern nations. Many nations reached bilateral (nation to nation) treaties and understandings.

Perhaps more important than the official summit at the remote and heavily guarded Rio Centrum conference center

was the "shadow assembly" or Global Forum of NGOs held in Flamingo Park along the beachfront of Guanabara Bay. Eighteen hours a day, the park pulsed and buzzed as thousands of activists debated, protested, traded information, and built informal networks. In one tent, a large television screen tracked nearly three dozen complex agreements being negotiated by official delegates at Rio Centrum. In other tents, delegates at minisummits discussed alternative issues such as the role of women, youth, indigenous people, workers, and the poor in protecting the environment. Specialized meetings focused on topics ranging from sustainable energy to endangered species. In contrast to Stockholm, where only a handful of citizen groups attended the meetings and almost all were from the developed world, more than 9,000 NGOs sent delegates to Rio. Brazil alone sent more than 700. The contacts made in these informal meetings may prove to be the most valuable part of the Earth Summit.

Will anything come of the official treaties and conventions? Perhaps so. Surely one nation cannot long defy the convictions of the whole world. As environmental problems become more pressing and global citizens become more vocal, perhaps the U.S. government will join the world majority in the struggle to protect our global environment.

estimates that more than 1 billion people live in poverty and that about the same number (mostly the same people) are malnourished and illiterate and lack adequate housing, sanitation, access to clean water, and medical care. The plight of these people is not just a humanitarian concern. Increasing evidence points to a cycle of worsening environmental conditions and depleted resources caused by and contributing to poverty. In an increasingly interconnected world, the environments and resource base damaged by poverty and ignorance are directly linked to those on which we depend.

Rich and Poor Countries

Where do these poor people live? Almost every country, even the richest (including the United States), has pockets of poverty. No doubt everyone reading this book has seen homeless people on city streets. In many countries, however, nearly everyone lives below the poverty line. The World Bank considers countries having an annual per capita gross domestic product (GDP) of less than $580 as being poor. Among the forty-one nations meeting this criteria, thirty-one are in sub-Saharan Africa (fig. 1.11). Nearly two-thirds of sub-Saharan Africa falls in this category. All the others

in the list of poorest nations, except Haiti in the Caribbean, are in Asia.

The ten economically poorest countries in the world are, in ascending order, Mozambique, Tanzania, Ethiopia, Somalia, Bangladesh, Guinea-Bissau, Malawi, Mali, Bhutan, and Chad. All of these countries have annual per capita gross domestic products of less than $200 per year. They also have low levels of social security and quality of life as indicated by table 1.2. By contrast, each of the ten richest countries in the world—Switzerland, Luxembourg, Japan, Finland, Norway, Sweden, Iceland, the United States, Denmark, and Canada (in descending order)—has an annual per capita GDP more than one hundred times that of the poorest countries. As you can see in the table, other conditions in the rich countries reflect this wide disparity in wealth.

Notice that the wealthiest countries tend to be in the north while the poorest countries tend to be located closer to the equator. It is common to speak of a North/South division of wealth and power even though many poorer nations such as India and China are in the Northern Hemisphere while some relatively rich nations like Australia and New Zealand are in the Southern Hemisphere. The rich nations are also highly

Figure 1.11 Three-quarters of the world's poorest nations are in Africa. Millions of people lack adequate food, housing, medical care, clean water, and safety.

TABLE 1.2 Average indicators of quality of life for the ten richest and poorest nations[1]

Indicator	Poor countries	Rich countries
GDP per capita	$176	$22,634
Life expectancy	49 years	77 years
Infant mortality[2]	122	6.4
Child deaths[3]	208	7.9
Share of calories for healthy life	95%	130%
Grams protein per day	50	95
Safe drinking water	36%	100%
Female literacy	20%	N.A.[4]
Birth rate[5]	45	12.7

[1]Averaged as a group.
[2]Per 1,000 live births.
[3]Per 1,000 children before age 5.
[4]Not available but close to 100 percent.
[5]Per 1,000 population.
Source: World Resources Institute, *World Resources*, *1992–1993*.

industrialized and are members of the Organization for Economic Cooperation and Development (OECD), made up of Western Europe, North America, Japan, Australia, and New Zealand. This group makes up about 20 percent of the world's population but consumes more than half of most resources. The other, poorer four-fifths of humanity are sometimes called the majority world.

Since poverty and social welfare are so closely linked with environmental degradation, in many instances in this book we will want to distinguish between conditions in groups of countries. In fact, income, social services, human development, and other indicators of quality of life don't fit neatly into just two categories. Nations actually fit a spectrum—or perhaps a three-dimensional matrix—of these indicators, making it difficult to designate categories.

Political Economies

Countries are sometimes classified according to their economic system. **First World** describes the industrialized, market–oriented democracies of Western Europe, North America, Japan, Australia, and New Zealand and their allies. **Second World** originally described the centrally planned, socialist countries such as the Soviet Union and its Eastern European allies. Several Asian socialist countries, such as China, Mongolia, North Korea, and North Vietnam, also once fit into this category.

The nonaligned, nonindustrial, ex-colonial nations such as India, Indonesia, Malaysia, Iran, Syria, and many African countries labeled themselves the **Third World** to show their independence from either of the superpower groups. This category was not intended to suggest that its members are third-rate or third-class. Some economists suggest that another category is necessary: The Fourth World, they say, is composed of the poorest nations with neither market economies nor central planning, as well as indigenous communities within wealthy nations.

The stunning political and economic upheavals of the past decade have thrown all of these categories into a turmoil. Few countries are either purely socialist or capitalist anymore. Nearly every government plans centrally and intervenes in its economy to some extent, and nearly every one has at least some market orientation. As old political alliances break down, these categories have less and less significance.

Development

We also categorize nations based on their level of economic development into less-developed countries (LDCs) and more-developed countries (MDCs). Distinguishing relative levels of affluence is useful, but remember that these are economic categories only. Countries that are poor economically may be highly developed culturally or in religion, art, or wisdom. Furthermore, countries that fall in between the richest and poorest groups also have distinguishing characteristics. There are the wealthy industrialized nations (WINs), such as those in Western Europe and the socially and environmentally ravaged poorer industrialized nations (PINs) of Eastern Europe. These latter nations probably belong in the LDC group, but they do have a history of industry and a trained work force.

In addition to the categories above, there are also recently industrialized countries (RICs) such as the "Asian tigers" of South Korea, Taiwan, Hong Kong, and Singapore, which have surpassed many European countries in wealth and industrial power. Some newly industrializing countries (NICs), such as Indonesia, Thailand, Malaysia, and Brazil, seem about to break into the international marketplace. In contrast to these now **developing countries,** some of the poorest countries seem likely to be never-to-be-developed countries (NDC). Economist Lester Brown of the World Watch Institute predicts that some of the countries in this last category may become so environmentally devastated that they will cease to function politically or economically and could be the world's first environmentally caused national fatalities (fig.1.12).

As you can see, it is difficult to find inclusive labels that are not judgmental. Many people from the majority world prefer First and Third World rather than North and South or more- or less-developed, so I will use those terms in this book. Please remember that I don't intend to be disparaging by doing so.

ENVIRONMENTAL FUTURES

In *A Christmas Carol* by Charles Dickens, Scrooge questions the ghost of the future after seeing the disparity between the rich and poor in London. "Answer me one question," says Scrooge. "Are these the shadows of the

Figure 1.12 The industrialized nations have set a poor example for the rest of the world.

By permission of Mike Luckovich and Creators Syndicate.

things that Will be, or are they shadows of things that May be, only?" We could ask something similar. Are the problems and dilemmas outlined in this book warnings of what will be, or only what may be if we fail to take heed and adjust our course of action?

What will be our environmental future? This is, perhaps, the most important question in environmental science. There are many opinions about what the data show and how we should interpret them. Think about the following worldviews as you read the subsequent chapters in this book.

Pessimism

You will find much in this book that justifies pessimism. A number of very serious environmental problems threaten us. Many environmental scientists see our world as one of scarcity and competition in which too many people fight for too few resources. This viewpoint is often called **neo-Malthusian** after Thomas Malthus, who predicted a dismal cycle of overpopulation, misery, vice, and starvation as a result of human fallibility. We will discuss Malthus and his predictions further in chapter 4.

Optimism

Science and technology have provided many benefits to humanity; they also have caused many difficulties, as you will see in the course of this book. **Technological optimists** believe that the beneficial aspects of technology will outweigh its problems. Proponents of this viewpoint see the world as one of abundance and opportunity. Critics call this a "cornucopian" fallacy—after the mythical horn of plenty—and see it as blind faith or an excuse for business as usual. Proponents argue that they simply expect historic patterns of progress to continue in the future.

Communal Cooperation and Social Justice

An alternative to both the neo-Malthusian view and technological optimism is to believe that we could experience gloom and doom if we continue patterns of ignorance, carelessness, and greed—but we don't have to. Technology cannot save us from ourselves, but neither are we preordained to breed and consume ourselves into an apocalyptic catastrophe. Stewardship (a sense of responsibility and care for nature) and social justice (a fair share of the benefits of society) might provide a decent, safe, free, and fulfilling life to everyone. This view is sometimes called **Gandhian sufficiency** in keeping with the spirit of Mahatma Gandhi (fig 1.13), who said that the world has "enough for everyone's need, but not enough for anyone's greed."

Is There Hope?

You will see, as you read this book, that I tend to favor communal cooperation and social justice. I believe that we can overcome obstacles if we face problems honestly, reasonably, and creatively. If we respect the needs of others—both humans and nonhumans—and strive for harmony, I believe that we can create a better world for everyone. This may be a leap of faith, because there is much evidence to suggest that humans are deeply flawed and fallible, and much reason to fear for the future. Our hopes and fears may become self-fulfilling prophecies, however. If we expect the worst, we will probably find it. As Aldo Leopold said, "I have no hope for a conservation based on fear."

You will have to decide for yourself as you study environmental science how much hope the future holds. If you feel overwhelmed and discouraged as you read about the many problems in the world, it might help to recall the disasters faced by the people who preceded us. Surely to those who have experienced the horrors of war, oppression, and terror, the world must have seemed a dreadful place. Many have found the strength to persevere and survive even in dreadful circumstances, however.

Consider the fourteenth century, when Europe was wracked by a terrible century-long war in which inno-

Figure 1.13 Mahatma Ghandi not only preached simplicity and self-sufficiency, but he practiced these virtues as well. Here he spins thread to be woven into cloth for locally-made clothing.

cent civilians were slaughtered by both sides. The climate changed inexplicably during this "little ice age," causing crops to fail across Europe and adding famine to the horrors of war. Bubonic plague, the "black death," swept repeatedly across the continent, killing at least one-third of the population in its first pass; as many as 70 percent died in some cities. Civilization crumbled as bands of brigands roamed the countryside. Parents abandoned their children; everything seemed lost. People interpreted this series of calamities as God's punishment for their sins. Many fully expected that the end of the world was at hand. Life went on, however, and things slowly got better. We haven't stopped killing each other yet, but at least those who kill and torture now try to hide their deeds. Perhaps there is still hope for the future.

Acting Locally: Box 1.4

What's Your Environmental Perspective?

*B*efore you go on with this book or your course of study, take a moment to think about where you stand on environmental issues. It might be interesting to compare how you feel at the beginning of this process with your attitudes when you reach the end. Take the following quiz now, and take it again when you finish this book.

Do you:	Every day	Sometimes	Never
Recycle paper, bottles, & cans	_____	_____	_____
Buy recycled products	_____	_____	_____
Car pool or use public transportation	_____	_____	_____
Belong to an environmental group	_____	_____	_____
Write to your elected representatives	_____	_____	_____
Use aerosol products	_____	_____	_____
Visit a nature center or engage in other outdoor activity	_____	_____	_____
Know where your food comes from	_____	_____	_____
Read a newspaper or watch world news on TV	_____	_____	_____
Agree that we should spend more on environmental protection	_____	_____	_____
Think that pollution control laws should be strictly enforced	_____	_____	_____
Have hope for the future	_____	_____	_____

Rank the importance of and your level of knowledge about the following environmental topics:

	Importance (1–14)	Knowledge Not Much	Some	Lots
Human population	_____	_____	_____	_____
Nuclear power	_____	_____	_____	_____
Sustainable energy	_____	_____	_____	_____
Toxics and pollution	_____	_____	_____	_____
Endangered species	_____	_____	_____	_____
Food supplies and soil	_____	_____	_____	_____
Global climate	_____	_____	_____	_____
Stratospheric ozone	_____	_____	_____	_____
Forests and timber	_____	_____	_____	_____
Parks and wilderness	_____	_____	_____	_____
Water supplies	_____	_____	_____	_____
Urban environment	_____	_____	_____	_____
Environmental ethics	_____	_____	_____	_____
International environmental politics	_____	_____	_____	_____

Summary

We live in a world of beauty and abundance. Our environment includes both the natural as well as the technological or built worlds in which we live. Many environmental problems warrant our immediate attention, but there is reason to hope that we can find solutions to protect and restore our environment. Humans have a long history of altering the environment. Among the technological revolutions that have changed both our history and that of the world at large are tool-making, agriculture, and industrialization.

Scientific research is a methodical, precise, objective way to study the world. It has given us power to improve our lives but also the potential to make bigger mistakes faster than ever before. Our attitudes toward nature and our place in creation influence how we use science and technology. A strong sense of responsibility toward nature is often found in tribal or indigenous societies, while assumptions that we are free to dominate nature and consume resources often permeate modern society. Many environmentalists advocate a return to a sense of stewardship of and cooperation with nature.

We have passed through three stages of nature protection and resource conservation in this century. The first stage focused on resource conservation and preservation as proposed by Gifford Pinchot and John Muir. Warnings by Rachel Carson about the dangers of indiscriminate pesticide use ushered in a new era of environmental activism and a new group of issues after the Second World War. Now we recognize that global environmental problems affect all of us. United Nations conferences such as the Earth Summit in Rio de Janeiro in 1992 have been organized to address these problems.

Today, the world is sharply divided into rich and poor nations, which can be described as more- or less-developed, North or South, or newly, rapidly, or never-industrializing. Nations are also divided into First, Second, and Third World economic systems. These disparities exacerbate environmental problems and resource scarcities. Still, although much remains to be done, there are signs of hope that we could resolve our environmental dilemmas and make a better life for everyone.

Review Questions

1. Define environment and list four central questions of environmental science that you will keep in mind as you read subsequent chapters.
2. List and describe seven attitudes and six steps required for reflective or critical thinking.
3. Describe the tool-making, agricultural, and industrial revolutions and how they affected the quality of human life and the environment. How have these revolutions affected human population sizes?
4. Draw a diagram showing the steps in experimental scientific research. Why are proofs from this technique always tentative?

5. Describe the attitudes toward nature embodied by ideas of stewardship, domination, and consumerism. What environmental and resource issues are related to these stages of development?
6. What percentage of the world's population lives in the United States? What proportion of the world's resources does the United States consume?
7. What philosophical differences separated John Muir and Gifford Pinchot? Define utilitarian conservation and biocentric preservation and explain what they mean.

8. Why was *Silent Spring* important? Who wrote it and what did it say?
9. Why was the Earth Summit in Rio de Janeiro a historic event? What happened there?
10. Describe the differences between the North and South, or rich and poor, or more-developed and less-developed nations. What do we mean by First, Second, and Third World?
11. Compare some indicators of quality of life between the richest and poorest nations.
12. Give some reasons for pessimism and optimism about our environmental future.

Exercises in Critical or Reflective Thinking

1. Review a topic in this chapter (historical stages in resource use or attitudes toward nature, for example) and summarize the arguments presented as a series of short statements. Determine whether each statement is a fact or opinion.
2. How would you verify or disprove the fact statements in exercise 1? Do the opinions or conclusions reached in the argument you summarized *necessarily* follow from the facts

given? What alternative conclusions could be proposed?
3. Reflect on the preconceptions, values, beliefs, and contextual perspective that you bring to the argument you summarized in exercise 1. Do your beliefs coincide with the values, beliefs, etc. in the textbook? What different interpretations does your perspective impose on the argument?
4. Try to put yourself in the place of a person from an

underdeveloped or Third World country in discussing questions of social justice and environmental quality. What preconceptions, values, beliefs, and contextual perspective would that person bring to the issue?
5. Do you think that environmental conditions are better now or worse than they were twenty or one hundred or one thousand years ago? Why?

Key Terms

agricultural revolution 6
biocentric preservation 11
critical thinking 4
deep ecology 12
developing countries 16
environment 3
environmentalism 11

environmental science 3
First World 15
Gandhian sufficiency 17
industrial revolution 7
neo-Malthusian 16
Second World 15
shallow ecology 12

stewardship 9
sustainable development 12
technological optimists 16
Third World 15
tool-making revolution 6
utilitarian conservation 11

Suggested Readings

Berry, Thomas. *The Dream of the Earth.* San Francisco: Sierra Club Books, 1988. An important theological exploration of humans and nature in an emergent earth.

Berry, Wendell. *The Unsettling of America.* New York: Avon Books, 1977. A poetic and passionate explanation of what agricultural mechanization is doing to rural culture and the land.

Brown, Lester, et al. *State of the World 1992.* Washington, D.C.: Worldwatch Institute, 1992. An overview of global environmental conditions and progress toward a sustainable future.

Carson, Rachel. *Silent Spring.* Boston: Houghton Mifflin, 1962. This landmark book awakened America to the dangers of indiscriminate pesticide use and initiated a new phase in environmental activism.

Commoner, Barry. *Making Peace with the Planet.* New York: New Press, 1992. An important analysis of the role of technology in culture and environment as well as how we can best control technological impacts.

Cronon, William. *Changes in the Land: Indians, Colonists, and the Ecology of New England.* New York: Hill and Wang, 1983. An intelligent and objective study of how humans have altered the New England environment.

Devall, Bill, and George Sessions. *Deep Ecology: Living as If Nature Mattered.* Salt Lake City: Perigrine Smith Books, 1985. A study of the philosophical roots of the modern environmental movement as well as alternative lifestyles more in harmony with nature.

Diamond, Jared. "The Golden Age that Never Was." *Discover* (December 1988):71-79. Debunks the myth that all indigenous people were master stewards of their environment.

Dubos, René. *Wooing of the Earth.* New York: Charles Scribner's Sons, 1980. A humane and compassionate discourse on how humans have both despoiled nature and embellished it.

Fox, Stephen. *John Muir and His Legacy: The American Conservation Movement.* Boston: Little Brown, 1981. An interesting and highly readable biography of Muir and the history of American environmental groups.

Glacken, Clarence J. *Traces on the Rhodian Shore: Nature and Culture in Western Thought from Ancient Times to the End of the Eighteenth Century.* Berkeley: University of California Press, 1967. An encyclopedic interdisciplinary compendium of the roots of Western attitudes toward nature and culture.

Leopold, Aldo. *Sand County Almanac and Sketches Here and There.* Oxford: Oxford University Press, 1949. This slim volume of essays is one of the most poetic and influential pieces of nature writing in American literature.

Marsh, George P. *Man and Nature: Or Physical Geography as Modified by Human Action.* New York: Charles Scribner, 1864. (Reprinted in 1965 by Harvard University Press.) Called the "fountainhead of the American conservation movement," this pioneering work is an enduring classic.

Meadows, Donnella, et al. *Beyond the Limits: Confronting Global Collapse and Envisioning a Sustainable Future.* Post Mills, Vt.: Chelsea Green Publishers, 1992. A prophetic warning that we are reaching the limits of nature's resources.

Merchant, Carolyn. *The Death of Nature.* New York: Harper & Row, 1980. Explores the connection between domination of nature, women, and minorities.

Tuan, Yi Fu. *Topophilia: A Study of Environmental Perception, Attitudes, and Values.* New York: Prentice Hall, 1974. A literate comparison of different cultures in time and place and their attitudes toward self and nature.

World Commision on Environment and Development. *Our Common Future.* Oxford: Oxford University Press, 1987. The report of the Brundtland Commission on sustainable development.

World Resources Institute. *World Resources 1992-93.* Washington, D.C.: World Resources Institute, 1992. An objective and comprehensive survey of global environmental conditions with extensive data tables. An excellent resource.

2
MATTER, ENERGY, AND LIFE

Statistically, the probability of any one of us being here is so small that you'd think the mere fact of existing would keep us all in a contented dazzlement of surprise. . . . Even more astounding is our statistical improbability in physical terms. The normal, predictable state of matter throughout the universe is randomness, a relaxed sort of equilibrium, with atoms and their particles scattered around in an amorphous muddle. We, in brilliant contrast, are completely organized structures, squirming with information at every covalent bond. We make our living by catching electrons at the moment of their excitement by solar photons, swiping the energy released at the instant of each jump and storing it up in intricate loops for ourselves. We violate probability, by our nature. To be able to do this systematically, and in such wild varieties of form, from viruses to whales, is extremely unlikely; to have sustained the effort without drifting back into randomness, was nearly a mathematical impossibility.

Lewis Thomas, *The Lives of a Cell*

Objectives

After studying this chapter you should be able to:

- Describe matter, atoms, and molecules and give simple examples of the roles of four major kinds of organic compounds in living cells.
- Define energy and explain the difference between kinetic and potential energy.
- Understand the principles of conservation of matter and energy and appreciate how the laws of thermodynamics affect living systems.
- Know how photosynthesis captures energy for life and how cellular respiration releases that energy to do useful work.

- Distinguish between species, populations, communities, and ecosystems.
- Discuss food chains, food webs, and trophic levels in biological communities. Why are there pyramids of energy, biomass, and numbers of individuals in the trophic levels of an ecosystem?
- Recognize the unique properties of water and explain why the hydrologic cycle is important to us.
- Compare the ways that carbon, nitrogen, and phosphorous cycle within ecosystems.

INTRODUCTION

Biological organisms come in a wonderful variety of sizes, shapes, and complexity (fig. 2.1). Besides providing us with the food, shelter, and oxygen essential for life, the living creatures with which we share the planet give us pleasure in their bright colors, lively songs, and graceful movements. No one who has experienced the beauty of wildflowers and songbirds, or the majesty of a grand old tree, can help but be awed by the elegant diversity and intricate structure of life on Earth. All living organisms use energy and matter from the environment to build organized structures and to carry out the dynamic processes that define life: sensing and responding to external stimuli, maintaining internal identity and integrity, and reproducing. How are these processes accomplished? This

question is among the most fundamental in biology and will be the focus of this chapter.

Modern biology covers a wide range of inquiry, from atoms to ecosystems to the whole planet Earth (fig. 2.2). This spectrum is usually divided into several levels of study. Molecular and cellular biology, along with associated fields such as biochemistry, study life at the microscopic level, while organismic biology examines the tissues and organs of organisms and how they function in the varied life cycles of the multitude of different species. Ecology is generally concerned with populations, communities, and ecosystems. Sometimes, because of the great differences in scale, these disciplines seem to be investigating different universes, and yet many of the underlying questions about how matter and energy are captured and used to maintain life are the same in all biological disciplines.

Figure 2.1 Matter and energy are essential for life. Matter is continuously recycled, but energy follows a one-way path through communities such as this magnificent redwood forest. How do organisms capture and use matter and energy? These are central questions for biologists.

FROM ATOMS TO CELLS

In a sense, every organism is a chemical factory that captures matter and energy from its environment and transforms them into structures and processes that make life possible. To understand how these processes work, we need to understand something about the fundamental properties of matter and energy. This section presents a survey of some of these principles.

Atoms, Molecules, and Compounds

Everything that takes up space and has mass is **matter.** All matter has three interchangeable physical forms or phases: gas, liquid, and solid. Water, for example, can exist as vapor (gas), fluid (liquid), or ice (solid). Matter also consists of unique chemical forms we call elements, molecules, and compounds. Each of the 106 known elements (92 natural and 14 synthetic) has distinct chemical characteristics. Among the more common elements in biology are carbon (represented by the symbol C), hydrogen (H), oxygen (O), nitrogen (N), and phosphorous (P).

All elements are composed of discrete units called **atoms,** which are the smallest particles that exhibit

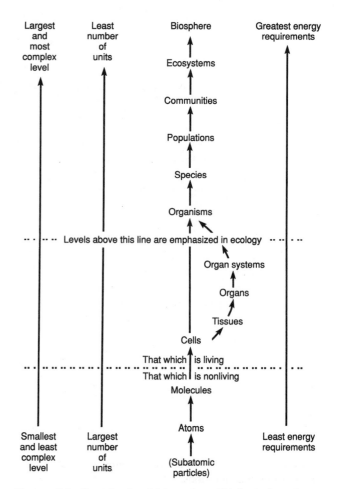

Figure 2.2 Organizational hierarchy of biological study.

the characteristics of the element. Each atom, in turn, is composed of several kinds of even smaller subunits (fig. 2.3). The three major kinds of subatomic particles are protons, neutrons, and electrons. Neutrons add mass to the central part, or nucleus, of an atom, but are electromagnetically neutral. Protons also add mass to the nucleus, but have a positive (+) charge. Electrons are negatively (−) charged particles that continually orbit the nucleus of the atom at the speed of light. Electrons are held in their paths, or orbitals, by attraction to the positive charge of the nucleus, which offsets their tendency to escape.

Atoms can join together to form **molecules.** A **compound** is a molecule containing different kinds of atoms. Water, for example, is a compound composed of two atoms of hydrogen attached to a single oxygen atom, shown by the formula H_2O. In a few cases, two atoms of the same element combine to form a molecule. Hydrogen gas (H_2), molecular oxygen (O_2), and molecular nitrogen (N_2) consist of such diatomic molecules. Most molecules are incredibly small, but some can be relatively large. The genetic information in your cells, for instance, is contained in deoxyribonucleic acid (DNA) molecules, each of

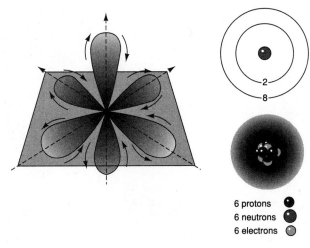

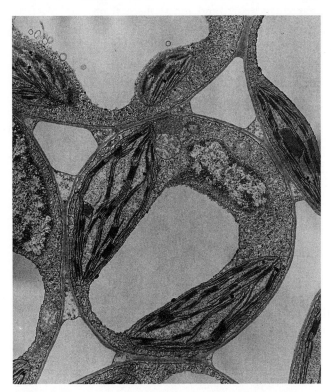

6 protons
6 neutrons
6 electrons

Figure 2.3 As difficult as it may be to imagine when you look at a solid object, all matter is composed of tiny, moving particles, separated by space and held together by energy. It is hard to capture these dynamic relationships in a drawing, so different models have been developed. Left and upper right models show two ways of depicting an atom having ten electrons. The inner two e^- are in a spherical orbital (not drawn in the first figure) and the other eight e^- are in three dumbbell-shaped orbitals at right angles to each other. The density of stippling in the first figure represents the relative probabilities of electrons being present in areas of the orbitals. The bottom model represents carbon[12], with a nucleus of six protons and six neutrons; the six electrons are represented as a fuzzy cloud of potential locations.

which contains billions of atoms and can be seen as long linear chains in very high-power microscopes.

Organic Chemicals

Organisms use some elements in abundance, others in trace amounts, and others not at all. Certain substances are hoarded within cells, while others are actively excluded. Carbon is particularly important because chains and rings of carbon atoms form the skeletons of **organic compounds,** the material of which biomolecules, and therefore living organisms, are made. At one time, scientists thought that organic chemicals had some mystical quality and were so complex that they could never be created outside of living cells. We now know this to be false. Many of the millions of known kinds of organic chemicals can be synthesized fairly easily in the laboratory.

The four major categories of bioorganic compounds are lipids, carbohydrates, proteins, and nucleic acids. Lipids are generally simple chains of carbon atoms, each of which has two hydrogens attached. Some common examples of lipids are fats and oils. They make up an important part of the membranes that surround cells and as their internal organelles. Carbohydrates, as their name suggests, are composed of carbon, hydrogen, and oxygen—$(CH_2O)n$—where n represents

many repeated units arranged in rings or linear chains. Some common examples of carbohydrates are sugars, starch, and cellulose.

Proteins are composed of long chains of subunits called amino acids, which are made primarily of carbon, hydrogen, and nitrogen (along with some other minor elements). Proteins make up much of both the structural and functional components of cells. Nucleic acids are composed of combinations of a sugar molecule, a nitrogen-containing ring structure, and a phosphate bridge that can link many subunits into long molecules such as DNA. Nucleic acids provide an immediate form of energy that drives many cellular processes. They also store the genetic information that directs the processes of life.

Cells, the Fundamental Units of Life

All living organisms are composed of **cells,** minute compartments within which the processes of life are carried out (fig. 2.4). Microscopic organisms such as bacteria, some algae, and protozoa are composed of single cells. Most higher organisms are multicellular, usually with many different cell varieties. Your body, for instance, is composed of several trillion cells of about two hundred distinct types. Every cell is surrounded by a thin but dynamic membrane of lipid and protein that receives information about the exterior world and regulates the flow of materials between the

Figure 2.4 A single plant cell is enclosed within a heavy cellulose cell wall. The bean-shaped structures are chloroplasts. Chlorophyll inside the chloroplasts captures light energy and uses it to carry out photosynthesis.

cell and its environment. Inside, cells are subdivided into tiny organelles and subcellular particles that provide the machinery for life. Some of these organelles store and release energy. Others manage and distribute information. Still others create the internal structure that gives the cell its shape and allows it to fulfill its role.

All of the chemical reactions required to create these various structures and provide them with energy and materials to carry out their functions, dispose of wastes, and perform other functions of life at the cellular level are carried out by a special class of proteins called **enzymes.** Enzymes are molecular catalysts; that is, they regulate chemical reactions without being used up or inactivated in the process. Think of them as tools such as hammers or wrenches that do jobs without being consumed or damaged as they work. There are generally thousands of different kinds of enzymes in every cell, all necessary to carry out the many processes on which life depends. Altogether, the multitude of enzymatic reactions performed by an organism is called its **metabolism.**

ENERGY AND MATTER

Energy and matter are essential constituents of both the universe and living organisms. Matter, of course, is the material from which structures are made. Energy provides the force to hold structures together, tear them apart, and move them from one place to another. In this section we will look at some fundamental characteristics of these components of our world.

Energy Types and Quality

Energy is the ability to move matter over a distance or to cause a heat transfer between two objects at different temperatures. Energy can take many different forms. Heat, light, electricity, gravity, and chemical energy are examples that we all experience. The energy contained in moving objects is called **kinetic energy.** A rock rolling down a hill, the wind blowing through the trees, water flowing over a dam, or electrons speeding around the nucleus of an atom are all examples of kinetic energy. **Potential energy** is stored energy that is latent but available for use. A rock poised at the top of a hill or water stored behind a dam are examples of potential energy (fig. 2.5). Chemical energy stored in the food that you eat or the gasoline that you put into your car are also examples of potential energy that can be released to do useful work.

Power is the rate of flow of energy from one object or one form to another. **Heat** describes the total kinetic energy of atoms or molecules in a substance not associated with bulk motion of the substance. **Temperature** is a measure of the speed of motion of

Figure 2.5 Water stored behind this dam represents potential energy. Water flowing over the dam has kinetic energy, some of which is converted to heat.

a typical atom or molecule in a substance. Note that heat and temperature are not the same. A substance can have a low temperature (low average molecular speed) but a high heat (much mass and many moving molecules or atoms). For example, the average temperature of the ocean is relatively low, but its total heat content is enormous. Conversely, the outer limits of the atmosphere contain gases that have very high kinetic energy (temperature), but there are so few of them per unit volume that their heat content is very low.

Energy that is diffuse, dispersed, and at low temperature is considered **low-quality energy** because it is difficult to gather and use for productive purposes. The heat stored in the oceans, for instance, is low quality. Conversely, energy that is intense, concentrated, and at high temperature is **high-quality energy** because of its usefulness in carrying out work. The intense flames of a very hot fire, for instance, or high-voltage electrical energy are high quality forms that are valuable to humans. These distinctions are important, because many of our most common energy sources are low quality and must be concentrated or transformed into high quality before they are useful to us.

Conservation of Matter

Under ordinary circumstances, matter is neither created nor destroyed but rather is recycled over and over again. Some of the molecules that make up your body probably contain atoms that once made up the body of a dinosaur and, most certainly, contain atoms that once made up the bodies of numerous smaller prehistoric organisms. Matter is transformed and combined in different ways, but it doesn't disappear; everything goes somewhere. These statements paraphrase the physical principle of **conservation of matter.**

How does this principle apply to human relationships with the biosphere? Particularly in affluent societies, we use natural resources to produce an incredible amount of "disposable" consumer goods. If everything goes somewhere, where do the things we dispose of go after the garbage truck leaves? As the sheer amount of "disposed-of stuff" increases, we are having greater problems finding places to put it. Another way of describing this principle simply is to say that there is no "away" where we can throw things we don't want any more.

Thermodynamics and Energy Transfers

How does the way organisms process energy differ from the way they process matter? Organisms use gases, water, and nutrients, then return them to the environment in altered forms as by-products of their metabolic processes. Year after year, century after century, the same atoms find endless reincarnation in new molecules synthesized by succeeding organisms as they feed, grow, and die. This exchange and continuity are made possible, however, by something that cannot be recycled: energy. Energy must be supplied from an external source to keep biological processes running. Energy flows in a one-way path through biological systems and eventually into some low-temperature sink such as outer space. It can be used to accomplish work in this process, and it can be stored temporarily in chemical bonds, but eventually it is released and dissipated.

The study of thermodynamics deals with how energy is transferred in natural processes. More specifically, it deals with the rates of flow and the transformation of energy from one form or quality to another. Thermodynamics is a complex, quantitative discipline, but you don't need a great deal of math to understand some of its broad principles that shape our world and our lives.

The **first law of thermodynamics** states that energy is *conserved;* that is, it is neither created nor destroyed under normal conditions. It may be transferred from one place or object to another, but the total amount of energy remains the same. Similarly, energy may be transformed, changed from one form to another (e.g., from the energy in a chemical bond to heat energy), but the total amount is neither diminished nor increased.

The **second law of thermodynamics** states that, with each successive energy transfer or transformation in a system, less energy is available to do work. This is not a contradiction of the first law; the energy is not lost or destroyed, merely degraded or dissipated from a higher quality form to a lower quality form. We can think of this process as an energy "expenditure," or the "cost in terms of useful energy" of doing work. The second law recognizes a tendency of all natural systems to go from a state of order (e.g., high-quality energy) toward a state of increasing disorder (e.g., low-quality energy, such as heat energy). Stated simply, the second law of thermodynamics says there is always less useful energy available when you finish a process than there was before you started. Because of this loss, everything in the universe tends to fall apart, slow down, and get more disorganized.

How does the second law of thermodynamics apply to organisms and biological systems? Organisms are highly organized, both structurally and metabolically. Constant care and maintenance is required to keep up this organization and a constant supply of energy is required to maintain these processes. Every time some energy is used by a cell to do work, some of that energy is dissipated or lost as heat. If cellular energy supplies are interrupted or depleted, the result, sooner or later, is death.

ENERGY FOR LIFE

Ultimately, most organisms depend on the sun for the energy needed to create structures and carry out the processes necessary for life. A few rare biological communities live in or near hot springs or thermal vents in the ocean where hot, mineral-laden water provides energy-rich chemicals that form the basis for a limited and unique way of life. For most of us, however, the sun is the ultimate energy source for life. In this section, we will look at how green plants capture solar energy and use it to create organic molecules that are essential for life.

Solar Energy: Warmth and Light

Our sun is a star, a fiery ball of exploding hydrogen gas. Its thermonuclear reactions emit powerful forms of radiation, including potentially deadly ultraviolet and nuclear radiation (fig. 2.6), yet life here is nurtured by and dependent upon this searing, energetic source. Solar energy is essential to life for two main reasons.

First, the sun provides warmth. Most organisms survive within a relatively narrow temperature range. In fact, each species tends to have its own range of temperatures within which it can function normally. At very high temperatures, biomolecules break or become distorted and are therefore nonfunctional. At very low temperatures, the chemical reactions of metabolism occur too slowly to enable organisms to grow and reproduce. Other planets in our solar system are either too hot or too cold to support life as we know it. The earth's water and atmosphere help to moderate, maintain, and distribute the sun's heat.

Second, organisms depend on solar radiation for life-sustaining energy, which is captured by green plants, algae, and some bacteria in a process called **photosynthesis** that converts radiant energy into

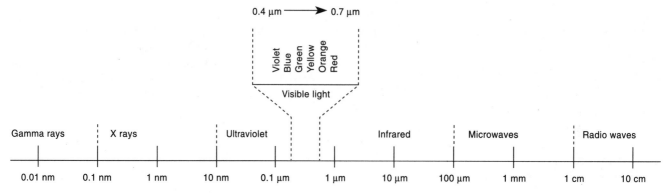

Figure 2.6 The electromagnetic spectrum.

useful, high-quality chemical energy in the bonds that hold together organic molecules.

How much potential solar energy is actually used by organisms? The amount of incoming, extraterrestrial solar radiation is incredible—about 2 gram-calories (gcal) per square centimeter per minute. (One gcal is defined as the amount of heat needed to raise the temperature of one gram or milliliter of water 1° C at 15° C.) However, not all of this radiation reaches the earth's surface. As much as half of the incoming sunlight may be reflected or absorbed by atmospheric clouds, dust, and gases. In particular, harmful, short wavelengths are filtered out by gases (such as ozone) in the upper atmosphere; thus, the atmosphere is a valuable shield, protecting lifeforms from harmful doses of ultraviolet and other forms of radiation. Even given these energy reductions, however, the sun provides much more solar energy than biological systems can harness, and more than enough for all our energy needs if technology can enable us to tap it efficiently.

On a clear day at noon, only about 67 percent of direct solar radiation may reach the earth's surface. Of that radiation, about 10 percent is ultraviolet, 45 percent is visible, and 45 percent is infrared. Most of the solar radiation that does reach the earth's surface is absorbed by land or water or is reflected into space by water, snow, and land surfaces. (Seen from outer space, the earth shines about as brightly as Venus.)

Fortunately for life on the earth, some radiation is captured by organisms through photosynthesis. Even then, however, the amount of energy that can be used to build organic molecules is further reduced. Photosynthesis can use only certain wavelengths of solar energy that are within the visible light range of the electromagnetic spectrum. These wavelengths are in the ranges we perceive as red and blue light. Furthermore, half or more of the light absorbed by leaves is consumed by evaporating water. Consequently, only about 1 to 2 percent of the sunlight falling on plants is captured for pho-

tosynthesis. This small percentage represents the energy base for virtually all life in the biosphere!

How Does Photosynthesis Capture Energy?

Photosynthesis occurs in tiny membraneous organelles called chloroplasts that reside within plant cells (see fig. 2.4). The most important key to this process is chlorophyll, a unique green molecule that can absorb light energy and use it to synthesize stable, highenergy compounds that serve as the fuel for all subsequent cellular metabolism. Chlorophyll doesn't do this important job all alone, however. It is assisted by a large group of other lipid, sugar, protein, and nucleic acid molecules. These components are arranged in two distinct subassemblies (fig. 2.7).

The first reactions in photosynthesis are called the light reactions because they continue only while light is being received by the chloroplast. During the light reactions, water molecules are split, releasing molecular oxygen. This process is the source of all the oxygen in the atmosphere on which animals, including humans, depend for life. The other products of the light reactions are small, mobile, high-energy nucleic acids called adenosine triphosphate (ATP) and reduced nicotinamide adenine diphosphate (NADPH). These nucleic acids serve as the fuel or energy source for the next set of processes: the dark reactions.

The dark reactions, as their name implies, can occur in the chloroplast after the light has been turned off. The enzymes in this complex use ATP and NADPH to add a carbon atom (from carbon dioxide) to a small sugar molecule. In most temperate-zone plants, this series of reactions can be summarized in the following equation:

$$6H_2O + 6CO_2 + energy \rightarrow C_6H_{12}O_6 \text{ (sugar)} + 6O_2$$

We read this equation as: water plus carbon dioxide plus energy produces sugar plus oxygen. The reason that the equation uses six water and six carbon dioxide molecules is that it takes six carbon atoms to make the sugar product. If you look closely, you

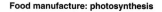

Food manufacture: photosynthesis

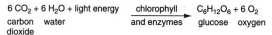

$6\,CO_2 + 6\,H_2O$ + light energy $\xrightarrow[\text{and enzymes}]{\text{chlorophyll}}$ $C_6H_{12}O_6 + 6\,O_2$

carbon water   glucose oxygen
dioxide

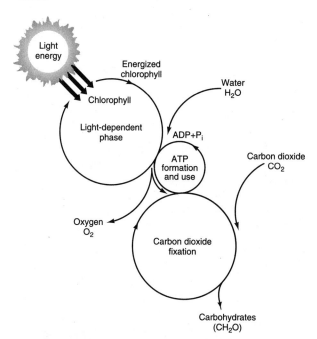

Figure 2.7 Photosynthesis, like other metabolic reactions, involves transfers of energy when chemical bonds are broken and formed. Also like other thermodynamic reactions, there is net loss of concentrated energy during these transfers. The flow of energy through photosynthesis, from *light units to food units* in the form of carbohydrate molecules, is shown here. The diagram shows input of energy, water, and carbon dioxide, and output of carbohydrates and oxygen. The three circles represent reaction cycles within the living photosynthetic cells.

will see that all the atoms in the reactants balance with the atoms in the products.

You might wonder how making a simple sugar benefits the plant. This particular sugar (glucose) doesn't even taste very sweet. The answer is that glucose is an energy-rich compound that serves as the central, primary fuel for all metabolic processes of cells. The energy in its chemical bonds—the ones created by photosynthesis—can be released by other enzymes and used to drive the synthesis of other molecules (lipids, proteins, nucleic acids, or other carbohydrates), or it can drive kinetic processes such as movement of ions across membranes, transmission of messages, changes in cellular shape or structure, or movement of the cell itself in some cases. This process of releasing chemical energy, called **cellular respiration**, involves splitting carbon and hydrogen atoms from the sugar molecule and recombining them with oxygen to recreate carbon dioxide and water. The net chemical reaction, then, is the reverse of photosynthesis:

$$C_6H_{12}O_6 + 6O_2 \rightarrow 6H_2O + 6CO_2 + \text{energy}$$

Note that in photosynthesis, energy is *captured,* while in respiration energy is *released.* Similarly, photosynthesis *consumes* water and carbon dioxide to *produce* sugar and oxygen, while respiration does just the opposite.

Animals do not have chlorophyll and cannot carry out photosynthetic food production. We do have the components for cellular respiration, however. In fact, this is how we get all our energy for life. We eat plants—or other animals that have eaten plants—and break down the organic molecules in our food through cellular respiration to obtain energy. In the process, we also consume oxygen and release carbon dioxide, thus completing the cycle of photosynthesis and respiration. In the next part of this chapter, we will see how these feeding relationships work.

FROM SPECIES TO ECOSYSTEMS

While cellular and molecular biologists study life processes at the microscopic level, ecologists generally study the interactions between species or between organisms and their environments. The word *species* literally means *kind.* In biology, **species** refers to all organisms that are similar enough genetically to breed in nature and produce live, fertile offspring. There are several qualifications and some important exceptions to this definition of species (especially among plants), but for our purposes this is a useful working definition.

Populations, Communities, and Ecosystems

An understanding of environmental interactions also requires an understanding of populations. A **population** consists of all the members of the same species that live in the same area at the same time. Chapter 4 deals with population dynamics. All of the populations of organisms that live and interact in the same area comprise a **biological community.** What populations make up the biological community of which you are a part? The population sign marking your city limits announces only the number of humans who live there, disregarding all the other populations of animals, plants, fungi, and microorganisms that are part of the biological community within the city's boundaries. Characteristics of biological communities are discussed in more detail in chapter 3.

An **ecosystem** is composed of a given biological community and its physical environment. This includes the abiotic factors or nonliving components of the environment such as climate, water, nutrients, sunlight, and soil as well as the biotic factors, which are the organisms and products of organisms (secretions, wastes, and remains) that are components of the environment. An ecosystem may be very large, such as a forest, or very small, such as a pond or even the surface of your skin. Many ecosystems have mechanisms that tend to stabilize composition and functions within certain limits. This has led some ecologists to suggest that ecosystems—or perhaps all of life—may be self-regulating and self-perpetuating superorganisms (box 2.1).

Most ecosystems are open in the sense that they lose or gain materials from other ecosystems. A stream ecosystem is a good example. Not only water and nutrients but also organisms enter from upstream and are lost downstream. The composition of the ecosystem may be relatively constant in terms of species and numbers of organisms present, but they are always different

Thinking Globally: Box 2.1

Is the Earth a Superorganism?

Your body is a marvelous system of self-regulating processes. Many of its subsystems at both the cellular level and at the level of tissues and organs have feedback mechanisms to maintain a stable, well-balanced whole. You modulate internal temperature by shivering when you are cold and sweating when you are hot. Hormones from your endocrine glands balance each other to regulate the functions of most organs in your body. We call this harmonious equilibrium **homeostasis** (from the Greek *homeo,* same, and *stasis,* stationary). Physiologist Walter Cannon, who coined the term, called it "the wisdom of the body."

Do these principles of self-regulation, stability, and self-perpetuation apply to ecological systems as well? Many people think so. The biogeographer F. E. Clements (1874–1945), for instance, considered ecological communities to be equivalent to superorganisms. He regarded the members of a community (individuals, species, populations) to be analogous to the organs of the body in the way they interact and the roles they carry out. Clements also saw a similarity between the growth and development of an individual and the regular successional stages of an ecosystem from bare ground or fallow field to mature forest (see chapter 3). He claimed that every landscape has a characteristic "climax" community toward which it will develop if free from external disturbances. The climax community, he believed, represents the maximum state of complexity and stability possible for a given set of environmental conditions.

Many people are attracted by this holistic view that the total system is greater than the sum of its parts. Most modern ecologists, however, are uneasy with the teleological (sense of purpose or design in nature) or deterministic implications of this Clemensian view. H. A. Gleason (1882–?), for instance, who was a contemporary of Clements, regarded biological communities as merely chance associations of species whose adaptations allowed them to live in a specific place at a particular time. The presence or absence of any one species, Gleason argued, is independent of all others. We see ecosystems as stable and uniform only because our lifetimes are so short and our view so limited.

If we look at landscape histories over thousands of years, for instance, what appears to be a stable forest or prairie community may have had dramatically different species compositions and ecosystem characteristics at different times. From this perspective, patchiness, variability, and randomness of species distribution in communities seem to be the rules rather than the exceptions; the self-regulation and self-determination of ecosystems seems much weaker than Clements imagined.

Recently, space research and cybernetics have awakened a new interest in the self-maintaining, equilibrating characteristics of biological systems. The Gaia hypothesis—named by James Lovelock after the ancient Greek goddess of the earth—was stimulated by speculations about the unique environmental conditions on the earth. In contrast to our neighboring planets, which are either much too hot or much too cold for life as we know it, most of the earth is just about right for our existence (some people call this the Goldilocks planet). The combination of gasses in the atmosphere is just what we need to breathe. In spite of the fact that our sun has gotten considerably stronger over the past several billion years, the temperature here has been remarkably constant.

This fortunate set of circumstances is not just a happy accident, according to the Gaia hypothesis. Rather, it is the result of active intervention by the living biota to create and sustain a livable environment. According to Lovelock, the entire earth operates as if it were a single superorganism: "The atmosphere, the oceans, the climate and the crust of the earth are regulated at a state comfortable for life because of the behavior of living organisms. This homeostasis is maintained by active feedback processes operated automatically and unconsciously by the biota." Although Lovelock dissociates himself from mysticism or teleological arguments about purpose or intention in this integrated system, many people find such holistic thinking gives great meaning to life.

One worry about these theories of intention or direction in life is that they may lull us into thinking that nature will take care of any environmental problems we create. If we believe that Gaia, like a patient and accommodating parent, has a purpose and a plan to keep the earth habitable for life, we may conclude that we can squander resources and cause havoc without thought for the future. However, Gaia's plan for a habitable earth may not include us, if we get too far out of line. Another view is that the fortuitous set of circumstances that we now enjoy may never occur again if we mess it up. What do you think?

Figure 2.8 Each time an organism feeds, it becomes a link in a food chain. In an ecosystem, food chains become cross-connected when predators feed on more than one kind of prey, thus forming a food web. How many food chains make up the food web shown here?

individuals. Some ecosystems are relatively closed in the sense that very little enters or leaves them. A balanced aquarium is a good example of a closed ecosystem. Aquatic plants, animals, and decomposers can balance material cycles in the aquarium if care is taken to balance their populations. Because of the second law of thermodynamics, however, every ecosystem must have a constant inflow of energy and a way to dispose of heat. Thus, at least with regard to energy flow, every ecosystem is open.

Food Chains, Food Webs, and Trophic Levels

Photosynthesis is the base of the energy economy of all but a few special ecosystems, and ecosystem dynamics are based on how organisms share food resources. In fact, one of the major properties of an ecosystem is its **productivity,** the amount of **biomass** (biological material) produced in a given area during a given period of time. For instance, a cornfield is a very simple ecosystem, and we measure its agricultural productivity by the biomass of corn produced, expressed in the United States as bushels of corn per acre per year. Why is this assessment of cornfield productivity incomplete in biological terms? Because it doesn't take into account the biomass of

other species, including grasshoppers, gophers, earthworms, and insectivorous birds that are also part of the ecosystem and therefore constitute a part of its ecological productivity.

Think about what you have eaten today and trace it back to its photosynthetic source. If you have eaten an egg, you can trace it back to a chicken, which ate corn. This is an example of a **food chain,** a linked feeding series. Now think about a more complex food chain involving you, a chicken, a corn plant, and a grasshopper. You could let the chicken eat grasshoppers that had eaten corn. You also could eat the grasshopper directly—some humans do. Or you could eat corn yourself, making the shortest possible food chain. Humans have several options of where we fit into food chains.

In ecosystems, some consumers are adapted to feed on a single species, but like humans, most consumers use more than one food species; similarly, most species in an ecosystem tend to have several predators. In this way, individual food chains become interconnected to form a **food web** (fig. 2.8). The matter of feeding becomes even more intricate when the energy needs and life histories of organisms that parasitize plants and animals in an ecosystem are considered. Perhaps you now can imagine the challenge

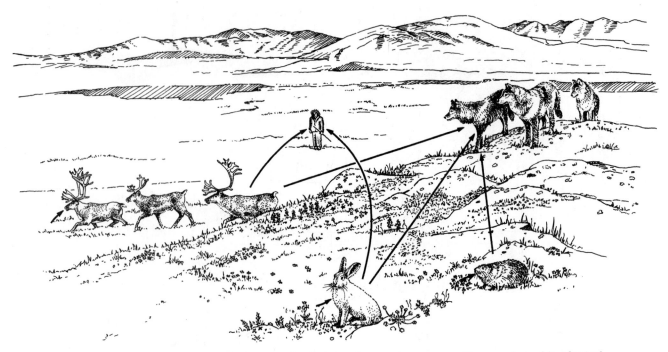

Figure 2.9 Harsh environments tend to have shorter food chains than environments with more favorable physical conditions. Compare the arctic food chains depicted here with the longer food chains in the food web in figure 2.8.

ecologists face in trying to quantify and interpret the precise matter and energy transfers that occur in a natural ecosystem!

An organism's feeding status in an ecosystem can be expressed as its **trophic level** (from the Greek *trophe,* food). In our first example, the corn plant is at the **producer** level; it transforms solar energy into chemical energy, producing food molecules. Other organisms in the ecosystem are **consumers** of the chemical energy harnessed by the producers. An organism that eats producers is a primary consumer. An organism that eats primary consumers is a secondary consumer, which may in turn be eaten by a tertiary consumer, and so on. Most terrestrial food chains are relatively short (seeds → mouse → owl), but aquatic food chains may be quite long (microscopic algae → copepod → minnow → crayfish → bass → osprey). The length of a food chain also may reflect the physical characteristics of an ecosystem (fig. 2.9).

Organisms can be identified both by trophic level at which they feed and by the *kinds* of food they eat (fig. 2.10). **Herbivores** are plant eaters, **carnivores** are flesh eaters, and **omnivores** eat both plant and animal matter. What are humans? We are natural omnivores, by history and by habit. Tooth structure is an important clue to understanding animal food preferences, and humans are no exception. Our teeth are suited for an omnivorous diet, with a combination of cutting and crushing surfaces that are not highly adapted for one specific kind of food, as are the teeth of a wolf or a horse.

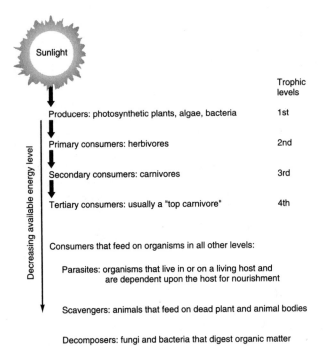

Figure 2.10 Organisms in an ecosystem may be identified by what they feed upon (produce, herbivore, carnivore, omnivore, scavenger, decomposer); by consumer level (producer; 1°, 2°, 3°, consumer); or by trophic level (1st, 2nd, 3rd, 4th).

Many kinds of organisms feed on the dead bodies of others. **Scavengers,** such as crows, ants, beetles, jackals, and many kinds of rodents, break apart dead plant and animal bodies into smaller pieces. This exposes inner tissues to small **decomposers,** organisms such as fungi and bacteria that grow on the tissues. A decomposer feeds by secreting digestive enzymes into its food source, then absorbing small molecules that are the products of digestion. It could be argued that decomposers are second in importance only to producers, because without their activity, nutrients would remain locked in the organic compounds of dead organisms and discarded body wastes, rather than being made available to successive generations of organisms.

Ecological Pyramids

Visualize the final predator species in a food chain, whom other species do not kill to eat. This species occupies the top trophic level. What does this position mean in terms of matter and energy? In practical terms, it means there is "less room at the top." There are fewer organisms at this level in an ecosystem than at previous levels. True to the second principle of thermodynamics, less food energy is available to the top trophic level than is available to preceding levels. This has an impact on other ecosystem characteristics. The energy and matter relationships of ecosystems suggest pyramidal models. The models are not perfect but are generally useful.

First, consider the energy pyramid (fig. 2.11). The energy that flows through an ecosystem is transferred and transformed many times in the metabolic pathways of organisms. Remember that at each successive link of the food chain, less energy is *available to do work* because at each previous level much of it is undigested, used, or released as heat. The losses often are about 90 percent from one level to the next, so as a general rule, only about 10 percent of the energy at one consumer level may be usable by the next level.

The amount of usable energy that flows through an ecosystem has a direct bearing on productivity and therefore on the composition of the biological community. As the energy at each successive consumer level decreases, less biomass is produced (fig. 2.12). This idea can be expressed as a biomass pyramid, based on the total amount of matter that is tied up in the bodies of organisms at each level of the pyramid. The biomass pyramid usually is related to a numbers pyramid, in which populations of species at each succeeding trophic level tend to be smaller (fig. 2.13).

Can you think of some discrepancies in the biomass and numbers models? Sometimes they can be turned upside down. The biomass pyramid, for instance, can be inverted by periodic fluctuations in producer populations (e.g., during winter in temper-

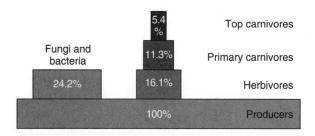

Figure 2.11 The classic example in this diagram is an aquatic ecosystem at Silver Springs, Florida, in which the producer energy base was calculated to be 20,810 kcal/m³/yr. The energy relationships between different trophic levels in an ecosystem can be depicted as a pyramid. Photosynthetic organisms are at the base of the pyramid and represent the maximal amount of primary food source for the ecosystem. Note that the numbers in each box represents the percentage of energy captured from the preceeding level. Thus, primary carnivores represent only 1.8 percent of primary production, and top carnivores have only 0.1 percent of the energy in the total system.

ate aquatic ecosystems there is low plant and algal biomass). The numbers pyramid also can be inverted. One coyote, for instance, can support numerous tapeworms. Numbers inversion also occurs at the lower trophic levels (e.g., one large tree can support numerous caterpillars).

It's tempting to think of ecosystems as being closed or self-contained, enabling us to make neat conclusions about their operation. In fact, most ecosystems are open to some influence from and exchange of materials with their neighbors, especially where they meet and intergrade. An obvious example is the range of effects a forest ecosystem has on the aquatic ecosystems with which it is associated. Streams are affected by soil, nutrients, and organic matter from the forest. Changes that affect the forest ecosystem, therefore, also affect the streams.

This example also emphasizes the interrelatedness of ecosystems within the biosphere. As conservationist John Muir said, "When we try to pick out anything by itself, we find it hitched to everything else in the universe."

MATERIAL CYCLES AND LIFE PROCESSES

To our knowledge, Earth is the only planet in our solar system that provides a suitable environment for life. Even our nearest planetary neighbors, Mars and Venus, do not meet the conditions required for life. Maintenance of these conditions requires a constant recycling of materials between the biotic (living) and abiotic (nonliving) components of ecosystems. In this section, we will examine some of the most important of those ecological cycles.

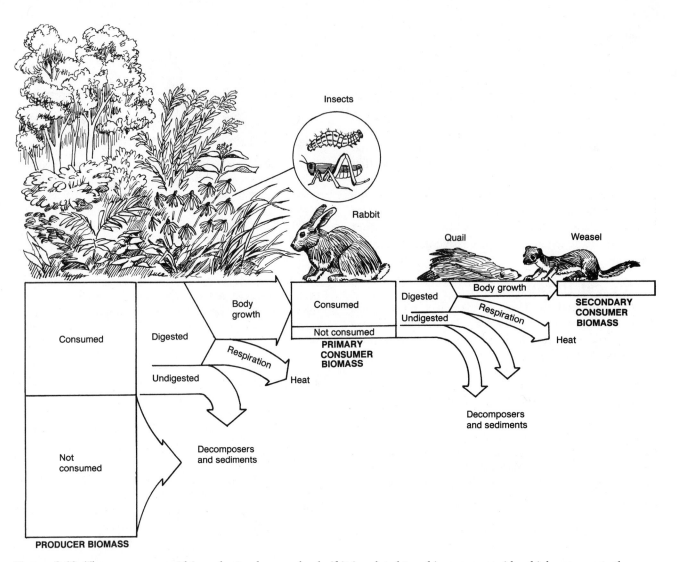

Figure 2.12 The energy pyramid is understood more clearly if it is related to a biomass pyramid, which represents the amount of biomass at each trophic level in a food chain. This figure illustrates how nutrients and energy become continuously less available to successive consumers.

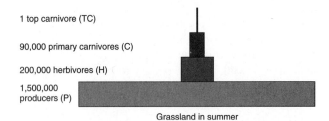

1 top carnivore (TC)

90,000 primary carnivores (C)

200,000 herbivores (H)

1,500,000 producers (P)

Grassland in summer

Figure 2.13 The pyramid of numbers of organisms in an ecosystem is even more exaggerated than the biomass pyramid. In this classic grassland study, 1.5 million primary producers were required to support one top carnivore.

The Water Cycle

"All rivers run into the sea, yet the sea is not full: Unto the place from which rivers come, thither they return again," says Ecclesiastes 1:7. Water has many marvelous and vital properties (box 2.2). The **hy-drologic cycle** (water cycle) describes the circulation of water as it evaporates from land, water, and organisms and enters the atmosphere; condenses and is precipitated to the earth's surface; and moves underground by infiltration or overland by runoff into rivers, lakes, and seas (fig. 2.14). This process supplies fresh water to the land masses while also playing a crucial role in creating a habitable climate and moderating world temperatures. Movement of water back to the sea in rivers and glaciers is a major geological force that shapes the land and redistributes material. Plants play an important role in the hydrologic cycle, absorbing groundwater and pumping it into the atmosphere by transpiration (transport plus evaporation) from leaf surfaces. In tropical forests, as much as 75 percent of annual precipitation is returned to the atmosphere by plants.

A "Water Planet"

*I*f travelers from other solar systems were to visit our lovely, cool, blue planet, they might call it Aqua rather than Terra because of its outstanding feature: the abundance of streams, rivers, lakes, and oceans of liquid water. This is the only place that we know of anywhere in the universe where water exists in liquid form in any appreciable quantity. Liquid water covers nearly three-fourths of the earth's surface, and during the winter snow and ice cover a good deal of the rest. Not only is water essential for cell structure and metabolism, but water's unique physical and chemical properties have a direct impact on the earth's surface temperatures, its atmosphere, and the interactions of life-forms with their environments. Water has many unique, almost magical qualities. Without the wonderful properties of water, life would not be possible here. Let's look in more detail at some of these properties and how they affect life.

1. Water is the primary component of cells and makes up 60 to 70 percent (on average) of the weight of living organisms. It plumps out cells, thereby giving form and support to many tissues. It is not just "filler," however, but has vital biological roles. Water is the medium in which all of life's chemical reactions occur, and it is an active participant in many of these reactions.

2. Water is the only inorganic liquid that exists in nature, and it is the solvent in which most substances must be dissolved before cells can absorb, use, or eliminate them. These substances include food molecules, mineral nutrients, gases, hormones and other chemical communicators, and waste by-products of metabolism. When molecules dissolve in water, they often have a tendency to break into positively and negatively charged parts, called ions. Ions are significant participants in cellular reactions.

3. Water molecules themselves can ionize, breaking into H^+ (hydrogen ions) and OH^- (hydroxyl ions). These ions help maintain the acid/base balance in cells, helping to offset (buffer) fluctuations caused by the release of other ions during metabolism. The term **pH** refers to the relative abundance of H^+ ions in a solution. On a pH scale from zero to fourteen, seven is neutral, values lower than seven are acidic, and those higher than seven are basic (box figure 2.2.1). A proper pH balance is critical to healthy cellular functioning.

4. Water molecules are cohesive, tending to stick together. You have experienced this property if you have ever done a bellyflop off a diving board. This cohesion causes capillary action, which is the tendency of water to be drawn into small channels. Without capillary action,

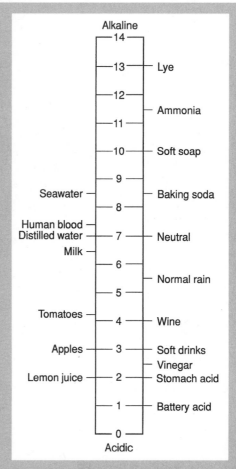

Figure 2.2.1 The pH scale. The numbers represent the negative logarithm of the hydrogen ion concentration in water.

movement of water and nutrients into groundwater reservoirs and through living organisms might not be possible. Because of its cohesiveness, water has the highest surface tension of any common, natural liquid. The surface of water, where it meets the air, acts like an elastic skin strong enough to support small insects (box figure 2.2.2).

5. Because water has a high heat capacity, it helps protect us from temperature fluctuations. It can absorb or release large amounts of heat energy before its own temperature changes. For this reason, large bodies of water such as the oceans and the Great Lakes have a moderating effect on their local climates. Without the presence of liquid oceans, the surface temperature of the earth would undergo wide temperature fluctuations between day and night as do the moon and Mars.

Figure 2.2.2 Surface tension is demonstrated by this water strider that dents but does not break the surface of the water as it walks across a pond.

6. Water exists as a liquid over a wide temperature range that for most of the world (at least during summer months) corresponds to the ambient temperature range. For most substances, the freezing point is only a few degrees lower than the boiling point. This means that they exist as either a solid or a gas, so that inorganic liquids are relatively rare. Organisms synthesize organic compounds such as oils and alcohols that remain liquid at ambient temperatures because they are

so valuable to life, but the original and predominant liquid in nature is water.

7. Water is unique in that it expands when it crystallizes. Most substances shrink as they change from liquid to solid. Ice floats because it is less dense than liquid water. When temperatures fall below freezing, the surface layers of lakes, rivers, and oceans cool faster and freeze before deeper water. Floating ice then insulates underlying layers, keeping most water bodies liquid throughout the winter in most places. Without this feature, lakes, rivers, and even oceans in high latitudes would freeze solid and never melt.

8. Water has a high heat of vaporization. Because of the amount of heat it absorbs in changing from a liquid to a vapor state, evaporating water is an effective way for organisms to shed excess heat. Many animals pant or sweat to moisten evaporative cooling surfaces. Why do you feel less comfortable on a hot, humid day than on a hot, dry day? Because the water vapor-laden air inhibits the rate of evaporation from your skin, thereby impairing your ability to shed heat. The heat absorbed when water vaporizes is released when condensation occurs. This accounts for a large transfer of heat from the oceans over the continents as water vapor turns into rain.

Altogether, these unique properties of water not only shape life at the molecular and cellular level, they also determine many features of both the biotic and abiotic components of our world.

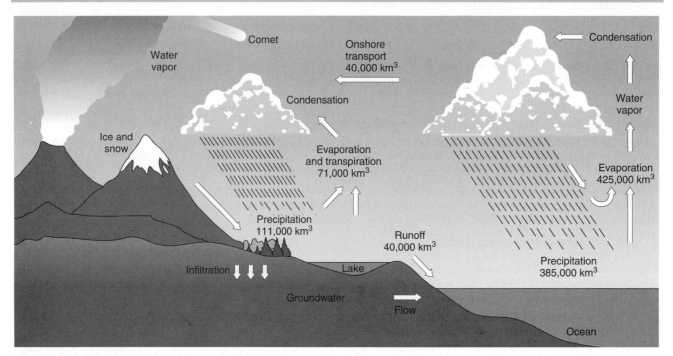

Figure 2.14 The hydrologic cycle moves water constantly among aquatic, atmospheric, and terrestrial compartments driven by solar energy and gravity. Total annual flows shown here are in thousands of cubic kilometers. The total annual runoff from land to the oceans is about 10.3×10^{15} gallons.

Solar energy drives the hydrologic cycle by evaporating surface water. As molecules of water vapor enter the atmosphere, they leave behind salts and other contaminants and thus create purified fresh water. This is essentially distillation on a grand scale. We once thought of rainwater as a symbol of purity, a standard against which pollution could be measured. Unfortunately, increasing amounts of atmospheric pollutants are picked up by water vapor as it condenses into rain. Chapter 11 has more to say about water pollution.

As air cools, water vapor may recondense into liquid droplets, or ice crystals if temperatures are below freezing. A cloud, then, is an accumulation of condensed water vapor in droplets or ice crystals. When cloud droplets and ice crystals become large enough, gravity overcomes uplifting air currents and precipitation occurs. Some precipitation never reaches the ground. As droplets fall through lower, warmer, and drier air layers, reevaporation occurs. Rising air currents lift this water vapor back into the clouds where it condenses again; thus, liquid water and ice crystals may exist for only a few minutes in this short cycle between clouds and air.

Everything about global hydrological processes is awesome in scale. Each year the sun evaporates approximately 496,000 cu km (km^3) of water from the earth's surface. More water evaporates in the tropics than at higher altitudes, and more water evaporates over the oceans than over land. Although the oceans cover about 70 percent of the earth's surface, they account for 86 percent of total evaporation. Ninety percent of the water evaporated from the ocean falls back on the ocean as rain. The remaining 10 percent is carried by prevailing winds over the continents where it combines with water evaporated from soil, plant surfaces, lakes, streams, and wetlands to provide a total continental precipitation of about 111,000 cu km.

What happens to the surplus water on land—the difference between what falls as precipitation and what evaporates? Some of it is incorporated by plants and animals into biological tissues. A large share of what falls on land seeps into the ground to be stored for a while (from a few days to many thousands of years) as soil moisture or groundwater. Eventually, all the water makes its way back downhill to the oceans. The 41,000 cu km carried back to the ocean each year by surface runoff or underground flow represents the renewable supply available for human uses and freshwater-dependent ecosystems.

The Carbon Cycle

Carbon serves a dual purpose for organisms: (1) It is a structural component of organic molecules, and (2) the energy-holding chemical bonds it forms represent energy "storage." The **carbon cycle** in an ecosystem begins with the intake of carbon dioxide (CO_2) by photosynthetic organisms (fig. 2.15). The carbon (and oxygen) atoms are incorporated into sugar molecules during photosynthesis. They are eventually released during respiration, closing the cycle.

The path followed by an individual carbon atom while it is cycling may be quite direct and rapid, depending on how it is used in an organism's body. Imagine for a moment what happens to a simple sugar molecule you swallow in a glass of fruit juice. The sugar molecule is absorbed into your bloodstream where it is made available to your cells for cellular respiration or for making more complex biomolecules. If it is used in respiration, you may exhale the same carbon atoms as CO_2 the same day.

Alternatively, the sugar molecule can be used to make larger organic molecules that become part of your cellular structure. The carbon atom could remain a part of your body until it decays after death. Similarly, carbon in the wood of a thousand-year-old tree will be released only when the wood is digested by fungi and bacteria that release carbon dioxide as a by-product of their respiration.

Can you think of examples in which carbon may not be recycled for even longer periods of time, if ever? Coal and oil are the compressed, chemically altered remains of plants or bacteria that lived millions of years ago. Their carbon atoms (and hydrogen, oxygen, nitrogen, sulfur, etc.) are not released until the coal and oil are burned. Enormous amounts of carbon also are locked up as calcium carbonate ($CaCO_3$) used to build shells and skeletons of marine organisms, from tiny protozoans to corals. Most of these deposits are at the bottom of the oceans. The world's extensive surface limestone deposits are biologically formed calcium carbonate from ancient oceans, exposed by geological events. The carbon in limestone has been locked away for millennia, which is probably the fate of carbon currently being deposited in ocean sediments. Eventually, even the deep ocean deposits are recycled as they are drawn down into deep molten layers and released via volcanic activity. Geologists estimate that every carbon atom on the earth has made about thirty such round trips over the past 4 billion years.

How does tying up so much carbon in the bodies and by-products of organisms affect the biosphere? Favorably. It helps balance CO_2 generation and utilization. Carbon dioxide is one of the so-called greenhouse gases because it blocks radiation of heat from the earth's surface, retaining it instead in the atmosphere. Photosynthesis and deposition of $CaCO_3$ remove atmospheric carbon dioxide; therefore, vegetation (especially large forested areas such as the tropical rainforests) and the oceans are very important **carbon**

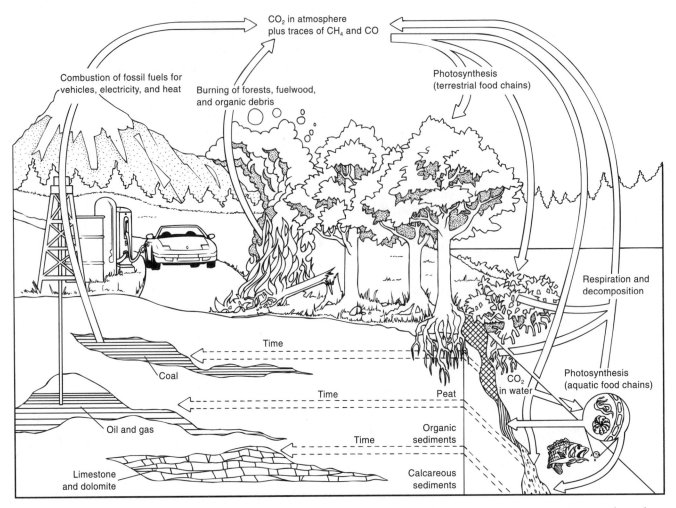

Figure 2.15 Atmospheric carbon dioxide is the "source" of carbon in the carbon cycle. It passes into ecosystems through photosynthesis and is captured in the bodies and products of living organisms. It is released to the atmosphere mainly by respiration and combustion. Carbon may be locked up for long periods in both organic and inorganic geological formations.

sinks (storage deposits). Cellular respiration and combustion both release CO_2, so they are referred to as carbon sources of the cycle.

Presently, natural fires and human-created combustion of organic fuels (mainly wood, coal, and petroleum products) release huge quantities of CO_2 at rates that seem to be surpassing the pace of CO_2 removal. Scientific concerns over the linked problems of increased atmospheric CO_2 concentrations, massive deforestation, and reduced productivity of the oceans due to pollution are discussed in later chapters.

The Nitrogen Cycle

Organisms cannot exist without amino acids, peptides, and proteins, all of which are organic molecules that contain nitrogen. The nitrogen atoms that form these molecules are provided by producer organisms. Plants assimilate (take up) inorganic nitrogen from the environment and use it to build their protein molecules, which are eaten by consumer organisms, digested, and used to build their bodies. This sequence

sounds very tidy, but there is one major problem. Even though N_2 is the most abundant gas (about 78 percent, by volume, of the atmosphere), plants cannot use N_2!

Where and how, then, do green plants get their nitrogen? The answer lies in the most complex of the gaseous cycles, the **nitrogen cycle.** Figure 2.16 summarizes the nitrogen cycle, emphasizing its biological aspects, though some nitrogen is converted to usable form by nonbiological processes. The nitrogen cycle has players of all kinds, each with a specific role. In many ways, the most critical players are the several categories of bacteria that make nitrogen available to plants. Nitrogen gas (N_2) is stable (chemically unreactive) and found in the air, water, and soil. Nitrogen-fixing bacteria (including some blue-green bacteria) have a highly specialized ability to "fix" nitrogen, meaning they change it to a less mobile, more useful form by combining it with hydrogen to make ammonia (NH_3), some of which combines with H^+ in water to become ammonium (NH_4^+). Nitrite-

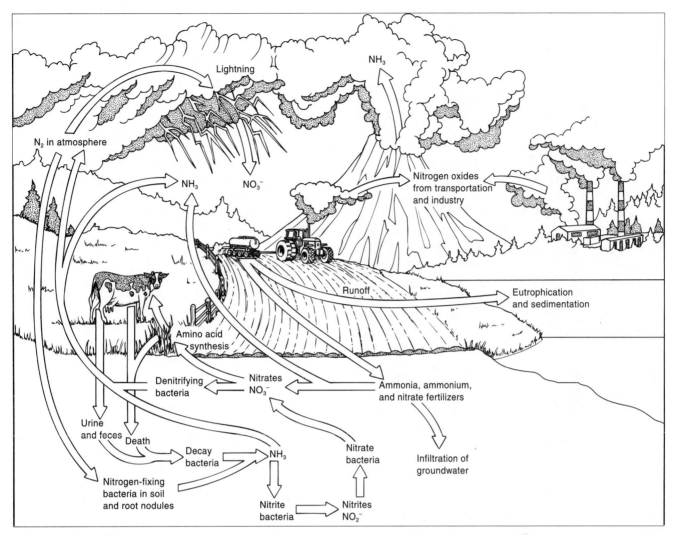

Figure 2.16 Nitrogen is incorporated into ecosystems when plants and bacteria use it to build their own amino acids. It is released from ecosystems by bacterial decomposition. Natural and human interactions with the nitrogen cycle are also depicted.

forming bacteria combine ammonia with oxygen, forming nitrites, which have the ionic form NO_2^-. Then n*itrate*-forming bacteria convert nitrites to nitrates, which have the ionic form NO_3^-. At this point nitrogen finally is in a form that can be absorbed and used by green plants—as nitrates! After nitrates have been absorbed into plant cells, they are broken down to ammonium, which is used to build amino acids, which become the building blocks for peptides and proteins.

You may have seen farmers injecting ammonium compounds into their fields to fertilize them. This treatment works because members of the bean family (legumes) and a few other kinds of plants have nitrogen-fixing bacteria actually living *in* their root tissues (fig. 2.17). The bacteria are clustered in nodules where they grow in a moist, nutrient-rich environment. This is an example of symbiosis, an intimate,

mutually beneficial "living-together" relationship between members of different species. Legumes and their symbiotic bacteria enrich the soil. Interplanting and rotating legumes with crops such as corn that use but cannot replace soil nitrates are beneficial farming practices that take practical advantage of the legume–bacteria relationship.

Nitrogen reenters the environment in several ways. The most obvious path is through the death of organisms. Their bodies are decomposed by fungi and bacteria, releasing ammonia and ammonium ions, which then are available for nitrate formation. Organisms don't have to die to donate proteins to the environment, however. Plants shed their leaves, needles, flowers, fruits, and cones; animals shed hair, feathers, skin, exoskeletons, pupal cases, and silk. Animals also produce excrement and urinary wastes that contain nitrogenous compounds. Urinary wastes are especially

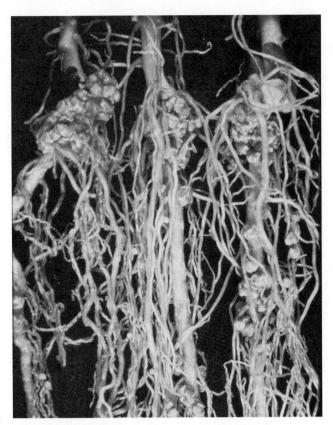

Figure 2.17 The roots of this adzuki bean plant are covered with bumps called nodules. Each nodule is a mass of root tissue containing many bacteria that help to convert nitrogen in the soil to a form the bean plant can assimilate and use to manufacture amino acids.

high in nitrogen because they contain the detoxified wastes of protein metabolism. All of these by-products of living organisms decompose, replenishing soil fertility.

How does nitrogen reenter the atmosphere, completing the cycle? Denitrifying bacteria break down nitrates into N_2 and nitrous oxide (N_2O), gases that return to the atmosphere; thus, it would seem that denitrifying bacteria compete with plant roots for available nitrates. However, denitrification occurs mainly in waterlogged soils, which have low oxygen availability and a high amount of decomposable organic matter. These are suitable growing conditions for many wild plant species in swamps and marshes, but not for most cultivated crop species, except for rice, which is a domesticated wetlands grass.

Phosphorus Cycles

Minerals become available to organisms after they are released from rocks. One mineral cycle of particular significance to organisms is that of phosphorus. Why do you suppose phosphorus is a primary ingredient in fertilizers? At the cellular level, ATP and other energy-rich, phosphorus-containing compounds are primary participants in energy-transfer reactions, as we have discussed. The amount of available phosphorus in an environment can, therefore, have a dramatic effect on productivity. Abundant phosphorus stimulates lush plant and algal growth, which makes it a major contributor to water pollution.

The phosphorus cycle (fig. 2.18) begins when phosphorus compounds are leached from rocks and minerals over long periods of time. Inorganic phosphorus is taken in by producer organisms, incorporated into organic molecules, and then passed on to consumers. It is returned to the environment by decomposition. An important aspect of the phosphorus cycle is the very long time it takes for phosphorus atoms to pass through it. Deep sediments of the oceans are significant phosphorus sinks of extreme longevity. Phosphate ores that now are mined to make detergents and inorganic fertilizers represent exposed ocean sediments that are millennia old. You could think of our present use of phosphates, which are washed out into the river systems and eventually the oceans, as an accelerated mobilization of phosphorus from source to sink. Aquatic ecosystems often are dramatically affected in the process because excess phosphates can stimulate explosive growth of algae and photosynthetic bacteria populations, upsetting ecosystem stability. Notice also that in this cycle, as in the others, the role of organisms is only one part of a larger picture.

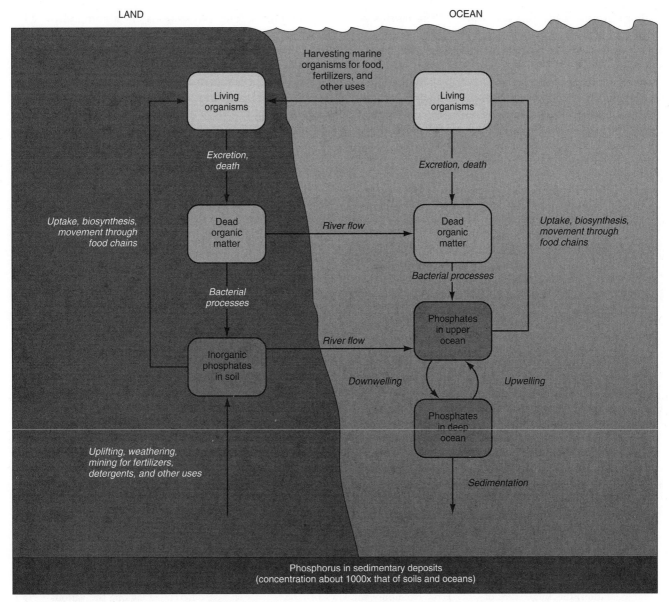

LAND   OCEAN

Harvesting marine
organisms for food,
fertilizers, and
other uses

Living organisms

Living organisms

Excretion, death

Excretion, death

Uptake, biosynthesis, movement through food chains

Uptake, biosynthesis, movement through food chains

Dead organic matter

River flow

Dead organic matter

Bacterial processes

Bacterial processes

Phosphates in upper ocean

Inorganic phosphates in soil

River flow

Downwelling   *Upwelling*

Phosphates in deep ocean

Uplifting, weathering, mining for fertilizers, detergents, and other uses

Sedimentation

Phosphorus in sedimentary deposits
(concentration about 1000x that of soils and oceans)

Figure 2.18 The phosphorus cycle includes relatively short-time biological components that involve the food chains of land and water. It also includes long-time geological processes of sedimentation, uplifting of landmasses, and weathering. Human activities such as mining and use of agricultural fertilizers, are now prominent factors in the phosphorus cycle. Bacterial processes noted in the diagram refer mainly to decomposition and formation of phosphates.

Summary

For life to exist on earth, certain conditions must be met. These include the availability of required chemical elements, a steady influx of solar energy, amenable surface temperatures, the presence of liquid water, and a suitable atmosphere. Life requires highly organized exchanges of matter and energy between organisms and their environment.

Matter is the observable material of which the universe is composed. It exists in three interchangeable phases: gas, liquid, and solid. Matter is made up of atoms, which are composed of particles called protons, neutrons, and electrons. The energy that holds atoms together forms the basis for energy transfers in the bodies of living organisms and, therefore, in the biosphere.

A steady influx of solar radiation provides the heat and light energy needed to support life in the biosphere. Water, which covers approximately three-fourths of the earth's surface, is readily available to life-forms. Because of its unique characteristics, it stabilizes the biosphere's temperature and provides the medium in which life processes occur. The earth's atmosphere provides gases necessary for life, helps maintain surface temperatures, and filters out dangerous radiation.

Ecosystem dynamics are governed by physical laws, including the law of conservation of matter and the first and second laws of thermodynamics. The recycling of matter is the basis of the cycles of elements that occur in ecosystems. Unlike matter, energy is not cycled. In ecosystems, solar energy enters the system and is converted to chemical energy by the process of photosynthesis. The chemical energy stored in the bonds that hold food molecules together is available for metabolism of organisms.

The populations of different species that live and interact within a particular area make up biological communities. The interactions of a community with the physical factors of its environment comprise the ecosystem level of organization and study.

Matter and energy are processed through the trophic levels of an ecosystem via food chains and food webs. At each energy transfer point, less energy is available to do work, so energy must be supplied to an ecosystem continuously. The relationships between producers and consumers in an ecosystem, often depicted as pyramids, demonstrate this principle. Most of the energy that enters an ecosystem comes, ultimately, from the sun.

The biosphere is a source of large quantities of essential elements, and in a given ecosystem these elements are constantly used and reused by living organisms. Water, carbon, nitrogen, and phosphorus, for instance, are recycled in ecosystems.

Review Questions

1. The number of board feet of lumber harvested per acre is a measure of the productivity of a forest. Why is this an incomplete measure of total biomass of a forest ecosystem?

2. Your body contains vast numbers of carbon atoms. How is it possible that some of these carbon atoms also may have been part of the body of a prehistoric creature?

3. In the biosphere, matter follows a circular pathway while energy follows a linear pathway. Explain.

4. The ocean stores a vast amount of heat, but (except for climate moderation) this huge store of energy is of little use to humans. Explain the difference between high quality and low quality energy.

5. Ecosystems require energy to function. Where does this energy come from? Where does it go? How does the flow of energy conform to the laws of thermodynamics?

6. Heat is released during metabolism. How is this heat useful to a cell and to a multicellular organism? How might it be detrimental, especially in a large, complex organism?

7. Photosynthesis and cellular respiration are complementary processes. Explain how they exemplify the laws of conservation of matter and thermodynamics.

8. Water cycles at highly variable rates. Describe a water cycle that occurs in a matter of several days and one that would take more than a year to complete.

9. The population density of large carnivores is always very small compared to the population density of herbivores occupying the same ecosystem. Explain this in relation to the concept of an ecological pyramid.

10. A species is a specific kind of organism. What general characteristics do individuals of a particular species share? Why is it important for ecologists to differentiate among the various species in a biological community?

Questions for Critical or Reflective Thinking

1. When we say that there is no "away" where we can throw things we don't want anymore, are we stating a premise or a conclusion? If you believe this is a premise, supply the appropriate conclusion. If you believe it is a conclusion, supply the appropriate premises. Does the argument change if this statement is a premise or a conclusion?

2. Suppose one of your classmates disagrees with the statement above, saying, "Of course there is an away. It's anywhere out of *my* ecosystem." How would you answer?

3. A few years ago laundry detergents commonly contained enzymes for added cleaning power. What do you suppose these enzymes were supposed to do? Can you imagine any disadvantages to adding enzymes to household products?

4. The first law of thermodynamics is sometimes summarized as "you can't get something for nothing." The second law is summarized as "you can't even break even." Explain what this means.

5. The ecosystem concept revolutionized ecology by introducing holistic, systems thinking as opposed to individualistic life history studies. Why was this a conceptual breakthrough?

6. Compare and contrast the views of F. E. Clements and H. A. Gleason concerning the concept of biological communities as superorganisms. How could these eminent biogeographers study the same communities and reach opposite interpretations? What evidence would be necessary to settle this question? Is lack of evidence the problem?

Key Terms

atoms   22
biological community   27
biomass   29
carbon cycle   35
carbon sinks   35-36
carnivores   30
cells   23
cellular respiration   27
compound   22
conservation of matter   24
consumers   30
decomposers   31
ecosystem   27
energy   24

enzymes   24
first law of thermodynamics   25
food chain   29
food web   29
heat   24
herbivores   30
high-quality energy   24
homeostasis   28
hydrologic cycle   32
kinetic energy   24
low-quality energy   24
matter   22
metabolism   24
molecules   22

nitrogen cycle   36
omnivores   30
organic compounds   23
pH   33
photosynthesis   25
population   27
potential energy   24
producers   30
productivity   29
scavengers   31
second law of thermodynamics   25
species   27
temperature   24
trophic level   30

Suggested Readings

Bent, Henry A. "Entropy and the Energy Crisis." *Journal of Science Teaching* 4, no. 44 (1977): 25-29. Implications of the second law of thermodynamics.

Bolin, B., and R. B. Cook. *The Major Biogeochemical Cycles and Their Interactions.* New York: Wiley, 1983. Detailed treatment of nutrient cycling.

Chapin, F. Stuart, III, et al. "Plant Responses to Multiple Environmental Factors." *BioScience* 1, no. 37(1987):49-57. Examination of resource allocations in plants, focusing on carbon and nitrogen, using a cost-benefit analysis approach.

Hanson, Robert W., ed. *Science and Creation—Geological, Theological and Educational Perspectives.* American Association for the Advancement of Science Series on Issues in Science and Technology. New York: Macmillan, 1986. Explores the false dichotomy in the creation/evolution controversy through working examples of how different teachers, scientists, and theologians deal with the topic.

Kormondy, E. J. *Concepts of Ecology.* 3d ed. New York: Harper & Row, 1984. A good, readable ecology text.

Kroschwitz, Jacqueline I., and Melvin Winokur. *Chemistry: General, Organic, Biological.* New York: McGraw-Hill, 1985. College textbook that gives clear explanations of basic chemistry principles plus practical examples.

Lovelock, James E. *Gaia: A New Look at Life.* Oxford: Oxford University Press, 1979. Presents the Gaia hypothesis, an intriguing but controversial hypothesis in which the biosphere is seen as a gigantic system with organismlike characteristics.

Odum, Eugene P. *Ecology and Our Endangered Life-Support Systems.* Sunderland, MA: Sinauer Assoc., 1993. A brief but compelling introductory ecology textbook.

Odum, Howard T., and Elisabeth C. Odum. *Energy Basis for Man and Nature.* New York: McGraw-Hill, 1980. Energy analysis, principles, and applications of information.

Osmond, C. B., et al. "Stress Physiology and the Distribution of Plants." *BioScience* 1, no. 37 (1987): 38-47. Good review of the subject.

Ricklefs, Robert E. *Ecology.* 3d ed. New York: Freeman, 1990. A comprehensive and well-written general ecology text with many compelling examples from nature and natural history.

3

BIOLOGICAL COMMUNITIES AND BIOMES

One hill cannot shelter two tigers.

Chinese Proverb

INTRODUCTION

Imagine yourself exploring a woods, bog, prairie, or some other natural area that you enjoy (fig. 3.1). If you are observant, you will see plants and animals of many different sizes and shapes, living out unique and varied life cycles in distinct parts of the ecosystem. Why are there so many different kinds of organisms? What determines who lives where and how? Distribution of and roles played by individuals in a biological community are determined by physical and chemical factors in the environment as well as by biological interactions. In chapter 2, we looked at some of the properties of matter and energy as well as ways in which they are cycled through ecosystems. In this chapter, we will study more closely the interactions among organisms in a biological community as they struggle for access to food, shelter, living space, and other resources necessary for life. Not all interactions within a community are competitive, however. Organisms also cooperate with, or at least tolerate, other individuals of their own or other species as they live out their daily lives.

Each biological community consists of several to many populations. Each population is made up of all the members of a single species in a given area. Species must adapt to the conditions they encounter if they are to persist in a particular place. They do this through **evolution** brought about by natural selection acting on inherent variation in populations. The physical environment is dynamic and constantly changing, so that evolution is never finished and complete. And yet, many biological communities are self-perpetuating, resilient, and relatively stable. Although changes in community type do occur through history, major transitions usually take time and are relatively rare. In this chapter, we'll look at how the characteristics of populations and communities are established.

SOME PROPERTIES OF COMMUNITIES

Four fundamental attributes of biological communities and ecosystems are of primary concern to ecologists: productivity, diversity, complexity, and organization. These properties of communities are interrelated, but a problem facing community ecologists is to measure and attempt to understand the exact nature of these relationships in a given setting.

Productivity

A community's **productivity** is measured as the rate of production of biomass, which also may be thought of as the rate of energy conversion. Photosynthetic rates are regulated by light levels, temperature, moisture, and nutrient availability. As you can see in figure 3.2, tropical forests, coral reefs, and estuaries (bays or inundated river valleys where rivers meet oceans)

Figure 3.1 Biological communities can be amazingly diverse and wonderfully abundant as this photograph of the Okefenokee Swamp in Georgia shows. Why are there so many kinds of organisms here?

have high levels of productivity because they have abundant supplies of all these resources. In deserts, lack of water limits photosynthesis. On the arctic tundra or in high mountains, low temperatures inhibit plant growth. In the open ocean, a lack of nutrients reduces the ability of aquatic plants to make use of plentiful sunshine and water.

Some agricultural crops such as corn (maize) and sugar cane grown under ideal conditions in the tropics approach the productivity levels of tropical forests. Because shallow-water ecosystems such as coral reefs, salt marshes, tidal mud flats, and other highly productive aquatic communities are relatively rare compared to the vast extent of open oceans—which are effectively biological deserts—marine ecosystems are much less productive *on average* than terrestrial ecosystems. Although oceans cover about 71 percent of the earth's surface, their total productivity is actually less than that of terrestrial areas.

Even in the most photosynthetically active ecosystems, only a small percentage of the sunlight available is captured and used to make energy-rich compounds. Between one-quarter and three-quarters of the light reaching plants is reflected by leaf surfaces. Most of the light that is absorbed by leaves strikes nonphotosynthetic pigments or other cellular components that convert the energy to heat that is either radiated away or dissipated by evaporation of water. Only one to two percent of the absorbed energy is used by chloroplasts to synthesize carbohydrates.

In a temperate-climate oak forest, for instance, about half the incident light available on a midsummer day is absorbed by the leaves. Ninety-nine percent of this energy is used to evaporate water. A large oak tree can transpire (evaporate) several thousand liters of water on a warm, dry, sunny day while it makes only a few kilograms of sugars and other energy-rich organic compounds. We describe the percentage of available light captured to make useful products the **photosynthetic efficiency.** As you can

imagine, this is an important factor in agriculture. Selecting environmental conditions or plant genetics to increase photosynthetic efficiency from, say, 1 percent to 2 percent could have a dramatic effect on crop yields.

Diversity

Species Diversity is a measure of the number of different species present in a community. Communities with a high level of diversity often have only a few members of any given species in a particular area. There is generally a gradient of diversity from the equator toward the poles. The arctic, for instance, has vast numbers of insects such as mosquitoes, but only a small number of species, while the tropics have vast numbers of species—some of which have incredibly bizarre forms and habits—but often only a few individuals of any particular species in a given area. Greenland is home to 56 species of breeding birds, New York State has 105 species, Guatemala boasts 469, and Columbia has 1,395. Why are so many species found in Columbia and so few in Greenland?

There are a number of reasons for this gradient of diversity. Climate and history are important factors. Greenland has such a harsh climate that the need to survive through the winter or escape to milder climates overwhelms all other factors and severely limits the ability of species to specialize or differentiate into new forms. Furthermore, Greenland was entirely covered by glaciers until relatively recently, so there has been little time for new species to develop.

Many areas in the tropics, by contrast, have relatively abundant rainfall and warm temperatures year-round. Moreover, these conditions have persisted for millions of years, making it possible for plants, animals, and microbes to specialize in the ways they obtain and use resources. This long, slow process of adaptation and evolution (discussed in more detail later in this chapter) has created the immense diversity of life in the tropics. Coral reefs are similarly stable, productive, and conducive to proliferation of diverse and exotic life-forms. The enormous abundance of brightly colored and fantastically shaped fish, corals, sponges, and arthropods in the reef community is one of the best examples we have of community diversity.

Complexity

Community complexity and compartmentalization are generally related to diversity. They are important because they help us visualize and understand community functions. **Complexity** in ecological terms refers to the number of species at each trophic level and the number of trophic levels in a community. We distinguish between diversity and complexity because a community might be diverse in the sense of having many different species, but all those organisms might

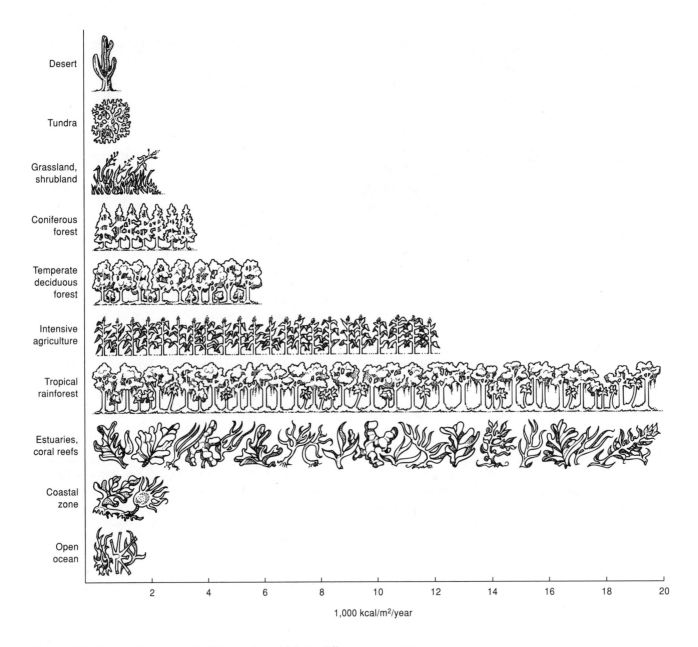

Desert

Tundra

Grassland, shrubland

Coniferous forest

Temperate deciduous forest

Intensive agriculture

Tropical rainforest

Estuaries, coral reefs

Coastal zone

Open ocean

2   4   6   8   10   12   14   16   18   20

1,000 kcal/m²/year

Figure 3.2 Gross primary productivity varies widely in different ecosystems.

be clustered in only two or three trophic levels. By contrast, a complex community might have not only many species but also many trophic levels. In addition, a very complex community might be compartmentalized into subdivisions within a single trophic level. In the tropical rainforest, for instance, the herbivores can be grouped into "guilds" based on the specialized ways they feed on plants. There may be fruit eaters, leaf nibblers, root borers, seed gnawers, and sap suckers, each composed of species of very different size, shape, and even biological kingdom, but that feed in related ways.

In 1955, Robert MacArthur, who was then a graduate student at Yale, proposed that the more complex a community is, the more stable and resilient it will be in the face of disturbance. This seems intuitively reasonable; if many different species occupy each trophic level, some can fill in if others are stressed or eliminated by external forces. The whole community should be able to recover relatively easily from perturbations and disruptions. This theory has been very controversial, however. Some studies, both theoretical and in the field, have supported it, while others have not. An alternate explanation is that in some communities, diversity creates time lags in population processes that destabilize the system, making it more—rather than less—sensitive to disturbances. Furthermore, specialization, which generally accompanies complexity, makes populations more dependent on a particular set of resources and less able to adapt quickly to changes. For instance, a field full of

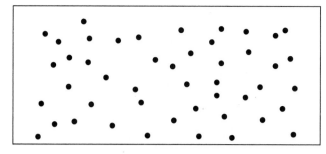

a.

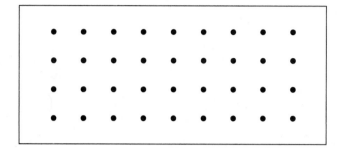

b.

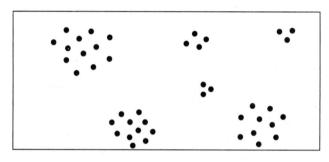

c.

Figure 3.3 Distribution of members of a population in a given space can be random (*a*), ordered (*b*), or clustered (*c*). These patterns are determined both by the physical environment and by biological interactions. They may produce a graininess or patchiness in community structure.

nothing but dandelions, which are generalists, may be better able to withstand some kinds of disturbance than a highly complex native prairie.

Structure

Structure refers to patterns of spatial distribution of individuals and populations within the community as well as the relation of a particular community to its neighbors. At the local level, even in a relatively homogeneous environment, individuals in a single population can be distributed randomly, clumped together, or exist in highly regular patterns (fig. 3.3). In randomly arranged populations, individuals live wherever resources are available. Ordered patterns may be determined by the physical environment but are more often the result of biological competition. For example, competition for nesting space in a penguin colony is often fierce. Each nest tends to be just out of reach of the neighbors sitting on their own nests. Constant squabbling produces a highly regular pattern. Similarly, sagebrush releases toxins from roots and shed leaves that inhibit the growth of competitors and create a circle of bare ground surrounding each bush. As neighbors fill in empty spaces up to the limit of this chemical barrier, a regular spacing results.

In contrast to these random or ordered patterns, some species cluster together (fig. 3.3c), often for protection or mutual assistance or to gain access to a particular environmental resource. For instance, dense schools of fish swim closely together in the ocean. By clustering together, they increase their chances of detecting predators and escaping if they are attacked. Similarly predators, whether sharks, wolves, or humans, often hunt in packs to catch their prey. A flock of blackbirds descending on a cornfield or a troop of baboons traveling across the African savanna band together both to avoid predators and to find food more efficiently.

Plants, too, cluster for protection. Wind-sheared evergreen trees are often found packed tightly together in groves at the crest of a high mountain or along the seashore. They offer protection from the wind not only to each other but also to other creatures that find shelter in or under their branches.

If we look closely at larger and more ecologically varied communities, we find they are often a mosaic of smaller units or subsets of the assemblage that composes the whole. These subunits derive from the fact that each species has a preference for specific, localized conditions. Ecologists describe these patterns as **patchiness** or graininess in spatial distribution. Some communities are coarse-grained with large patches; others are so fine-grained as to be almost indistinguishable. Even though individual patches may be small and not very distinct from adjacent areas, the specific interactions in each patch may be more important in determining functions of the community than the overall composition.

Distribution in a community can be vertical as well as horizontal. The tropical forest, for instance, has many layers, each with different environmental conditions and combinations of species (box 3.1). While the larger trees are part of each layer, the communities of smaller plants, animals, and microbes that live in each part of this complex community generally are distinct and very different from each other. Similarly, aquatic communities are often stratified into layers based on light penetration in the water, temperature, salinity, pressure, or other factors.

Finally, community structure also refers to a community's relation to adjacent communities. In many cases, the boundary between communities is relatively sharp and distinct. In moving from a forest out onto a grassland, you sense a dramatic change from one to the other. In the forest, you are surrounded by dense

Tropical Rainforests: Life in Layers

The richest and most productive biological communities in the world are in the tropical forests. These forests have been reduced to less than half of their former extent by human activities and now cover only about 7 percent of the earth's land area. In this limited area, however, lives about two-thirds of the vegetation mass and about half of all the living species in the world!

The largest, lushest, and most biologically diverse of the remaining tropical moist forests are in the Amazon River basin of South America, the Congo River basin of central Africa, and the large islands of Southeast Asia (Sumatra, Borneo, and Papua New Guinea). Whereas the forests of mainland Southeast Asia, western Africa, and Central America are strongly seasonal, with wet and dry seasons, the South American and central African forests are true rainforests. Rainfall is generally more than 400 cm (160 in) per year and falls more or less evenly throughout the year. It is said that such rainforests make their own rain, because about half the rain that falls in the forests comes from condensation of water vapor released by transpiration from the trees themselves. Rainforests at lower elevations are hot and humid year-round. At higher elevations, tropical mountains intercept moisture-laden clouds, so the forests that blanket their slopes are cool, wet, and fog-shrouded. They are aptly and poetically called "cloud forests."

Tropical forests are mostly very old. Unlike temperate forests, they haven't been disturbed by glaciation or mountain building for hundreds of millions of years. This long period of evolution under conditions of ample moisture and stable temperatures has created an incredible diversity of organisms of amazing shapes, colors, sizes, habits, and specialized adaptations.

Habitats in a tropical rainforest are stratified into three to five distinct layers from ground level to the tops of the tallest trees (box fig. 3.1.1). Let's start at the top.

Hundreds of tree species grow together in lush profusion, their crowns interlocking to form a dense, dappled canopy about 40 m (120 ft) above the forest floor. These unusually tall trees are supported by relatively thin trunks that are reinforced by wedge-shaped buttresses, instead of having thick trunks and deep roots. A few emergent trees rise above the seemingly solid canopy into a world of sunlight, wind, and open space. Numerous species of birds, insects, reptiles, and small mammals live exclusively in the forest canopy, never descending below the crowns of the trees.

The forest understory is composed of small trees and shrubs growing between the trunks of the major trees, as well as climbing woody vines (lianas) and many epiphytes—mainly orchids, bromeliads, and arboreal ferns—that attach themselves to the trees. Some of the larger trees may support fifty to one hundred different species of epiphytes and an even larger population of animals that are specialized to live in the many habitats they create. These understory layers are a world of bright but filtered light abuzz with animal activity.

By contrast, the forest floor is generally dark, humid, quiet, and rather open. Few herbaceous plants can survive in the deep shade created by the layered canopy of the forest trees and their epiphytes. The most numerous animals are ants and termites that scavenge on the detritus raining down from above. A few rodent species gather fallen fruits and nuts. Rare predators such as leopards, jaguars, smaller cats, and large snakes hunt both on the ground and in the understory.

What happens at the soil level? The productivity of a tropical rainforest can be as high as 90 tons per hectare per year. You might think that the soil that supports this incredible growth is rich and fertile. Instead, however, it is old, acidic, and nutrients poor. Ages of incessant tropical rains and high temperatures have depleted minerals, leaving an iron- and aluminum-rich ultisol. Tropical forests have only about 10 percent of their organic material and nutrients in the soil, compared to boreal

trees and the species they harbor. As you move out into the sun, wind, and open space of the grassland, you find an entirely different group of organisms and environmental conditions. Few species from one community extend very far into the adjacent community. We call this boundary an **ecotone.** A community that is sharply divided from its neighbors is called a closed community (fig. 3.4). In contrast, communities with indistinct boundaries across which many species cross are called open communities. Many ecologists, however, argue that the distinctions we perceive across ecotones are less real and less important than we may tend to believe. They argue that many species cross even apparently sharp ecotones. Birds feed on the

grassland but nest in the forest. Mammals, reptiles, insects, and other groups spend some time in both and have important influences on both communities. Nutrients are carried from one to the other.

One important result of openness of communities is that it may allow species to migrate from one area into another in which they might not otherwise be able to survive. The presence of individuals of a particular species in a community does not necessarily prove that a self-sustaining, breeding population is present. There could be occasional or continual recruitment from an adjacent community where conditions are more suitable for the species and where the population is self-perpetuating.

forests, which may have 90 percent of their organic material in litter and sediments.

The interactions of decomposers and living plant roots in the soil are, literally, the critical base that maintains the rainforest ecosystem. Tropical rainforests are able to maintain high productivity only through rapid recycling of nutrients. As you might suspect, the constant rain of detritus and litter that falls to the ground is quickly decomposed by populations of fungi and bacteria that flourish in the warm, moist environment. Some of these decomposers have symbiotic relationships with the roots of specific trees. Trees have broad, shallow root systems to capitalize on this surface nutrient source; an individual tree might create a dense mat of superficial roots 100 meters in diameter and 1 meter thick. In this way, nutrients are absorbed quickly and almost entirely and are resued almost immediately to build fresh plant growth, the necessary base to the trophic pyramid of this incredible ecosystem.

Figure 3.1.1 The layered communities of a tropical rainforest are directly related to the gradual extinction of light, from the brightness of the uppermost canopy to the deep shade of the forest floor.

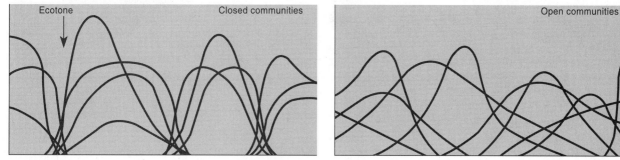

Figure 3.4 Hypothetical distribution of species along two gradients of environmental factors. In closed communities (left), species are grouped in distinct assemblages separated by sharp ecotones (boundaries). In open communities (right), species are distributed at random along the gradient.

WHO LIVES WHERE, AND WHY?

"Why" questions often are the stimulus for scientific research, but the research itself centers on "how" questions. Why, we wonder, does a particular species live where it does? More to the point, how is it *able* to live there? How does it use the physical resources of its environment, and are some of its techniques unique? How does it interact with other species present? What gives one species an edge over another species in a particular habitat?

In this section we will emphasize some specific ways species are limited by the physical aspects of their environment. We then will discuss some ways members of a biological community interact, pointing out some of the difficulties ecologists encounter when they attempt to discern patterns and make generalizations about community interactions and organization.

Limiting Factors and Tolerance Limits

Victor Shelford, a pioneer in American ecology, developed the principle of **limiting factors.** This principle states that for each physical factor in the environment, there are both minimum and maximum limits, called **tolerance limits,** beyond which a particular species cannot survive (fig. 3.5). Furthermore, the one factor that is closest to a tolerance limit for a given species at a given time is the *critical factor* that determines, more than any other physical factor, the abundance and distribution of a species in a given area. Water, light or food, and temperature are the most common critical factors for plants and animals, although something as specific as a particular pH or mineral requirement also can be a critical factor for a species.

Sometimes the requirements and tolerances of species can be useful indicators of specific environmental characteristics. The absence or presence of such species can tell us something about the community and the ecosystem as a whole. Locoweeds, for instance, are small legumes that grow where soil concentrations of selenium are high. Because selenium often is found with uranium deposits, locoweeds have an applied economic value as environmental indicators. Such indicator species also may demonstrate the effects of human activities. Lichens and eastern white pine are less restricted in habitat than locoweeds, but are indicators of air pollution, as they are extremely sensitive to sulfur dioxide and acid precipitation. Bull thistle is a weed that grows on disturbed soil but is not eaten by cattle; therefore, an abundant population of bull thistle in a pasture is a good indicator of overgrazing. Anglers know that trout species require clean, well-oxygenated water, so the presence or absence of trout is an indicator of water quality.

Natural Selection and Adaptation

What do genetics and evolution have to do with ecology? The answer is, a great deal! Each organism carries a specific genetic heritage, the legacy of generations of predecessors. Organisms inherit specific nutritional requirements and tolerances for environmental factors. Cultivated plants provide numerous good examples: citrus

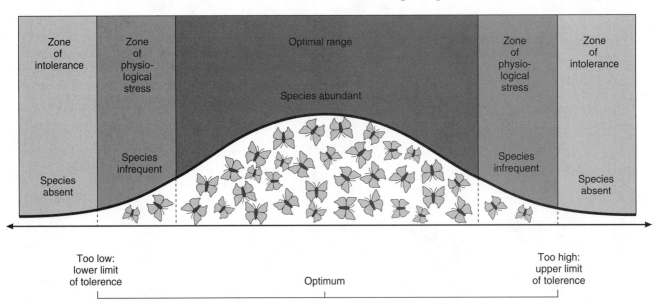

Figure 3.5 The concepts of limiting factors and tolerance ranges recognize that the presence of a species often is related to its ability to survive within a specific range of variation for each factor of the environment. The factor may be, for instance, the amount of shade, a particular mineral concentration, or the presence of a required food source. A species does best within an optimum range but is responsive to stresses created by higher or lower levels of the factor, as depicted by the hypothetical population curve in this drawing.

trees have a low tolerance for cold winters, whereas apple trees require winter chilling; most ornamental shrubs prefer soil that is neutral in pH, but azaleas and rhododendrons thrive in acidic soil.

How have species accumulated their particular structural and metabolic characteristics, their specific sets of genetic characteristics? Overwhelming scientific evidence supports the theory of evolution developed by Charles Darwin and Alfred Wallace. This theory states that species have developed by a long series of small, random mutations (changes in genetic characteristics), some of which prove to be better adapted to a particular set of environmental conditions. **Natural selection** is the complex process by which environmental pressures cause those individuals with genetic combinations that make them more "fit"—that is, better able to acquire or use resources to survive and reproduce—to become more abundant than others in the population whose genetics make them less fit in that particular environment.

Some environmental factors that can influence fertility or survivorship are (1) physiological stress due to inappropriate levels of some critical environmental factor, such as moisture, light, temperature, pH, or a specific nutrient; (2) predation, including parasitism and disease; and (3) competition. Be sure you understand that evolution and adaptation work at the population level. Each individual is locked in by genetics to a particular way of life. Most plants, animals, or microbes can do little to modify their physical makeup or behavior to better suit a particular environment. Over time, however, a whole species can adapt, as individuals that happen to be better suited for existing conditions survive and are more successful in reproduction than others who are less fit.

Natural selection and adaptation can cause organisms with a similar origin to become very different in appearance and habits over time, but it can also result in unrelated organisms coming to look and act very much alike. We call this latter process convergent evolution. If a community lacks woodpeckers, for instance, some other species will likely take on the role of eating insects that live under tree bark and in dead wood. Over time, these ecological equivalents may come to look much like woodpeckers even though they are genetically unrelated. Some examples of ecological equivalents are shown in figure 3.6.

We often make the mistake of believing that organisms develop certain characteristics because they want or need them (box 3.2). This is incorrect. The genetic mutations (changes) that give rise to evolution occur at random. A variety of different genetic types are always present in any population. Natural selection favors those best suited for particular conditions. Over many generations this results in adaptation and

Figure 3.6 Ecological equivalents are unrelated species that fill similar niches in similar ecosystems in different places in the world. The Old World animals on the left (gerbil, jackal, gazelle) have North American equivalents (kangaroo rat, coyote, pronghorn).

creation of new species. Whether there is a purpose or direction to this process is a theological rather than a scientific question.

The Ecological Niche

Habitat describes the place or set of environmental conditions in which a particular organism lives. An **ecological niche** is a functional description of the role a species plays in a community—how it makes its living. Well-established communities often demonstrate **resource partitioning,** in which populations share limiting resources through specialization (fig. 3.7). By specializing in their choice of seeds, for instance, two populations of seed-eating birds

How the Camel Got Its Hump

How did the many different kinds of plants and animals acquire their unique characteristics? How did the camel get its hump, the giraffe its long neck, and the elephant its remarkable trunk? These questions have intrigued people for a long time, as is shown by the many myths in nearly every culture that attempt to explain the amazing variation we see in nature.

Many people remain confused about how evolution works. It seems natural to suppose that creatures acquire specialized features in order to adapt to their particular environments. French biologist Jean Baptiste de Lamarck (1744–1829) developed a theory of acquired characteristics in 1809. He stated that traits could be determined by use or disuse of an organ; that is, a body part not used by an organism atrophies and is lost to future generations, while a body part used extensively is amplified. For example the Lamarckian explanation for the long neck of the giraffe is that the ancestors of modern giraffes constantly reached into the trees to feed on high-growing vegetation. This repeated stretching over time resulted in long necks that were inherited by offspring, eventually leading to the extreme length we see today. Similar explanations have been given for camels' humps, elephants' trunks, and many other amazing features of living organisms.

It's true that organisms can modify their body shape to some extent through use or disuse, but the ability to do so is determined by the genes each individual inherits from its parents. Your potential for building muscles through exercise was fixed at conception. Whether you exercise or not won't have any effect on your genes, nor will exercise influence the genes you pass on to your offspring. Of course, if you got genes for big muscles from your parents, there is a good chance that your offspring will inherit this trait as well.

How then does change come about? How does evolution occur? The answer is that genes are shuffled and recombined in each generation. Some of your offspring are likely to have genes for bigger muscles or a longer neck than you have; others will have genes for smaller or shorter or otherwise different features than you. This variation exists in every large population for every genetically determined characteristic. The way in which the average features of the population change over time is through reproduction.

If the trait passed from one generation to another affects survival or reproduction in a way that makes those that inherit it more successful in producing offspring, it will gradually become dominant in a population. We call this process natural selection. It is driven by the struggle for access to resources, survival, and reproduction, but it is made possible by random mutations (changes) in the genetic materials that serve as blueprints for life. These changes occur first and then are selected by environmental or social factors.

It's tempting to think that organisms develop characteristics because they need to or want to. The truth is, however, that features are acquired only through a slow process of small, random changes selected for through reproductive success. A duck doesn't have webbed feet *in order* to swim; rather, a duck *is able* to swim because some ancestor happened to have a gene for webbed feet that it could pass on to future generations. Similarly, giraffe ancestors with long necks had better access to food than giraffes with shorter necks, and camels have humps because their ancestors with large fat deposits survived better and had more offspring than those without fat deposits.

---

can share the seed resources of a community with less competition. A community can be thought of as a collection of interacting, niche-differentiated species adapted for the existing environmental conditions and complementing each other in the use of space, resources, and time.

Perhaps you haven't thought of time as an ecological factor, but in a community, niche specialization is a twenty-four-hour phenomenon. Flowers of different species open at different times of the day and night, attracting pollinators that are active at different times. Swallows and insectivorous bats both catch insects as they fly, but some insect species are active during the day and others are active at night, providing noncompetitive feeding opportunities for day-active swallows and night-active bats.

If the presence of a resource is predictable, a population may specialize to exploit it, reducing the level of interspecific competition. This phenomenon is demonstrated in the **vertical stratification** of communities, whereby specific populations form subcommunities at different layers, or strata. Terrestrial forests and most aquatic ecosystems provide good examples of vertical stratification of niches within a community (fig. 3.8).

Some species have narrow, well-defined niches, whereas others do not. Compare, for instance, the European starling and the golden-cheeked warbler. The starling, now a common resident in the United States, is a niche generalist. It has wide tolerance limits and is found in many habitats, from natural areas to inner cities. It is flexible in its choice of foods and nesting sites, choosing a tree nest hole or electrical

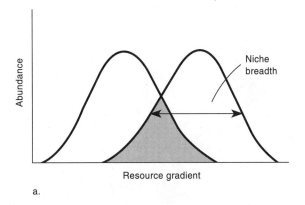

a.

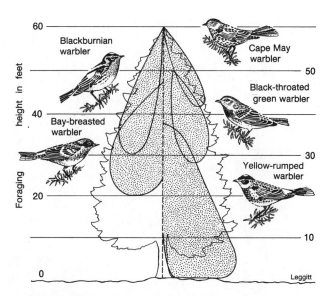

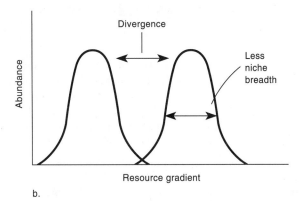

b.

Figure 3.7 Resource partitioning and niche specialization caused by competition. Where niches of two species overlap along a resource gradient, competition occurs (shaded area in *a*). Individuals occupying this part of the niche are less successful in reproduction so that characteristics of the population diverge to produce more specialization, narrower niche breadth, and less competition between species (*b*).

Figure 3.8 Resource partitioning and the concept of the ecological niche are demonstrated by the way several species of wood warblers use different strata of the same forest. This is a classic example of the principle of competitive exclusion.

transformer box with equal ease. By contrast, the golden-cheeked warbler is a niche specialist, with a very restricted habitat preference. The species is making its last stand on a juniper-covered limestone hillside in central Texas, its habitat threatened by development of vacation homes. Species with such narrow tolerance limits are likely to be sensitive to small differences in microclimate, soil characteristics, food, nesting sites or materials, or some other critical factor. They lack flexibility to respond to changing environmental conditions.

SPECIES INTERACTIONS AND COMMUNITY DYNAMICS

The world is dynamic. Changes constantly occur in topography, climate, moisture levels, and other environmental factors. In response to these changes and in the struggle for existence, populations of organisms also change. Not all biological interactions are competitive, however. Organisms also cooperate or at least tolerate individuals of their own species as well as individuals of other species in order to survive and reproduce.

Predation

All organisms need food to live. Producers make their own food, and consumers get theirs by eating organic molecules produced by other organisms. In most communities, photosynthetic organisms are the producers. Consumers include herbivores, carnivores, omnivores, scavengers, and decomposers. With which of these categories do you associate the term "predator"? Ecologically, the term has a much broader meaning than you might expect. A **predator** in community ecology is an organism that feeds directly upon another living organism, whether or not it kills the prey to do so. This definition separates live-feeders from scavengers and decomposers. By this definition, herbivores, carnivores, and omnivores all are predators (fig. 3.9). Also, in this broad sense, predation in a community includes feeding by **parasites,** organisms that live in or upon the body of a host organism and take nourishment from it, usually without killing it. **Pathogens,** disease-causing organisms, also may be included.

Predation is a potent and complex influence on the population balance of communities. It involves (1) all stages of the life cycles of predator and prey species, (2) many specialized food-obtaining mechanisms, and

Figure 3.9 Insect herbivores are predators as much as are lions and tigers. In fact, insects consume the vast majority of biomass in the world. Often, complex patterns of predation and defense have evolved between insect predators and their plant hosts.

(3) specific prey–predator adaptations that either resist or encourage predation.

For instance, predation throughout the life cycle is very pronounced in marine intertidal animals. Many crustaceans, mollusks, and worms release eggs directly into the water, and the eggs and free-living larval and juvenile stages are part of the floating community, or **plankton** (fig. 3.10). Planktonic animals feed upon each other and are food for successively larger carnivores, including small fish. As prey species mature, their predators change. For instance, barnacle larvae are planktonic and are eaten by fish. Adult barnacles build hard shells that protect them from fish but can be crushed by limpets and other mollusks. Also, predators change their feeding targets. Adult frogs, for instance, are carnivores, but the tadpoles of most species are grazing herbivores. To sort out the trophic levels in these communities can be very difficult.

Obviously, direct feeding is one way predators affect a community; however, predators also have indirect effects on nonprey populations they do not feed upon. For instance, sea otters eat sea urchins, and sea urchins, in turn, feed on kelp. When California sea otters were eliminated by fur hunters early in this century, sea urchin populations exploded and the kelp forests that sheltered many other species were decimated. Large areas were turned into impoverished "urchin barrens." Now that sea otters are protected and populations are rebounding, sea urchins are being controlled, and kelp forests, fish, and many other spe-

Figure 3.10 Adult forms of many marine invertebrates—such as crabs, snails, and starfish—have planktonic larval stages. Thus, adults and larvae do not compete for food resources. They participate in different food chains.

cies are also recovering. Ecologists call a species that determines the essential characteristics of a community the keystone species. This may be a dominant plant such as kelp that feeds and shelters other species, or it may be the key predator that shapes the community by how it feeds.

Predation is an important factor in evolution. Predators prey most successfully on the slowest, weakest, and least fit members of the prey population, thus allowing successful traits in the prey population to become dominant. Prey species have evolved many protective or defensive adaptations to avoid predation. In plants this often takes the form of thick bark, spines, thorns, or chemical defenses. Animal prey may become very clever at hiding, fleeing, or fighting back against predators. Predators, in turn, evolve mechanisms to overcome the defenses of their prey. This mutual molding of interacting partners is called **coevolution**.

Competition

Competition is another kind of antagonistic relationship within a community. For what do organisms compete? To answer this question, think again about what all organisms need to survive: energy and matter in usable forms, space, and specific sites for life activities. Plants compete for growing space for root and shoot systems so they can absorb and process sunlight, water, and nutrients. Animals compete for living, nesting, and feeding sites, food, water, and mates. Competition among members of the same species is called *intraspecific* competition, whereas competition between members of different species is called *interspecific* competition.

Intraspecific competition can be especially intense because members of the same species have the same space and nutritional requirements; therefore, they compete directly for these environmental resources. How do plants cope with intraspecific competition? The inability of seedlings to germinate in the shady conditions created by parent plants acts to limit intraspecific competition by favoring the mature, reproductive plants. Many plants have adaptations for dispersing their seeds to other sites by air, water, or animals. You've seen dandelion plumes and probably have had sticky or burred seeds attach themselves to your clothing (fig. 3.11). Some plants secrete leaf or root exudates that inhibit the growth of seedlings near them, including their own and those of other species. This strategy is particularly significant where water is a limiting factor.

Animals also have developed adaptive responses to intraspecific competition. Two major examples are varied life cycles and territoriality. The life cycles of many invertebrate species have juvenile stages that are very different from the adults in habitat and feeding. Compare, for instance, a leaf-munching caterpillar to a nectar-sipping adult butterfly, a herbivorous tadpole to an insectivorous frog, or a planktonic crab larva to its bottom-crawling adult form. In these examples, the adults and juveniles do not compete, even though they are members of the same species and live in the same habitats.

You may have observed robins chasing other robins during the mating and nesting season. Robins and many other vertebrate species demonstrate **territoriality**, in which they define an area surrounding their home site or nesting site and defend it, primarily against other members of their own species. Territoriality helps to allocate the resources of an area by spacing out the members of a population. It also promotes dispersal into adjacent areas by pushing grown offspring outward from the parental territory.

Interspecific competition can affect community composition as well as the characteristics and habits of competing species through natural selection. The role of competition as a selective force is expressed in the classic principle of **competitive exclusion.** This principle states that two species will not occupy exactly the same ecological niche in a community for very long. Natural selection will favor those individuals that use resources in ways or at times that avoid competition. This will tend to cause niches to diverge (see fig. 3.7) in ways that reduce rivalry.

The process of niche specialization not only separates existing species but also allows subpopulations within a species to evolve into distinct species. Why doesn't this proceed until there is an infinite number of species? The answer is that a given resource can be partitioned only so far. Populations must be maintained at a minimum size to avoid genetic problems and to survive bad times. This puts an upper limit on the number of different niches—and therefore the number of species—that a given community can support.

Symbiosis

In contrast to predation and competition, symbiotic interactions between organisms are nonantagonistic. **Symbiosis** is the intimate living together of members of two or more species. Lichens are a combination of a fungus and an alga or a bacterium (box 3.3). Their association is a type of symbiosis called mutualism, because both members of the partnership benefit. **Commensalism** is a type of symbiosis in which one member clearly benefits and the other apparently is neither benefited nor harmed. Parasitism, described earlier as a form of predation, also may be considered a type of symbiosis, where one species benefits and the other is harmed. All of these relationships have a bearing on such ecological factors as resource utilization, niche specialization, diversity, predation, and competition. Symbiotic relationships often enhance the survival of one or both partners.

In symbiotic relationships, the partners often show some degree of coadaptation in which their structural and behavioral characteristics are shaped, at least in part, by the relationship. One of the most interesting cases of mutualistic coadaptation involves Central and South American swollen thorn acacias and acacia ants (fig. 3.12). Acacia ant colonies live within the swollen thorns on the acacia tree branches and feed on two

Figure 3.11 Pioneer plant species usually have some mechanism for seed dispersal over long distances so they can reach new areas for growth.

Partners for Life

*L*ichens are symbiotic associations between a filamentous fungus (usually an ascomycete) and a photosynthetic partner, which may be either a green alga or a cyanobacterium or both. They are an excellent example of obligate mutualism in which both partners not only benefit from the association but cannot survive without it. Lichens come in many sizes, colors, and shapes. Some form a thin, flaky crust on boulders, statues, and gravestones. Others are more leafy in appearance and grow on tree trunks or directly on the ground as well as on rocks. The largest species form wispy filaments—sometimes called "old man's beard"—that hang from branches in moist forests. In the northern tiaga and tundra the ground is often covered by thick, puffy mats of gray "reindeer moss"—really a lichen—that are a major food source for caribou and musk oxen.

Most of the visible body of the lichen is fungus whose thin hyphae (filaments) wrap around and encase the individual cells of the algal or bacterial partner (box fig. 3.3.1). Enough light penetrates the translucent body of the fungus to make photosynthesis possible. Thus the alga or bacterium synthesizes energy-rich compounds that sustain both partners. The fungus, in turn, forms a thick, leathery body that shelters the photosynthetic cells. By absorbing and storing water, the fungus helps keep both partners alive during dry times. Apparently, biochemical signals they exchange to coordinate their growth and metabolism have become essential for normal development of both partners.

With an estimated 1,500 species, lichens can be found from tropical rainforests to the tops of high mountains. The tough, durable body of the fungus coupled with the photosynthetic ability of its partner allows them to survive in the harshest environments. They are often the first pioneers in succession, colonizing places where no other plant can grow. Clinging to bare rock or other inhospitable substrate, they survive on rainwater, dustfall, and the few minerals they obtain by secreting organic acids to etch away surfaces to which they adhere. Where a cyanobacterium is the photosynthetic partner, the lichen can fix nitrogen from the air. Although lichens usually grow very slowly, they generally have little competition for the rugged habitats they colonize. Most lichens are unpalatable and have few predators. After years or even centuries of growth, they may build up enough soil so that grasses and other faster-growing plants can get a foothold and develop a more productive community.

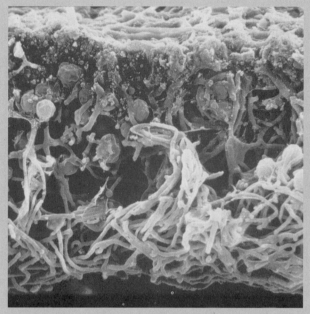

Figure 3.3.1 Mutualistic association of a fungus and photosynthetic partner in the body of a lichen are shown in this scanning electron microscope photograph. The fungus forms a filamentous matrix within which the spherical, single-celled photosynthetic partners live.

Although lichens are very tough and resistant to the harsh environmental conditions to which they are so well adapted, they are very sensitive to air pollution. Because they often are totally dependent on rainwater, which is stored in the absorptive fungal filaments, they are very susceptible to sulfur dioxide, sulfuric acid, and other air pollutants. Even in remote and seemingly pristine areas such as Acadia National Park in Maine and Big Bend National Park in Texas, research has shown that lichens can be used as indicator species for long-range transport of air pollutants.

Lichens also take up and store radioactive materials and pesticides. In the 1950s it was shown that DDT from agricultural areas thousands of kilometers away was accumulating in arctic reindeer moss and was then being passed up the food chain from caribou to indigenous Inuit people who accumulated high levels in fat deposits. Similarly, after the explosion and fire at the Chernobyl Nuclear Reactor in 1985, radioactive materials accumulated in the tissues of reindeer owned by the Lapp people in northern Scandinavia so that thousands of animals had to be killed and buried.

Figure 3.12 The swollen thorn acacia has a complex symbiotic relationship with certain ant species, shown here feeding on nutritious fluids produced in the row of nectaries on the leaf petiole.

kinds of food provided by the acacias—nectar produced in nectaries at the leaf bases and special protein-rich structures (Beltian bodies) produced on leaflet tips. The acacias thus provide shelter and food for the ants. Although they expend energy to provide these services, they are not physically harmed by ant feeding.

What do the acacias get in return and how does the relationship relate to community dynamics? Ants tend to be aggressive defenders of their home areas, and acacia ants are no exception. They kill herbivorous insects that attempt to feed on their home acacia, thus reducing predation. They also trim away vegetation that grows around its base, thereby reducing competition. This is a fascinating example of how a symbiotic relationship fits into community interactions. It is also an example of coevolution based on mutualism rather than competition or predation.

COMMUNITIES IN TRANSITION

So far our view of communities has focused on the day-to-day interactions of organisms with their environments, set in a context of survival and selection. In this section, we'll step back to look at some transitional aspects of communities, including where communities meet and how communities change over time.

Ecological Succession

Biological communities have a history in a given landscape. The process by which organisms occupy a site and gradually change environmental conditions so that other species can replace the original inhabitants is called ecological **succession** or development (fig. 3.13). **Primary succession** occurs when a community begins to develop on a site that has not been occupied previously by living organisms, such as an island, a sand or silt bed, a body of water, or a new volcanic flow. **Secondary succession** occurs when an existing community is disrupted by some natural catastrophe such as fire or flooding, or by a human activity such as deforestation, plowing, or mining, and a series of communities subsequently develops at the site.

In primary succession on a terrestrial site, the new site first is colonized by a few hardy **pioneer species,** often microbes, mosses, and lichens that can withstand harsh conditions and lack of resources. Their bodies create patches of organic matter in which protists and small animals can live. Organic debris accumulates in pockets and crevices, providing soil in which seeds can become lodged and grow. The community of organisms becomes more diverse and increasingly competitive as development continues and new niche opportunities appear. The pioneer species gradually disappear as the environment changes and new species combinations replace the preceding community.

Figure 3.14 depicts succession in an aquatic ecosystem, showing how a developing community changes its own environment. The amount of open water in the lake or pond gradually decreases as vegetation encroaches from the margins, resulting in gradual community replacement progressing from the edges of the pond toward the center. Succession proceeds from open lake, to shallow pond with highly vegetated edges, to marshy area with rooted, emergent vegetation, and finally to grassland or forest.

Examples of secondary succession are easy to find. Observe an abandoned farm field or burned-over forest in a temperate climate. The bare soil first is colonized by rapidly growing annual plants (those that grow, flower, and die the same year) that have light, wind-blown seeds and can tolerate full sunlight and exposed soil (see fig. 3.11). They are followed and replaced by perennial plants (those that live for several to many years), including grasses, various non-woody flowering plants, shrubs, and trees. As in primary succession, plant species progressively change the environmental conditions. Biomass accumulates and the site becomes richer, better able to capture and store moisture, more sheltered from wind and climate change, and biologically more complex.

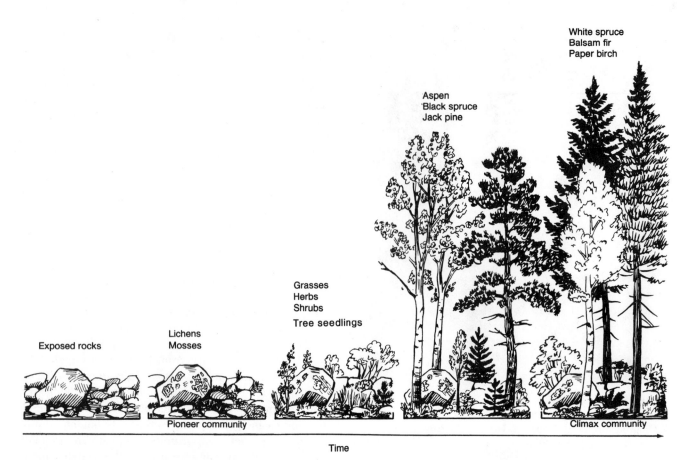

White spruce
Balsam fir
Paper birch

Aspen
Black spruce
Jack pine

Grasses
Herbs
Shrubs

Tree seedlings

Exposed rocks

Lichens
Mosses

Pioneer community

Climax community

Time

Figure 3.13 Primary succession on a terrestrial site is shown in five stages, from left to right, beginning with rocks that are initially colonized by a pioneer community of lichens and mosses and ending with a climax forest community.

Figure 3.14 Succession occurs in a pond as vegetation gradually encroaches from the margins toward the center. Eventually, the pond may fill in completely.

Species that cannot survive in a bare, dry, sunny, open area find shelter and food as the field turns to prairie or forest.

Eventually in either primary or secondary succession a community develops that resists further change. Ecologists call this a **climax community** because it seems to be the climax of the successional process. An analogy is often made between community succession and organism maturation. Beginning with a primitive or juvenile state and going through a complex developmental process, each progresses until a stable or mature form is reached. It is dangerous to carry this analogy too far, however, because no mechanism is known to regulate communities in the same way that genetics and physiology regulate development of the body.

Mature or highly developed communities tend to be resilient and stable over long periods of time because they can resist or recover from external disturbances. Many are characterized by high species diversity, narrow niche specialization, well-organized community structure, good nutrient conservation and recycling, and a large amount of total organic matter. Community functions, such as productivity and nutrient cycling, tend to be self-stabilizing or self-perpetuating. The concept of the climax community is not as dogmatic as it once was because additional research indicates that what once were regarded as "final" climax communities still are changing. It's probably more accurate to say that the rate of succession is so slow in a climax community that, from the perspective of a single human lifetime, it seems a stable community.

Some landscapes never reach a stable climax in the traditional sense because they are characterized by and adapted to periodic disruption. They are called equilibrium communities or disclimax communities. Grasslands, the chapparral shrubland of California, and some kinds of coniferous forests, for instance, are shaped and maintained by periodic fires that have been a part of their history. They are, therefore, often referred to as fire-climax communities. Plants in these communities are adapted to resist fires or to reseed quickly after fires or both. In fact, many of the plant species we recognize as dominants in these communities *require* fire to eliminate competition, to prepare seedbeds for germination of seedlings, or to open cones or thick seed coats. Without fire, community structure might be quite different.

Introduced Species and Community Change

Succession requires the continual introduction of new community members and the disappearance of previously existing species. New species move in as conditions become suitable; others die or move out as the community changes. New species also can be introduced after a stable community already has become established. Some cannot compete with existing species and fail to become established. Others are able to fit into and become part of the community, defining new ecological niches. If, however, an introduced species preys upon or competes more successfully with one or more populations that are native to the community, the entire nature of the community can be altered.

Some human introductions of European plants and animals to non-European communities have been disastrous to native species because of competition or overpredation. Oceanic islands offer classic examples of devastation caused by rats, goats, cats, and pigs liberated from sailing ships. All these animals are prolific, quickly developing large populations. Goats are efficient, nonspecific herbivores; they eat nearly everything vegetational, from grasses and herbs to seedlings and shrubs. In addition, their sharp hooves are hard on plants rooted in thin island soils. Rats and pigs are opportunistic omnivores, eating the eggs and nestlings of sea birds that tend to nest in large, densely packed colonies, and digging up sea turtle eggs. Cats prey upon nestlings of both ground-nesting and tree-nesting birds. Native island species are particularly vulnerable because they have not evolved under circumstances that required them to have defensive adaptations to these predators.

Sometimes, in an attempt to solve problems created by previous introductions, we introduce new species but end up making the situation worse. In Hawaii and on several Caribbean Islands, for instance, mongooses were imported to help control rats that had escaped from ships and were destroying indige-

nous birds. Since the mongooses were diurnal (active in the day), however, while rats are nocturnal, they tended to ignore each other. Instead the mongooses also killed native birds and further threatened endangered species. Our lessons from this and similar introductions have a new technological twist. Some of the ethical questions currently surrounding the release of genetically engineered organisms are based on concerns that they are novel organisms, and we might not be able to predict how they will interact with other species in natural ecosystems—let alone how they might respond to natural selective forces. Many people argue that we cannot predict either their behavior or their evolution.

BIOMES

Many places on Earth share similar climatic, topographic, and soil conditions, resulting in generally similar communities in widely separated locations. These major community types are called **biomes**. Although the communities within a particular biome type may not be identical, these broad generalizations are useful in describing what kind of community is likely to occur in a particular place under a given set of environmental conditions (fig. 3.15). Furthermore, comparing and contrasting related biomes can help us understand how ecosystems work.

Biome classifications are usually based on the dominant plant species in each community, both because plants are conspicuous and migrate slowly and because plants play such an important role in determining what other species can inhabit an area. Water availability is often a critical limiting factor in plant distribution, and water availability, in turn, is dependent on temperature and precipitation. By looking at the interaction of these two factors, you can make a general prediction about the biome type that might be expected in a particular area (fig. 3.16). Let's look in more detail at some of these world biomes.

Deserts

Deserts are harsh environments where only drought-adapted species of plants and animals thrive (fig. 3.17). Precipitation is often sporadic and unpredictable, while daily and seasonal temperature variations can be extreme. Antarctica, the tops of very high mountains, and the arctic islands near the North Pole are classified as cold deserts because the small amount of precipitation that does occur is locked up for most of the year in ice and snow and is not available for plant growth. Belts of seasonally hot, dry deserts girdle the earth at about 30° north and south of the equator including the Sahara of North Africa, the Gobi of China, and the deserts of southwestern United States and northern Mexico. Most plant species in these areas are either ephemeral (appearing only when moisture

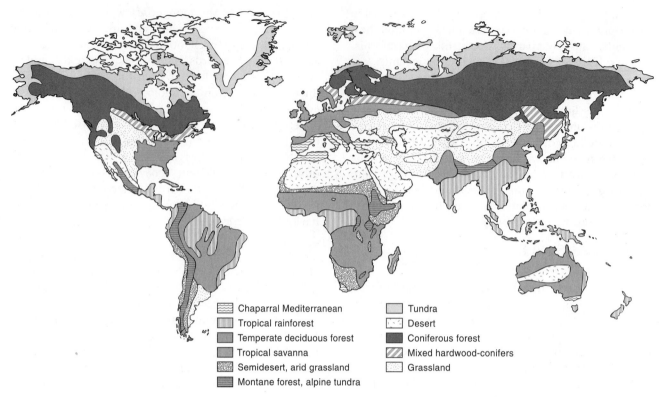

Figure 3.15 This generalized map of world biomes gives an overview of the vegetation types on different landmasses. At this scale, much detail cannot be shown, so such maps should be interpreted with that limitation in mind.

Legend:
- Chaparral Mediterranean
- Tropical rainforest
- Temperate deciduous forest
- Tropical savanna
- Semidesert, arid grassland
- Montane forest, alpine tundra
- Tundra
- Desert
- Coniferous forest
- Mixed hardwood-conifers
- Grassland

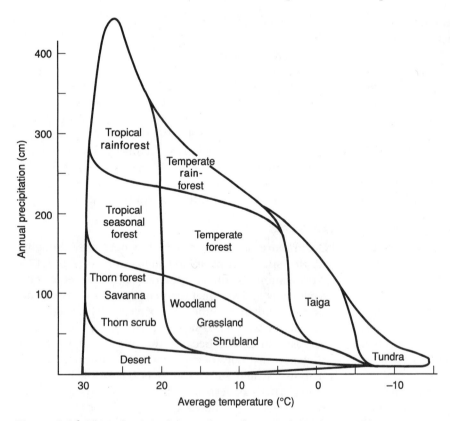

Figure 3.16 This schematic diagram shows the general distribution of biome types along two gradients: temperature and moisture. Other things such as topography, soils, and history being equal, you can use this diagram to predict the vegetation type in an area given a particular combination of temperature and moisture.

is available and completing their life cycles quickly) or have mechanisms to conserve water, endure the harsh climate, and discourage predation. Seasonal leaf production, water-storing tissues, thick epidermal layers, spines, thorns, and toxic chemicals help them survive. Animals, too, have both structural and behavioral adaptations to meet their three most critical needs: food, water, and heat survival. Many desert animals have remarkable abilities to reduce water loss, and some can survive without ever drinking liquid water.

Grasslands

Grasslands generally occupy areas with cold winters and dry, hot summers (fig. 3.18). Continental climates in the centers of the largest continents create vast grasslands such as the Great Plains of North America, the steppes of Central Asia, the Serengeti Plain of East Africa, and the pampas of South America. These broad open spaces once supported enormous herds of

Figure 3.17 The hot deserts of the American Southwest are dominated by large cacti, creosote bush, ocotillo, and paloverde trees—all perennial plants with adaptations to survive extreme temperatures and drought. Many species of annual plants also live in the hot desert. Their survival strategy involves germinating and producing seed during short periods of adequate rainfall, plus having seeds that can remain dormant for years.

Figure 3.18 Grasslands, such as this area in Badlands National Park, dominated much of the North American interior before they were converted to croplands and rangelands.

migratory mammals, but hunting and conversion of land for crop production have reduced many of the largest herds to a small fragment of their former size. Periodic fires are often important in maintaining grasses and suppressing trees. Native prairie soils are famous for their fertility, and many former grasslands have been converted into the breadbaskets of the world. Tundra and paramo are specialized grasslands that occur at high latitudes or high altitudes where temperatures are too cold and growing seasons too short to support woody species.

Forests

Temperate deciduous forests occupy regions with warm, humid summers and cold winters (fig. 3.19). This climate supports lush summer growth when water is plentiful but requires survival adaptations when temperatures are below freezing. Many plants adapt by producing leaves in the summer and then shedding them at the end of the growing season—a deciduous pattern. Small animals in these forests often hibernate during cold spells or migrate to warmer climates. A wide variety of tree species are found in these temperate forests, and their crowns often meet overhead to form a closed canopy. Most of the eastern United States, western Europe, and southern Siberia were once covered by deciduous forests. Savannas and woodlands are intermediate between these closed forests and prairies with individual trees or small groves scattered across open grasslands. The prairie-forest border of the central United States was originally an oak savanna.

Conifer forests are dominated by pine, spruce, fir, hemlock, and redwood, evergreen trees that are cone-bearing and have needles or scales rather than broad leaves. These forests have a variety of temperature, soil, moisture, and fire conditions that favor drought or fire-resistant needle-leaved species over broad-leaved deciduous species. Boreal (northern) and montane (mountain) forests occupy regions where winter temperatures are too low and snows too deep for hardwoods to survive (fig. 3.20). Southern pine forests occur in regions with sandy soil where seasonal droughts and frequent fires favor hardy evergreen species. Huge, ancient redwoods, Douglas firs,

Figure 3.19 Deciduous forests are stratified assemblages of tall trees, understory shrubs, and small, lovely, ground-covering species. In turn, a rich fauna is supported by the variety of niche opportunities that are present.

Figure 3.20 Montane ecosystems are distributed in vertical zones because of the effects of altitude on physical factors such as temperature, length of growing season, availability of soil moisture, and wind exposure. This photograph shows the transition from valley floor to alpine tundra.

hemlocks, Sitka spruce, and western cedars flourish in the temperate rainforests of the northwest coast of North America, which are cool, cloudy, and moist year-round. The small remnants still remaining of these ancient, cathedral forests are among the most productive and biologically diverse of any biological communities in the world.

Tropical forests can be either very wet, as in the lush, evergreen tropical rainforests of Amazonia, or seasonally dry, as in the arid deciduous forests of the Pacific coast of Central America and the Atlantic coast of South America. Moist tropical forests are probably the most productive and biologically diverse communities in the world (see box 3.1). Tropical forests are

threatened everywhere, but dry forests have generally been destroyed to a much greater extent than wet forests because they are easier to clear and more conducive to human habitation. Less than 1 percent of the Central and South American dry tropical forests have been preserved, while large tracts of moist forest remain, especially in the Amazon River basin.

Aquatic Biomes

Wetlands also come in a variety of community types depending on the depth and duration of water coverage, the water source, flow, and chemistry, and the climate and other conditions. Freshwater wetlands range from wet meadows that are only occasionally inundated; to marshes, bogs, and fens where shallow standing water covers the soil for most of the growing season and grasses, sedges, cattails, reeds, rushes, and other emergent herbaceous species dominate; to open water in lakes, rivers, and ponds. Swamp forests are dominated by tree species that can survive occasional or even permanent flooding. These include river valley flood plains as well as the cypress bayous and swamps of the deep South (see fig 3.1). Saltwater wetlands occupy a similar range from salt meadows and tidal mud flats to deep marshes, eel grass shallows, mangrove swamps, coral reefs, banks, and shoals. Although all these wetland habitats amount to only a small percentage of the total area occupied by biological communities, they account for a large percentage of biological productivity and reproductive activity. They are the nurseries for many species. Without these vital communities, biological diversity is severely reduced.

Summary

Four fundamental properties of biological communities are productivity, diversity, complexity, and organization. Productivity is a measure of the rate at which photosynthesis produces biomass made of energy-rich compounds. Tropical rainforests are generally the most productive of all terrestrial communities; coral reefs and estuaries are generally the most productive aquatic communities. Diversity is a measure of the number of different species in a community, while abundance is the total number of individuals. Generally the most productive and stable communities are both the most diverse and most abundant. Community complexity refers to the number of species at each trophic level as well as the total number of trophic levels in a community. Structure concerns the patterns of organization, both spatial and functional, in a community.

The principle of limiting factors states that for every physical factor in the environment there are both maximum and minimum tolerable limits beyond which a given species cannot survive. Furthermore, the factor closest to the tolerance limit for a particular species at a particular time is the critical factor that will determine the abundance and distribution of that species in that ecosystem. Natural selection is the process by which environmental pressures affect survival and reproduction of organisms. Over a very long time, given a large enough population, natural selection works on the naturally occurring random variation in a population to allow evolution of species and adaptation of the population to a particular set of environmental conditions.

Habitat describes the place or set of conditions in which an organism lives; niche describes the role an organism plays: how it makes a living. Natural selection often leads to niche specialization and resource partitioning that reduce competition between species. Organisms interact within communities in many ways. Predation—feeding on another organism—involves pathogens, parasites, and herbivores as well as carnivorous predators. Competition is another kind of antagonistic relationship in which organisms vie for space, food, or other resources. The principle of competitive exclusion states that no two species will remain in direct competition for very long because natural selection and adaptation will cause niche divergence and specialization to minimize competition. Symbiosis is the intimate living together of two species. Mutualism means that both species benefit; commensalism means that one species benefits while the other is indifferent.

Ecological succession is a series of changes through which communities progress as organisms alter the environment in ways that allow some species to replace others. Primary succession starts with a previously unoccupied site. Secondary succession occurs on a site that has been disturbed by external forces. In many cases, succession proceeds until a stable, climax community is established. Introduction of new species by natural processes, such as the opening of a land bridge, or through human intervention can upset the natural relationships in a community and cause catastrophic changes for indigenous species.

Biomes consist of broad regional groups of related ecosystems. Their distribution is determined primarily by climate, topography, and soils. Often, similar niches are occupied by different but similar species in geographically separated biomes. Some of the major biomes of the world are deserts, grasslands, wetlands, forests of various types, and tundra.

Review Questions

1. Productivity, diversity, complexity, and organization are characteristics of all communities and ecosystems. Describe how these characteristics apply to the ecosystem in which you live.

2. Describe the general niche occupied by a bird of prey, such as a hawk or an owl. How can hawks and owls exist in the same ecosystem and not adversely affect one another?

3. All organisms within a biological community interact with each other. The greatest interaction, however, occurs between individuals of the same species. What concept discussed in this chapter can be used to explain this fact?

4. Predator/prey relationships play an important role in the energy transfers that occur in ecosystems. They also influence the process of natural selection. Explain how predators affect the adaptations of their prey. This relationship also works in reverse. How do prey species affect the adaptations of their predators?

5. Competition for a limited quantity of resources occurs in all ecosystems. This competition can be between members of different species (interspecific) or between members of the same species (intraspecific). Explain some of the ways an organism might deal with interspecific competition. How might it deal with intraspecific competition?

6. Each year fires burn large tracts of forestland. Describe the process of succession that occurs after a forest fire destroys the existing biological community. Is the composition of the final successional community likely to be the same as that which existed before the fire? What factors might alter the final outcome of the successional process?

7. Explain the concept of climax community. Why does the climax community exhibit a higher level of stability than that found in other successional stages?

8. Discuss the dangers to existing community members of introducing new species into ecosystems where they did not previously exist. What type of organism would be most likely to survive and cause problems in a new habitat?

Questions for Reflective or Critical Thinking

1. Ecologists debate whether biological communities have self-sustaining, self-regulating characteristics or are highly variable, accidental assemblages of individually acting species. What outlook or worldview might lead scientists to favor one or another of these theories?

2. Natural selection and evolution are central to how most scientists understand and interpret the world, and yet this theory is an anathema to many religious groups. Why do you think this theory is so important to science and so strongly opposed by others? What evidence would be required to convince opponents of evolution?

3. What is the difference between saying that a duck has webbed feet because it needs them to swim and saying that a duck is able to swim because it has webbed feet?

4. It is often argued that complexity in a biological community creates stability and justifies preservation of native communities. What would be the affect on our government's environmental policy if complexity was not related to stability?

5. Some scientists look at the boundary between two biological communities and see a sharp dividing line. Others looking at the same boundary see a gradual transition with much intermixing of species and many interactions between communities. How can they have such different interpretations of the same landscape?

6. The absence of certain lichens is used as an indicator of air pollution in remote areas such as national parks. How can we be sure that air pollution is really responsible? What evidence would be convincing?

7. We tend to regard generalists or "weedy" species as less interesting and valuable than rare and highly specialized endemic species. What values or assumptions underlie this attitude?

8. What part of this chapter do you think is most likely to be challenged or modified in the future by new evidence or new interpretations?

Key Terms

biomes 57
climax community 56
coevolution 52
commensalism 53
competition 52
competitive exclusion 53
complexity 43
conifer forests 59
deserts 57
diversity 43
ecological niche 49
ectone 46
evolution 42
grasslands 58
habitat 49
limiting factors 48
natural selection 49
parasites 51
patchiness 45

pathogens 51
photosynthetic efficiency 43
pioneer species 55
plankton 52
predator 51
primary succession 55
productivity 42
resource partitioning 49
secondary succession 55
structure 45
succession 55
symbiosis 53
temperate deciduous forests 59
territoriality 53
tolerance limits 48
tropical forests 60
vertical stratification 50
wetlands 60

Suggested Readings

Davis, Margaret B. "Pleistocene Biogeography of Temperate Deciduous Forests." *Geoscience and Man* 13 (1976): 13-26. A fascinating historical account of the migration of forests after the last glacial period.

Ehrlich, Paul R., and Jonathan Roughgarden. *The Science of Ecology.* New York: Macmillan, 1987. The community ecology segments of this textbook are particularly appropriate for expanded reading in regard to this chapter.

Elton, C. *Animal Ecology.* New York: Macmillan, 1927. A pioneering analysis of community ecology.

Holldobler, B., and E. O. Wilson. *The Ants.* Cambridge, Mass.: Harvard University Press, 1990. A Pulitzer Prize-winning monograph on the evolution, ecology, and social biology of the most diverse of all social species.

Krebs, Charles J. *The Message of Ecology.* New York: Harper & Row, 1988. A concise summary of ten essential principles of ecology. Recommended.

Krebs, Charles J., and N. B. Davies. *An Introduction to Behavioral Ecology.* Sunderland, Mass.: Sinauer, 1991. Perhaps the best source for information on the evolution and adaptiveness of behavior.

May, Robert M. *Stability and Complexity in Model Ecosystems.* Princeton, N.J.: Princeton University Press, 1973. Largely responsible for influencing our present understanding of the relationships between the number of species in a community and its stability in fluctuating or stable environments. May's work incorporates use of mathematical models for analysis and prediction.

Odum, Eugene P. *Basic Ecology.* Philadelphia: Saunders, 1983. An excellent source of additional and more detailed information about biological communities and ecosystems.

Pimm, Stuart L. *Balance of Nature?* Chicago: University of Chicago Press, 1991. A good discussion of modern concepts of ecological issues in the conservation of species and communities.

Rennie, John. "Living Together." *Scientific American* 266, no. 1 (1992): 122-133. Parasites and their hosts have devised many odd strategies—perhaps even using sex—in their endless game of adaptive one-upmanship. Yet sometimes they seem to cooperate.

Ricklefs, Robert E. *Ecology.* New York: Freeman, 1990. A good, comprehensive, general ecology textbook that expands on many of the topics in this chapter. Ricklefs gives many interesting examples to illustrate ecological principles.

Smith, Robert Leo. *Ecology and Field Biology* 4th ed. New York: Harper & Row, 1990. A thorough overview of field ecology by a pioneer in the field.

Soulé, M. E. Conservation Biology: The Science of Scarcity and Diversity. Sunderland, MA: Sinauer Assoc., 1986. A pioneering anthology in the emerging field of conservation biology.

Terborgh, J. *Diversity and the Rain Forest.* Scientific American Library. New York: Freeman, 1992. An interesting examination of why tropical rainforests are so rich in biological diversity.

Wilson, E. O. *The Diversity of Life.* Cambridge: Harvard University Press, 1992. A fascinating and well-written account of the experiences of this world leader in field biology and his reflections on biodiversity and what we are doing to the environment.

4
HUMAN POPULATIONS

*Imagine your life a few short years from now: You're living in a world where fuel,
food, jobs, housing, and health care are at a premium—where natural resources are
severely depleted and cities horribly polluted, uncomfortably crowded, and plagued by
crime. Right now, we're heading on a collision course towards the population crisis that
could prompt this nightmare scenario, not only in distant, developing countries of
Africa and Latin America, but right here in the United States.*

Zero Population Growth, Inc.

*Population growth is the only force powerful enough to bring about a long-range
transformation to more advanced and productive societies. The world has immense
physical resources for agriculture and mineral production. In industrial countries the
benefits of large expanding markets are abundantly clear. The principal problems
created by expanding populations are not those of poverty, but of exceptionally rapid
increase of wealth and its attractions to migration and unmanageable spread of cities.*

Collin Clark

Objectives

*After studying this chapter, you should be
able to:*

■ Describe the causes and consequences of exponential
(J curve) population growth that overshoots or
oscillates around the carrying capacity of the
environment.

■ Explain why some species stabilize their populations
while others do not.

■ Trace the history of human population growth.

■ Discuss Malthusian theories of population growth as
well as why technological optimists and supporters of
social justice oppose these theories.

■ Understand pressures for and against family planning in
traditional and modern societies.

■ Compare modern birth control methods and plan a
personal family planning agenda.

INTRODUCTION

Human populations have grown rapidly during the
past three hundred years. By mid-1992, the world
population reached 5.4 billion, and it is doubling
about every forty-one years. About 92 million more
people are added to the world each year, making us
now the most numerous vertebrate species on earth
(fig. 4.1). We have good reason to fear that this population explosion, unless checked immediately, will
bring disaster of an unknown scale. Many people call
for immediate, worldwide birth control programs to
reduce population growth and eventually stabilize the
total number of humans on Earth.

Optimists argue, on the other hand, that humans
will not destroy life-support systems on which we all
depend because our ingenuity and effort can expand
the ability of the environment to support us. According to this view, population growth could result in

Figure 4.1 This crowd in Chapultepec Park in Mexico
City gives some impression of the population density in
some areas of the world. Ninety percent of the population
growth in the next century will be in developing
countries, and most of that growth will be in urban areas.

economic growth and technological development that might bring a higher standard of living for everyone.

Whether human populations will continue to grow at present rates and what that would imply for environmental quality and human life are among the most central and pressing questions in environmental science. In this chapter, we will look at some causes of population growth as well as how populations are measured and described. We also will look at some opinions of what the optimum human population is and how we might reach that level. Family planning and birth control are essential for stabilizing populations. The number of children that a couple decides to have and the methods they use to regulate fertility, however, are strongly influenced by culture, religion, politics, and economics, as well as basic biological and medical considerations. We will look at how some of these factors impinge on human demographics.

POPULATION BIOLOGY

In the 1930s, biologist G. F. Gause carried out pioneering experiments in population dynamics that warn us of the dangers of unlimited reproduction. After inoculating a small sample of microorganisms into a culture flask, he measured total population size at regular intervals. A typical result is shown in figure 4.2. The population grows slowly at first but then accelerates rapidly. We describe this accelerating pattern of growth as a J curve because of its shape. It is also called **exponential growth** because the number of individuals increases by an exponential factor or a compound interest rate. A bank account also grows in an exponential fashion. When the account is small, a 5 percent annual addition doesn't seem like much. If you have a million dollars in your account, however, 5 percent interest gives you $50,000 more each year.

Biotic Potential

Biological organisms have the potential to increase at an exponential rate because they can produce more than one offspring per generation. In fact, some very fertile species can produce many offspring in each generation. If all were to survive, the rate of growth would be phenomenal. We call the maximum possible reproductive rate of an organism its **biotic potential.**

Consider houseflies, for example. A female housefly (*Musca domestica*) can lay up to 120 eggs in its lifetime. In 56 days, the eggs hatch and mature into sexually active adults that lay their own eggs. In one year (seven generations), if all its offspring lived long enough to reproduce, a single female could give rise to 5.6 trillion flies! If this rate of reproduction continued for ten years, the whole earth could be covered several meters deep with the offspring of a single female. Fortunately, only a small

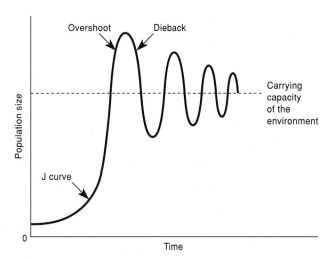

Figure 4.2 Population oscillations. Some species demonstrate a pattern of cyclic overshoot and dieback.

number of eggs and flies survive in each generation, and we are spared this unappetizing fate.

Catastrophic Declines and Population Oscillations

In the real world, there are limits to growth (fig. 4.2). Eventually, a rapidly growing population reaches the limits of its environment to furnish food, water, living space, and other essential resources. The **carrying capacity** is the maximum number of organisms that a given environment can sustain on a stable basis. When a population **overshoots** (exceeds) its carrying capacity, death rates will begin to surpass additions to the population. We call this a population crash or **dieback.**

A population may go through repeated oscillating cycles of overshoot and dieback as shown in figure 4.2. These cycles can be regular if they depend on a single factor such as the amount of nutrients or space available in an ecosystem. They also may be quite irregular if they depend on complex environmental and biotic relationships. For instance, the population explosions of migratory locusts in the desert or tent caterpillars and spruce budworms in northern forests occur with highly variable timing. If the dieback is very severe, a given species may become locally extinct.

Growth to a Stable Population

Not all biological populations go through the boom and bust cycles shown in figure 4.2. Some species' reproductive and survival rates are regulated by both external and internal factors so that they come into equilibrium with their environmental resources. These species may grow exponentially when resources are abundant, but their growth slows as they approach the carrying capacity of their environment (fig. 4.3). Together, the factors that limit growth rates and produce an equilibrium at the carrying

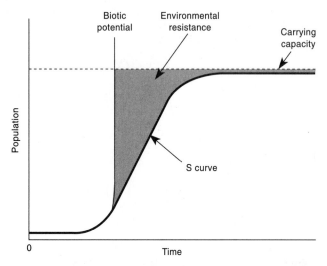

Figure 4.3 Growth to a stable population size follows an S-shaped curve that levels off as it approaches the carrying capacity of the environment. The difference between the theoretically possible J curve of biotic potential and the actual S curve of growth is called environmental resistance and is the cumulative effect of all growth-slowing factors in the ecosystem.

capacity are called **environmental resistance.** We describe this pattern of population growth as an S curve because of its shape.

Generally, reproduction rates of species at or near the bottom of the food web are regulated by external factors such as food availability or predation. Species such as clams, mice, or dandelions can do little to help their offspring survive dangers in their environment. These species compensate for high mortality rates by producing large numbers of offspring, only a few of which usually survive to reproduce. If external restraints are relaxed, however, these organisms will exhibit boom and bust cycles of exponential growth. Predators may actually benefit these populations by stabilizing their size (box 4.1).

The top carnivores in an ecosystem generally have no predators and few competitors except their own species. Organisms such as lions, wolves, and elephants often have behavior patterns or other intrinsic restraints on reproduction that make populations grow to stable numbers in the S curve pattern. Which type of growth best describes humans? Are we more like mice or lions?

Mortality and Survivorship

The mortality rates (or death rates) that cause diebacks and slowing of population growth do not affect all ages or all species equally. **Survivorship** is a measure of the percentage of maximum lifespan attained by members of a population. Figure 4.4 shows four idealized survivorship curves for species at different trophic levels in ecosystems. Curve A represents species such as whales that have few natural preda-

tors. Most individuals live to old age and die of disease or senility. Curve B represents species such as sea gulls in which the main causes of death are accidents, poisoning, or diseases that strike at any age. Mortality in this case is linear and random with respect to age.

Curve C represents species such as deer that have high mortality when young but have a good chance of living out the maximum lifespan once they attain adulthood. Curve D represents organisms such as rabbits or mice that are at the bottom of the food chain, where most babies are eaten by predators or die from diseases soon after birth. Interestingly, humans in primitive societies have survivorship patterns like curve B, with high infant mortality rates due to diseases. Pathogens are our most deadly predators. People with access to modern sanitation and medicine are much less vulnerable to diseases and have survivorship patterns more like those in curve A.

Density-dependent and Density-independent Factors

Some of the controlling factors in growth regulation are dependent on population size or density. **Density-dependent factors** are those for which mortality or reproductive failure are linked to population size. For example, starvation was rare on Isle Royale when moose numbers were small but increased catastrophically as the moose population grew.

Density-independent factors are those that cause mortality or reproductive failure regardless of population size. These are often abiotic environmental factors such as temperature, rainfall (or lack thereof), and accidents. Accidental poisoning of sea gulls, for instance, occurs by chance and has no relationship to population density. Similarly, thrips (small leaf-eating insects) grow well in the spring but are killed by high temperatures and drought in the summer. The total number is determined more by these density-independent factors than by living space, food supply, or some other density-dependent factor.

HUMAN POPULATION HISTORY

For most of our history, humans have not been very numerous compared to other species. Studies of hunting and gathering societies suggest that the total world population was probably only a few million people before the invention of agriculture and the domestication of animals around ten thousand years ago. The larger and more secure food supply made available by the agricultural revolution allowed the human population to grow, reaching perhaps 50 million people by 5000 B.C. For thousands of years, the number of humans continued to increase very slowly. Archaeological evidence and historical descriptions suggest

Wolves and Moose on Isle Royale: A Case Study

*I*sle Royale National Park occupies the largest island in Lake Superior, the largest freshwater lake (in surface area) in the world. The island presents a spectacular wilderness setting of high rocky ridges covered by a dense boreal forest. Cut off from the mainland of Minnesota and Ontario by 30 km (20 mi) of rough, deep, very cold water, the island is a nearly closed ecosystem that is a unique laboratory for studying large animal population dynamics.

Until this century, the island was populated by woodland caribou, which disappeared about 1900 due to hunting and human-caused fires. Moose first appeared about the time that caribou became extinct on the island. The first moose probably swam from the mainland or crossed on ice that occasionally forms a bridge to the island. The island was an ideal place for moose: shrubs and aquatic plants on which they prefer to browse were plentiful, and no major predators were present to harass them.

The moose population increased slowly at first. In 1915, Isle Royale had only about two hundred moose, or one animal for each 2.6 sq km (1 sq mi). In the 1920s, however, the moose population exploded. Their number jumped more than tenfold in ten years, reaching a peak of more than five thousand animals in the summer of 1929. How could this increase occur so rapidly? Each cow moose reaches sexual maturity at age three or four. She then can produce one or two calves each year for the next eight to ten years if she remains healthy and has enough to eat. This means that the moose population can double every two years under optimum conditions, a 35 percent annual growth rate.

Pioneering wildlife biologist Adolph Murie went to the island to study the moose situation in the summer of 1929. He reported that the moose had overshot the carrying capacity of their environment. All the tender branches on which the moose browse in the winter were eaten back as high as the animals could reach. Much of the summer food (aquatic plants and annuals) was also badly depleted. He predicted that disease and starvation would soon cause an extensive dieback. As you can see in box figure 4.1.1, his prediction came true. By 1941, only 171 moose were found on the island—fewer than were found there twenty-five years earlier.

In the early 1940s, shortly after the moose population on Isle Royale crashed, wolves appeared on the island in pairs and small groups, presumably having crossed on the ice during previous winters. Wolves are shy and diffi-

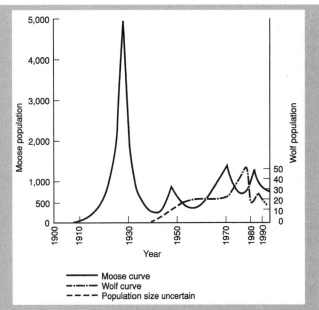

Figure 4.1.1 Growth of moose and wolf populations on Isle Royale National Park, 1900 to 1985. Moose first appeared on the island shortly after 1900. By 1929, the moose population had grown to about 5,000, clearly overshooting the carrying capacity of the ecosystem. A catastrophic dieback occurred during the 1930s, when 97 percent of the moose died. The appearance of wolves after 1940 at first reduced the moose population but then established a dynamic equilibrium around what appears to be the carrying capacity.

Source: David Mech, *The Wolves of Isle Royale*, 1966, published by National Park Service.

cult to spot, so we don't know exactly when the first ones arrived or how many there were in those early years. By 1957, however, when the first systematic census was taken, twenty-one wolves were on the island.

You might think that the wolves, arriving as they did when the moose were weakened by starvation, could have exterminated their prey completely. Instead, the wolves and moose established a dynamic balance, oscillating between six hundred to twelve hundred moose and twelve to fifty wolves at any given time (box fig. 4.1.1). Wolf predation prevented both excess population growth and catastrophic decline of moose on the island. Weaker moose were eaten by the wolves while wolf populations were regulated by social behavior and food supply. Only the dominant wolves in each pack (the alpha male and female) mate and produce pups. In years when moose are plentiful, more wolf pups survive; if moose become scarce, few wolves reach adulthood.

In the 1980s, an alarming decline in the wolf population on Isle Royale was observed. The 1988 winter survey revealed only twelve wolves in three small packs, down from fifty wolves only six years earlier. Although three pups were born in 1990 and two more in 1992, mortality has been high and the population has remained around ten animals. Scientists proposed several possible causes for this population crash. Although moose are plentiful, they are young and healthy, and perhaps too difficult for the wolves to catch. Diseases might have been introduced by dogs or stray wolves from the mainland. Some scientists believe the problem may be genetic. When a population starts with only a few founding individuals and is highly inbred, as are the wolves of Isle Royale, defective genes are likely to be expressed, resulting in a high rate of reproductive failures and infant mortality.

Ten wolves may not be a viable population over the long term, no matter how ideal the environment or how carefully the species is protected. This problem has important implications in managing other endangered species and in determining the size of protected habitat and wildlife preserves (chapter 9). Researchers are continuing studies to determine why reproductive rates are so low and what that means for the future of both wolves and moose on the island.

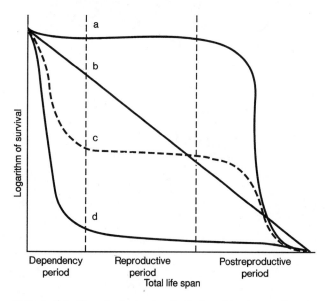

Figure 4.4 Four basic types of survivorship curves for organisms with different life histories. Curve (*a*) represents organisms such as humans or whales that tend to live out the full physiological life span if they survive early growth. Curve (*b*) represents organisms such as sea gulls in which the rate of mortality is fairly constant at all age levels. Curve (*c*) represents organisms such as white-tailed deer, moose, or robins that have high mortality rates in early and late life. Curve (*d*) represents organisms such as clams and redwood trees that have a high mortality rate early in life but live a full life if they reach adulthood.

TABLE 4.1 World population growth and doubling times		
Date	**Population**	**Doubling time**
5000 B.C.	50 million	?
800 B.C.	100 million	4,200 years
200 B.C.	200 million	600 years
1200 A.D.	400 million	1,400 years
1700 A.D.	800 million	500 years
1900 A.D.	1,600 million	200 years
1965 A.D.	3,200 million	65 years
1990 A.D.	5,300 million	38 years
2020 A.D. (estimate)	8,230 million	55 years

Source: Population Reference Bureau, Inc., Washington, D.C.

that only about 300 million people were living at the time of Christ (table 4.1).

Populations grew slowly up until the Middle Ages (fig. 4.5) because diseases, famines, and wars made life short and uncertain for most people. Furthermore, there is evidence that many early societies regulated their population size through cultural taboos and practices such as infanticide. Among the most destructive of natural population checks were bubonic plagues that periodically swept across Europe between 1348 and 1650. During the worst plague years, between 1348 and 1350, it is estimated that at least one quarter of the European population perished. Notice, however, that this did not retard population growth for very long. In 1650, at the end of the last great plague, there were about 600 million people in the world.

As you can see in figure 4.5, human population began to increase rapidly after 1600 A.D. Many factors contributed to this rapid growth. Better sailing and navigating skills stimulated commerce and communication between nations. Agricultural developments, better sources of power, and better health care and hygiene also played a role. Together, all these advances can be considered part of the scientific and technological revolution brought about by the Renaissance and the Age of Reason. We are now in a J curve pattern of growth. Will this population explosion continue until we overshoot the carrying capacity of our environment (perhaps we have done so already) and experience a catastrophic dieback? Or will our population stabilize at some optimum level? These are among the most important questions in environmental science.

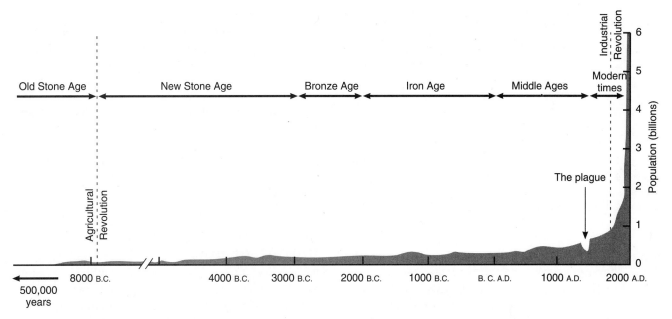

Figure 4.5 Human population levels through history. Since about A.D. 1000, our population curve has assumed a J shape. Are we on the upward slope of a population overshoot? Will we be able to adjust our population growth to an S curve? Or can we just continue the present trend indefinitely?

LIMITS TO GROWTH: SOME OPPOSING VIEWS

Few people believe that human populations can continue to grow at current rates for very long. Clearly, the physical world has finite limits that must restrict growth sooner or later. But how will our numbers be brought into balance with resources? Will we recognize when we are approaching the limits of our world and control growth ourselves, or will we be controlled by external forces?

Malthusian Checks on Population

In 1798, the Reverend Thomas Malthus wrote *An Essay on the Principle of Population as It Affects the Future Improvement of Society, with Remarks on the Speculations of Mr. Godwin, M. Condorcet, and Other Writers* to refute the views of progressives and optimists—including his father—who were inspired by the egalitarian principles of the French Revolution to predict a coming utopia. Malthus argued that human populations tend to increase at an exponential or compound rate of interest, while food production either remains stable or increases only slowly. The result, he predicted, was that human populations would outstrip food supply, resulting inevitably in starvation, crime, and misery. Malthus's theory might be summarized by the equation in figure 4.6.

According to Malthus, the only ways to stabilize human populations are "positive checks," such as diseases or famines that kill people, and "preventative checks," including all the factors that prevent human birth. One preventative check he advocated was

"moral restraint," which included later marriage and celibacy until a couple could afford to support children. Many social scientists and biologists have been influenced by Malthus. Charles Darwin, for instance, derived his theories about the struggle for scarce resources and survival of the fittest after reading Malthus's essay.

If Malthus's views of the consequences of exponential population growth were dismal, the corollary he drew was even more bleak. He believed that most people are too lazy and immoral to regulate birth rates voluntarily. Consequently, he opposed efforts to feed and assist the poor in England because he feared that more food would simply increase their fertility and thereby perpetuate the problem of starvation and misery.

Not surprisingly, Malthus's ideas provoked a great social and economic debate. Karl Marx was one of his most vehement critics, claiming that Malthus was a "shameless sycophant of the ruling classes." According to Marx, population growth is a symptom rather than a root cause of poverty, resource depletion, pollution, and other social ills. The

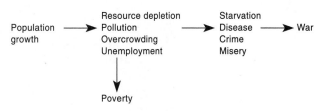

Figure 4.6 Thomas Malthus argued that excess population growth is the ultimate cause of many other social and environmental problems.

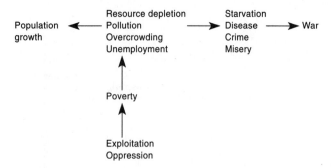

Population growth ← Resource depletion / Pollution / Overcrowding / Unemployment → Starvation / Disease / Crime / Misery → War

Poverty

Exploitation / Oppression

Figure 4.7 Karl Marx argued that oppression and exploitation are the real causes of poverty and environmental degradation. Population growth in this view is a symptom or result of other problems, not the source.

real causes of these problems, he believed, are exploitation and oppression (fig. 4.7). Marx argued that workers always provide for their own sustenance given access to the tools of production and a fair share of the fruits of their labor. According to Marx, the way to slow population growth *and* alleviate crime, disease, starvation, misery, and environmental degradation is through social justice.

Malthus and Marx Today

Both Marx and Malthus developed their theories about human population growth in the nineteenth century when understanding of the world, technology, and society were much different than they are now. Still, the questions they raised are relevant today. While the evils of racism, classism, and other forms of exploitation that Marx denounced still beset us, it is also true that at some point available resources must limit the numbers of humans that the earth can sustain.

Those who agree with Malthus that we are approaching—or may have already surpassed—the carrying capacity of the earth are called **neo-Malthusians**. In their view, we should address the issue of surplus population directly by promoting birth control programs. Neo-Marxians, on the other hand, believe that only eliminating oppression and poverty through technological development and social justice will solve population problems.

Can Technology Make the World More Habitable?

Technological optimists argue that Malthus was wrong two hundred years ago in his predictions of famine and disaster because he failed to account for scientific progress. In fact, food supplies have increased faster than population growth since Malthus's time. Terrible famines have occurred in the past two centuries, but they were caused more by politics and economics than by lack of resources or sheer population size. Whether this progress will continue remains to be seen, but technological advances have increased human carrying capacity more than once in our history.

Figure 4.8 shows historical human population growth plotted on a logarithmic scale. Seen this way, the data show several episodes in which technological progress has made it possible to support larger populations. As

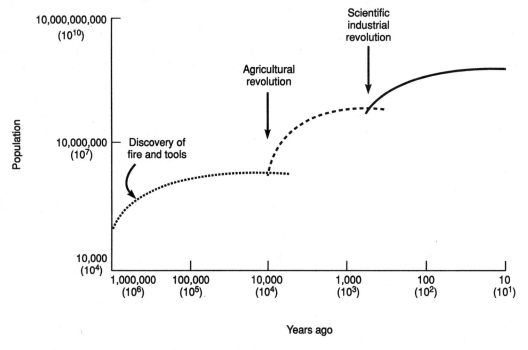

Figure 4.8 Human population plotted on a log-log scale. Plotted this way, population growth is seen as occurring in three bursts associated with revolutions in culture, agriculture, science, and technology.

we discussed in chapter 1, the first of these revolutions was the discovery of fire and the invention of tools about a million years ago that enabled our ancestors to be more effective predators. The second population increase corresponds to the domestication of plants and animals in the agricultural revolution about ten thousand years ago.

The third burst of growth, of which we are a part, was stimulated by the scientific and industrial revolution. Progress in agricultural productivity, engineering, information technology, commerce, medicine, sanitation, and other achievements of modern life have made it possible to support approximately one thousand times as many people per unit area as was possible ten thousand years ago.

Much of our progress in the past three hundred years has been based on availability of easily acquired natural resources, especially cheap, abundant fossil fuels. Whether we can develop alternative, renewable energy sources in time to avert disaster when current fossil fuels run out is a matter of great concern (chapter 12). Furthermore, we have traditionally exploited resources rather than managed them. Many biologists fear that the basic life-supporting systems of the biosphere, which are threatened by overuse and global pollution, may limit the number of humans the earth can sustain.

Can More People Be Beneficial?

There can be benefits as well as disadvantages in larger populations. More people mean larger markets, more workers, and efficiencies of scale in mass production of goods. More people also provide more intelligence and enterprise to overcome problems such as underdevelopment, pollution, and resource limitations. Human ingenuity and intelligence can create new resources by substituting new materials for old materials and new ways of doing things for old ways. For instance, utility companies are finding it cheaper and more environmentally sound to finance insulation and energy-efficient appliances for their customers rather than build new power plants. The effect of saving energy that was formerly wasted is comparable to creating a new fuel supply. We will return to the question of what constitutes a resource and which resources are most likely to limit further growth of human populations in subsequent chapters.

HUMAN DEMOGRAPHY

Demography is derived from the Greek words *demos* (people) + *graphos* (to write or to measure). It encompasses vital statistics about people such as births, deaths, and marriages as well as total population size. In this section, we will survey ways human populations are measured and described and discuss demographic factors that contribute to population growth.

How Many of Us Are There?

Even in this age of information technology and communication, we do not know exactly how many people there are in the world. Some countries still have such difficult and confusing conditions that estimates from different sources vary widely. Governments may overstate or understate their population for a variety of social, political, and economic reasons. Homeless people, refugees, and illegal aliens may not want to be counted or identified.

Although there is considerable uncertainty about the exact number of people in some places, most demographers agree that about 5.5 billion people were alive in mid-1993. Only about 20 percent of that population lives in the more-developed or rich countries of the world. Four out of five humans live in the poor countries of the less-developed world. Demographers estimate that 90 percent of the population growth expected to occur in the next century will take place in the Third World (fig. 4.9). Figure 4.10 shows the distribution of human populations around the world. Notice the high densities supported by fertile river valleys of the Nile, Ganges, Yellow, Yangtze, and Rhine rivers and the well-watered coastal plains of India, China, and Europe.

Fertility and Birth Rates

Fecundity is the physical ability to reproduce, while **fertility** describes the actual production of offspring. Those without children may be fecund but not fertile. The most accessible demographic statistic of fertility is usually the **crude birth rate,** the number of births in a year per thousand persons. It is statistically "crude" in the sense that it is not adjusted for population characteristics such as the number of women in reproductive age.

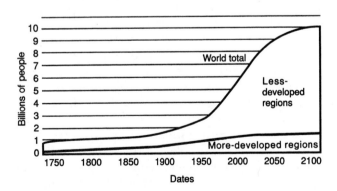

Figure 4.9 Human population growth, 1750–2100, in less-developed and more-developed regions. The human population is now in an exponential growth phase. More than 90 percent of all growth in this century and all growth projected for the next is in the less-developed countries. Total population is expected to stabilize at about 10 billion in the next century.

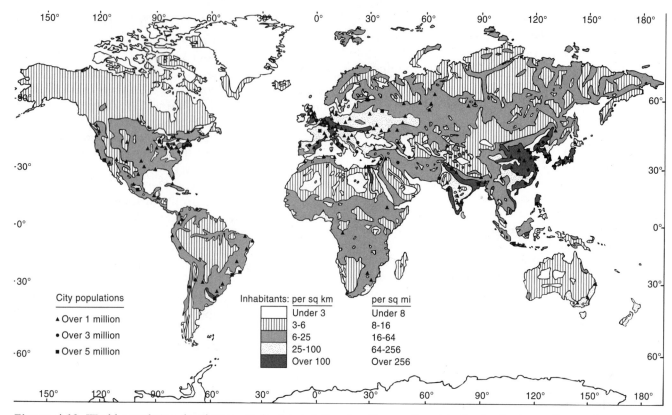

Figure 4.10 World population distribution. Highest population densities are found in the fertile river valleys and well-watered coastal plains of temperate zones.

The **total fertility rate** is the number of children born to an average woman in a population during her entire reproductive life. Upper-class women in seventeenth and eighteenth century England, whose babies were given to wet nurses immediately after birth and who were expected to produce as many children as possible, often had twenty-five or thirty pregnancies. The highest recorded total fertility rates for working-class people is among some Anabaptist agricultural groups in North America that have averaged up to twelve children per woman. In most tribal or traditional societies, food shortages, health problems, and cultural practices limit total fertility to about six or seven children per woman even without modern methods of birth control.

When analyzing population growth, we usually express birth rate in terms of the number of children per couple. The **zero population growth** (ZPG) rate (also called the replacement level of fertility) is the number of births at which people are just replacing themselves. In the more highly developed countries, where infant mortality rates are low, this rate averages 2.2 children per couple. It takes slightly more than two children per couple to stabilize the population because some people are infertile, have children who do not survive, or choose not to have children. In the less-developed countries, the replacement birth rate is often six children per couple.

The nations with the highest birth rates are currently in Africa and the Middle East. In 1992, the Population Reference Bureau listed Gaza as having a crude birth rate of 55 children per 1,000 population while Mali, Niger, Malawi, and Uganda all had 52. By comparison, the United States and Canada have crude birth rates of 16 and 15 per 1,000 population. Rwanda had the highest total fertility rate in the world of 8.0 children per woman, compared to 1.8 in Canada and 2.0 in the United States in 1992.

Mortality and Death Rates

A traveler to a foreign country once asked a local resident, "What's the death rate around here?" "Oh, the same as anywhere," was the reply, "about one per person." In demographics, however, **crude death rates** (or crude mortality rates) are expressed in terms of the number of deaths per thousand persons in any given year. Countries in Africa where health care and sanitation are limited often have mortality rates of twenty or more per thousand people. Wealthier countries generally have mortality rates around ten per thousand. The number of deaths in a population is sensitive to the age structure of the population, however. Rapidly growing, less-developed countries such as Belize or Costa Rica have lower crude death rates (5 per 1,000) than do the more-developed, slowly growing countries such as Denmark (12 per

1,000). This is because there are proportionately more youths and fewer elderly people in a rapidly growing country than in a more slowly growing one.

Population Growth Rates

Crude death rate subtracted from crude birth rate gives the **natural increase** of a population. We distinguish natural increase from the **total growth rate,** which includes immigration and emigration as well as births and deaths. Both these growth rates are usually expressed as a percent (number per hundred people) rather than per thousand. A useful rule of thumb is that if you divide seventy by the annual percentage growth, you will get the approximate doubling time in years. Gaza, which has the world's highest current growth rate of 4.6 percent per year, is doubling its population every fifteen years. The United States and Canada, which have natural increase rates of 0.8 percent per year, are doubling in eighty-nine years. Actually, because of immigration, total U.S. growth is about twice as fast as natural increase. Denmark, with a natural increase rate of 0.1 percent, is doubling in about seven hundred years. The world growth rate is now 1.7 percent, which means that the population will double in 41 years.

China, now the world's most populous country, has a growth rate of only 1.3 percent. India, the world's second most populous country, is expected to surpass China in about fifty years because Indian population control programs have not been as successful as China's (box 4.2).

Life Span and Life Expectancy

Life span is the oldest age to which a species is known to survive. Although there are many claims in ancient literature of kings living for one thousand years or more, the oldest age that can be certified by written records is that of Jeanne Louise Calment of Arles, France, who was believed to be 118 in 1993. The aging process is still a medical mystery, but it appears that cells in our bodies have a limited ability to repair damage and produce new components. At some point they simply wear out, and we fall victim to disease, degeneration, accidents, or senility.

Life expectancy is the average age that a newborn infant can expect to attain in any given society. It is another way of expressing the average age at death. We believe that for most of human history life expectancy in most societies has been between thirty-five and thirty-nine years. This doesn't mean that no one lived past age forty but rather that so many deaths at earlier ages (mostly early childhood) balanced out those who managed to live longer. It once was widely believed that differences in life expectancy between ethnic groups were biological and therefore difficult to change. We now know that social and environmental, not biological, factors are responsible for most variations in mortality. In the highly developed

Thinking Globally: Box 4.2

"Now Because of My Large Family, I am a Rich Man"

*I*n the 1950s, the Rockefeller Foundation invested more than $1 million in a major birth control study in India. The Khana study, as it was called, focused on seven test villages in the Punjab region of northwest India housing about eight thousand people. Trained family planning professionals visited villagers regularly for six years and distributed free birth control supplies. While birth rates decreased in the villages during the study (from 40 to 35 per 1,000), the rates were no different than those in nearby villages that did not receive birth control advice and materials.

This lack of positive results is puzzling since 90 percent of the residents in the test villages claimed that they favored contraception and had used the methods offered to them. Fakir Singh, for instance, insisted that his wife had taken birth control pills regularly; yet she had four children during the study. Rockefeller officials suggested that villagers failed to understand how birth control works or rejected it be-

cause of fear of technology. In a follow-up study, however, Mahmood Mamdani found that the villagers understood very well what they were doing when they threw away the birth control pills and had children. Ignoring the advice of the Rockefeller team was a rational economic choice; people had large families because they wanted and needed them (box fig. 4.2.1).

In tropical, agricultural societies, the cost of having a child is relatively little compared to the contribution that child can make to the family. In times of rapid change when parents are displaced from traditional jobs by changing technology, children may be more able than their parents to adapt and learn new ways to earn money. Listen to the words of Thaman Singh, an unemployed water carrier.

You were trying to convince me in 1960 that I shouldn't have any more sons. Now, you see, I have six sons and two daughters and I sit home in leisure. They are grown up and they bring me money. One even works outside the village as a laborer. You told me I was a poor man and couldn't support a large family. Now, you see, because of my large family, I am a rich man.

Because there is little public assistance for the elderly in India, children are the only source of support for their parents in old age. This makes lack of medical care and high infant mortality levels important determinants of birth rates. In poor countries where one child in four dies before age five, and most girls leave home when they get married, a couple needs to have six or seven children to be sure that at least one son will survive to adulthood and support them when they are old. This, in fact, is a common total fertility rate in many Third World countries. Furthermore, in patriarchal societies where only boys own property and carry on the family name, it becomes even more important for family security to have a living son.

Birth control programs in India and other similar Third World countries will probably never be very successful until the forces that require people to want large families are reduced. Try to put yourself in the situation they face. What would you do in similar circumstances?

Figure 4.2.1 A large family can be an advantage to poor people in less-developed countries. Each child shares in family chores or earns a little money working part-time. Collectively, the family gets by because each member helps the others.

countries, the average life expectancy has nearly doubled in the past two centuries. The maximum life span, however, does not appear to have been changed at all by modern medicine.

Declining mortality, not rising fertility, is the primary cause of most population growth in the past three hundred years. Crude death rates began falling in Western Europe during the late 1700s. Most of this advance in survivorship came long before the advent of modern medicine and is due primarily to better food and better sanitation. Figure 4.11 shows the change in life expectancy at birth in major regions of the world between 1950 and 1992. The greatest gains have been in the developing countries, especially in China and East Asia, where life expectancies are approaching those of the developed world.

Figure 4.12 shows how life expectancy has changed during this century in the United States. Life expectancy for white men and women in the United States has been increasing at about the same rate as in Western Europe, but nonwhites continue to have significantly lower life expectancies. Differences between the sexes are also becoming more pronounced.

Some demographers believe that life expectancy is approaching a plateau while others predict that advances in biology and medicine might extend longevity markedly. If it is true that an infant born in 1990 can expect to live at least one hundred years, on average, our society will be profoundly affected. In 1970, the median age in the United States was thirty. By 2100 the median age could be over sixty. If workers continue to retire at sixty-five, half of the population

could be retired, and retirees might be facing thirty-five or forty years of retirement. We may need to find new ways to structure our lives.

Living Longer: Demographic Implications

A population that is growing rapidly by natural increase has more young people than does a stationary population. One way to show these differences is to graph age classes in a histogram as is shown in figure 4.13. In Mexico, which is growing at a rate of 2.5

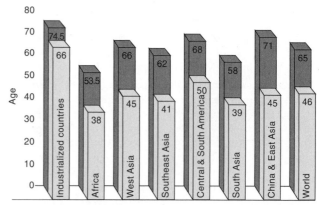

Life expectancy at birth, by regions of the world, 1950 and 1992
☐ 1950 ■ 1992

Figure 4.11 Life expectancy at birth by regions of the world, 1950 and 1992.

Sources: Data from UNICEF, *The State of the World's Children*, 1987, p. 110; 1992 data from Population Reference Bureau.

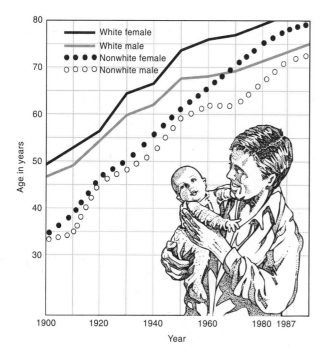

Figure 4.12 Life expectancy in the United States during the twentieth century.

percent per year, 42 percent of the population is in the prereproductive category (below age fifteen). Even if total fertility rates were to fall abruptly, the total number of births and population size would continue to grow for some years as these young people enter reproductive age. This phenomenon is called **population momentum.**

A population that has recently entered a lower growth rate pattern, such as the United States, will have a bulge in the age classes for the last high-birth-rate generation (fig. 4.13). A country that has had a stable population for many years, such as Sweden, will have approximately the same numbers in all age classes (fig. 4.13). Notice that there are more females than males in the older age groups in both the United States and Sweden because of differences in longevity between the sexes. These countries also have a high percentage of retired people (17 percent in Sweden, for instance) because of long life expectancy.

Countries with a high percentage of children (such as Mexico) and countries with a high percentage of

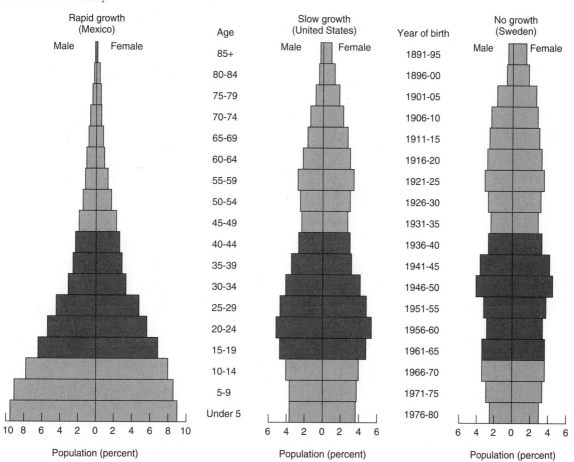

Figure 4.13 Population by age and sex in rapidly growing, slowly growing, and no-growth countries in 1980. Colored region represents individuals in reproductive ages. Note the high proportion of children in the rapidly growing population and the high proportion of elderly in the stable, no-growth population.

Source: Data from Carl Haub, "Understanding Population Projection," *Population Bulletin*, Vol. 42, no.4, Dec. 1987.

China's One-Child Family Program

*I*t is difficult to comprehend the number of people in China. In 1992, the population was estimated to be 1.166 billion, more than one-fifth of all the people in the world. The 22 million Chinese babies born each year are nearly equal to the combined metropolitan populations of New York, Los Angeles, and Chicago. More than four-fifths of China's population are rural peasants. The per capita income is about $370 per year, placing it among the least developed countries in the world, and yet it appears to be achieving population stabilization under conditions that experts had predicted were impossible.

For many centuries, China followed a cycle of disasters, famines, and political upheavals. It was widely believed that China was so large and unmanageable that this cycle could never be broken. In the 1950s, as a result of the misguided Great Leap Forward Program, some 20 million people died of starvation. Some experts predicted that China would never be able to feed itself. Now however, the country provides an adequate, if spartan, nutritional level for all of its people. Medical care, housing, education, and social security have also improved markedly. Between 1950 and 1980, the death rate dropped from twenty per thousand to eight per thousand, and average life expectancy increased from forty-seven to seventy years. During the same period, the United States had a death rate of nine per thousand and a life expectancy of seventy-four years.

Profamily traditions and public policies caused explosive population growth following the Socialist Revolution, especially in the postfamine recovery in the early 1960s (box fig. 4.3.1). Former Chairman Mao Zedong believed, as do many Marxists, that labor is the

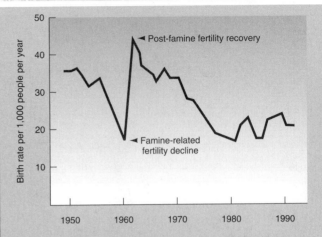

Figure 4.3.1 China's crude birth rate, 1950–1990.
Source: Population Reference Bureau, Inc.

source of all wealth. He proclaimed that "revolution plus production will solve all problems." After Mao's death in 1976, new leaders reversed those policies and sought to bring population growth under control. Premier Deng Xiaoping saw rampant population growth as the main obstacle to an improved standard of living. He also worried that modernization of agriculture would make hand labor obsolete, displacing as much as half of the rural population and creating a monumental unemployment problem.

Chinese demographers believe that scarcity of resources (only 11 percent of the land is arable) will limit the population that China can support to a maximum of 1.25 billion people, or only about 100 million more than the present population. Because of the high proportion of young people in the society now, China's population has an internal momentum that will carry it well past the

old people (such as Sweden) have a problem with their **dependency ratio,** or the number of nonworking compared to working individuals in a population. Mexico has a high number of children to be supported by each working person. In Sweden, although there are fewer children, a large number of retired persons must be supported by a small working population. There is considerable worry that not enough workers will be available to support retirement systems in countries where population growth has slowed sharply and life expectancies have increased. Retirement might have to be postponed or eliminated altogether.

Emigration and Immigration

Humans are highly mobile, so emigration and immigration play a larger role in human population dynamics than they do in those of many species. Currently, about 570,000 people immigrate legally to the United States each year, but at least twice that number enter illegally. In 1992, the American Refugee Committee estimated that at least 18 million people left their countries for political or economic reasons while another 50 million fled their homes but remained in their countries.

Immigration is a controversial issue in most wealthy countries. "Guest workers" often perform heavy, dangerous, or disagreeable work that citizens are unwilling to do. Many of these people are of a different racial or ethnic background than the majority in the country to which they go in search of work. The treatment of migrants and guest workers often is appalling. They are paid low wages and given substandard housing, poor working conditions, and few rights.

target of 1.25 billion in the year 2010 unless the average number of children per couple immediately falls below the replacement level of 2.2. For the past decade, therefore, China has carried on an extensive campaign to persuade couples to have only one child. Although there have been problems, the program has made dramatic progress. Between 1968 and 1988, the crude birth rate fell from about forty per thousand to twenty per thousand. The Chinese birth rate is now only a little higher than that of the United States, which has more than fifty times the per capita income of China. In Shanghai, Beijing, and other urban centers, 80 to 90 percent of all births are first children.

The success of this campaign undoubtedly lies in the unique social organization in China. Communes and production teams are responsible for jobs, health care, and political organization. Officials in these units pursue the national birth control goals through a combination of social pressure, incentives, and disincentives that many of us would find repressive. Newly married couples are asked to pledge that they will have only one child. Birth control specialists in the production team monitor all married women, and couples must be granted permission to be part of the quota of births allowed in any year. Free contraceptives and medical advice are provided. All methods of birth control are used, including pills, condoms, and abortion, but major reliance is placed on the intrauterine device (IUD), in spite of serious questions about its health effects.

Demographers point out that millions of girl babies "disappear" during their first year of life in China. At birth, the sex ratio is a normal 105 boys for every 100 girls. After one year, however, the ratio is 110 boys per 100 girls. No one knows what happens to the missing girls. They may be neglected or even killed by families who prefer their one child

to be a male, or they may simply be hidden so that parents can try again for a boy. At any rate, they appear to be causalities of the one-child policy.

Couples who sign the one-child pledge are guaranteed free delivery of that child when they give birth. The child receives preference in education and job placement as he or she grows up. The parents get better housing, longer vacations, and an extra month's pay each year. Because of a traditional preference for boys to carry on the family name and a tendency of couples to want to try again if they have a girl, special privileges are accorded to parents of a girl. Penalties for unsanctioned pregnancies include official reprimands, pay cuts, and public censure. Because each cadre is anxious to make a good showing in meeting its birth quotas, zealous local officials may use coercive techniques to demand abortions in cases of unapproved pregnancies and to control the behavior of individuals.

In contrast to most countries where family planning is effective only in wealthy, educated, urban upper- or middle-class families, birth control in China is nearly as effective among the working class as it is in the more affluent classes. The plan has met with resistance, however, in rural areas and among ethnic minorities. Allowing peasant families to farm private plots has created a demand for large families to work the fields. In many areas couples have been permitted to have a second or even third child.

In spite of some disturbing aspects, the relative success of China's one-child program suggests that a demographic transition can occur without concomitant industrialization and a shift to a high standard of living. If other developing countries can achieve similar results by more humane techniques, the human population might stabilize much sooner and at lower levels than demographers have previously thought possible.

People already established in the area often complain that immigrants take away jobs, overload social services, and ignore established rules of behavior or social values. There are often undertones of racism and xenophobia in these claims. Ironically, those who are themselves immigrants or descendants of immigrants are sometimes the most adamant about closing doors to future migration. Proponents of an open door policy claim that immigrants make a valuable contribution to society, bringing new energy, ingenuity, cultural diversity, and vigor.

POPULATION GROWTH: OPPOSING FACTORS

A number of social and economic pressures affect decisions about family size, which in turn affects the population at large.

Pronatalist Pressures

Factors that increase people's desires to have babies are called **pronatalist pressures.** Many people consider raising a family the most enjoyable and rewarding part of their lives. Children can be a source of pleasure, pride, and comfort. Also, children may be the only source of support for elderly parents in countries without a social security system. Where infant mortality rates are high, couples may have many children in order to be sure that at least a few will survive to take care of them when they are old. In places where people have little opportunity for upward mobility, children may provide status in society, express parental creativity, and provide a sense of continuity and accomplishment otherwise missing from life. Our response to babies (especially our own) and our

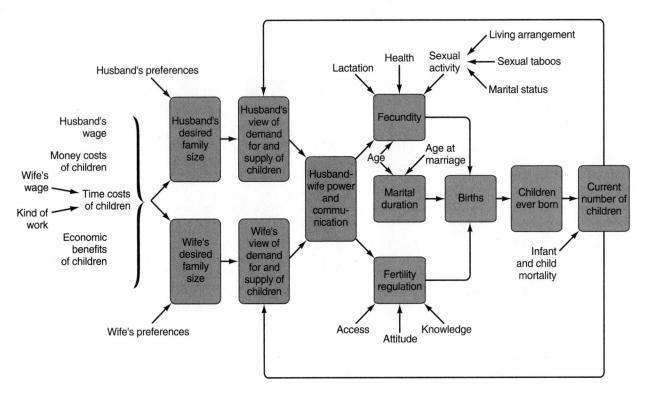

Figure 4.14 A model for the variables determining fertility. Three major factors interact in this model: the biological supply of children, the demand for children, and the regulation of fertility. Other intervening factors have either positive or negative effects on the interaction of these factors.

urge to reproduce must have at least some basis in an instinctive need to ensure survival of the species.

Society also has an impetus to survive, and the need to replace members who die or become incapacitated has centuries-old historic roots. This need often is codified in cultural or religious values that encourage bearing and raising children. In some societies families with few or no children are looked upon with pity or contempt. The idea of deliberately controlling fertility may be shocking, even taboo. Women who are pregnant or have small children may be given special status and protection. Frequently, boys are more valued than girls because they carry on the family name and are expected to support their parents in old age. Couples may have more children than they really want in an attempt to produce a son.

Males often derive a sense of pride from having as many children as possible. In some cultures, they accomplish this by having multiple wives. In societies where women have little control over their own lives, they may be subjected to the demands of their husbands, families, religion, or social groups in determining how many children they will have. Even though women might desire fewer children, they may be powerless to make that decision for themselves. In many societies, women have no status outside of their roles as wives and mothers. Without children, they have no source of support.

Figure 4.14 shows a model for the variables determining fertility. Three primary factors interact in this model: the biological supply of children, determined by fecundity and infant mortality; the demand for children, determined by economics and social values; and the regulation of fertility, determined by knowledge of, attitudes toward, and access to birth control. The combined interaction of these factors and their intervening variables determine fertility.

Birth Reduction Pressures

In highly developed countries, many pressures tend to reduce fertility. Higher education and more personal freedom for women often result in decisions to limit childbearing. The desire to have children is offset by a desire for goods and activities that compete with childbearing and childrearing for time and money. When women have opportunities to earn a salary, they are less likely to stay home and have children. Not only are the challenge and variety of a career attractive to many women, but the money that they can earn outside the home becomes an important part of the family budget. Thus, education and socioeconomic status are usually inversely related to fertility in richer countries. In less-developed countries, however, fertility is likely to be positively related to educational levels and socioeconomic status. As income increases, families are better able to afford the

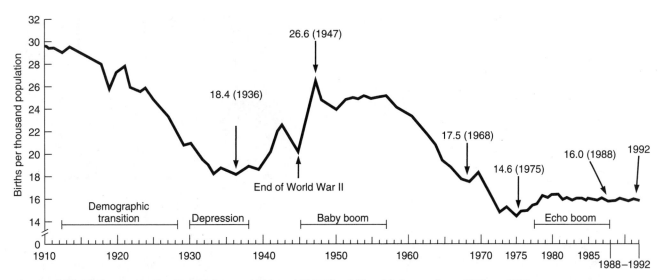

Figure 4.15 Birth rates in the United States, 1910 to 1992. The falling birth rate from 1910 to 1929 represents a demographic transition from an agricultural to an industrial society. Note that this decline occurred before the start of the Great Depression. The "baby boom" following the Second World War lasted from 1945 to 1957. A much smaller "echo boom" occurred around 1980 when the baby boomers started to reproduce, but it produced far fewer births than anticipated.

Sources: Data from Population Reference Bureau, Inc., and U.S. Bureau of Census.

children they want; higher income means that women are likely to be healthier and therefore better able to conceive and carry a child to term.

In less-developed countries, where feeding and clothing children can be a minimal expense, adding one more child to a family usually doesn't cost much. By contrast, raising a child in the United States can cost hundreds of thousands of dollars by the time he or she is through school and is independent. Under these circumstances, parents are more likely to choose to have one or two children on whom they can concentrate their time, energy, and financial resources.

Figure 4.15 shows U.S. birth rates between 1910 and 1992. As you can see, birth rates have fallen and risen in a complex pattern. The period between 1910 and 1930 was a time of industrialization and urbanization. Women were getting more education than ever before and entering the work force. The Great Depression in the 1930s made it economically difficult for families to have children, and birth rates were low. The birth rate increased at the beginning of World War II (as it often does in wartime). For reasons that are unclear, a higher percentage of boys are usually born during war years.

At the end of the war, a baby boom occurred as couples were reunited and new families were started. This high birth rate persisted through the prosperity and optimism of the 1950s but began to fall in the 1960s. Part of this decline was caused by the small number of babies born in the

1930s. This meant that fewer young adults were giving birth in the 1960s. Part was due to changed perceptions of the ideal family size. Whereas in the 1950s women typically wanted four children or more, in the 1970s the norm dropped to one or two (or no) children. A small echo boom occurred in the 1980s as people born in the 1950s began to have babies, but changing economics and attitudes seem to have permanently altered desired ideal family sizes in the United States.

Birth Dearth?

Most European countries now have birth rates below replacement rates, and Germany, Bulgaria, and Hungary are experiencing negative rates of natural increase. In Asia, Japan, Singapore, Hong Kong, and Taiwan are also facing a "child shock" as fertility rates have fallen well below the replacement level of 2.2 children per couple. There are concerns in all these countries about falling military strength (lack of soldiers), declining economic power (lack of workers), and declining social systems (not enough workers and taxpayers) if low birth rates persist or are not balanced by immigration.

Economist Ben Wattenberg warns that this "birth dearth" might seriously erode the powers of western democracies in world affairs. He points out that Europe and North America accounted for 22 percent of the world's population in 1950. By the 1980s, this number had fallen to 15 percent, and by the year 2030, Europe and North America will make up only 9 percent of the world's population. Germany, Hungary, Denmark, and Russia now offer incentives to encourage women to bear children. Some Asian countries are also worried about a birth dearth. Japan offers financial support to new parents, and Singapore provides a dating service to encourage marriages among the upper classes as a way of increasing population.

On the other hand, if the population of European people falls to about 10 percent of the world population, it will be nearly the same as it was before the population explosion of the eighteenth and nineteenth centuries. Furthermore, since Europeans and North Americans consume so much more of the world's resources per capita than most other people in the world, a reduction in the population of these countries will do more to spare the environment than would a reduction in population almost anywhere else.

DEMOGRAPHIC TRANSITION

In 1945, demographer Frank Notestein pointed out a typical pattern in which falling death rates and birth rates due to improved living conditions accompany economic development. He called this pattern the **demographic transition** from high birth and death rates to lower birth and death rates. An idealized version of this transition is shown in figure 4.16.

Development and Population

In figure 4.16, phase I represents the conditions in a premodern society. Food shortages, malnutrition, accidents, lack of sanitation and medicine, and other hazards keep death rates high. Birth rates are correspondingly high to keep population densities relatively constant.

Phase II represents the conditions in a developing country where better jobs and higher incomes improve the standard of living. More food is available, and sanitation and medicine increase longevity. As a result, death rates fall, often very rapidly. Birth rates also fall, but only after a lag of several decades, because people take a generation or so before they believe that social improvements are permanent. It also takes some years to change attitudes and values. One generation clings to the values it grew up with, and it takes a new generation to adapt to changing realities. During phase II, the population grows rapidly as the rate of natural increase (birth rate minus death rate) rises. The population will grow exponentially and, depending on how long it takes to complete this phase, may go through one or more rounds of doubling before coming into balance again.

Phase III represents the conditions in a developed country where both birth rates and death rates are low, often one-third or less of those in the predevel-

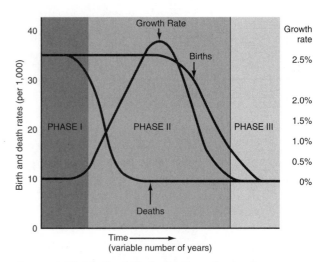

Figure 4.16 Idealized birth, death, and growth rate curves during a demographic transition. Phase I—pre-modern society with high birth and death rates and stable population size. Phase II—developmental stage in which death rate has fallen, but birth rate remains high. Population grows rapidly. Phase III—modern society in which birth and death rates come into equilibrium again but at a much lower level than before. Population size stabilizes in this phase but at a much higher level than before.

opment era. During this phase the population comes into a new equilibrium but at a much higher density than before. Most of the countries of northern and western Europe went through a demographic transition in the nineteenth or early twentieth century, although the shifts were not as uniform or pronounced as in the theoretical model.

The most rapidly growing countries in the world such as Gaza, Syria, and Zambia are now in Phase II of this demographic transition. Their death rates have fallen close to the rates of the fully developed countries, but their birth rates have not fallen correspondingly. In fact, their birth rates are higher than those in most European countries three hundred years ago when industrialization began. The large disparity between birth and death rates means that many developing countries are growing at 3 to 4 percent per year. This high rate of growth in the Third World will boost total world population to 10 billion or more before the end of the next century. Perhaps the most important questions in this whole chapter are "Why are birth rates not yet falling in these countries?" and "What can be done about it?"

An Optimistic View

Some demographers claim that a demographic transition is already in progress in most developing nations and that the world population will stabilize sometime in the next century. Some evidence supports this view. The total fertility rate in developing countries dropped from 6.1 in 1970 to 3.9 in 1990. What factors might facilitate transition to a stable world population?

- The growing prosperity and social reforms that typically accompany development should reduce the need and desire for large families in most countries.
- Technology is available to bring advances to the developing world much more rapidly than was the case a century ago, and the rate of technology transfer is much faster than it was when Europe and North America were developing.
- Less-developed countries have historic patterns to follow: They can benefit from the mistakes of developed nations and can chart a course to stability more quickly than they might otherwise be able to.
- Modern communications (especially television) have caused a revolution of rising expectations that act as a stimulus to spur change and development.

A Pessimistic View

Economist Lester Brown of the Worldwatch Institute presents a more pessimistic view of the demographic situation. He warns that many of the poorer countries of the world appear to be caught in a "demographic trap" that prevents them from escaping from phase II of the demographic transition. When populations expand to the point where human demands exceed the sustainable yields of local forests, grasslands, croplands, or water resources, the likely result is environmental deterioration, economic decline, and political instability.

Many people argue that the only way to break out of the demographic trap is to immediately and drastically reduce population growth by whatever means are necessary. They argue strongly for birth control education and bold national policies to encourage lower birth rates.

Some agree with Malthus that helping the poor will simply increase their reproductive success and further threaten the resources on which we all depend. Author Garret Hardin, for example, calls for lifeboat ethics. "Each rich nation," he says, "amounts to a lifeboat full of comparatively rich people. The poor of the world are in other much more crowded lifeboats. Continuously, so to speak, the poor fall out of their lifeboats and swim for a while, hoping to be admitted to a rich lifeboat, or in some other way to benefit from the goodies on board. . . . We cannot risk the safety of all the passengers by helping others in need. What happens if you share space in a lifeboat? The boat is swamped and everyone drowns. Complete justice, complete catastrophe."

A Social Justice View

A third view is that **social justice** is the key to successful demographic transitions. According to this view, the world has enough resources for everyone, but inequitable social and economic systems cause maldistributions of those resources. Hunger, poverty, violence, environmental degradation, and overpopulation are symptoms of a lack of social justice rather than a lack of resources. The way to solve these problems is to establish fair systems, not to blame the victims. The cartoon in figure 4.17 expresses the opinion of many people in the less-developed countries about the relation between resources and population.

An important part of this view is that many of the rich countries are or were colonial powers, while the poor, rapidly growing countries are or were colonies. The wealth that paid for progress and security for developed countries was often extracted from colonies, which now suffer from exhausted resources, exploding populations, and chaotic political systems. Some of the world's poorest countries such as India, Ethiopia, Mozambique, and Haiti had rich resources and adequate food supplies before they were impoverished by colonialism. Those in developed nations who now enjoy abundance may need to help the poorer countries not only as a matter of justice but because we all share the same environment.

Ecojustice

In addition to considering the rights of all humans, we should also consider the rights of other species. Rather than ask only how many humans the world can support, perhaps we should think about the needs of other creatures. As we convert natural landscapes into agricultural or industrial areas, species are crowded out that may have just as much right to exist as we do. Perhaps we should seek the optimum human population at which we can provide a fair and decent life for all people while causing a minimum impact on our nonhuman neighbors.

Infant and Child Mortality

Survival of children is one of the most critical factors in stabilizing population. When infant and child mortality rates are high, as they are in much of the developing world, parents tend to have high numbers of children to ensure that some will survive to adulthood. There has never been a sustained drop in birth rates that was not first preceded by a sustained drop in infant and child mortality. One of the most important distinctions in our demographically divided world is the high proportion of deaths under five years of age in the less-developed countries. Improved health care, simple oral rehydration therapy, and immunization against infectious diseases (chapter 6) have

Figure 4.17 Controlling our population and resources—there may be more than one side to the issue.

Source: Asian Cultural Forum on Development.

brought about dramatic reductions in child mortality rates that have been accompanied in most regions by falling birth rates. It has been estimated that saving 5 million children each year from easily preventable communicable diseases would prevent 20 or 30 million extra births.

FAMILY PLANNING AND FERTILITY CONTROL

Family planning allows couples to determine the number and spacing of their children. It doesn't necessarily mean fewer children—people may use family planning to have the maximum number of children possible—but it does imply that people will control their reproductive lives and make rational, conscious decisions about how many children they will have and when those children will be born, rather than leaving it to chance. As the desire for smaller families becomes more common, birth control becomes an essential part of family planning in most cases. In this context, **birth control** usually means any method used to reduce births, including celibacy, delayed marriage, contraception, methods that prevent implantation of embryos, and induced abortions.

Traditional Fertility Control

Evidence suggests that people in every culture and every historic period have used a variety of techniques to control population size. Studies of hunting and gathering people, such as the Kung! or San of the

Kalahari Desert in southwest Africa, indicate that our early ancestors had stable population densities not because they killed each other or starved to death regularly but because they controlled fertility.

For instance, San women breast-feed children for three or four years. When calories are limited, lactation depletes body fat stores and suppresses ovulation. Coupled with taboos against intercourse while breast-feeding, this is an effective way of spacing children. Other ancient techniques to control population size include celibacy, polygamy, folk medicines, abortion, and infanticide. We may find some or all of these techniques backward, unpleasant, or morally unacceptable, but we shouldn't assume that other people are too ignorant or too primitive to make decisions about fertility.

Current Birth Control Methods

Modern medicine gives us many more options for controlling fertility than were available to our ancestors. Some of these techniques are safer, easier, and more pleasant to use. The major categories of birth control techniques include (1) avoidance of sex during fertile periods (celibacy, using changes in body temperature or cervical mucous color and viscosity to judge when ovulation will occur); (2) mechanical barriers that prevent contact between sperm and egg (condoms, spermicides, diaphragm, cervical cap, and vaginal sponge; figs. 4.18 and 4.19c); (3) surgical methods that prevent release of sperm or egg (sterilization: tubal ligation or use of the Filshie clip in females; vasectomy in males; figs. 4.19a

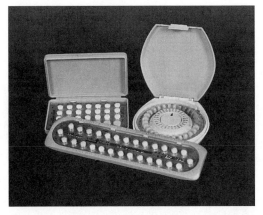

a.

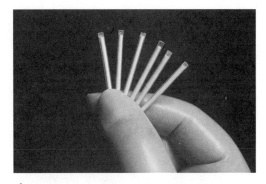

d.

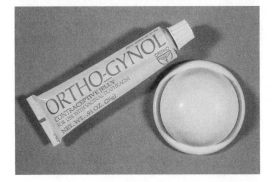

b.

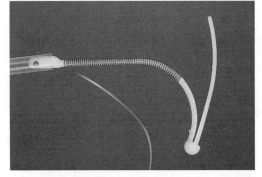

e.

Figure 4.18 (*a*) Daily birth control pills; (*b*) spermicidal jelly and diaphram; (*c*) spermicidal foam; (*d*) subdermal slow-release progestin; (*e*) intrauterine device; (*f*) condom.

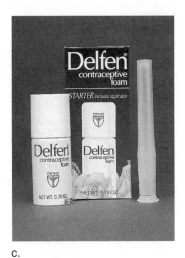

c.

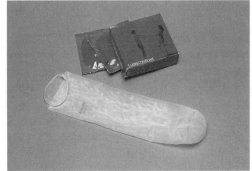

f.

and *b*); (4) chemicals that prevent maturation or release of sperm or eggs or implantation of the embryo in the uterus (the pill: estrogen + progesterone, progesterone alone for females; gossypol for males); (5) physical barriers to implantation (intrauterine device, or IUD); and (6) abortion.

None of these methods is perfect and none suits every contraceptive need. The best choice for you depends on your life situation and plans for the future (box 4.4).

New Developments in Birth Control

In 1991, Norplant, the trade name for flexible, matchstick-sized, silicon-rubber implants containing slow-release analog of progesterone was approved for use in the United States. The implants are inserted under the skin where they will release hormones for up to five years (fig. 4.20). In 1992, Depo-Provera, an injectable form of progesterone, was approved by the Food and Drug Administration after many years of research and controversy. The injections are given four

Acting Locally: Box 4.4

Preparing a Personal Fertility Plan

What are the options for personal reproductive decisions? More alternatives are available now than were a generation ago, but we face more concerns about health, sexuality, and parenthood as well. How can an individual know what to do? Each of us must examine our own values, situation, and choices to design a personal fertility plan.

First, examine your own life and your hopes for the future. How old are you? Are you sexually active now? Will you be in the future? How many children—if any—will you want, and when would you like them to be born? Timing and number of births can critically effect education, jobs, travel, and other life aspirations. How many children will you be able to support and educate, given projections for future costs and income? As global citizens we also should ask how many children it is responsible for us to bring into the world, given the impact we have.

Next, examine the alternatives available to meet your goals. Remember that you don't need to commit yourself to a single option for life. Among the choices are

- *Abstinence:* One hundred percent effective in birth control and prevention of sexually transmitted diseases. Under some circumstances, this may be the best and safest course of action. As a long-term strategy, however, celibacy may inhibit intimate and fulfilling relationships for many people.

- *Oral contraceptives (pills):* Ninety-nine percent effective. Easy to use and readily reversible, the pill is not recommended for women over age thirty-five who smoke because of the risk of strokes and other cardiovascular problems. Pills with low estrogen levels or progesterone alone can reduce these risks. Some increase in breast cancer is associated with prolonged use of synthetic hormones, but on the other hand they seem to give some protection from endometrial and ovarian cancer. Overall, cancer risks seem slightly lower with birth control pills than without.

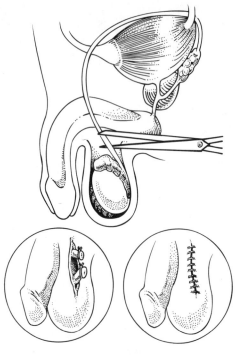
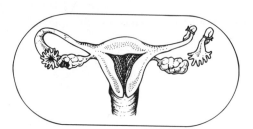
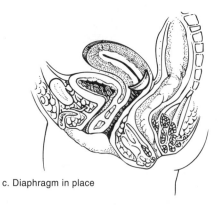

b. Oviduct cut and tied

a. Vas deferens cut and tied

c. Diaphragm in place

Figure 4.19 Surgical birth control procedures include vasectomy for men and tubal ligation for women. (*a*) Vasectomy requires only a small incision under local anesthesia to cut and tie the sperm-conducting *vas deferens*. (*b*) Tubal ligation in women is a more difficult operation and usually requires general anesthesia. Neither procedure affects sex life or sexual characteristics, but neither is easily reversible. (*c*) The cervical cap or diaphragm is an example of a mechanical barrier. It covers the mouth of the cervix and prevents sperm from entering the uterus. While not as dependable as surgical methods, mechanical barriers are easily reversible.

- *Implantable or injectable progesterone analogs:* Ninety-nine percent effective and highly reversible. Doesn't require daily attention or approval of partner. Implants may be suitable for women who cannot take oral contraceptives. Some women experience dizziness, fatigue, headaches, vaginal bleeding, or absent periods, and implants or injections may increase the risk of osteoporosis. They have about the same breast cancer risk and other cancer benefits as oral pills.

- *Barrier methods:* Condoms, diaphragms, cervical caps, and vaginal sponges are generally 85 to 95 percent effective. They help prevent sexually transmitted diseases, but don't bet your life on them. They require some planning ahead but not the daily attention of the pill. This may be a good option for someone who cannot use hormone analogs or has sex only occasionally.

- *Spermicidal creams, jellies, douches:* Not very effective by themselves in either birth control or disease prevention. These methods may help, however, when used together with barrier methods.

- *Intrauterine device:* Ninety-four percent effective. Doesn't require daily attention or alter natural hormone balance. IUDs may increase frequency or severity of sexually transmitted diseases as well as vaginal bleeding, cramps, and other side effects. They are generally recommended only for mature women in monogamous relationships.

- *Natural methods:* Careful record keeping, temperature measurements, and examination of cervical mucus can be 70 to 90 percent effective in birth control. This method meets ethical or religious scruples but may be difficult and demanding to follow. Prohibits sex on some days.

- *Vasectomy, tubal ligation, Filshie clip:* These surgical methods are more than 99 percent effective in birth control but are not easily reversed. These may be good options for people who are sure they don't want any more children.

As you can see, no birth control method is perfect for everyone. All have advantages and disadvantages. By thinking carefully and communicating clearly about values and desires, couples can reach responsible decisions about their fertility and sexual behavior.

times a year. Both the implants and the injections are as effective as daily oral contraceptives (about 1 percent failure rate) but eliminate the need to keep track of and take daily pills. They also can be used without the knowledge of one's partner, who may oppose birth control. Both the injections and the implants cause prob-

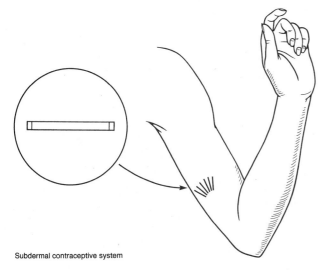

Subdermal contraceptive system

Figure 4.20 In a subdermal contraceptive system, six flexible, match-sized capsules are implanted under the skin in the woman's arm. They release a slow but steady supply of birth control hormones over about five years, or they can be removed if the woman wants to become pregnant.

lems for some women who experience increased vaginal bleeding or the absence of menstrual periods, and some increased breast cancer and osteoporosis (bone thinning) have been associated with their use.

A condom for women called Reality has recently been introduced. It consists of two flexible plastic rings connected by a strong, clear polyurethane sheath. One ring fits over the cervix much like a diaphragm, while the other remains outside the vagina. Each condom costs about two dollars and is designed to be discarded after a single use. The six-month failure rate is about 12 percent, which means that 12 percent of women using only this device will get pregnant in six months. An important advantage of condoms is the protection they offer against sexually transmitted diseases. A female condom gives women control over their own reproduction, but some women may find them difficult or unpleasant to use.

Some new methods of birth control presently being studied may have great promise. The "morning-after" drug, RU486 (mifepristone or mifegyne), can be taken up to ten days after a missed period. It blocks the effects of progesterone in maintaining the lining of the uterine wall and prevents the embryo from implanting. It is reported to cause fewer physical and psychological side effects than abortion after implantation has occurred. Interestingly, RU486 also appears to have promise in treating breast cancer, brain cancer, diabetes, and hypertension. Some groups consider this antipregnancy agent to be the same as abortion, however, and oppose its use. In 1993 the nonprofit Population Council signed an agreement to carry out clinical trials of RU486 on 2,000 women in Oregon. More than 120,000 women have already received the drug from clinics in France, Great Britain, and Sweden, but approval for general use in the United States is probably years away.

Development of simple, inexpensive, do-it-yourself tests for levels of estrogen and progesterone in urine may make the rhythm method easier to follow and more reliable for women who cannot or would rather not use other methods of birth control. Some antipregnancy and antisperm vaccines are being tested that would use the immune system to prevent fertilization or pregnancy, but they are several years away from the market. The sperm suppressant gossypol is being tested as a male contraceptive, but side effects and male reluctance to take responsibility for reproduction inhibit its introduction. Finally, potent new spermicides are being tested, as are new methods of using them.

THE FUTURE OF HUMAN POPULATIONS

How many people will be in the world a century from now? Most demographers believe that world population will stabilize sometime during the next century. The total number of humans, when we reach that equilibrium, probably will be between 6 and 16 billion people, depending on the success of family planning programs and the multitude of other factors affecting human populations (fig. 4.21). Will the earth be able to support that many humans on a sustainable basis? As you have seen in this chapter, there are good reasons to fear that we may be headed for a population crash. Even if technology can increase the carrying capacity of the earth, however, it clearly will be easier to feed, clothe, and house 6 billion people rather than 16 billion people; thus family planning programs are a good idea.

In 1992, the World Health Organization estimated that nearly 1 million conceptions occur daily around the world as a result of some 100 million sex acts. At least half of those conceptions are unplanned or unwanted or both. About 380 million people in the Third World used contraceptives in 1990, up nearly

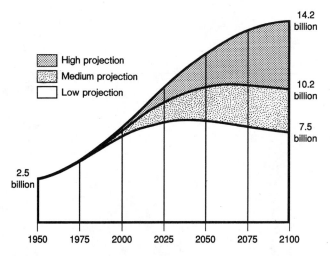

Figure 4.21 Long-term projections of world population result in different scenarios, depending on assumptions about the pace of development, birth control, and subsequent demographic transitions. The United Nations projection, shown in the middle scenario above, assumes that world fertility will decline to two children per woman in 2035, resulting in a population of 10.2 billion or a little less than twice that of today.

Source: Data from *1988 World Population Data Sheet*, Population Reference Bureau, Inc.

tenfold from 1960, but another 300 million people say they want but do not have access to family planning. Coordinated, effective family planning requires governmental action, ranging from policies to assistance. The methods used to promote family planning and the ethics of social intervention and population control are often controversial.

Deep societal changes are often required to make family planning programs successful. Among the most important of these are (1) improved social, educational, and economic status for women (birth control and women's rights are often interdependent); (2) improved status for children (fewer children are born as parents come to regard them as valued individuals rather than possessions); (3) acceptance of calculated choice as a valid element in life in general and in fertility in particular (belief that we have no control over our lives discourages a sense of responsibility); (4) social security and political stability that give people the means and the confidence to plan for the future; and (5) knowledge, availability, and use of effective and acceptable means of birth control.

S u m m a r y

Human populations have grown at an unprecedented rate over the past three centuries. By 1992, the world population stood at 5.4 billion people. If the current growth rate of 1.7 percent per year persists, the population will double in thirty-nine years. Most of that growth will occur in the less-developed countries of Asia, Africa, and Latin America. There is a serious concern that the number of humans in the world and our impact on the environment will overload the life-support systems of Earth.

The crude birth rate is the number of births in a year divided by the average population. A more accurate measure of growth is the general fertility rate, which takes into account the age structure and fecundity of the population. The crude birth rate minus the crude death rate gives the rate of natural increase. When this rate reaches a level at which people are just replacing themselves, zero population growth is achieved.

In the more highly developed countries of the world, growth has slowed or even reversed in recent years so that without immigration from other areas, populations would be declining. The change from high birth and death rates that accompanies industrialization is called a demographic transition. Many less-developed countries have already begun this transition. Death rates have fallen, but birth rates remain high. Some demographers believe that as infant mortality drops and economic development progresses so that people in these countries can be sure of a secure future, they will complete the transition to a stable population. Others fear that excessive population growth and limited resources will catch many of the poorer countries in a demographic trap that could prevent them from ever achieving a stable population or a high standard of living.

While larger populations bring many problems, they also may be a valuable resource of energy, intelligence, and enterprise that will make it possible to overcome resource limitation problems. A social justice view argues that a more equitable distribution of wealth might reduce both excess population growth and environmental degradation.

We have many more options now for controlling fertility than were available to our ancestors. Some techniques are safer than those available earlier; many are easier and more pleasant to use. Sometimes it takes deep changes in a culture to make family planning programs successful. Among these changes are improved social, educational, and economic status for women; higher values placed on children; accepting responsibility for our own lives; social security and political stability that give people the means and confidence to plan for the future; and knowledge, availability, and use of effective and acceptable means of birth control.

Review Questions

1. At what point in history did the world population pass its first billion? What factors restricted population before that time, and what factors contributed to growth after that point?
2. How might growing populations be beneficial in solving development problems?
3. Why do some economists consider human resources more important than natural resources in determining the future of a country?
4. Where will most population growth occur in the next century? What conditions contribute to rapid population growth in some countries?
5. Define crude birth rate, total fertility rate, total growth rate, crude death rate, and zero population growth.
6. What is the difference between life expectancy and longevity?
7. What is dependency ratio, and how might it affect the United States in the future?
8. What pressures or interests make people want to or not want to have babies?
9. Describe the conditions that lead to a demographic transition.
10. Describe the major choices in modern birth control.

Questions for Critical or Reflective Thinking

1. What do you think is the optimum human population, and what do you think is the maximum human population? Are the numbers different? If so, why?
2. Some people argue that technology can provide solutions for environmental problems; others believe that a "technological fix" will make our problems worse. What personal experiences or worldviews do you think might underlie these positions?
3. Karl Marx called Thomas Malthus a "shameless sycophant of the ruling classes." Why would the landed gentry of the eighteenth century be concerned about population growth of the lower classes? Are there comparable class struggles today?
4. Try to imagine yourself as a person your age in a Third World country. What family planning choices and pressures would you face? How would you choose between these options?

5. Some demographers claim that population growth has already begun to slow; others dispute this claim. How would you evaluate the competing claims of these two camps? Is this an issue of uncertain facts or differing beliefs? What sources of evidence would you accept as valid?

6. What role do race, ethnicity, and culture play in our immigration and population policies? How can we distinguish between prejudice and selfishness on one hand and valid concerns about limits to growth on the other?

Key Terms

biotic potential 65
birth control 82
carrying capacity 65
crude birth rate 71
crude death rate 72
demographic transition 80
demography 71
density-dependent factors 66
density-independent factors 66
dependency ratio 76
dieback 65
environmental resistance 66
exponential growth 65
family planning 82

fecundity 71
fertility 71
life expectancy 73
life span 73
natural increase 73
neo-Malthusian 70
overshoot 65
population momentum 75
pronatalist pressures 77
social justice 81
survivorship 66
total fertility rate 72
total growth rate 73
zero population growth 72

Suggested Readings

Berreby, D. "The Numbers Game." *Discover* 11, no. 4 (April 1990):42. An interesting comparison of the views of Paul Ehrlich and Julian Simon.

Blaxter, K. L. *People, Food and Resources.* Cambridge: Cambridge University Press, 1986. An economist looks at demographics and resources. Well written.

Brown, L. R., and J. L. Jacobson. "Our Demographically Divided World." *Worldwatch Paper 74.* Washington, D.C.: Worldwatch Institute, December 1986. A sobering analysis of our demographic future.

Caldwell, J. C., and P. Caldwell. "High Fertility in Sub-Saharan Africa." *Scientific American* 262, no. 5 (May 1990):118. Birth rates and population growth have begun to decline everywhere else in the world. What makes this region different?

Cole, H., et al. *Models of Doom: A Critique of the Limits to Growth.* New York: Universe Books, 1973. An answer to the computer forecasts of Meadows, et al.

Eberstadt, N. *Fertility Decline in the Less Developed Countries.* New York: Praeger Scientific, 1981. Excellent survey of demographic transitions.

Ehrlich, A. H., and P. R. Ehrlich. "Why Do People Starve?" *The Amicus Journal* 9, no. 2 (Spring 1987):42. What would be the carrying capacity of a world full of saints?

Ehrlich, P. R. *The Population Bomb.* New York: Ballantine Books, 1968. A landmark in public concern about population problems.

Greenhalgh, S., and J. Bongaarts. "Fertility Policy in China: Future Options." *Science* 235, no. 4793 (March 6, 1987):1167. An excellent overview of current conditions in China.

Gustlee, C. "The Coming World Labor Shortage." *Fortune* (April 1990):71. Describes the baby bust facing industrialized economies. Warns that retirement may be impossible in the future.

Hardin, G. *Exploring Ethics for Survival: The Voyage of the Spaceship Beagle.* New York: Viking Press, 1972. Controversial ideas from one of our most challenging thinkers.

Jacobsen, J. "Promoting Population Stabilization: Incentives for Small Families." *Worldwatch Paper 54.* Washington, D.C.: Worldwatch Institute, June 1983.

Keyfitz, Nathan. "The Growing Human Population." *Scientific American* 261, no. 3 (September 1989):119. This distinguished population biologist predicts that human population will stabilize but that the planet's life-support capacity may not support all of us.

Kolata, G. "Wet-nursing boom in England explored." *Science* 235, no. 4790 (February 13, 1987):745. A fascinating study of why seventeenth- and eighteenth-century English upper-class women had as many as thirty children.

McKeown, T. R., G. Brown, and R. G. Record. "An Interpretation of the Modern Rise of Population in Europe." *Population Studies* 26 (1972):3. An interesting historical analysis of possible causes of accelerating population growth in Europe in the eighteenth and nineteenth centuries.

Mamdani, M. *The Myth of Population Control: Family, Caste and Class in an Indian Village.* New York: Monthly Review Press, 1972. An intriguing study of why externally imposed population policies don't work.

Mass, B. *Population Target: The Political Economy of Population Control in Latin America.* Toronto: Canadian Women's Educational Press, Latin American Working Group, 1976. Demographics in a social and political perspective.

Meadows, D. H., et al. *Beyond the Limits.* Post Mills, Vt: Chelsea Green Publishing, 1992. A computer model of population–resource interactions.

Menken, J., ed. *World Population and U.S. Policy: The Choices Ahead.* New York: Norton, 1986. A thorough but accessible anthology of articles on demographics and family planning.

Population Reference Bureau. *1992 World Population Data Sheet of the Population Reference Bureau, Inc.* Washington, D.C., 1992. An excellent, annually updated source of demographic data.

Preston, S. H. "Population Growth and Economic Development." *Environment* 28, no. 2 (March 1986):6. Links these two important factors.

Repetto, R. *The Global Possible.* New Haven: Yale University Press, 1985. An excellent overview of the current situation with an emphasis on ways we can improve.

Simon, J. L. *The Ultimate Resource.* Princeton, N.J.: Princeton University Press, 1981. A cornucopian view that more people means more geniuses, bigger markets, and a better life for all.

Tierney, J. "State of the Species: Fanisi's Choice." *Science 86* 7, no. 1 (January/February 1986):26. A personal view of why women in less-developed countries might choose to have eight children.

Ulmann, A., et al. "RU486." *Scientific American* 262, no. 6 (June 1990):42. An excellent description of both the science and the politics of this controversial method of birth control.

PART II: BASIC CONCEPTS

Air, water, life these above all
—these that in their huge cycles constantly renew themselves—
these, with the sun's light, form the great conditions of Man's being"

Ansel Adams and Nancy Newhall, *This Is the American Earth*

5

ENVIRONMENTAL RESOURCE ECONOMICS

The ecologist lays down an environmental imperative that requires an end to economic growth as the price of biological survival. The economist counters with a socioeconomic imperative that requires the continuation of growth as the price of social survival . . . and restoring the environment.

Walter Heller

Objectives

After studying this chapter you should be able to:

- Define natural and human resources and distinguish between economic resource categories.

- Describe frontier, industrial, and postindustrial economies and discuss how each uses and affects its environment.

- Diagram the relationship between supply and demand at different stages of economic or technological development.

- Understand how technology can mitigate scarcity and increase the carrying capacity of our environment.

- Appreciate the limits to growth and the features of a steady-state system.

- Explain internal and external costs, market approaches to pollution control, and cost/benefit ratios.

INTRODUCTION

Ecologists warn that if human populations continue to grow and we continue to consume resources at current rates, we will surpass the carrying capacity of our environment and suffer disastrous collapses in ecological and social systems. Many economists, on the other hand, argue that economic growth and development are essential to raise the standard of living for all humans and provide resources for restoring our environment. Where many environmentalists contend that all growth must stop, economists tend to regard the structure rather than the mere fact of growth as our main problem.

A major source of disagreement in this argument is the nature and extent of natural resources. Are we using up irreplaceable assets that will cause crippling shortages in our own lifetime and condemn our children to lives of poverty, misery, and privation? Or can technological development create substitutes for now essential resources that will give everyone a richer and more fulfilling life? Questions such as these about limits to economic development, resource scarcity, and population growth underlie much of our global debate about environmental policy.

Understanding how our environment works and how we should interact with it requires some appreciation of both ecology and economics. This chapter surveys the field of natural resource economics to show how it applies to environmental science. We will look at some contrasting views about the relationship between resource values, scarcity, economic development, and limits to growth. Deciding which of these theories is correct (wholly or in part) poses a difficult, controversial question that each of us must examine in planning for the future.

Ironically, ecology and economy are derived from the same root words and concerns. *Oikos* or *ecos* is the Greek word for household. Economy is the *nomos* or counting of the household goods and services. Ecology is the *logos* or logic of how the household works. In both disciplines, household is expanded to include the whole world—the household of humans. Although ecologists and economists study the same household, their approaches and interpretations are so different and the language they use to describe what they see is so specialized and mutually incomprehensible that a gulf of mistrust and misunderstanding often separates the two disciplines.

ECONOMIC CONTEXT

Money and politics are the languages of most policy planners and decision makers. They ask, "How much will it cost?" and "What are the benefits?" Economists try to answer those questions. Basically, economics deals with resource allocation or tradeoffs, either on the "micro" scale of buying and selling by individuals and businesses, or on the "macro" scale of national policy and world economic systems. Economics is a description of how valuable goods and services are to us as we make decisions about how to use time, energy, creativity, or physical resources. If resources were unlimited, there would be no necessity to choose between alternatives and no need for a system of economics.

In the real world, however, we usually face decisions about tradeoffs between the goods and services we desire and the resources to produce them (fig. 5.1). Economists ask, what shall we produce, for whom or for what purpose? Furthermore, in what manner, and when, shall we produce these goods and services? What level of pollution and environmental or social disruption is acceptable to obtain the things we want? For those of us in the affluent countries of the world, as most of our basic needs are satisfied, a new question becomes important. We should ask ourselves, how much is enough? How much luxury, convenience, or simple acquisition of stuff do we need? Would we—and the world—be better off if we were satisfied with less?

A classic expression of tradeoffs in economics is the question, What shall we produce: guns or butter? This assumes a reciprocal relationship between these two symbolic sets of products, shown in figure 5.2: If we have more guns, we will have less butter, and vice versa. The curve describing the maximum output at a given stage of technological or capital development is called the **production frontier.** A point inside this curve, such as point U in this figure, represents resources and production capacity that are not being fully employed in the most efficient manner. However, there may be very good societal or ecological reasons to not fully use resources at a given time.

Economic growth is an increase in the total wealth of a nation regardless of the population. If the population grows faster than the economy, there may be real growth, but the share per person will decline. **Economic development** is a rise in the average real income (adjusted for inflation) *per person.* It does not guarantee that everyone is better off, however. Development often involves the introduction of labor-saving machines that create greater profits but displace workers. As a result, the per capita income may rise, but the actual standard of living for much of the population may decline.

Figure 5.3 shows how growth in an economic system expands the production frontier. With better access to resources, improved technological efficiency, more workers, or more investment in production facilities (all considered to be forms of capital), we can produce more of both guns and butter (or whatever else people want). In most economic systems, continued growth is thought to be essential to maintain full employment and

Figure 5.1 Are natural resources like canned goods on a shelf, awaiting our use until they are consumed, or can they be extended and created by technology and enterprise? Economics deals with resource values. What should we produce, for whom, and at what cost?

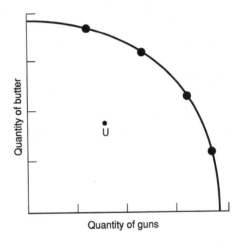

Figure 5.2 Production possibilities: "guns" or "butter." An economy can produce all "guns" or all "butter," or some combination of the two at a given level of technology. The curve represents balanced production possibilities. Any point inside the curve, such as *U,* indicates that resources are not being fully employed in the most efficient manner possible.

prevent class conflict that arises from inequitable distribution. As President John F. Kennedy said, "a rising tide lifts all boats." Recently, however, many people have become concerned that incessant growth of both human populations and productive systems soon will exhaust natural resources and will surpass the capacity of natural systems to withstand disruption, either by effluents and wastes or by the harvesting of more stock than can be replaced.

Many ecologists and a few unorthodox economists call for a transition to a **steady-state economy**

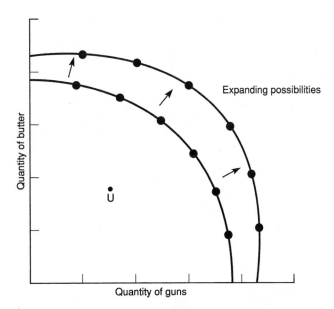

Figure 5.3 Compare this figure to figure 5.2. Economic growth brings expanding possibilities for both "guns" and "butter" or any other pair of competing goods and services.

characterized by low birth and death rates, use of renewable energy sources, recycling of materials, and an emphasis on durability, efficiency, and stability rather than high throughput of materials. Perhaps the first economist to advocate this new way of looking at the world was John Stuart Mill, whose *Principles of Political Economy,* published in 1857, preceded Karl Marx. More recently, the eccentric Romanian economist Nicholas Georgescu-Roegen spent his career exploring the limitations imposed on economics by the laws of thermodynamics.

Kenneth Boulding's 1966 essay, "The Economics of the Coming Spaceship Earth" and Herman E. Daly's 1991 *Steady-State Economics* both built on the concept that continual growth cannot be sustained. We will discuss the implications of steady-state economic theory further after looking at some of the questions of resource scarcity that make it seem necessary.

RESOURCES AND RESERVES

What are the resources that economic systems are set up to manage and allocate? How can we determine the available amount of a specific resource, given a particular economic system and technology? These are vital questions in the field of environmental studies because much of our concern about population growth hinges on a continual supply of resources.

Defining Resources

Simply defined, a resource is any useful information, material, or service. Within this broad generalization, we can differentiate between **natural resources**

(goods and services supplied by our environment) and **human resources** (human wisdom, experience, skill, labor, and enterprise). It is also useful to distinguish between exhaustible, renewable, and intangible resources.

In general, **exhaustible resources** are the earth's geologic endowment: the minerals, nonmineral resources, fossil fuels, and other materials present in fixed amounts in the environment (fig. 5.4). In theory, these exhaustible resources place a strict upper limit on the number of humans and the amount of industrial activity our environment can support. Predictions that we are in imminent danger of running out of one or another of these exhaustible resources are abundant. In practice, however, the available supplies of many commodities such as metals can be effectively expanded by more efficient use, recycling, substitution of one material for another, or better extraction from dilute or dispersed supplies.

Renewable resources include sunlight—our ultimate source of energy—and the biological organisms and biogeochemical cycles powered by solar energy. In contrast to minerals or fossil fuels, biological organisms are self-renewing (fig. 5.5). With careful management, we can harvest surplus plants and animals indefinitely without reducing the available supply. Unfortunately, our stewardship of these resources is often less than ideal. Most species have thresholds of population size, habitat, or other critical factors below which populations can suddenly crash. Once vast populations of species such as passenger pigeons, American bison (buffalo), or Atlantic cod, for instance, were exhausted by overharvesting in only a

Figure 5.4 The geological resources of the earth's crust, such as oil from this well, are nonrenewable. Often the limit to our use of these resources is not so much the absolute amount available as the energy required to extract the resources and the environmental consequences of doing so.

Figure 5.5 Biological resources are renewable in that they replace themselves by reproduction, but if overused or misused, populations die. When a whole species is lost, it is unlikely that it will ever be re-created. These northern fur seals were almost totally exterminated by overhunting in the nineteenth century.

few years (chapter 9). Ironically, human mismanagement may make these renewable resources more ephemeral and limited than fixed geological resources.

Abstract or **intangible resources,** including open space, beauty, serenity, genius, information, diversity, and satisfaction, are also important to us (fig. 5.6). Strangely, these resources can be both infinite *and* exhaustible. There is no upper limit to the amount of beauty, knowledge, or satisfaction that can exist in the world, yet they can be easily destroyed. A single piece of trash can ruin a beautiful vista, or a single mean-spirited remark can spoil an otherwise perfect day. On the other hand, unlike tangible resources that usually are reduced by use or sharing, intangible resources often are increased by use and multiplied by sharing. These nonmaterial resources can be important economically. Information management and tourism—both based on intangible resources—have become two of the largest and most powerful industries in the world.

Economic Categories

We have defined a resource as anything useful, but we should distinguish between economic usefulness and the total amount of a material or service that is available. Vast supplies of potentially important materials are present in the earth's crust, for instance, but they are useful to us only if we can recover them in reasonable amounts with available technology and with acceptable environmental and economic costs. Within the aggregate total of any natural resource, we can distinguish categories on the basis of economic and technological feasibility as well as the resource location and quantity (fig. 5.7).

Figure 5.6 Many of the intangible resources of nature enrich the quality of our lives. Our enjoyment of these resources may be enhanced by sharing them with others. On the other hand, too many people, or people engaged in conflicting activities, can destroy the values we seek.

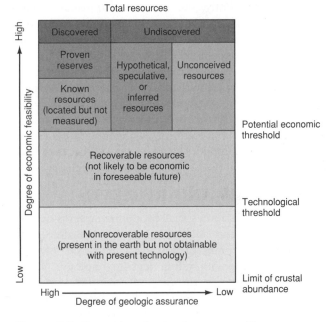

Figure 5.7 Categories of natural resources. These categories, based on degree of geological assurances and economic feasibility, were intended to describe mineral resources. With some modification, these categories can describe many types of nonrenewable resources.

You've probably read in newspapers or magazines about fossil fuel resources or proven reserves of oil and natural gas. What do these terms mean? *Proven reserves* of a resource are those that have been thoroughly mapped and are economical to recover at current prices with available technology. *Known resources* are those that have been located but are not completely mapped. Their recovery or development may not be economical now, but they are likely to become economical in the foreseeable future. *Undiscovered resources* are only speculative or inferred from similarities between known deposits or conditions and unexplored ones. There are also *unconceived resources*—sometimes called unknown-unknowns—that may be economic but have not even been thought of yet. *Recoverable resources* are accessible with current technology but are not likely to be economic in the foreseeable future, whereas *nonrecoverable resources* are so diffuse or remote that they are not ever likely to be technologically accessible.

There is a vast difference in amount between the first and last of these categories. The U.S. Geological Survey estimates, for instance, that only 0.01 percent (one hundredth of one percent) of the minerals in the upper one kilometer of the earth's crust will ever be economically recoverable. That means that ten thousand times as much is not economically available, and billions or even trillions of times as much is neither economically nor technically feasible to recover. Chapter 12 gives some practical examples of proven, known, and ultimately recoverable fuel resources.

ECONOMIC DEVELOPMENT AND RESOURCE USE

The supplies of economically accessible resources and their use by society depend on economic and technological development. In chapter 1, we discussed changing attitudes toward nature in various historical periods. We can see similar changes in our valuing of resources in different phases of economic development.

Frontier Economy

Reflect for a moment on conditions European settlers found as they moved across the North American continent. Vast forests, abundant wildlife, and rich soils seemed to be in unlimited supply and were easily exploited during the frontier era. Natural resources were regarded as expendable and of little value relative to capital and labor, which tended to be in short supply. Given these conditions, what now seems like a terrible waste of resources could be tolerated or even encouraged. People who could open up the wilderness, extract food, fiber, or minerals, and get them to market were regarded as heroes. What mattered was how

much you could deliver, not how much you used in the process, since the supply of raw materials was not seen as a problem.

This attitude may not be irrational, given the circumstances. Perhaps you have experienced an abundance of garden produce such as tomatoes or zucchini during harvest season. You may have more than you can possibly use before it spoils. Time and energy to process your harvest are the limiting factors. As you sort through piles of produce, you might hack away half a tomato to remove a slight blemish or simply throw it out and concentrate your efforts on the best fruit. Later, however, after the abundance of the harvest has passed, that same fruit would have much more value. You might spend much more time and energy trying to rescue it for use.

Pollution levels tend to be relatively low in a frontier economy because population density and industrial activities tend to be lower than the absorptive capacity of the environment. Natural systems can process wastes without becoming overloaded. Pioneers could let the smoke from their campfires drift away in the wind or leave organic wastes to decay on the ground because they were unlikely to bother anyone else. These are practices we cannot tolerate now that our populations are more dense.

Industrial Economy

The Industrial Revolution was one of the most significant developments in human history. The substitution of water power and fossil fuels for muscle power and the introduction of machines and mass production techniques brought great changes to both human society and the environment.

In an industrial economy, greater production capacity increases the rate of resource use and depletes stocks of readily available resources. This, in turn, causes rising extraction and shipping costs because poorer and more remote sources are used. Factory owners are willing to pay more for raw materials because they can produce more valuable goods from those materials than they could have produced in a frontier society. Workers receive higher hourly wages because technology and economy of scale make them more productive. This means that they can afford to pay more for goods and services. Pollution levels tend to be high in an industrial society because of the high rate of materials that flow through the system.

Unfortunately, we in the industrial world often retain what economist Kenneth Boulding called "cowboy economics" in our attitudes toward resource use. Because we assume that we can always find more resources and more places to discard our wastes, we measure our economic success by the rate of flow—or throughput—of materials through our industrial network. Our term for this is the gross national product (GNP), the sum total of all goods and services produced in a country. Sometimes the term gross domestic product (GDP) is used to distinguish economic activity within a country from that of off-shore corporations.

The more resources we harvest and consume, the higher our GDP will be, and presumably the higher our standard of living, happiness, and satisfaction. Natural resources are assigned no value until they are harvested, extracted, or otherwise used. Saving natural capital or building up equity for future generations does not figure into our GDP and is considered inefficient or backward. Clear-cutting a forest or exhausting our oil supplies shows as a positive contribution to our national balance sheet. The fact that future generations will have to do without those resources isn't taken into consideration.

Postindustrial Economy

Most of the developed world has undergone a dramatic change in its sources of wealth and power in the past century. More than 50 percent of the real economic growth in the United States in that period has been in organizational, educational, and technological areas. We have gone beyond an industrial-based economy to an information-based economy. A century ago, 90 percent of the American work force was engaged in farming, harvesting resources (logging, mining, etc.), or manufacturing. Now only 2 percent of the work force farms and 20 percent harvests resources or produces material goods. The rest is involved in "service" work, at least two-thirds of which involves transferring, storing, or processing information of some kind.

Information has become our most important resource and is, therefore, the main source of wealth and power. Unlike most resources, it is not depleted through use. It is expandable, shareable, transportable, substitutable, and storable (fig. 5.8). The more widely information is disseminated and used, the more valuable it becomes, for the most part, and the more we have. Although some people may try to restrict information flow to gain power or financial benefits, there is no scarcity rent (charge for using a depletable resource) on information that reflects diminishing stocks. Since few materials are used in an information and service economy, pollution levels are lower than those in an industrial economy.

It is becoming apparent that our previous presumption of infinite natural resources no longer applies. The earth can neither provide endless supplies of materials nor absorb unceasing wastes. We now must move to what Kenneth Boulding calls a mature or "spaceship" economy in which throughput of materials is minimized rather than maximized. As in a spaceship on a long flight, our resource supply is limited

Figure 5.8 In a postindustrial society, information, education, research, communication, and technology replace material goods or natural resources as the main sources of wealth and power. Service also becomes a major segment of the economy.

and our ability to store or process wastes is reaching maximum capacity. Our success will no longer be measured by rates of production and consumption but rather by the extent, quality, and complexity of our natural stocks. Efficiency in getting more benefits from fewer resources will become our goal.

POPULATION, TECHNOLOGY, AND RESOURCE SCARCITY

As the preceding discussion shows, the supply of a particular natural resource available for human use is not determined so much by the absolute amount present on the earth as by economic, social, and technological factors. Let's look more closely at these relationships.

Supply, Price, and Demand Relationships

Supply depends on (1) which raw materials can supply a service using present technology; (2) the availability of those materials in various quantities; (3) the costs of extracting, shipping, and processing them; (4) competition for those materials by other uses and processes; (5) feasibility and cost of recycling already used material; and (6) social and institutional arrangements in force.

In a market system, most of the considerations just mentioned are reflected in the market price for a good or service—the amount it sells for. The available quantity of a resource or opportunity usually increases as the price rises. For example, in 1978, the Congressional Office of Technology Assessment estimated that at $11 per barrel some 21 billion barrels of oil were available. If the price were to double (as it soon did) to $22 per barrel, the supply also would double to 42 billion barrels. The reason for this increase was not that new oil was being created but that it becomes worthwhile to drill into lower quality and more remote oil fields as prices rise. If the price were to go even higher, substitute fuels such as oil shales and tar sands that are not now economical to extract might become competetive with oil. The effect would be as if a whole new resource had been created.

In economic terms, the relationship between available supply of a commodity or service and its price is described by a **supply/demand curve.** Demand is the amount of a product that consumers are willing and able to buy at various possible prices assuming they are free to express their preferences. Supply is the quantity of that product being offered for sale at various prices, other things being equal. The inverse relationship between supply and demand is shown in the supply/demand graph in figure 5.9. As the price rises, the supply increases and the demand falls. The reverse holds as the price decreases. The intersection between these two curves is called the **market equilibrium.** In a competitive free market of willing buyers and sellers, this is the price that should be maintained by natural market forces. If sellers increase the quantity available faster than prices increase, or if buyers increase their purchases more rapidly than prices are falling, we say that the product has **price elasticity.**

Market Efficiencies and Technological Development

In a frontier economy, procedures for gaining access to resources and turning them into useful goods and services tend to be primitive and inefficient. As markets develop, however, producers gain experience in obtaining and working with particular resources. Specialization and experimentation lead to discovery of new, more efficient technologies, making it possible to produce larger quantities of goods at lower prices. The supply curve shifts and the market moves to a new equilibrium point, as is shown in figure 5.10. At each successive stage in this development process, a larger quantity of product is available at a lower price. The effect is that the standard of living increases—at least in economic terms.

Population Effects

Growing populations can offset advances in science and technology. As the number of workers increases, a point may be reached at which there are not enough jobs to employ everyone efficiently. As a result, the productivity per person declines and wages fall. This predicament is intensified by the pressure on resources created by the growing population. More people use more resources, and we must look to less accessible or desirable supplies. Raw materials, therefore, become more expensive, as do the prices of goods and services provided by those resources; thus, workers have less money to buy more expensive

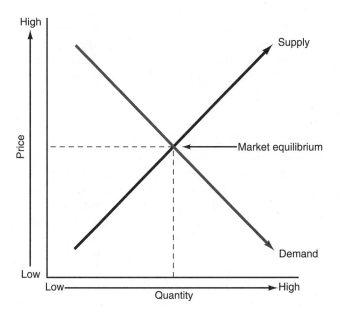

Figure 5.9 Classic supply-demand curves. When price is low, supply is low and demand is high. As prices rise, supply increases, but demand falls. The market equilibrium is the price at which supply and demand are equal.

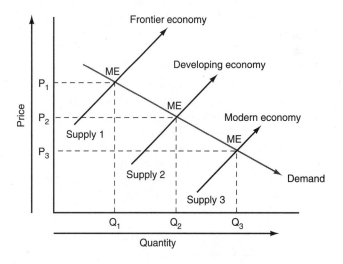

Figure 5.10 Supply and demand curves at three different stages of economic development. At each stage there is an equilibrium point at which supply and demand are in balance. As the economy becomes more efficient, the equilibrium shifts so there is a larger quantity available at a lower price than before. (P = price, Q = quantity, ME = market equilibrium.)

items, and the standard of living declines. This "iron **law of diminishing returns**" led Thomas Malthus (chapter 6) to predict that unrestrained population growth would inevitably cause the standard of living to decrease to a subsistence level where poverty, misery, vice, and starvation would make life permanently drab and miserable. This dreary prophecy has led some people to call economics "the dismal science."

Growing populations also place a strain on economic development by diverting the capital necessary for growth. In a rapidly growing country, much of the population is made up of children who require social overhead expenditures such as housing, schools, and roads that contribute little to development. Also, the creation of new jobs to employ a growing population traps capital in conventional industries and lessens investments in new technologies that might provide a real improvement in the standard of living. This diversion of investment capital is called the **population hurdle.**

Growing populations, however, also can create markets that encourage specialization, innovation, and capital investment that result in efficiency. They can bring young, energetic, and better-trained workers into the work force and make changes possible in traditional ways of doing things. Some demographers argue that while growing populations cause problems, they also result in more human ingenuity, energy, and cooperation to solve those problems. Where are we now in the process of economic development and population growth? Are we on a curve of diminishing returns, or are we benefiting from economy of scale in terms of human populations and environmental problems?

Factors that Mitigate Scarcity

Human social systems can adapt to resource scarcity in a number of ways. Some economists point out that scarcity can be a catalyst for innovation and change (fig. 5.11). As materials become more expensive and difficult to obtain, it becomes cost-efficient to discover new supplies or to use the ones we have more carefully, and we may be better off in the long run because of these developments.

Several factors can alleviate the effects of scarcity.

- Technological inventions can increase the efficiency of extracting, processing, using, and recovering materials.
- Substituting new materials or commodities for scarce ones can extend existing supplies or create new ones. For instance, substituting aluminum for copper, concrete for structural steel, grain for meat, and synthetic fibers for natural ones all remove certain limits to growth.

Figure 5.11 Scarcity/development cycle. Paradoxically, resource use and depletion of reserves can stimulate research and development, substitution of new materials, and the effective creation of new resources.

- Trade makes remote supplies of resources available and may also bring unintended benefits in information exchange and cultural awakening.
- Discovery of new reserves through better exploration techniques, more investment, and looking in new areas becomes rewarding as supplies become limited and prices rise.
- Recycling becomes feasible and accepted as resources become more valuable. Recycling now provides about 37 percent of the iron and lead, 20 percent of the copper, 10 percent of the aluminum, and 60 percent of the antimony that we consume each year in the United States.

Increasing Environmental Carrying Capacity

Economist Julian Simon says that in spite of recurring fears of natural resource scarcity, the mitigating factors listed above have made every commodity cheaper in real terms as far back as we can find records. In fact, responding to the growing scarcity of resources actually enables us to increase the carrying capacity of the environment for humans. Figure 5.12 shows the change in real gross national product (as a percent of that in 1890) during the period of industrial growth and transformation to a postindustrial society. There has been about a 500 percent increase in per capita GNP during this century even though average working hours have declined, population has tripled, and the easily accessible local resources largely have been used up.

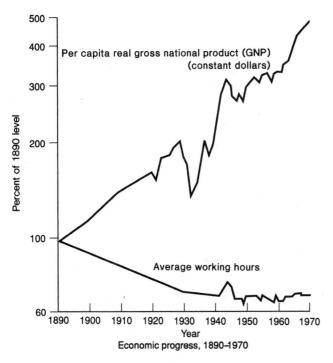

Figure 5.12 Technological improvements, capital investments, and more highly trained labor have raised production faster than population growth in the United States. Between 1890 and 1970, real per capita GNP rose fourfold even though average working hours fell nearly 40 percent. In other words, our increased productivity has given us both more output and more leisure.

Source: U.S. Department of Commerce.

Will this economic progress be sustained, however? Ecologist Paul Ehrlich contends that increasing levels of population and consumption will inevitably lead to scarcity and rising prices as more of us try to share less and less. In 1980, Ehrlich made a wager with Simon. They bet $1,000 on five metals—chrome, copper, nickel, tin, and tungsten. If the 1990 combined prices, corrected for inflation, were higher than $1,000, Simon would pay the difference. If prices had fallen, Ehrlich would pay. In 1990 Ehrlich sent Simon a check for $576.07; prices of these five metals had fallen 47.6 percent. Perhaps, however, the time was just too short. Would you care to bet on resource scarcity fifty or a hundred years from now?

Most countries have not shared in the rapid increase of wealth enjoyed by the United States in the past century. Figure 5.13 shows GNP and population by region. As you can see, North America and Europe have 65 percent of the world GNP but only 18 percent of the world population. Asia, Latin America, and Africa have 74 percent of the population but only 16 percent of the GNP. This may be a very real example of the population hurdle discussed in chapter 4.

LIMITS TO GROWTH

We return again and again in environmental studies to the underlying question of whether continued population growth, economic growth, or both would be good or bad. At what point will the number of people or the extent of economic activities that impinge on our environment bring disaster, not only to humans, but to the whole life-supporting system of the biosphere? On the other hand, how will we improve the standard of living in less-developed countries and clean up damage already done to the environment if growth stops? The crux of this argument is whether resources are finite or can be effectively expanded through human ingenuity and enterprise. Let's look at what some economic models predict might be the result of further economic and population growth.

Computer Models of Resource Use

In the early 1970s, an influential study of resource limitations funded by the Club of Rome, a group of wealthy industrialists, was undertaken by a team of scientists from the Massachusetts Institute of Technology headed by Donella H. Meadows. The results of this study were published in a 1972 book entitled *Limits to Growth*. Several computer models of world economy were used to examine various scenarios of resource depletion, growing population, pollution, and industrial output. Given the Malthusian assumptions built into these models, catastrophic social and environmental collapse seemed inescapable. Increasing the resources available at the beginning of the computer run simply resulted in a faster rate of consumption and a more disastrous situation eventually.

Figure 5.14 is a run of one of the world models given a doubling of present resources but no change in public policies or attitudes. Population, resources, industrial output, and food supplies all crash precipitously sometime about the middle of the next century. Pollution continues to rise after everything else collapses, presumably because of high rates of death and destruction. Note the strong similarity between these curves and the boom and bust cycles observed when natural populations exceed environmental carrying capacity (chapter 4).

Some authors criticized *Limits to Growth* because the computer models used by the Meadows group discounted factors that might mitigate the effects of scarcity. More recently, in *Beyond the Limits,* Meadows et al. modified their model to include technical progress, pollution abatement, population stabilization, and new public policies that work for a sustainable future. If we adopt these changes sooner rather than later, the computer shows an outcome like that in

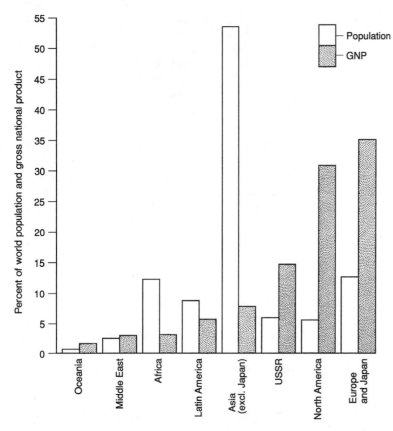

Figure 5.13 Percent of world population and gross national product (GNP) by region. The more developed countries (MDC) have 65 percent of the world GNP, but only 18 percent of the world's population. The less developed countries (LDC) have 74 percent of the world's population but only 16 percent of the total GNP.

Source: Data from Population Reference Bureau

figure 5.15, in which all factors stabilize at an improved standard of living for everyone. Neither of these computer models show what *will* happen, only what some possible outcomes *might* be depending on what we do.

Why Not Conserve Resources?

Even if large supplies of resources are available, or technological advances to mitigate scarcity are possible, wouldn't it be better to reduce our current use of natural resources so they will last as long as possible? Will anything be lost if we're frugal now and leave more to be used by future generations? This makes sense if the assumption is correct that resources are limited and cannot be replenished, or that their extraction inevitably degrades the environment. Using them more slowly and sharing them with fewer people will give each of us a larger share, will be kinder to the environment, and will make our supply last longer.

Most economists, however, look at resources as means to an end rather than having inherent value in themselves. Resources have to be used to be of value. If you bury your savings in a jar in the backyard, it will last a long time but may not be worth much when you dig it up. If you

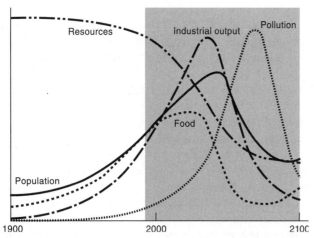

Figure 5.14 A run of one of the world models in *Limits to Growth*. This model assumes business-as-usual for as long as possible until Malthusian limits cause industrial society to crash. Notice that pollution continues to increase well after industrial output, food supplies, and population have all plummeted.

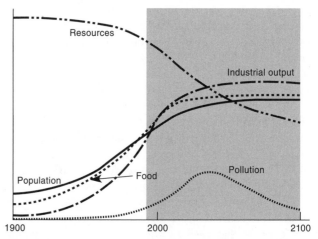

Figure 5.15 A run of the world model from *Beyond the Limits*. This model assumes that population and consumption are curbed, new technologies are introduced, and sustainable environmental policies are introduced immediately rather than after resources are exhausted.

invest it productively, you will have much more in the future than you do now. Furthermore, a window of opportunity for investment may be open now but not later.

RESOURCE ECONOMICS

Some of the most crucial commodities that may limit our future growth are not represented by monetary values in the marketplace. Groundwater, sunlight, clean air, biological diversity, and other common resources are treated as public goods (benefits) that anyone can use freely. Our economic system typically has not charged for using the absorptive capacity of the environment to dispose of wastes. In theory, these resources are self-renewing, but as we will see in other chapters of this book, many of these vital environmental goods and services are threatened by human activities. If we damage basic life-support systems of the biosphere, we cannot simply substitute another material or service for the ones that have become limited. The crux of this question is the way we manage resources in a market system. Let's look now at how our economic system handles internal and external costs and intergenerational justice.

Internal and External Costs

Internal costs are the expenses (monetary or otherwise) that are borne by those who use a resource. Often, internal costs are limited to the direct out-of-pocket expenses involved in gaining access to the resource and turning it into a useful product or service.

External costs are the expenses (monetary or otherwise) that are borne by someone other than the individuals or groups who use a resource. External costs often are related to public goods and services derived from nature. Some examples of external costs are the environmental or human health effects of using air or water to dispose of wastes. Since these effects usually are diffuse and difficult to quantify, they do not show on the ledgers of the responsible parties. They are likely to be ignored in private decisions about the costs and benefits of a purchase or a project. One way to use the market system to optimize resource use is to make sure that those who reap the benefits of resource use also bear all the external costs. This is referred to as **internalizing costs.**

A controversial provision of the 1990 revision of the Clean Air Act allows companies to market emission quotas as a way of reducing pollution in the most efficient and least costly way possible (box 5.1). This provision would have the effect of internalizing external costs, but it is regarded by its opponents as merely a license to pollute.

Intergenerational Justice and Discount Rates

"A bird in the hand is worth two in the bush." All of us are familiar with sayings that suggest it is better to

Understanding Principles and Issues: Box 5.1

Market-Based Incentives for Environmental Protection

What is the most efficient and economical way to eliminate pollution? Some people argue that we should simply say to polluters, "Stop it! You can't dump garbage into the air or water anymore." While this approach has a certain moral appeal, it tends to force all businesses to adopt uniform standards and methods of pollution control regardless of cost or effectiveness. This approach also can lead to an adversarial climate in which resources are used in litigation rather than pollution control.

Furthermore, the command and control approach tends to freeze technology by eliminating incentives for continued research and development. Industry is discouraged—even prohibited—from trying new technologies or alternative production methods. These problems can be overcome, many economists believe, by using market mechanisms rather than rigid rules and regulations to reduce pollution. Since there may be a one hundredfold variation in the cost of eliminating a specific pollutant from different sources due to the age of equipment in use, environmental factors, and other considerations, market-based incentives such as pollution charges or tradable permits can be more cost-effective and flexible than simply saying, "Thou shalt not."

Pollution charges are fees assessed per unit of pollution. They could be the same for a given pollution type regardless of source or location, or they could be adjusted to reflect relative amounts of damage. In either case, the more you reduce pollution, the more you save. The charges might have to be set quite high to discourage some types of pollution, however, and could exaggerate inequities between the rich and the poor (fig. 5.1.1).

Five types of pollution charges are now being considered or are already in place for some industries: (1) effluent charges based on the quantity of discharge, (2) user charges based on the cost of public treatment facilities such as sewage treatment plants that clean up effluents, (3) production charges based on potential damage caused by a product, (4) administrative charges based on government monitoring services, and (5) differential taxes to encourage "green" products.

Figure 5.1.1 Allowing firms to buy and sell pollution allowances or "rights to pollute" may be the most efficient way to reduce pollution on a regional level. It could have unfortunate local impacts, however.

Tradable permits are based on an assumption of thresholds below which some types of pollution are acceptable. If we can agree on those acceptable levels, industries can be given permits to emit a fair share of pollution. Any company below its limit can sell or lease the excess amount to another company. In theory, this should allow us to reach pollution control goals in the most cost-effective manner. Companies that can lower effluents most cheaply will do so because they can make money by selling credits to others. If what we want is a given result—clean air for instance—and we don't care how that goal is attained, then this may be a good approach.

Like pollution charges, permits tend to encourage some innovation and technological improvements. The more efficiently pollutants are removed, the more money you can make by selling the excess. While this should make pollution control cheaper, it doesn't create incentives to lower pollution below established targets. Also, where charges require a large bureaucracy to handle paperwork, measure effluents, and collect fees, permits tend to be handled in the private sector.

Both permits and effluent charges are being considered or are already being used to regulate several types of pollution. The 1990 Clean Air Act contained provisions for marketing sulfur dioxide permits for power plants as a way of reducing acid precipitation. A carbon tax on fossil fuels is being studied as a way of reducing greenhouse gases. A tax is considered more effective than emission permits or effluent standards because the large number of carbon dioxide sources makes emission measurements and enforcement difficult. It is estimated that a $100 per ton tax on fossil fuels could reduce carbon dioxide emissions by 36 percent by the year 2000 and raise $120 billion per year for pollution control.

Similarly, a "gas-guzzler" tax of 50 cents per gallon on gasoline is being considered to encourage fuel-efficient automobiles and to improve urban air quality. Both of these taxes would fall most heavily on the poor, but the revenues generated could be used to fund social programs that would mitigate these negative effects. For instance, the $40 billion per year raised by the gas guzzler tax could offset the social security taxes paid by the lowest one-third of all wage earners.

have something now than in the distant future. **Discount rates** are the economist's way to introduce a time factor in accounting. It is a recognition that a ton of steel delivered today is worth more than the same ton delivered a year from now; the difference is an extra year's worth of use of products made from the steel. In theory the discount rate should be equal to the interest rate on borrowing money.

The choice of discount rates to apply to future benefits becomes increasingly problematic with intangible resources or long time frames. How much will a barrel of oil or a four-thousand-year-old redwood tree be worth a couple of centuries from now? Maybe there will be substitutes for oil by then so the barrel won't be highly valued. On the other hand, the oil or the tree may be priceless. This valuation is compli-

cated by the fact that we are making decisions not only for ourselves but also for future generations.

Although having access to clean groundwater or biological diversity one hundred years from now isn't worth much to me—assuming that I will be long gone by that time—those resources might be quite valuable to people alive at the time. Future citizens will be affected by the choices we make today, but they don't have a vote. Our decisions about how to use resources raise difficult questions about justice between generations. How shall I weigh their interests in the future against mine right now?

Which discount rate should be used for future benefits and costs is often a crucial question in large public projects such as dams and airports. Proponents may choose discount rates that make a venture seem attractive while opponents prefer rates that show the investment to be unwise. These questions are especially difficult when comparing future environmental costs to immediate financial returns. Environmental activists need to understand economic nuances when they are fighting environmentally destructive schemes.

Cost/Benefit Ratios

One way to evaluate the outcomes of large-scale public projects is to analyze the costs and benefits that accrue from them. The assumption that marginal cost/benefit analysis can be applied to present and future values of a resource, given proper criteria and procedures, is one of the main conceptual frameworks of resource economics. This process is controversial, however, because it deals with vague and uncertain values and compares costs and benefits that are as different as apples and oranges. Yet we continue doing these analyses because we don't have better ways to allocate resources.

Figure 5.16 presents a flowchart for preparing a cost/benefit analysis. As you can see, several different tributary paths come together to determine the final outcome. The easiest parts of the equation to quantify

are the direct costs and benefits to the developer or agent who has proposed the project, i.e., the out-of-pocket expenses and the immediate profits that will result from this investment. These direct monetary costs and benefits are usually the most concrete and accurate components in the analysis. It is important that they not outweigh other factors more difficult to ascertain but of equal importance.

The other branch of the flowchart involves analysis of more diffuse, nonmonetary factors, such as environmental quality, ecosystem stability, human health impacts, historic importance of the area to be affected, scenic and recreational values, and potential future uses. These are difficult values to quantify. It is even more difficult to express them in monetary terms. How much are beauty or tranquility worth? What are the benefits of ethical behavior? How much would you pay for good health?

Some costs and benefits simply cannot be expressed in monetary terms. These invaluable (in a positive sense) factors bypass the mathematical stages of comparison and are considered—we hope—in the final decision-making process, which is more political than mechanical. Also factored in at

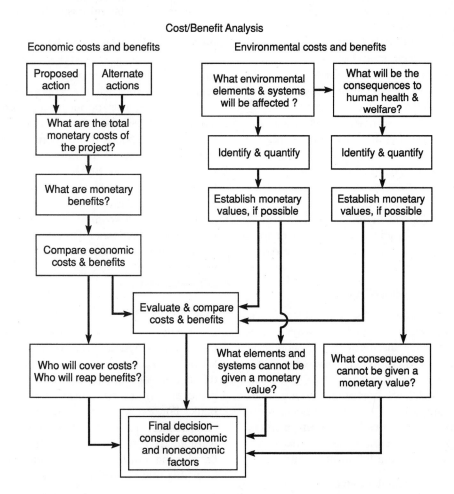

Figure 5.16 A flowchart for a cost/benefit analysis.

this stage are distributional considerations, i.e., who will bear the costs of the project and who will reap the benefits. If these are different groups, as they usually are, questions of justice arise that must be resolved, perhaps in some other venue.

Criticisms and complications of this process include the following:

1. *Absence of standards.* Each person assesses costs and benefits by his or her own criteria, often leading to conflicting conclusions about the comparative values of a project. It has been suggested that an agency or influential group set specifications for how factors should be evaluated.

2. *Inadequate attention to alternatives.* To really understand the true costs of a project, including possible loss of benefits from other uses, it is essential to evaluate alternative uses for a resource and alternative ways to provide the same services. This is often slighted.

3. *Assigning monetary values to intangibles and diffuse or future costs and benefits.* Some critics of this process claim that we should not even try. They believe that attempting to express all values in monetary terms suggests that only monetary gains and losses are important. This can lead to the "slippery slope" argument that everything has a price and that any behavior is acceptable as long as we can pay for it.

4. *Acknowledging the degree of effectiveness and certainty of alternatives.* Sometimes speculative or even hypothetical results are given specific numerical values and treated as if they were hard facts. We should use risk-assessment techniques to evaluate and compare uncertainties in the process.

5. *Justification of the status quo.* Agencies may make decisions to go ahead with a project for political reasons and then manipulate the data to correspond to preconceived conclusions.

Distributing Intangible Assets

Can the market set priorities for use and protection of intangible assets in a socially responsible way? Sometimes it does, but in many instances it does not. It is important to recognize when we can and cannot depend on market forces to manage natural resource economics. Market systems are notoriously poor at handling distributional questions of equity or justice. Most economists recognize that we need to regulate markets to achieve social goals that market mechanisms alone cannot accomplish. Consider the following points:

1. In theory, environmental services such as pollution absorption are available to everyone and are not bought and sold. There is, therefore, no market for them and no price-setting mechanism.

2. In deciding whether it would be more profitable to use a resource now or save it for the future, private interests almost always choose to use it quickly because they alone benefit directly from immediate use, whereas the benefits from future use may be spread among many users.

3. Private discount rates for future benefits do not consider social objectives and therefore do not reflect complete values of *in situ* resources.

4. The costs of destroying a resource are often external; that is, they accrue to individuals other than those who benefit from using the resource.

Since cost/benefit ratios are often the deciding factors in embarking on public works projects and in setting priorities regarding natural resources, students of environmental science should understand how these ratios are determined. First, we have to decide who benefits and who pays. Then, we have to set prices on the goods and services provided by the resource. For some benefits and costs this may be fairly straightforward, but the value of intangible assets such as opportunity or existence can be difficult to determine.

For instance, a wild and scenic river flows not far from my home. Even though I don't go there often, it is important to me that the opportunity to do so exists. How much is this opportunity worth? A cost/benefit study might assume that, since the owner of a large cabin cruiser spends much more money per day on the river than I spend in my homemade canoe, I don't value the opportunity as highly. (I don't agree.) It is even more difficult to determine existence values. I like to know that some grizzly bears still live in Alaska, and that great whales still swim in the ocean, even though I may never see either one. How much is this knowledge worth? Does a diffuse value to many of us outweigh a very specific economic gain to the hunter who wants to kill these animals?

International Development

No single institution has more influence on the financing and policies of developing countries than the World Bank. Of some $25 billion loaned by multinational development banks each year for Third World projects, about two-thirds comes from the World Bank. For every dollar invested by the World Bank, two dollars are attracted from other sources. If you want to have an impact on what is happening in the developing countries, it is imperative to understand how this huge enterprise works.

The World Bank was founded in 1945 to provide aid to war-torn Europe and Japan. In the 1950s, its emphasis shifted to development aid for Third World countries. This aid was justified on humanitarian grounds, but providing markets and political support for Western capitalism was an important by-product.

The Bank is jointly owned by 150 countries, but one-third of its support comes from the United States. The Bank president has always been a U.S. citizen. The bulk of its $66.8 billion capital comes from private investors who buy instruments (bonds and debentures) on the open market. Loan applications are first screened by the professional staff and then voted on by a council of all member countries. No loan has ever been turned down once it arrives at the full council. So far, one hundred countries have borrowed some $140 billion from the World Bank for a wide variety of development projects.

Many World Bank projects have been environmentally destructive and highly controversial. In Botswana, for example, $18 million was provided to increase beef production for export by 20 percent, despite already severe overgrazing on fragile grasslands. The project failed, as did two previous beef production projects in the same area. In Ethiopia, rich floodplains in the Awash River Valley were flooded to provide electric power and irrigation water for cash export crops. More than 150,000 subsistence farmers were displaced and food production was seriously reduced. In India, the sacred Narmada River is being transformed by thirty large dams and 135 medium-sized ones financed by the World Bank. About 1.5 million hill people and farmers will be displaced as a result.

These are only a few of the projects that have aroused concern and protest. They typify the environmental and social costs of many development programs. Loans from the World Bank tend to favor large-scale agricultural, transportation, and energy projects, and these types of loans make up more than half of all the funds it distributes. In part, this is because both the World Bank and debtor nations find it easier to manage one big project than many small ones. Furthermore, both the World Bank and the governments borrowing money prefer large, impressive, modern projects to show how the money is being spent. Small cooperatives or cottage industries might do more for the people than a big dam, but they don't look as impressive. Recently, however, a new, effective model has been demonstrated (box 5.2).

Thinking Globally: Box 5.2

Microlending at the Grammeen Bank

Development projects in poorer nations have generally been financed through loans from international development banks. The megaprojects funded by these banks are highly political and often show little sensitivity to local culture or environment. Furthermore, they fail to help the informal sector (street vendors, household industries, or unorganized service workers), even though this sector may account for more than half of all economic activity in many less-developed countries.

The greatest barrier to productive self-employment for poor people often is a lack of access to capital. Since they generally have few assets to use for collateral and no credit record, the poor can't go to traditional banks for loans to buy tools and materials to start a small business. An inspiring new approach to banking called microlending has been shown to be both successful and profitable, however, in making loans to the poor. This promises to be a model for grass-roots economic development.

Pioneered by economist Muhammad Yunus of Bangladesh, the Grammeen (village) Bank provides credit directly to the poorest people, those who have no collateral and no steady source of income. Started in 1976, the Grammeen Bank now has nearly one thousand local offices and some 1.2 million customers, 90 percent

Figure 5.2.1 Community development banks make "micro loans" to individuals and small companies to buy tools, such as this spinning wheel used by this woman in India. Collective management of these projects provides mutual support and education in financial management. Grass roots development is often both more democratic and more effective than huge multinational projects.

of whom are women who could never have borrowed money from an ordinary bank. Their loans are small, averaging only $67. This is enough, however, to buy a used sewing machine, a bicycle, a loom, a cow, some garden tools—a start in providing needed family income (fig. 5.2.1).

Compared to a Bangladesh national average of only 30 percent repayment rate on loans, the recovery rate on Grammeen accounts is an astonishing 98 percent. The key to this success is peer lending. Borrowers are organized into five-member peer groups that act both as mutual aid societies and collection agencies. Payments must be made in regular weekly installments. If one member of the group defaults, the others must repay the loan. Having some dignity, respect, and independence encourages responsibility and self-reliance.

Microlending and self-help programs are now being used around the world. More than a hundred organizations in the United States currently assist microenterprises by providing loans, grants, or training. The Women's Self-Employment Project in Chicago, for instance, is teaching skills and empowerment to single mothers in public housing projects. The repayment rate on its loans is nearly 100 percent. In 1992, the Small Business Administration earmarked $15 million for microloan projects. Interestingly, the richest country in the world may have learned a valuable economic lesson from one of the poorest.

In a response to criticisms of World Bank policies, the U.S. Congress now insists that all loans for international development be reviewed for environmental and social effects before being approved. It asks for assurance that each project (a) use renewable resources and not exceed the regenerative capacity of the environment, (b) not cause severe or irreversible environmental deterioration, and (c) not displace indigenous people.

As recently as 1986, only six of the nearly seven thousand World Bank staff members were assigned to do ecological assessments on $17 billion in development projects. Pressure from environmental groups has forced the World Bank to add about a hundred ecologists and environmental specialists to the staff. Critics charge that many of these people are merely recycled economists and that business is going on as usual. Loans for road building that would lead to tropical forest destruction have been canceled, however, and loans for environmental restoration projects in Brazil have been made. Environmental groups continue to criticize the World Bank, at least partly because it's easy to aim at a large target. A fundamental question remains whether technological and economic growth can bring a better standard of living for everyone without causing unacceptable environmental disruption.

International Trade

International issues further complicate questions of resource management. Much of the vast discrepancy between richer and poorer nations is related to their economic and political histories, as well as to current international trade relations. The banking and trading systems that control credit, currency exchange, shipping rates, and commodity prices were set up by the richer and more powerful nations in their own self-interest. These systems tend to keep the less-developed countries in a perpetual role of resource suppliers to the more-developed countries. The producers of raw materials, such as mineral ores or agricultural products, get very little of the income generated from international trade (fig. 5.17). Furthermore, they suffer both from low commodity prices relative to

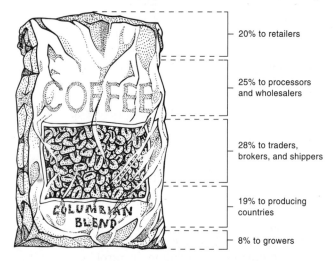

20% to retailers

25% to processors and wholesalers

28% to traders, brokers, and shippers

19% to producing countries

8% to growers

Figure 5.17 What do we really pay for when we purchase a pound of coffee?

manufactured goods and from wild "yo-yo" swings in prices that destabilize their economies and make it impossible to either plan for the future or to accumulate capital for further development.

In 1974, the United Nations General Assembly adopted Resolution Number 3281 calling for a "new international economic order" in which "every State has and shall freely exercise full permanent sovereignty, including possession, use and disposal over all its wealth, natural resources and economic activities." This idea of a new relationship between the richer nations of the North and the poorer nations of the South was developed further by the United Nations Commission, chaired in 1980 by Willy Brandt, former mayor of West Berlin, and at the Rio Earth Summit (see box 1.3). As you might suppose, most rich nations have been less enthusiastic than the poor ones have been about changing present economic relationships.

Jobs and the Environment

For years business leaders and politicians have portrayed environmental protection and jobs as mutually exclusive. Pollution control, protection of natural areas and endangered species, and limits on use of nonrenewable resources, they claim, will strangle the economy and throw people out of work. A new brand of environmental economists dispute this claim, however. Environmental protection, they argue, is not only necessary for a healthy economic system, it actually creates jobs and stimulates business.

Recycling, for instance, takes much more labor than extracting virgin raw materials. This doesn't necessarily mean that recycled goods are more expensive than those from virgin resources. We're simply substituting labor in the recycling center for energy and huge machines used to extract new materials in remote places.

Japan, already a leader in efficiency and environmental technology, has recognized the multibillion dollar economic potential of "green business." The Japanese government is investing 4 billion U.S. dollars per year on research and development that targets seven areas ranging from utilitarian projects such as biodegradable plastics and heat-pump refrigerants to exotic schemes such as carbon-dioxide fixing algae and hydrogen-producing microbes.

By the year 2000, Japan expects to be selling $12 billion worth of equipment and services per year worldwide. The country is already marketing advanced waste incinerators, pollution control equipment, alternative energy sources, and water-treatment systems. Unfortunately, the United States has been resisting international pollution control conventions rather than recognizing the potential for economic growth *and* environmental protection in the field of green business.

SUSTAINABLE DEVELOPMENT: THE CHALLENGE

One of the most difficult, and as yet unanswered, questions in this chapter is whether further growth in either population or economic activity is tolerable. On one hand, many environmental scientists warn that human activities are overwhelming the basic life-support systems of the biosphere. They call for a steady-state economic system that will minimize our impact on the environment. On the other hand, many people see economic growth as a way to bring the underdeveloped nations of the world up to a higher standard of living without calling on those of us in the richer nations to give up the wealth we now enjoy.

Economic growth is also advocated as necessary to provide funds to clean up the environmental damage caused by earlier, more primitive technologies and misguided resource uses. It is estimated that the world will need $350 billion per year to control population growth, develop renewable energy sources, stop soil erosion, protect ecosystems, and provide a decent standard of living for the world's poor. This is a great deal of money but is small compared to the $1 trillion per year spent on wars and military equipment.

An intermediate position between the extremes of no growth and unlimited growth is **sustainable development** based on use of renewable resources in harmony with ecological systems. Perhaps the best statement of this principle is that of *Our Common Future* (see box 1.3), which defined sustainable development as meeting the needs of the present without compromising the ability of future generations to meet their own needs. Some goals of sustainable development are:

- A demographic transition to a stable world population of low birth and death rates.
- An energy transition to high efficiency in production and use, coupled with increasing reliance on renewable resources.
- A resource transition to reliance on nature's "income" without depleting its "capital."
- An economic transition to sustainable development and a broader sharing of its benefits.
- A political transition to global negotiation grounded in complementary interests between North and South, East and West.
- An ethical or spiritual transition to attitudes that do not separate us from nature or each other.

Some strategies for attaining these goals are listed in table 5.1.

Will we accomplish the transition to an economy based on sustainable development, intergenerational justice, an equitable distribution of resources, and a harmony with nature? We will succeed only if we bring the wisdom and knowledge of both ecology *and* economics to bear on our problems. The future will be what we make it.

It is scarcely necessary to remark that a stationary condition of capital and population implies no stationary state of human improvement. There would be just as much scope as ever for all kinds of mental culture and moral and social progress; as much room for improving the art of living and much more likelihood of its being improved when minds cease to be engrossed by the art of getting on.

John Stuart Mill
Principles of Political Economy, 1857

TABLE 5.1 Strategies for sustainable development

1. *Greater attention to the problems of the poor people of the world.* In many cases, environmental damage is caused by people who have no other alternatives in their struggle for existence. Modest improvements in their income, health status, political freedom, and access to education, capital, and technology could have major impacts on society and the environment.

2. *Local input in planning and managing resource development.* Local people often have valuable ecological knowledge that is overlooked by planners. Giving them a better share of proceeds from resource development provides an incentive to protect and conserve resources.

3. *Proper resource pricing.* Internalizing external costs shows the real trade-offs in resource development and gives resource users the incentive to minimize *all* costs, not just those that directly affect themselves.

4. *Demand management.* Dramatic and far-reaching improvements in resource availability can be obtained by reducing or eliminating wasteful or unnecessary resource uses. This can be accomplished either by pricing mechanisms or by changes in regulations and institutional mechanisms. Sometimes it is faster and more effective to pay for new efficient equipment (insulation, new furnaces for the poor) than to depend on market forces or laws.

5. *Better management capability.* We need technical personnel, information, and legal and administrative systems to plan and guide resource use so that market forces or legal mechanisms will work as they should to protect and sustain vital resources.

6. *Minimize throughput.* By emphasizing durable goods, recycling and reusing materials, efficiency, and lower consumption, we can lower our consumption of resource "capital" and reduce our impact on the environment.

7. *Redistribute wealth and power.* We can no longer justify inequality as necessary for savings, investment, and growth. As economist Herman Daly points out, "sustainable development will make fewer demands on our environmental resources, but greater demands on our moral resources."

8. *Develop nondestructive resource uses.* By focusing on activities that use intangible resources, such as information, creativity, communications, leisure, and art, we can have a life rich in values but with minimal impact on environmental resources.

From Robert Repetto, *World Enough in Time*, p. 18, World Resources Institute, Washington D.C. Published by Yale University Press.

Summary

In this chapter, we have reviewed some economic theories of the effects of natural resource scarcity. A resource is defined as any useful material or service. This includes tangible, physical assets and intangible services of the environment. Resources are defined by the economic and technological feasibility of extracting them, as well as by their location and physical size. We distinguished between known, proven, inferred, and unconceived resources, and between economically important and technically accessible resources.

Two main mechanisms determine who shall benefit from natural resources. In the case of private goods, we depend on the marketplace to set prices through the interplay of supply and demand. For public goods, where costs and benefits are widely spread and difficult to evaluate in terms of a market price, we use the political process to reflect social values and to distribute resources fairly. The market may fail for a number of reasons to optimally set priorities in conserving or utilizing natural resources. Among the most important of these reasons are inadequate reflection of external costs, the public good, and future values in the price system.

Questions about the scarcity of resources and their effects on economic development are important in determining what kind of society we have. Different theories about economics and the role of resources form the basis for market or centrally planned economic systems. Resources are also treated differently in various stages of development such as frontier, industrial, and postindustrial economies. Some people have called for a transition to a steady-state economic system, one in which there is sustainable development.

It is important for us to decide, individually as well as collectively, how we should use our resources and how we can best reach the goal of a just, sustainable society.

1. Define a resource and distinguish between tangible and intangible resources.
2. What is the difference between economic growth and economic development?
3. List four economic categories of resources and describe the differences among them.
4. Describe the relationship between supply and demand.
5. What causes diminishing returns in natural resource use? How does population growth affect this phenomenon?
6. Describe how cost/benefit ratios are determined and how they are used in natural resource management.
7. Distinguish between a material-based and an information-based economy.
8. Describe how frontier, industrial, and postindustrial economies use resources, capital, and labor.
9. Why does the marketplace sometimes fail to optimally allocate natural resource values?
10. What are some characteristics of a sustainable economic system?

1. If you could retroactively stabilize economic growth or population growth at some point in the past, what point in time would you choose? What assumptions or values shape your choice?
2. When the ecologist warns that we are using up irreplaceable natural resources and the economist rejoins that ingenuity and enterprise will find substitutes for most resources, are they talking about the same things? What underlying premises and definitions shape these arguments?
3. How can intangible resources be infinite and exhaustible at the same time? Isn't this a contradiction? Can you find other similar paradoxes in this chapter?
4. What is the difference between hypothetical and unconceived (sometimes called unknown-unknown) resources? How can we plan for resources that we haven't even thought of yet? Are there costs in assuming that there are no unknown-unknowns?
5. What would be the effect on the developing countries of the world if more-developed countries were to change to a steady-state economic system? How could we achieve a just distribution of resource benefits while still protecting environmental quality and future resource use?
6. Resource use policies bring up questions of intergenerational justice. Suppose you were asked, "What has posterity ever done for me?" How would you answer?
7. If you were doing a cost/benefit study, how would you assign a value to the opportunity for good health or the existence of rare and endangered species in faraway places? Is there a danger or cost in simply throwing up our hands and saying some things are immeasurable and priceless and therefore off-limits to discussion?
8. What does it really mean to say that sustainable development meets the needs of the present without compromising the ability of future generations to meet their own needs? Is this possible? What is meant here by needs?

discount rates 102
economic development 92
economic growth 92
exhaustible resources 93
external costs 101
human resources 93
intangible resources 94
internal costs 101
internalizing costs 101
law of diminishing returns 98

market equilibrium 97
natural resources 93
population hurdle 98
price elasticity 97
production frontier 92
renewable resources 93
steady-state economy 92
supply/demand curve 97
sustainable development 107

Adams, William, M. *Green Development: Environment and Sustainability in the Third World.* London: Routledge, 1990. An excellent overview of sustainable development, poverty, and environmental activism.

Boulding, K. "The Economics of the Coming Spaceship Earth." In *Environmental Quality in a Growing Economy.* Baltimore: Johns Hopkins University Press, 1966. A comparison of "cowboy" economics and the "spaceship earth model" of steady-state economics.

Brant Commission. *Common Crisis: North-South Cooperation for World Recovery.* Cambridge: MIT Press, 1983. A plan for reducing Third World poverty and disparities between rich and poor nations.

Carlin, Alan, et al. "Environmental Investments: The Cost of Cleaning Up." *Environment* 34, no. 2 (March 1992):12. The U.S. pays hundreds of billions of dollars each year to clean up pollution. This summary of an EPA study compares different kinds of pollution control costs.

Costanza, Robert, ed. *Ecological Economics: The Science and Management of Sustainability.* New York: Columbia University Press, 1991. An excellent compendium of articles in the new field of ecological economics.

Daly, H. E., *Steady-State Economics.* Washington, D.C.: Island Press, 1991. A collection of essays on steady-state systems by economists, ecologists, and others.

George, Susan. *A Fate Worse than Debt: The World Financial Crisis and the Poor.* New York: Grove Weidenfeld, 1990. An award-winning examination of the link between national debt, poverty, and environmental destruction.

Georgescu-Roegen, N. "The Steady-state and Ecological Salvation: A Thermodynamic Analysis." *BioScience* 27, no.4 (1977): 266. An insightful analysis on the ultimate limits to growth and the necessity for steady-state systems.

Gilman, Robert. "Economics, Ecology, and Us." *In Context* 26 (Summer 1990): 10. Lead article in a special issue devoted to asking "What is enough?" in the context of humane sustainable culture.

Larson, E. D., M. H. Ross, and R. H. Williams. "Beyond the Era of Materials." *Scientific American* 254, no. 6 (June 1986): 34. Economic growth in industrial nations is no longer accompanied by increased consumption of basic materials.

Leonard, H. Jeffrey, et al. *Environment and the Poor: Development Strategies for a Common Agenda.* New Brunswick, N.J.: Transaction Books, 1989. A powerful argument that development strategies must take on the twin challenges of environment and poverty together to be successful. Special report #11 of the Overseas Development Council on U.S.–Third World Policy Perspectives.

MacNeill, Jim, et al. *Beyond Interdependence: The Meshing of the World's Economy and the Earth's Ecology.* New York: Oxford University Press, 1991. Builds on the work of the Brundtland Commission to argue the critical relationships between the global environment, the world economy, and international order.

Meadows, Donella H., et al. *Beyond the Limits.* Post Mills, Vt.: Chelsea Green, 1992. An encouraging addition to computer projections of world resources that suggests we can acheive sustainability if we act soon.

Meeker-Lowry, Susan. *Economics as If the Earth Really Mattered: A Catalyst Guide to Socially Conscious Investing.* Santa Cruz, Calif.: New Society Publishers, 1988. A valuable examination of how we can use money, both individually and collectively, to better society and the environment.

Mill, J. S. *Principles of Political Economy.* Vol. 2. London: J. W. Parker & Son, 1857. Contrary to most economists of his time, Mill questioned the need for and wisdom of continual growth.

Neher, Philip. *Natural Resource Economics: Conservation and Exploitation.* Cambridge: Cambridge University Press, 1990. A detailed mathematical presentation of economics with a social conscience.

Pearce, David W., and R. Kerry Turner. *Economics of Natural Resources and the Environment.* Baltimore: Johns Hopkins University Press, 1990. A good overview of basic economic theory and its application to natural resources and the environment.

Plant, Christopher, and Judith Plant, eds. *Green Business: Hope or Hoax?* Santa Cruz, Calif.: New Society Publishers, 1991. A well-researched critique of "shallow-green" consumerism and businesses that claim their products are "environmentally friendly" when they are really not. Includes many useful suggestions for authentic conversion of our economy.

Renner, Michael. *Jobs in a Sustainable Economy.* Paper 104. Washington, D.C.: Worldwatch Institute, 1992. Converting to a sustainable economy will create jobs rather than eliminate them.

Repetto, Robert. "Accounting for Environmental Assets." *Scientific American* 266, no. 6 (June 1992):

94. The national system of accounts used by the United Nations badly needs overhaul to consider depletion of natural resource stocks.

Repetto, Robert. "Earth in the Balance Sheet: Incorporating Natural Resources in National Income Accounts." *Environment* 34, no. 7 (September 1992): 12. Similar to the article above but with more data from case studies.

Richie, Mark. "Free Trade Versus Sustainable Agriculture: The Implications of NAFTA." *The Ecologist* 22, no. 5 (September/October 1992): 221. The North American Free Trade Agreement threatens both jobs and environmental protection under the guise of economic growth.

Schramm, Gunter, and Jeremy J. Warford, eds. *Environmental Management and Economic Development.* Baltimore: Johns Hopkins University Press, 1989. A World Bank view of national resource accounting, environment, and resource management with case studies from many countries.

Schumacher, E. F. *Small Is Beautiful: Economics as If People Mattered.* New York: Harper & Row, 1973. A highly readable and insightful discussion on Buddhist economics and good work from the father of appropriate technology.

Shrybman, S. "International Trade and the Environment." *Ecologist* 20, no. 1 (1990): 30. A useful and timely environmental assessment of the General Agreement on Tariffs and Trade (GATT).

Speth, James G. "Coming to Terms: Toward a North-South Compact for the Environment." *Environment* 32, no. 5 (1990): 16. Meeting the challenge of global environmental problems will require unprecedented cooperation and understanding between nations.

Stavins, Robert N., and Bradley W. Whitehead. "Market-Based Incentives for Environmental Protection." *Environment* 34, no. 7 (September 1992): 7. A good discussion of using market mechanisms for pollution control with many examples and case studies.

Trainer, F. E. "Environmental Significance of Development Theory." *Ecological Economics* 2, no. 4 (1990): 124. A cogent review of the effects of development on the environment.

Wachtel, Paul. *The Poverty of Affluence: A Psychological Portrait of the American Way of Life.* Santa Cruz, Calif.: New Society Publishers, 1992. Examines the reasons and consequences of our seemingly insatiable desire for growth. Asks, How much is enough?

World Bank. *World Development Report 1992.* Oxford: Oxford University Press, 1992. A good overview of world resources and natural resource economics with a special emphasis on the relationship between sustainable development, poverty, and the environment.

6

*E*NVIRONMENTAL HEALTH AND TOXICOLOGY

The best measure we have for designing our future technologies is human health. There is nothing that seems more immediate and important than personal health, our own and that of our loved ones. It is here that we feel the greatest urgency to solve problems of environmental pollution, and it is here that the consequences of our actions are most dramatically demonstrated.

Mike Samuels and Hal Zina Bennett

O b j e c t i v e s

After studying this chapter, you should be able to:

- Define health and disease in terms of some major environmental factors that affect humans.
- Identify some major infectious organisms and hazardous agents that cause environmental diseases.
- Distinguish between toxic and hazardous chemicals and between chronic and acute exposures and responses.

- Compare the relative toxicity of some natural and synthetic compounds as well as report on how such ratings are determined and what they mean.
- Be aware of the major environmental risks we face and how risk assessment and risk acceptability are determined.

INTRODUCTION

If you read a newspaper or watch television news, you undoubtedly have seen many stories about toxic and hazardous chemicals in the environment. There are scares about pesticide residues on fruits and vegetables; cancer-causing radon in our homes; neurotoxic heavy metals in fish; pathogenic bacteria in eggs, milk, and cheese; and dangerous synthetic chemicals released into the environment by industrial accidents or deliberate dumping (fig. 6.1).

What are these toxic and hazardous agents? Why are they dangerous, and how are we exposed to them? In this chapter, we will survey some principles of toxicology and environmental health that will help answer these questions.

TYPES OF ENVIRONMENTAL HEALTH HAZARDS

What is health? The World Health Organization defines **health** as a state of complete physical, mental, and social well-being, not merely the absence of disease or infirmity. By that definition, we all are ill to some extent. Likewise, we all can improve our health to live happier, longer, more productive, and more satisfying lives if we pay attention to what we are doing.

What is a disease? A **disease** is a deleterious change in the body's condition in response to an environmental factor that could be nutritional, chemical,

Figure 6.1 Environmental health studies the ways that infectious agents, toxic chemicals, or physical factors in the environment affect our well-being.

biological, or psychological. Diet and nutrition, infectious agents, toxic chemicals, physical factors, and psychological stress all play roles in the onset or progress of human diseases. To understand how these factors affect us, let's look at some of the major categories of environmental health hazards.

TABLE 6.1 Some major environmental health problems

Disease	New cases each year	Yearly deaths
Diarrhea	1 billion	10 million
Malaria	800 million	5-10 million
Parasitic worms, including flukes	1 billion	*
Anemia	375 million	*
Respiratory diseases†	500 million	5-6 million
Trachoma	300 million	*
Goiter and cretinism	200 million	*
Tetanus	5 million	800,000
Polio	2 million	200,000

*Few people die directly from these diseases, but debilitation can be severe and can contribute to other diseases.

†Respiratory diseases include tuberculosis, influenza, pneumonia, and whooping cough.

Infectious Organisms

For most people in the world, the greatest environmental health threat continues to be pathogenic (disease-causing) organisms (table 6.1). Although much of our attention is focused on toxic synthetic chemicals, we also should be aware of the biological hazards to which we are exposed. In the less-developed countries, where nearly 80 percent of the world population lives, infectious agents, parasites, and nutritional deficiencies still are the main cause of **morbidity** (illness) and mortality (death).

Gastrointestinal infections (diarrhea, dysentery, and cholera) probably cause more deaths worldwide than any other group of diseases. Diarrhea can be caused by either bacteria or protozoans (fig. 6.2). At least 1

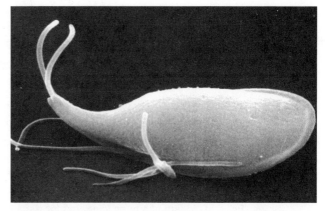

Figure 6.2 Giardia, a parasitic intestinal protozoan, is reported to be the largest single cause of diarrhea in the United States. It is spread from human feces through food and water. Even pristine wilderness areas have giardia outbreaks due to careless campers.

billion new cases of diarrhea occur each year—mostly among young children—and around 10 million deaths result from a combination of malnutrition and diarrhea. That means that one child dies every three seconds from these highly preventable diseases. In the time that it will take you to read this paragraph, about ten children will die. In many of the less-developed countries, one child in four dies before the age of five from infectious diseases.

Malnutrition and diarrhea create a vicious cycle. Poor nutrition makes people more susceptible to infection, and infections, in turn, make it more difficult to obtain, absorb, and retain food. Improved sanitation and better food could prevent most, if not all, gastrointestinal infections. Simple oral rehydration therapy (ORT), in which patients are given an inexpensive mixture of sugar and salts in water, is highly effective in treating diarrhea, and costs only a few cents per patient (box 6.1).

Thinking Globally: Box 6.1

The Child Survival Revolution

Every year in the developing countries of the world, some 14 million children under the age of five die of common infectious diseases (box fig. 6.1.1). Most of these children could be saved by simple, inexpensive, preventative medicine. How we treat the most vulnerable members of society is a good measure of our level of civilization. Many public health officials argue that it is as immoral and unethical to allow children to die of easily preventable diseases as it would be to allow them to starve to death or to be murdered. In 1986, the United Nations announced a worldwide campaign to prevent unnecessary child deaths. Called the "child

survival revolution," this campaign is based on four principles designated by the acronym GOBI.

G is for growth monitoring. A healthy child is a growing child. Underweight children are much more susceptible to infectious diseases, retardation, and other medical problems than children who are better nourished. Regular growth monitoring is the first step in health maintenance.

O is for oral rehydration therapy (ORT). About one-third of all deaths under five years of age are caused by diarrheal diseases. A simple solution of salts, glucose or rice powder, and boiled water given orally is almost miraculously effective in preventing death from dehydration shock in these diseases. The cost of treatment is only a few cents per child. The British

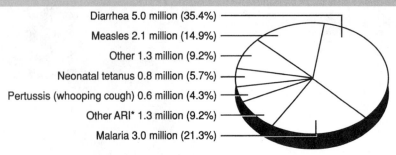

Diarrhea 5.0 million (35.4%)

Measles 2.1 million (14.9%)

Other 1.3 million (9.2%)

Neonatal tetanus 0.8 million (5.7%)

Pertussis (whooping cough) 0.6 million (4.3%)

Other ARI* 1.3 million (9.2%)

Malaria 3.0 million (21.3%)

Estimated total annual child deaths: 14.1 million

Notes: For purposes of this chart, one cause of death has been allocated for each child death when, in fact, children die of multiple causes.

*Other acute respiratory infections (ARI): Tuberculosis, diphtheria, pneumonia, influenza, pleurisy, acute bronchitis, otitis media, and other respiratory tract diseases.

Figure 6.1.1 Estimated annual deaths of children under five by cause. Most of these deaths could be prevented by better sanitation and inexpensive health care.

From WHO and UNICEF estimates, *State of the World's Children.* Copyright © 1987 Oxford University Press, Oxford, England. Used by permission of Oxford University Press.

medical journal *Lancet* called ORT "the most important medical advance of the century."

B is for breast-feeding. Babies who are breast-fed get natural immunity to diseases from antibodies in their mothers' milk, but infant formula companies have been persuading mothers in many developing countries that bottle-feeding is more modern and healthful than breast-feeding. Unfortunately, these mothers usually don't have access to clean water to combine with the formula, and they can't afford enough expensive synthetic formula to nourish their babies adequately. Consequently, the mortality among bottle-fed babies is much higher than among breast-fed babies in developing countries.

I is for universal immunization against the six largest, preventable, communicable diseases of the world: measles, tetanus, tuberculosis, polio, diphtheria, and whooping cough. In 1975, less than 10 percent of the developing world's children had been immunized. By 1990, this number had risen to over 50 percent. Although we have not yet reached our goal of full immunization for all children, many lives are being saved every year. In some countries, yellow fever, typhoid, meningitis, cholera, and other diseases also urgently need attention.

Burkina Faso provides an excellent example of how a successful immunization campaign can be carried out. Although this West African nation is one of the poorest in the world (with an annual GNP per capita of only $140), and roads, health care clinics, communication, and educational facilities are either nonexistent or woefully inade-

quate, a highly successful "vaccination commando" operation was undertaken in 1985. In a single three-week period, one million children were immunized against three major diseases (measles, yellow fever, and meningitis) with only a single injection per child. This represents 60 percent of all children under age fourteen in the country. The cost was less than $1 per child.

In addition to being an issue of humanity and compassion, reducing child mortality may be one of the best ways to stabilize world population. There has never been a reduction in birth rates that was not preceded by a reduction in infant mortality. When parents are confident that their children will survive, they tend to have only the number of children they actually want, rather than "compensating" for likely deaths by extra births. In Bangladesh, where ORT was discovered, a children's health campaign in the slums of Dacca has reduced infant mortality rates 21 percent since 1983. In that same period, the use of birth control increased 45 percent and birth rates decreased 21 percent.

Sri Lanka, China, Costa Rica, Thailand, and the Republic of Korea have reduced child deaths to a level comparable to those in many highly developed countries. This child survival revolution has been followed by low birth rates and stabilizing populations. The United Nations Children's Fund estimates that if all developing countries had been able to achieve similar birth and death rates, there would have been 9 million fewer child deaths in 1987, and nearly 22 million fewer births.

Malaria is an infection of red blood cells by a parasitic protozoan (*Plasmodium* sp.). It is reported to be the second leading disease-caused source of mortality in the world and is especially common in the moist tropical countries of Africa, where the *Anopheles* mosquitoes that spread the disease thrive. About 160 million people have malaria at any given time, and there are around 800 million new cases each year.

Once the malaria parasite becomes established in the blood, the disease can recur every few months for many years. Insecticide use greatly reduced malaria incidence (as much as 90 percent in India and Sri Lanka) in the 1950s and 1960s, but pesticide-resistant mosquito populations have developed, allowing the disease to reappear—in some cases at higher levels than before. Malaria has now become the most rapidly growing disease in the world, in many areas increasing at a much higher rate than AIDS.

Parasitic nematodes (roundworms) and flatworms (flukes and tapeworms) are very common in less-developed countries where sanitation is primitive. People rarely die directly from the infections, but they can be extremely debilitated. Intestinal worms, such as tapeworms and some roundworms, are especially serious in persons who are already malnourished.

Schistosomiasis is a disease caused by waterborne blood flukes. About 200 million people are infected worldwide, and about 1 million deaths are caused by complications of this disease. The adult flukes live in blood vessels of the human digestive tract where they cause dysentery, anemia, general weakness, and greatly reduced resistance to other infections. They have a complex life cycle involving an aquatic snail as an intermediate host (fig. 6.3). Rice paddy farming, the most common type of agriculture in many tropical countries, creates a nearly perfect environment for transmission of these flukes because human feces are used for fertilizer, the water is shallow and warm—just right for the snails—and people spend hours wading in the water tending rice plants. Large irrigation projects, such as those made possible by building the Aswan Dam in Egypt, bring about increased crop yields but also increase problems with schistosomiasis.

Nematodes are among the most numerous animals in the world, and several parasitic species cause serious human diseases. Onchocerciasis (river blindness) is caused by worms transmitted by the bite of black flies. Masses of dead nematodes accumulate in the eyeball, destroying vision. This disease affects 18 million people in the world and permanently blinds 500,000 each year. In some African villages, nearly every adult over thirty is blind from onchocerciasis (fig. 6.4). The World Health Organization (WHO) undertook a massive pesticide spraying campaign in Africa during the early 1980s that reduced the incidence of river blindness in many areas, but the flies have become pesticide resistant, and many nontarget species were destroyed. The drugs amocarzine and ivermectin can effectively control onchocerciasis and a number of other parasitic infections in humans and domestic livestock. The WHO predicts that river blindness, and perhaps other parasitic infections, could be wiped out by the year 2000 with continued efforts in public education and early treatment.

Some other parasitic worms also cause dreadful diseases. Filariasis (one form of which is elephantiasis) is transmitted by mosquitoes. The worms block the lymphatic system, causing large fluid accumulations and swellings in various body areas. Guinea worms (*Dracunculus*) live as larvae inside small aquatic crustaceans. They infect people who drink unfiltered water containing the crustaceans. Adult worms can be

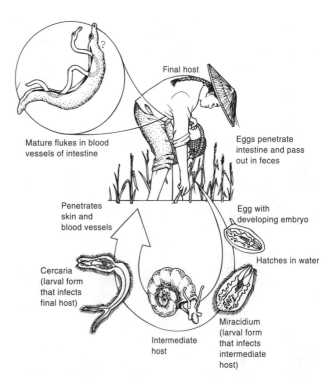

Figure 6.3 The blood fluke life cycle. Sexual forms (inset) live in human tissues. The mature female fits in a groove running the length of the larger male's body. Eggs pass out of the body in feces. They hatch in fresh water and infect the intermediate host (snails) where they reproduce asexually to produce free-swimming larvae that reinfect humans.

up to 1 m (3 ft) long. After a year of growth, the worm emerges through the skin to lay its eggs. This painful ordeal may last several weeks, and the open sores it creates become infected, causing further illness.

Tuberculosis and other respiratory diseases (influenza and pneumonia) are the leading cause of death in many subtropical countries, especially in Latin America. Until 1900, tuberculosis was the major cause of death in the United States. This dreaded disease had been largely eliminated by good sanitation and inexpensive inoculations, but ''Super-strains'' of tuberculosis bacteria that are resistant to multiple antibiotics are appearing in many countries. We may have to go back to quarantines and isolating patients. In the United States, pneumonia and influenza are now the leading causes of deaths from infection, ranking sixth among all causes. This statistic reflects the fact that these diseases are often the last, fatal complication of a variety of other ailments.

Trachoma is another widespread eye disease. It is a contagious inflammation of the inner eyelid, tear glands, and cornea caused by viruses. This disease is found where sanitation is poor. If not treated, it can cause blindness. Several hundred million people, mostly children, suffer from trachoma.

Figure 6.4 A child leads a blind adult in West Africa. River blindness, a parasitic disease transmitted by biting flies, affects 18 million people. In some African villages, nearly every adult over age thirty is blind due to this disease.

Although sexually transmitted diseases don't kill nearly as many people worldwide as malaria, diarrhea, and tuberculosis do, they may become more serious problems in the future. Gonorrhea and syphilis bacteria have developed resistance to many antibiotics and are spreading rapidly. The acquired immune deficiency syndrome (AIDS) virus causes an invariably fatal disease because it attacks the immune system itself. Currently, AIDS kills about as many people each year as tetanus does, but it is spreading rapidly in Asia and may infect tens of millions of people there.

You might suppose that all these terrible diseases will reduce or reverse population growth rates. If you look at the history of population growth in figure 4.5, however, you will notice that even pandemics such as the bubonic plague (Black Death), which killed about one-third of the population of Europe in the fourteenth century, didn't slow population growth for very long. The great influenza contagion of 1919,

which killed at least 20 million people worldwide, doesn't even show up as a ripple on the curve. Fortunately, there are more humane and more effective ways than disease to solve population problems.

Chemicals

Toxic chemicals in the environment are becoming a source of increasing concern to people in industrialized countries. Humans probably have always been subjected to a variety of toxic materials from natural sources; now we are exposed to an increasing variety and quantity of dangerous synthetic chemicals as well. In this section, we will look at some of these materials and why they are of concern.

Chemical agents are divided into two broad categories: those that are hazardous and those that are toxic. **Hazardous** means dangerous. This category includes flammables, explosives, irritants, sensitizers, acids, and caustics. Many chemicals that are hazardous in high concentrations are relatively harmless when dilute. **Toxins** are poisonous; they react with specific cellular components to kill cells. Because of this specificity, they often are harmful even in dilute concentrations. Toxins can be either general poisons that kill many kinds of cells, or they can be extremely specific in their target and mode of action. Ricin, for instance, is a protein found in castor beans. It is one of the most toxic organic compounds known. Three hundred picograms (trillionths of a gram) injected intravenously is enough to kill an average mouse. A single molecule can kill a cell. This is about two hundred times the acute toxic dose for dioxin, which sometimes is claimed to be the most toxic substance known. Table 6.2 shows some of the environmental toxins of greatest concern to the Environmental Protection Agency. This group of chemicals includes heavy metals, inorganic chemicals, and both natural and synthetic organic compounds.

Irritants are corrosives (strong acids), caustics (alkaline reagents), and other substances that damage biological tissues on contact. Some examples are sulfuric and nitric acid, ammonia, sodium hydroxide, toxic metal fumes (e.g., beryllium or nickel), ozone, chlorine, sulfur or nitrogen oxides, formaldehyde, benzene hexachloride, and dioxin. These agents not only damage cells directly but also make them susceptible to infections and can trigger transformations to a cancerous state. Skin diseases caused by irritants (dermatoses) are the most common occupational disease.

Respiratory fibrotic agents are a special class of irritants that damage the lungs, causing scar tissue formation that lowers respiratory capacity. This group includes both chemical reagents and particulate materials. Some conditions are common enough to be given specific names: silicosis (caused by silica dust), black lung (caused by coal dust), brown lung (caused by

TABLE 6.2 Toxic chemicals causing the greatest risk to human health

Benzene	Methyl ethyl ketone
Cadmium	Methyl isobutyl ketone
Carbon tetrachloride	Nickel
Chloroform	Tetrachloroethylene
Chromium	Toluene
Cyanides	Trichloroethane
Dichloromethane	Trichloroethylene
Lead	Xylene(s)
Mercury	

Source: Environmental Protection Agency, 1991.

cotton fibers), asbestosis (caused by asbestos fibers), and farmer's lung (caused by moldy hay), among others. Some of these health problems are simply obstructive diseases in which the lungs fill with residue and tissue that interfere with breathing. Some also lead to cancer.

Asphyxiants are chemicals that exclude oxygen or actively interfere with oxygen uptake and distribution. Pure nitrogen, methane, and carbon dioxide are all passive asphyxiants. Under normal circumstances they are relatively inert, but they can be deadly when they fill enclosed spaces like mines, caves, or farm silos. By contrast, active asphyxiants react chemically with blood or lung tissue to prevent oxygen uptake. Some examples are carbon monoxide, hydrogen cyanide, hydrogen sulfide, and aniline. The effects of these chemicals tend to be relatively irreversible and are toxic even in low concentrations.

Allergens are substances that activate the immune system. Some allergens act directly as **antigens;** that is, they are recognized as foreign by white blood cells and stimulate the production of specific antibodies. Other allergens act indirectly by binding to other materials and changing their structure or chemistry so they become antigenic and cause an immune response.

Formaldehyde is a good example of a widely used synthetic chemical that is a powerful immune system activator. It is both directly and indirectly allergenic. Some people who are exposed to formaldehyde in plastics, wood products, insulation, glue, fabric dyes, and a variety of other products become hypersensitive not only to formaldehyde itself but also to many other materials in their environment. This is sometimes called "sick house" syndrome. Victims may have to go to great lengths to protect themselves from these allergenic substances.

Immune system depressants are pollutants that seem to suppress the immune system rather than activate it. Little is known about how this occurs or which chemicals are responsible. Immune system failure is thought to have played a role, however, in widespread deaths of seals in the North Atlantic and of dolphins in the Mediterranean in recent years. The bodies of these dead animals generally contained high levels of pesticide residues, PCBs, and other contaminants, perhaps making the animals susceptible to a variety of opportunistic infections. Similarly, some humans with sick house syndrome or other environmental illnesses seem to have defective immune responses. The evidence for a pollution link is mostly anecdotal, however, and little hard scientific data supports these claims.

Neurotoxins are a special class of metabolic poisons that specifically attack nerve cells (neurons). The nervous system is so important in sensing information from the environment and regulating body activities that disruption of its activities is especially fast-acting and devastating. Different types of neurotoxins act in different ways. Anesthetics (ether, chloroform, halothane), chlorinated hydrocarbons (DDT, Dieldrin, Aldrin), and heavy metals (lead, mercury) disrupt the ion transport across cell membranes necessary for nerve action. Organophosphates (Malathion, Parathion) and carbamates (Sevin, Zeneb, Maneb) inhibit acetylcholinesterase, an enzyme that regulates nerve signal transmission between nerve cells and the tissues or organs they innervate (e.g., muscle). You could lose many cells in most tissues and organs with little ill effect, but the loss of a few critical neurons that regulate essential functions such as breathing can quickly be fatal. In contrast to most cells in the body, neurons are generally not replaced when they die.

Mutagens are agents such as chemicals and radiation that damage or alter genetic material (DNA) in cells. If the damage occurs during embryonic or fetal growth, it can lead to birth defects. Later in life, genetic damage may trigger neoplastic (tumor) growth. When damage occurs in reproductive cells, the results can be passed on to future generations. Cells have repair mechanisms to detect and restore damaged genetic material, but some changes may be hidden, and the repair process itself can be flawed. It is generally accepted that there is no "safe" threshold for exposure to mutagens. Any exposure has some possibility of causing damage.

Teratogens are chemicals or other factors that specifically cause abnormalities during embryonic growth and development. Some compounds that are not otherwise harmful can cause tragic problems in these sensitive stages of life. One of the most well-known examples of teratogenesis is that of the widely used sedative, thalidomide. In the 1960s, thalidomide (marketed under the trade name Cantergan) was the most widely used sleeping pill in Europe. It seemed to have no unwanted effects and was sold without prescription. When used by pregnant women, however, it caused abnormal fetal development resulting in phocomelia, meaning seal-like limbs, in which children have a hand or foot but no arm or leg (fig. 6.5). There is evidence that taking a single thalidomide pill in the first weeks of pregnancy is sufficient to cause these tragic birth defects. Altogether, about ten thousand children were affected before this drug was withdrawn from the market. Fortunately, thalidomide was not approved for sale in the United States because the Food and Drug Administration was not satisfied with the laboratory tests of its safety.

We don't know how many birth defects are due to mutagens, teratogens, or other environmental factors. Between 5 and 10 percent of all children born alive have birth defects that require medical attention in the first few years of life. It is thought that as many as 75 percent of all human embryos fail to implant in the uterus, or undergo spontaneous abortions, often without the mother's knowledge that conception has occurred. Cell damage and errors in development caused by environmental factors may play a large role in these failed pregnancies.

Carcinogens are substances that cause **cancer**, invasive, out-of-control cell growth that results in **malignant tumors**. Cancer rates have been rising rapidly in the United States and most other industrialized countries in recent years, and cancer is now the second leading cause of death in the United States, killing 510,000 people in 1990. Some investigators warn that we are entering an era of "cancer epidemic" due to the proliferation of toxic pollutants in the environment. They point out that 30 percent of Americans now living eventually will have cancer if present rates continue. A part of this rapid rise is due to increasing longevity. Since cancer often takes twenty or thirty years to develop, it is especially associated with old age. If we adjust for an aging population and better diagnosis, the rate for most cancers has remained steady or even declined. The only major types that have increased significantly are skin cancer and respiratory cancers, most of which are associated with excessive sun exposure and smoking (fig. 6.6). Still, in spite of improving preventive measures and increasingly successful therapy, cancer eventually will strike in approximately three of four families. Few of us will escape contact with this dreaded disease.

There are many different kinds of cancer and probably many different environmental causes. Some viral infections may trigger cancer. Mutagenic agents, such as radiation, heavy metals, and organic chemicals, also initiate this process. It may be that every mutagen is a potential carcinogen. Asbestos fibers and other crystalline minerals can cause cancer when they are ingested or inhaled. Foreign material, such as plastic in the body, also can trigger tumor formation. Repeated damage to cells by toxic agents like alcohol also can result in cancer, especially in the liver. Some agents such as the phorbal esters in the oil from croton plants are not carcinogenic themselves but assist in the progression and spread of tumors. These factors are called cocarcinogens or **promoters.**

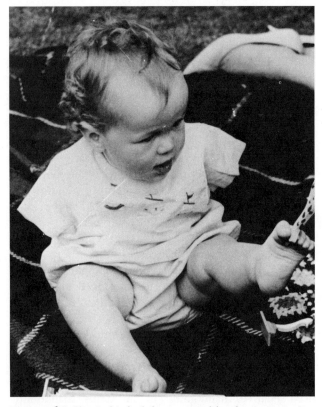

Figure 6.5 Tragic birth defects caused by the teratogenic sedative thalidomide. Fortunately, thalidomide was not approved for use in the United States.

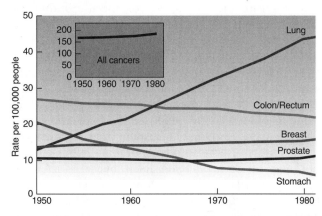

Figure 6.6 U.S cancer death rates. Although the total cancer rate in the United States has increased in recent years, the age-adjusted rates have been constant or falling for all but respiratory cancers, most of which are caused by smoking.

Natural and Synthetic Toxins

There has been so much bad news lately about the dangers of industrial chemicals that some people assume that all human-made compounds are poisonous while all natural materials must be benign and innocuous. In fact, many natural chemicals are just as dangerous as synthetic ones. Since most plants and many animal species can't escape from predators or defend themselves by fighting back, many of them have evolved a kind of chemical warfare, secreting or storing in their tissues a vast armamentarium of irritants, toxins, metabolic disrupters, and other chemicals that discourage competitors and predators. Some of the chemical defenses employed by organisms are sophisticated and specific. Both plants and animals make chemicals similar to—or even identical with—neurotransmitters, hormones, or regulatory molecules of predators or potential enemies. Our cells do not distinguish between different origins of these chemicals.

Toxicologist Bruce Ames claims that there are ten thousand times as many natural pesticides in our diets as synthetic ones. He argues that our fear of synthetic chemicals may divert our attention from more important issues. For instance, he has found natural pesticides in crops such as potatoes, tomatoes, coffee, celery, and mushrooms that are more carcinogenic than some commercial products. When plants are attacked by insects, for instance, they may synthesize natural toxins that are more dangerous than the residues left from protective treatment with synthetic pesticides. Similarly, treatment of crops with fungicides (which are carcinogenic) may prevent growth of molds that are even more carcinogenic. Simply because food is raised "organically" may not necessarily make it safer than food raised by current commercial practices.

Other environmental health specialists argue that the effects of toxic chemicals in our diet are significantly different when mixed with fiber and a multitude of other substances than they are as pure chemicals in laboratory tests. Broccoli, for instance, which can be shown to contain toxins in a bacterial test, clearly reduces the risk of cancer when added to the diet of laboratory animals.

Physical Agents, Trauma, and Stress

Physical agents, such as radiation, also are serious environmental health hazards. Radiation associated with nuclear power is discussed in chapter 12. Although the data are not totally conclusive, ubiquitous low-frequency electromagnetic fields associated with power lines and ordinary household appliances may pose a health risk (box 6.2). Noise, another important physical health threat, is discussed in chapter 14.

Acting Locally: Box 6.2

Electromagnetic Fields and Your Health

Many forms of technology seem scary and mysterious, but perhaps none are quite as insidious as potential dangers from invisible, unfelt electric and magnetic fields associated with our use of electricity. These fields are generated by power lines, household appliances, video display terminals, or any other device in which electricity flows through a wire or a motor. Although the data are vague and often contradictory, there appears to be some increased risk of cancers, miscarriages, and birth defects associated with exposure to these fields. Epidemiological studies generally implicate only the magnetic fields in human health risks, but most studies use the term electromagnetic field (EMF) because of the difficulty in separating electric and magnetic effects.

The first published report of adverse health effects from EMF was a 1966 study of electrical switchyard workers in the Soviet Union who experienced a variety of rather vague symptoms including headaches, fatigue, and reduced fertility. A more alarming study published in 1979 reported that children living near power lines in Denver, Colorado, had two or three times higher rates of childhood leukemia than matched controls. Like the Soviet study, this report was greeted with skepticism because no direct measurement of exposure was available. Instead, researchers estimated field strength based on distances between homes and power lines. Exposure to other possible sources of cancer could not be determined. Furthermore, no cellular mechanism by which EMF could cause cancer is known.

These studies have stimulated further research. Some investigators have found evidence of adverse health effects associated with EMF, while others have failed to show a link. A recent EPA study found a causal link between EMF exposure and leukemia, lymphoma, and brain cancer, but stopped short of listing EMF as a probable carcinogen. A survey of workers in power plants or telephone switching stations found evidence for breast cancer among both men and women. Analysis of childhood leukemias in Los Angeles found links between use of electric hair dryers and black-and-white televisions. Canadian research showed evidence for miscarriages, brain tumors, and birth defects among children whose mothers worked at video display terminals or used electric blankets while pregnant. The statistical significance of these studies is generally weak, and they often fail to show clear dose/response relationships expected for a direct and unequivocal link. In 1992, however, Swedish studies reported that electrical workers and children who live near power lines have twice the normal leukemia rate of one case in twenty thousand people. This research does show a statistically significant dose/response relationship between field strength and cancer incidence.

What does this evidence suggest for the average person? First of all, homes and schools should be at least one kilometer away from high-voltage power lines. Electric distribution lines that bring power into homes create much less powerful fields but should still be shielded and routed away from the parts of houses where people spend the most time. An electric blanket generates only minute fields, but it lies right on top of you for many hours each night. It might be advisable to use a quilt instead, especially if you are pregnant. People who watch television or work at a video display terminal (computer screen) for many hours each day should back up at least one meter (three feet) from the screen. Children, especially, should be discouraged from sitting close to television screens.

Bedside appliances such as electric clocks, telephone answering machines, or anything with an electric motor that runs continuously should be placed at least a meter away from your head. Even better, why not place them across the room? Other electric appliances such as hair driers, curling irons, electric shavers, can openers, and microwave ovens should be used as briefly as possible and at the greatest distance from your person as is feasible. Do not stand right in front of the microwave door watching your food cook. Consider using a towel to dry your hair or a nonmotorized razor to shave.

We also should keep relative risks in mind. If it is true that cancer risks are doubled by exposure to EMFs, remember that smoking increases cancer risks twenty times. Riding in an automobile, being overweight, eating a high-fat diet, engaging in unsafe sex, excessive drinking, risky jobs, radon in your home, and stress all probably pose much greater threats to your health than EMF. Still, prudent avoidance makes sense; if you can reduce your exposure to EMF at little cost, why not do it?

Trauma, injury caused by accidents and violence, has surely always been a life-threatening environmental factor for humans. There is probably less danger from physical trauma in the more-developed countries of the world now than ever before, even though modern media coverage might make it seem otherwise. The death rate from accidents in the United States is about half what it was in 1900. Still, accidents, homicide, and suicide are the principal causes of death for people between the ages of one and thirty-eight in the United States, and trauma is the leading cause of years of life lost before age sixty-five in all industrialized countries. Every year about 100,000 premature deaths from trauma in the United States and twice as many cases of permanent disability occur. More than half of these deaths and injuries are caused by motor vehicle accidents. About 90 percent of the accidents involve private automobiles, trucks, or motorcycles, and at least half are caused by drivers under the influence of alcohol or other drugs.

Stress and lifestyle once were considered forces that might cause unhappiness, but they were not considered life-threatening. Now we know, however, that stress is clearly related to physical diseases such as heart attack, stroke, and atherosclerosis, which are the leading causes of nontrauma-related death in the United States and most other industrialized countries. Since stress is a component of our cultural environment, it deserves a place in a discussion of environmental health.

In medical terms, **stress** refers to physical, chemical, or emotional factors that place a strain on an organism for which there is inadequate adaptation. This can result in physical responses that contribute to disease. Adverse stress responses are not unique to humans. Plants show signs of environmental stress, as do most animals. When laboratory or zoo animals are kept in cages next to especially aggressive members of the same species, they often show many signs of anxiety and stress. Even though they are separated by glass walls so that no physical contact is possible, the stressed animals often will die prematurely of cardiovascular diseases. Gastrointestinal disturbances, such as ulcers, are common human responses to stress. Stress also contributes to susceptibility to infectious diseases, as many students discover at exam time.

Diet

Diet also has an important effect on health. For instance, there are correlations between cardiovascular disease and cancer with the amount of salt and animal fat in one's diet (fig. 6.7). Highly processed foods, fat, and smoke-cured, high-nitrate meats also are associated with cancer. Fruits, vegetables, whole grains, complex carbohydrates, and dietary fiber (plant cell walls), on the other hand, often have beneficial health effects. Certain dietary components, such as pectins; vitamins A, C, and E; substances produced in cruciferous vegetables (cabbage, broccoli, cauliflower, brussel sprouts); and selenium (in low levels) seem to have anticancer effects.

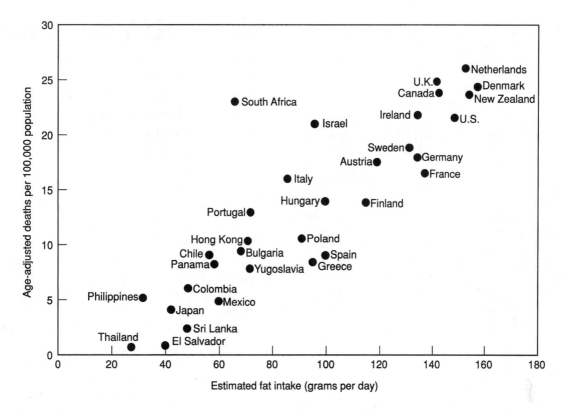

Figure 6.7 Strong linear correlation between dietary fat intake and deaths from breast cancer. The exact cause/effect relationship is unknown.

Eating too much food is a significant dietary health factor in developed countries and among the well-to-do everywhere. At least one-fourth of all Americans are considered overweight. Cutting back on the number of calories consumed reduces the strain on bones, muscles, and other organs and has additional beneficial effects, including reducing cardiovascular disease, diabetes, and—perhaps—cancer.

In some areas of the world, people seem to live exceptionally long lives. The Abkhasian people in the Caucasus Mountains of Soviet Georgia, the Hunzans in the mountains of Pakistan, and the Vilcabama villagers in Ecuador, for instance, are among the longest-lived people in the world. Many claim to be 120 to 140 years old, although it is difficult to substantiate when they were born. Still, they do appear to live longer and to be more physically active later in life than do most of us.

These people share several factors that probably contribute to longevity. They live at moderately high elevations where the climate is cool, dry, and sunny. They lead active, vigorous lives in low-pressure, nonindustrialized settings. Life in their small villages is uniform and predictable. Stress levels are low. People of all ages work together in the fields and in the home, doing practical, physical work that directly benefits their community. The elderly are respected and live a useful, active life. The whole family shares in decision making, recreation, and religion. Their diet is usually simple, with low fat and salt, high fiber content, little meat, and lots of fruits and vegetables. They enjoy clean air and pure water. Their culture and uncomplicated lifestyles reduce conflict and anxiety. We might benefit from incorporating some aspects of their lives into our own (table 6.3).

TABLE 6.3 National health recommendations and diet goals

Eat only enough calories to meet body needs (fewer if overweight).

Eat less fat and cholesterol.

Eat less salt.

Eat less sugar.

Eat more whole grains, cereals, fruits, and vegetables.

Eat more fish, poultry, beans, and peas.

Eat less red meat.

Eat less additives and processed foods.

Source: The Surgeon General's Report: *Healthy People*, 1980.

MOVEMENT, DISTRIBUTION, AND FATE OF TOXINS

There are many sources of toxic and hazardous chemicals in the environment. The danger of each chemical can be determined by factors related to the chemical itself, its route or method of exposure, and its persistence in the environment, as well as characteristics of the target organism (table 6.4). We can think of an ecosystem as a set of interacting compartments among which a chemical moves, based on its molecular size, solubility, stability, and reactivity (fig. 6.8). The routes used by chemicals to enter our bodies also play important roles in determining toxicity (fig. 6.9). In this section, we will consider some of these characteristics and how they affect environmental health.

Solubility

Solubility is one of the most important characteristics in determining how, where, and when a toxic material will move through the environment or through the body to its site of action. Chemicals can be divided into two major groups: those that dissolve more readily in water and those that dissolve more readily in oil. Water-soluble compounds move rapidly and widely through the environment because water is ubiquitous. They also tend to have ready access to most cells in the body because aqueous solutions bathe all our cells. Molecules that are oil- or fat-soluble (usually organic molecules) generally need a carrier to move through the environment and into and within the body. Once inside the body, however, oil-soluble toxins penetrate readily into tissues and cells

TABLE 6.4 *Factors in environmental toxicity*
Factors related to the toxic agent
1. Chemical composition and reactivity
2. Physical characteristics (e.g., solubility, state)
3. Presence of impurities or contaminants
4. Stability and storage characteristics of toxic agent
5. Availability of vehicle (e.g., solvent) to carry agent
6. Movement of agent through environment and into cells
Factors related to exposure
1. Dose (concentration and volume of exposure)
2. Route, rate, and site of exposure
3. Duration and frequency of exposure
4. Time of exposure (time of day, season, year)
Factors related to organism
1. Resistance to uptake, storage, or cell permeability of agent
2. Ability to metabolize, inactivate, sequester, or eliminate agent
3. Tendency to activate or alter nontoxic substances so they become toxic
4. Concurrent infections or physical or chemical stress
5. Species and genetic characteristics of organism
6. Nutritional status of subject
7. Age, sex, body weight, immunological status, and maturity

because the membranes that enclose cells are themselves made of similar oil-soluble chemicals. Once they get inside cells, oil-soluble materials are likely to

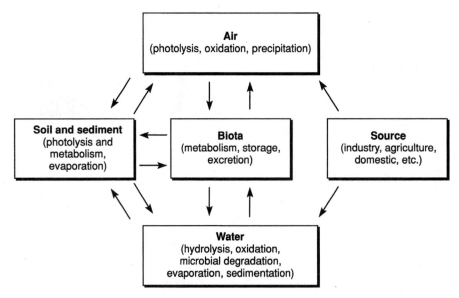

Figure 6.8 Movement and fate of chemicals in the environment. Mechanisms that modify, remove, or sequester compounds are shown in parentheses.

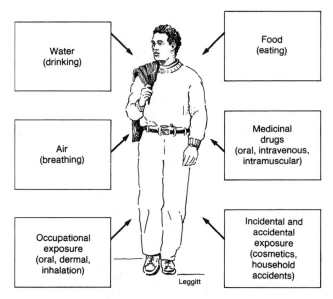

Figure 6.9 Routes of exposure to toxic and hazardous environmental factors.

be accumulated and stored in lipid deposits where they may be protected from metabolic breakdown and persist for many years.

Bioaccumulation and Biomagnification

Cells have mechanisms for **bioaccumulation,** selectively absorbing and storing a great variety of molecules. This allows them to accumulate nutrients and essential minerals, but they also may absorb and store harmful substances through these same mechanisms. Toxins that are rather dilute in the environment can reach dangerous levels inside cells and tissues through this process of bioaccumulation.

The effects of toxins are magnified in the environment through food chains. **Biomagnification** occurs when the toxic burden of a large number of organisms at a lower trophic level is accumulated and concentrated by a predator at a higher trophic level. Phytoplankton and bacteria in aquatic ecosystems, for instance, take up heavy metals or toxic organic molecules from water or sediments (fig. 6.10). Their predators—zooplankton and small fish—collect and retain the toxins from many prey organisms, building up higher concentrations of toxins. The top carnivores in the food chain—game fish, fish-eating birds, and humans—can accumulate such high toxin levels that they suffer adverse health effects. One of the first known examples of bioaccumulation and biomagnification was DDT, which accumulated through food chains so that by the 1960s it was shown to be interfering with reproduction of peregrine falcons, brown pelicans, and other predatory birds at the tops of their food chains.

Figure 6.10 Bioaccumulation and biomagnification. Lower organisms take up and store toxins from the environment. They are eaten by larger predators, who are eaten, in turn, by even larger predators. The highest members of the food chain can accumulate very high levels of the toxin.

Persistence

Some chemical compounds are very unstable and degrade rapidly under most environmental conditions so that their concentrations decline quickly after release. Some of the modern herbicides, for instance, quickly lose their toxicity. Other substances are more persistent and last for a long time. Some of the most useful chemicals, such as chlorofluorocarbons, plastics, chlorinated hydrocarbons, and asbestos, are valuable because they are resistant to degradation. This stability also causes problems because these materials persist in the environment and have unexpected effects far from the sites of their original use. DDT, for instance, is a useful pesticide because it breaks down slowly and doesn't have to be reapplied often. Its toxic effects may spread to unintended victims, however, and it may be stored for long periods of time in organisms that lack mechanisms to destroy it.

MECHANISMS FOR MINIMIZING TOXIC EFFECTS

A fundamental concept in toxicology is that every material can be poisonous under some conditions, but most chemicals have some safe level or threshold below which their effects are undetectable or insignificant. Each of us consumes lethal doses of many chemicals in our lifetime. One hundred cups of strong coffee, for instance, contain a lethal dose of caffeine. Similarly, a bottle of one hundred aspirin tablets, or twenty pounds of spinach or rhubarb, or a fifth of Scotch, vodka, or gin would be deadly if consumed all at once. Taken in small doses, however, most toxins can be broken down or excreted before they do much harm. Furthermore, damage they cause can be repaired. Sometimes, however, mechanisms that protect us from one type of toxin or that protect us at one stage in the life cycle can become harmful with a different substance or at a different stage of development. Let's look at how these processes help protect us from harmful substances as well as how they can go awry.

Metabolic Degradation and Excretion

Most organisms have enzymes that process waste products and environmental poisons to reduce their toxicity and biological activity. In mammals, many of these enzymes are located in the liver, the primary site for detoxifying both natural wastes and introduced poisons. Sometimes, however, these enzymes work to our disadvantage. Compounds such as benzopyrene, for example, that are not toxic in their original form, are processed by these same enzymes into cancer-causing carcinogens. Why would our bodies contain a system that makes a chemical more dangerous? Evolution and natural selection are expressed through reproductive success or failure. Defense mechanisms that protect us from toxins and hazards early in life are "selected for" by evolution. Factors or conditions like cancer or premature senility that affect us at postreproductive ages usually don't affect reproductive success or exert "selective pressure."

We also reduce the effects of waste products and environmental toxins by eliminating them from our body through excretion. Some volatile molecules, such as carbon dioxide, hydrogen cyanide, and ketones, are excreted via breathing. Some excess salts and other substances are excreted in sweat. Primarily, however, excretion is a function of the kidneys, which can eliminate significant amounts of soluble materials through urine formation. Toxins accumulated in the urine can damage this vital system, however, and the kidneys and bladder often are subjected to harmful levels of toxic compounds. In the same way, the stomach, intestine, and colon often suffer damage from materials concentrated in the digestive system and may be afflicted by diseases and tumors.

Repair Mechanisms

In the same way that individual cells have enzymes to repair damage to DNA at the molecular level, tissues and organs that are exposed regularly to physical wear and tear or to toxic or hazardous materials often have mechanisms for damage repair. Our skin and the epithelial linings of the gastrointestinal tract, blood vessels, lungs, and urogenital system have high cellular reproduction rates to replace injured cells. With each reproduction cycle, however, there is a chance that some cells will lose normal growth controls and run amok, creating a tumor. Thus any agent such as smoking or drinking that irritates tissues is likely to be carcinogenic. And tissues with high cell-replacement rates are among the most likely to develop cancers.

MEASURING TOXICITY

In 1540, the German scientist Paracelsus said, "the dose makes the poison," by which he meant that almost everything is toxic at some level. This remains the most basic principle of toxicology. Sodium chloride (table salt), for instance, is essential for human life in small doses. If you were forced to eat a kilogram of salt all at once, however, it would make you very sick. A similar amount injected into your blood stream would be lethal. How a material is delivered—at what rate, through which route of entry, and in what medium—plays a vitally important role in determining toxicity.

This does not mean that all toxins are identical, however. Some are so poisonous that a single drop on your skin can kill you. Others require massive amounts injected directly into the blood to be lethal. Determining and comparing the toxicity of various materials is difficult because not only do species differ in sensitivity, but individuals within a species respond differently to a given exposure. In this section, we will look at methods of toxicity testing and at how results are analyzed and reported.

Animal Testing

The most commonly used and widely accepted toxicity test is to expose a population of laboratory animals to measured doses of a specific substance under controlled conditions. This procedure is expensive, time-consuming, and often painful and debilitating to the animals being tested. It commonly takes hundreds—or even thousands—of animals, several years of hard work, and hundreds of thousands of dollars to thoroughly test the effects of a toxin at very low doses. More humane toxicity tests using computer simulation of model reactions, cell cultures, or other substitutes for whole living animals are being developed. However, conventional large-scale animal testing is the method in which we have the most confidence and on which most public policies about pollution and environmental or occupational health hazards are based.

In addition to humanitarian concerns, there are several problems in laboratory animal testing that trouble both toxicologists and policy makers. One problem is that members of a specific population will have differences in sensitivity to a toxin. Figure 6.11 shows a typical dose/response curve for exposure to a hypothetical toxin. Some individuals are very sensitive to the toxin, while others are insensitive. Most, however, fall in a middle category forming a bell-shaped curve. The question for regulators and politicians is whether we should set pollution levels that will protect everyone, including the most sensitive people, or only aim to protect the average person. It might cost billions of extra dollars to protect a very small number of individuals at the extreme end of the curve. Is that a good use of resources?

Dose/response curves are not always symmetrical, making it difficult to compare toxicity of unlike chemicals or different species of organisms. A convenient way to describe toxicity of a chemical is to determine the dose to which 50 percent of the test population is sensitive. In the case of a lethal dose (LD), this is called the **LD50** (fig. 6.12).

Unrelated species can react very differently to the same toxin, not only because body sizes vary but also because of differences in physiology and metabolism. Even closely related species can have very dissimilar reactions to a particular toxin. Hamsters, for instance, are nearly five thousand times less sensitive to some dioxins than are guinea pigs. Of 226 chemicals found to be carcinogenic in either rats or mice, 95 caused cancer in one species but not the other. These variations make it difficult to estimate the risks for humans since we can't perform controlled experiments in which we deliberately expose people to toxins.

Toxicity Ratings

It is useful to group materials according to their relative toxicity. A moderate toxin takes about one gram per kilogram of body weight (about an ounce for an average human) to make a lethal dose. Very toxic materials take about one-tenth that amount, while extremely toxic substances take one-hundredth as much (only a few drops) to kill most people. Supertoxic chemicals are extremely potent; for some, a few micrograms (millionths of a gram—an amount invisible to the naked eye) make a lethal dose. These materials are not all synthetic (human-made). As mentioned earlier, ricin, a protein found in castor bean seeds, is one of the most toxic chemicals known. It is so toxic that 0.3 billionths of a gram given intravenously will generally kill a mouse. If aspirin were this toxic, a single tablet, divided evenly, could kill one million people.

Many carcinogens, mutagens, and teratogens are dangerous at levels far below their direct toxic effect because abnormal cell growth exerts a kind of biological amplification. A single cell, perhaps altered by a

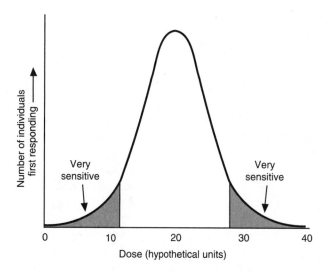

Figure 6.11 Variations in sensitivity to a toxin within a population. Some members of a population are very sensitive to a given toxin, while others are much less sensitive. The majority of the population falls somewhere between the two extremes.

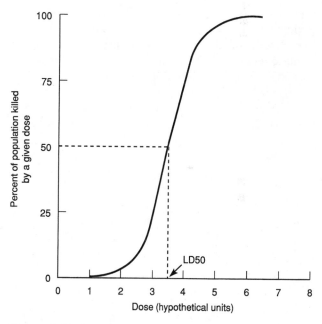

Figure 6.12 Cumulative population response to increasing doses of a toxin. The LD50 is the dose that is lethal to half the population.

single molecular event, can multiply into millions of tumor cells or an entire organism. Just as there are different levels of direct toxicity, however, there are different degrees of carcinogenicity, mutagenicity, and teratogenicity. Methanesulfonic acid, for instance, is highly carcinogenic while the sweetener saccharin is a suspected carcinogen whose effects may be vanishingly small.

Acute versus Chronic Doses and Effects

Most of the toxic effects that we have discussed so far have been **acute effects.** That is, they are caused by a single exposure to the toxin and result in an immediate health crisis of some sort. Often, if the individual experiencing an acute reaction survives this immediate crisis, the effects are reversible. **Chronic effects,** on the other hand, are long-lasting, perhaps even permanent. A chronic effect can result from a single dose of a very toxic substance or it can be the result of a continuous or repeated sublethal exposure.

We also describe long-lasting *exposures* as chronic, although their effects may or may not persist after the toxin is removed. It usually is difficult to assess the specific health risks of chronic exposures because other factors, such as aging or normal diseases, act simultaneously with the factor you would like to study. It often requires very large populations of experimental animals to obtain statistically significant results for low-level chronic exposures. Toxicologists talk about "megarat" experiments in which it might take a million rats to determine the health risks of some supertoxic chemicals at very low doses. Such an experiment would be terribly expensive for even a single chemical, let alone for the thousands of chemicals and factors suspected of being dangerous.

An alternative to enormous studies involving millions of animals is to give massive doses of a toxin being studied to a smaller number of individuals and then to extrapolate what the effects of lower doses might have been. This is a controversial approach because it is not clear that responses to toxins are linear or uniform across a wide range of doses.

Figure 6.13 shows three possible results from low doses to a toxin. Curve A shows a baseline level of response in the population, even at zero dose of the toxin. This suggests that some other factor in the environment also causes this response. Curve B shows a straight-line relationship from the highest doses to zero exposure. Many carcinogens and mutagens show this kind of response. Any exposure to such agents, no matter how small, carries some risk. Curve C shows a threshold for the response where some minimal dose is necessary before any effect can be observed. This generally suggests the presence of some defense mechanism that prevents the toxin from reaching its target in an active form or repairs the damage that it causes. Low levels of exposure to the toxin in question may have no deleterious effects, and it might not be necessary to try to keep exposures to zero.

Which, if any, environmental health hazards have thresholds is one of the most important questions in environmental science. The Delaney Clause of the

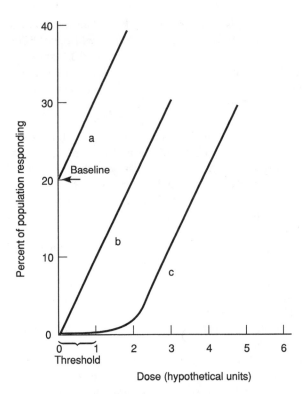

Figure 6.13 Three possible dose-response curves at low doses. (*a*) Some individuals respond, even at zero dose, indicating that some other factor must be involved. (*b*) Response is linear down to the lowest possible dose. (*c*) Threshold must be passed before any response is seen.

U.S. Food and Drug Act forbids the addition of *any* amount of a known carcinogen to foods or drugs. This is based on the assumption that *any* exposure to these substances will cause some increased risk of cancer. This may not be true in every case. Holding exposures to absolute zero may be impossible and unnecessary; however, attempting to do so seems prudent until we learn more.

Detection Limits

You may have seen or heard dire warnings about toxic materials detected in samples of air, water, or food. A typical headline announced recently that twenty-three pesticides were found in sixteen food samples. What does that mean? The implication seems to be that the mere presence of dangerous materials is equivalent to risk and that counting the numbers of compounds detected is a reliable way to establish danger. We have seen, however, that the dose makes the poison. It matters not only what is there, but how much, where it is located, how accessible it is, and who is exposed. At some level, the mere presence of a substance is insignificant.

Toxins and pollutants may seem to be more widespread now than in the past, and this is surely a valid perception for many substances. The daily reports we hear of new materials found in new places, however, are also due in part to our more sensitive measuring techniques. Twenty years ago, parts per million were generally the limits of detection for most chemicals. Anything below that amount was often reported as zero or absent rather than simply undetected. A decade ago, new machines and techniques were developed to measure parts per billion. Suddenly, chemicals were found where none had been suspected. Now we can detect parts per trillion or even parts per quadrillion in some cases. Increasingly sophisticated measuring capabilities may lead us to believe that toxic materials have become more prevalent. In fact, our environment may be no more dangerous; we are just better at finding trace amounts.

RISK ASSESSMENT AND ACCEPTANCE

Even if we know with some certainty how toxic a specific chemical is in laboratory tests, it still is difficult to determine how dangerous that chemical will be if it is released into the environment. As you already have seen, many factors complicate the movement and fate of chemicals both around us and within our bodies. Furthermore, public perception of relative dangers from environmental hazards can be skewed so that some risks seem much more important than others.

Assessing Risk

A number of factors influence how we perceive relative risks associated with different situations.

1. People with social, political, or economic interests tend to downplay certain risks and emphasize others that suit their own agendas. We do this individually, as well, building up the dangers of things that don't benefit us while diminishing or ignoring the negative aspects of activities we enjoy or profit from.

2. Most people have difficulty understanding and believing probabilities. We feel that there must be patterns and connections in events, even though statistical theory says otherwise. If the coin turned up heads last time, we feel certain that it will turn up tails next time. In the same way, it is difficult to understand the meaning of a 1-in-10,000 risk of being poisoned by a chemical.

3. Our personal experiences often are misleading. When we have not personally experienced a bad outcome, we feel it is more rare and unlikely to occur than it actually may be. Furthermore, the anxieties generated by life's gambles make us want to deny uncertainty and to misjudge many risks.

4. We have an exaggerated view of our own abilities to control our fate. We generally consider ourselves above-average drivers, safer than most when using appliances or power tools, and less likely than others to suffer medical problems, such as heart attacks. People often feel they can avoid hazards because they are wiser or luckier than others.

5. News media give us a biased perspective on the frequency of certain kinds of health hazards, overreporting the frequency of some accidents or diseases while downplaying or underreporting others. Sensational, gory, or especially frightful causes of death like murders, plane crashes, fires, or terrible accidents occupy a disproportionate amount of attention in the public media. Heart disease, cancer, and stroke kill nearly fifteen times as many people in the United States as do accidents and seventy-five times as many people as do homicides, but the emphasis placed by the media on accidents and homicides is nearly inversely proportional to their relative frequency compared to either cardiovascular disease or cancer. This gives us an inaccurate picture of the real risks to which we are exposed.

6. We tend to have an irrational fear or distrust of certain technologies or activities that leads us to overestimate their dangers. Nuclear power, for instance, is viewed as very risky, while coal-burning power plants seem to be familiar and relatively benign; in fact, coal mining, shipping, and combustion cause an estimated 10,000 deaths each year in the United States, compared to none known so far for nuclear power. An old, familiar technology seems safer and more acceptable than does a new, unknown one.

Accepting Risks

How much risk is acceptable? How much is it worth to minimize and avoid exposure to certain risks? Most people will tolerate a higher probability of occurrence of an event if the harm caused by that event is low. Conversely, harm of greater severity is acceptable only at low levels of frequency. A 1-in-10,000 chance of being killed might be of more concern to you than a 1-in-100 chance of being injured. For most people, a 1-in-100,000 chance of dying from some event or some factor is a threshold for changing what we do. That is, if the chance of death is less than 1 in 100,000, we are not likely to be worried enough to change our ways. If the risk is greater, we will probably do something about it. The Environmental Protection Agency generally assumes that a risk of 1 in 1 million is acceptable for most environmental hazards. Critics of this policy ask, acceptable to whom?

For activities that we enjoy or find profitable, however, we are often willing to accept far greater risks than this general threshold. Conversely, for risks that benefit someone else we demand far higher protection. For instance, your chances of dying in a motor vehicle accident in any given year are about 1 in 5,000, but that doesn't deter many people from riding in automobiles. Your chances of dying from lung cancer if you smoke one pack of cigarettes per day are about 1 in 1,000. By comparison, the risk from drinking water with the EPA limit of trichloroethylene is about 2 in 1 *billion*. Strangely, many people demand water with zero levels of trichloroethylene, while continuing to smoke cigarettes.

Table 6.5 lists some activities estimated to increase your chances of dying in any given year by 1 in 1 million. These are statistical averages, of course, and there clearly are differences in where one lives or how one rides a bicycle that affect the danger level of these activities. Still, it is interesting how we readily accept some risks while shunning others.

Our perception of relative risks is strongly affected by whether risks are known or unknown, whether we feel in control of the outcome, and how dreadful the results are. Risks that are unknown or unpredictable and results that are particularly gruesome or disgusting seem far worse than those that are familiar and socially acceptable. Figure 6.14 shows the relative acceptability of a variety of technologies and activities judged by their familiarity, controllability, and consequences. The relative undesirability of these risks is indicated by the location of the circle that marks its position. Note that factors in the upper right quadrant tend to be much more feared than those in the lower left quadrant, even though the actual numbers of deaths or disease from automobile accidents, smoking, etc. are thousands of times higher than those from pesticides, nuclear energy, or genetic engineering.

ESTABLISHING PUBLIC POLICY

A problem in setting environmental standards is that we are dealing with many sources of harm to which we are exposed simultaneously or sequentially. It is difficult to separate the effects of all these different hazards and to evaluate their risks accurately, especially when the exposures are near the threshold of measurement and response. In spite of often vague and contradictory data, public policy makers must make decisions.

The case of the sweetener saccharin is a good example of the complexities and uncertainties of risk assessment in public health. Studies in the 1970s at the University of Wisconsin and the Canadian Health Protection Branch suggested a link between saccharin

TABLE 6.5 *Activities estimated to increase your chances of dying in any given year by 1 in 1 million**

Activity	Resulting death risk
Smoking 1.4 cigarettes	Cancer, heart disease
Drinking 0.5 liter of wine	Cirrhosis of the liver
Spending 1 hour in a coal mine	Black lung disease
Living 2 days in New York or Boston	Air pollution
Traveling 6 minutes by canoe	Accident
Traveling 10 miles by bicycle	Accident
Traveling 150 miles by car	Accident
Flying 1,000 miles by jet	Accident
Flying 6,000 miles by jet	Cancer caused by cosmic radiation
Living 2 months in Denver	Cancer caused by cosmic radiation
Living 2 months in a stone or brick building	Cancer caused by natural radioactivity
One chest X ray	Cancer caused by radiation
Living 2 months with a cigarette smoker	Cancer, heart disease
Eating 40 tablespoons of peanut butter	Cancer from aflatoxin
Living 5 years at the site boundary of a typical nuclear power plant	Cancer caused by radiation from routine leaks
Living 50 years 5 miles from a nuclear power plant	Cancer caused by accidental radiation release
Eating 100 charcoal-broiled steaks	Cancer from benzopyrene

*From William Allman, "Staying Alive in the Twentieth Century," *Science 85,* 5(6):31, October 1985. Copyright 1985 by the American Association for the Advancement of Science.

and bladder cancer in male rats. Critics of these studies pointed out that humans would have to drink eight hundred cans of diet soda *per day* to get a saccharin dose equivalent to that given to the rats. Furthermore, they argued that people are not just very large rats. Although the Food and Drug Act forbids the addition of any substance that causes cancer in any amount in any animal, Congress has repeatedly exempted saccharin from being banned from food because of the uncertainty about its risks.

Experiments testing the toxicity of saccharin in rats merely give a range of probable toxicities in humans.

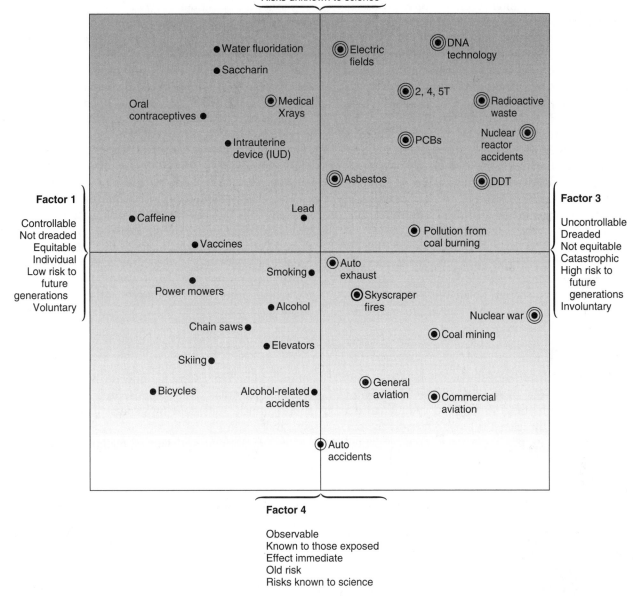

Factor 2

Factor 2

Not observable
Unknown to those exposed
Effect delayed
New risk
Risks unknown to science

• Water fluoridation

• Saccharin

Oral
contraceptives •

• Medical Xrays

• Intrauterine
device (IUD)

◉ Electric fields

◉ DNA technology

◉ 2, 4, 5T

◉ Radioactive waste

◉ PCBs

◉ Nuclear reactor accidents

◉ Asbestos

◉ DDT

Factor 1

Controllable
Not dreaded
Equitable
Individual
Low risk to
  future
generations
Voluntary

• Caffeine

Lead
•

• Vaccines

◉ Pollution from coal burning

Factor 3

Uncontrollable
Dreaded
Not equitable
Catastrophic
High risk to
  future
generations
Involuntary

Smoking •

• Power mowers

• Alcohol

Chain saws •

• Elevators

Skiing •

• Bicycles

Alcohol-related • accidents

◉ Auto exhaust

◉ Skyscraper fires

Nuclear war ◉

◉ Coal mining

◉ General aviation

◉ Commercial aviation

◉ Auto accidents

Factor 4

Observable
Known to those exposed
Effect immediate
Old risk
Risks known to science

Figure 6.14 Relative perception of various risks. Size of circle indicates perception of risk; location on graph indicates contribution of different factors in apprehension of risk. Each factor is derived from combinations of characteristics.

The lower end of this range indicates that only 1 person in the United States would die every 1,000 years from using saccharin. That is clearly inconsequential. The higher estimate, however, indicates that 3,640 people would die each year from the same exposure. Is that too high a cost for the benefits of having saccharin available to people who must restrict sugar intake? How does the cancer risk compare to the dangers of obesity, cardiovascular disease, etc.? What other alternatives might there be to saccharin? A

popular but more expensive alternative (aspartame, derived from the amino acid aspartic acid) that bears the trade name NutraSweet™ also is controversial because of uncertainties about its safety.

In 1989, a near-panic swept the United States after the Natural Resources Defense Council issued a report claiming that children were being exposed to dangerous chemicals in fruits and vegetables. The worst of these was believed to be Alar (daminozoide), used to treat apples to promote even ripening and reduce surface blemishes. The apple industry claimed that you would have to eat 28,000 pounds of treated apples every day for 70 years to approach the exposures that caused cancer in laboratory animals.

In setting standards for environmental toxins, we need to consider (1) combined effects of exposure to many different sources of damage, (2) different sensitivities of members of the population, and (3) effects of chronic as well as acute exposures. Some people argue that pollution levels should be set at the highest amount that does *not* cause measurable effects. Others demand that pollution be reduced to zero if possible, or as low as is technologically feasible. It may not be reasonable to demand that we be protected from every potentially harmful contaminant in our environment, no matter how small the risk. As we have seen, our bodies have mechanisms that enable us to avoid or repair many kinds of damage so that most of us can withstand some minimal level of exposure without harm.

On the other hand, each challenge to our cells by toxic substances represents stress on our bodies. Although each individual stress may not be life-threatening, the cumulative effects of all the environmental stresses, both natural and human-caused, to which we are exposed may seriously shorten or restrict our lives. Furthermore, some individuals in any population are more susceptible to those stresses than others. Should we set pollution standards so that no one is adversely affected, even the most sensitive individuals, or should the acceptable level of risk be based on the average member of the population?

Finally, policy decisions about hazardous and toxic materials need to be based also on information about how such materials affect the plants, animals, and other organisms that define and maintain our environment. In some cases, pollution can harm or destroy whole ecosystems with devastating effects on the life-supporting cycles on which we depend. In other cases, only the most sensitive species are threatened. The 1992 budget of the Environmental Protection Agency reflected a concern that our exclusive focus on reducing pollution to protect human health has neglected risks to natural ecological systems. Former EPA head William K. Reilly pointed out that while there have been many benefits from a case-by-case approach in which we evaluate the health risks of individual chemicals, we have often missed broader ecological problems that may be of greater ultimate importance.

Summary

Health is a state of physical, mental, and social well-being, not merely the absence of disease or infirmity. The cause or development of nearly every human disease is at least partly related to environmental factors. For most people in the world, the greatest health threat in the environment is now, as always, from pathogenic organisms. Bacteria, viruses, protozoans, parasitic worms, and other infectious agents probably kill more people each year than any other cause of death.

Stress, diet, and lifestyle also are important health factors. Our social or cultural environment may be as important as our physical environment in determining the state of our health. People in some areas in the world live exceptionally long and healthful lives. We might be able to learn from them how to do so as well.

Estimating the potential health risk from exposure to specific environmental factors is difficult because information on the precise dose, length and method of exposure, and possible interactions between the chemical in question and other potential toxins to which the population may have been exposed is often lacking. In addition, individuals have different levels of sensitivity and response to a particular toxin and are further affected by general health condition, age, and sex.

The distribution and fate of materials in the environment depend on their physical characteristics and the processes that transport, alter, destroy, or immobilize them. Uptake of toxins into organisms can result in accumulation in tissues and transfer from one organism to another.

Estimates of health risks for large, diverse populations exposed to very low doses of extremely toxic materials are inexact because of biological variation, experimental error, and the necessity of extrapolating from results with small numbers of laboratory animals. In the end, we are left with unanswered questions. Which are the most dangerous environmental factors that we face? How can we evaluate the hazards of all the natural and synthetic chemicals that now exist? What risks are acceptable? We have not yet solved these problems or answered all the questions raised in this chapter, but it is important that these issues be discussed and considered seriously.

1. What is the difference between toxic and hazardous? Give some examples of materials in each category.
2. What are some of the most serious infectious diseases in the world? How are they transmitted?
3. How do stress, diet, and lifestyle affect environmental health? What diseases are most clearly related to these factors?
4. How do the physical and chemical characteristics of materials affect their movement, persistence, distribution, and fate in the environment?
5. Define LD50. Why is it more accurate than simply reporting toxic dose?
6. What is the difference between acute and chronic toxicity?
7. Define carcinogenic, mutagenic, teratogenic, and neurotoxic.
8. What are irritants, sensitizers, allergens, caustics, acids, and fibrotic agents?
9. How do organisms reduce or avoid the damaging effects of environmental hazards?
10. What are the relative risks of smoking, driving a car, and drinking water with the maximum permissible levels of trichloroethylene? Are these relatively equal risks?

Questions for Critical or Reflective Thinking

1. What consequences (positive or negative) do you think might result from defining health as a state of complete physical, mental, and social well-being? Who might favor or oppose such a definition?
2. How would you feel or act if your child were dying of diarrhea? Why do we spend more money on AIDS or cancer research than childhood illnesses?
3. Some people seem to have a poison paranoia about synthetic chemicals. Why do we tend to assume that natural chemicals are benign while industrial chemicals are evil?
4. Analyze the claim that in our diet we are exposed to thousands of times more natural carcinogens than industrial ones. Is this a good reason to ignore pollution?
5. Describe what is shown in figure 6.7. Find the two countries with similar fat intake that have the greatest difference in deaths from breast cancer. How would you explain this difference?
6. What are the premises in the discussion of assessing risk? Could conflicting conclusions be drawn from the facts presented in this section? What is your perception of risk from your environment?
7. Table 6.5 equates activities such as smoking 1.4 cigarettes, having one chest X ray, and riding ten miles on a bicycle. How was this equation derived? Do you agree with it? Do any items on this list require further clarification?
8. Who were the stakeholders in the saccharine controversy, and what were their interests or biases? Was Congress justified in refusing to ban saccharine? Should soft drink cans have warning labels similar to those on cigarettes?
9. Should pollution levels be set to protect the average person in the population or the most sensitive? Why not have zero exposure to all hazards?
10. What level of risk is acceptable to you? Are there some things for which you would accept more risk than others?

Key Terms

acute effects 126
allergens 117
antigens 117
asphyxiants 117
bioaccumulation 123
biomagnification 123
cancer 118
carcinogens 118
chronic effects 126
disease 112
hazardous 116
health 112

irritants 116
LD50 125
malignant tumor 118
morbidity 113
mutagens 117
neurotoxins 117
promoters 118
respiratory fibrotic agents 116
stress 120
teratogens 117
toxins 116
trauma 120

Albert, A. *Selective Toxicity: The Physico-Chemical Basis of Therapy.* 5th ed. London: Chapman and Hall, 1985. A technical but thorough introduction to toxicology with much interesting and useful information.

Allman, W. October 1985. ''Staying Alive in the Twentieth Century.'' *Science 85* 5, no. 6 (October 1985):31. A highly readable summary of risks and risk perception.

Ames, B., R. Magaw, and L. Gold. ''Ranking Possible Carcinogenic Hazards.'' *Science* 236, no. 4799 (April 17, 1987): 271. Startling new conclusions about natural versus synthetic products.

Campt, D. ''Reducing Dietary Risk.'' *E.P.A. Journal* 16, no. 3 (1990): 18 There is a crisis in public confidence about the safety of food. How can food be made more safe?

Castleman, M. ''Toxics and Male Infertility.'' *Sierra* 70, no. 2 (March/April 1985): 49. Is there a connection?

Efron, E. *The Apocalyptics: Cancer and the Big Lie—How Environmental Politics Controls What We Know about Cancer.* New York: Simon & Schuster, 1984. A blistering attack on the cancer establishment.

Guthrie, F., and J. Perry. *Introduction to Environmental Toxicology.* New York: Elsevier, 1980. A good introductory textbook in toxicology.

Hester, Gordon L. ''Electric and Magnetic Fields: Managing an Uncertain Risk'' *Environment* 34, no. 1 (January/February 1992): 6. A thorough but inconclusive discussion of EMF and health risks.

Hirschhorn, N., and W. B. Greenough III. ''Progress in Oral Rehydration Therapy.'' *Scientific American* 264, no. 5 (1991): 50. Treatment of diarrhea-induced dehydration with simple electrolyte solutions now saves 1 million children a year from death.

National Institute of Occupational Safety and Health. *Registry of Toxic Effects of Chemical Substances.* Lewis, R., and D. Sweet, eds. Washington, D.C.: Government Printing Office, (1985). The official toxic substances list.

Reilly, W. K. ''Why I Propose a National Debate on Risk.'' *E.P.A. Journal* 17, no. 2 (1991): 2. The EPA proposes to weigh the relative importance of different risks in setting environmental policies.

Russell, M., and M. Gruber. ''Risk Assessment in Environmental Policy-Making.'' *Science* 236, no. 4799 (April 17, 1987): 286. How should we measure environmental hazards?

Slovic, P. ''Perception of Risk.'' *Science* 236, no. 4799 (April 17, 1987): 280. An excellent discussion of risk assessment.

Trunkey, D. ''Trauma.'' *Scientific American* 249, no. 2 (August 1983): 28. An excellent overview of accidents and other hazards.

Tschirley, F. ''Dioxin.'' *Scientific American* 254, no. 2 (February 1986): 29. Argues that dioxins may be less dangerous to humans than previously thought.

PART III: THE EARTH AND BIOLOGICAL RESOURCES

▼

*Here, in this rich wilderness, we dreamed
that over the next ridge, beyond the next stream
freedom lay forever.*

Ansel Adams and Nancy Newhall, *This Is the American Earth*

7

FOOD AND AGRICULTURE

The battle to feed all of humanity is over. In the 1970s the world will undergo famines—hundreds of millions of people are going to starve to death in spite of any crash programs embarked upon now. At this late date nothing can prevent a substantial increase in the world death rate, although many lives could be saved through dramatic programs to stretch the carrying capacity of the earth by increasing food production. But these programs will only provide a stay of execution unless they are accompanied by determined and successful efforts at population control.

Paul Ehrlich

Properly managed, the Earth's more fertile lands and its forests could meet everyone's food and wood needs abundantly and indefinitely. The persistent undernourishment of some half a billion people today does not stem from a global scarcity of resources; even as tens of thousands of babies die each day from diseases exacerbated by malnutrition, over one third of the world's grain is fed to livestock to supply the meat-rich diet of the affluent.

Erick Eckholm

Objectives

After studying this chapter, you should be able to:

- Understand the sources and distribution of world hunger.

- See how recent increases in food supply have occurred and what the prospects are for future increases.

- Analyze the role of cash crops and food trade policies in world hunger.

- Discuss the causes and consequences of soil erosion, nutrient depletion, waterlogging, salinization, and other abuses that decrease soil fertility and crop production.

- Describe the agricultural inputs necessary for sustained food production.

- Evaluate pesticide benefits, dangers, and uses.

- Explain techniques of soil conservation and low-input, sustainable agriculture.

INTRODUCTION

We live in perplexing times. In the past forty years, world food supplies have increased at spectacular rates. Between 1950 and 1990, world agricultural output rose more than 2.5-fold. Food supplies have risen faster than population on every continent except Africa, yet there are more chronically undernourished people now—both in total numbers and as a percentage of the world population—than there were in 1950. Many countries that have had the biggest gains in food production also have the greatest numbers and largest percentages of malnourished people. Some places that have the most heart-rending scenes of starvation and wasted human potential also grow billions of dollars worth of luxury foods to be exported to the industrialized countries of Europe and North America.

Even the richest countries have high numbers of hungry people (fig. 7.1). The United States, which pays farmers around $25 billion a year not to grow crops and to store mountains of surplus food, has at least 25 million undernourished people, mostly children. How can this occur when we have an embarrassing surplus of food? If we cannot feed everyone now, in this time of relative plenty, what will happen in the future if populations continue to grow and resources become more limited?

Fertile, tillable soil for growing crops is an indispensable resource for agricultural crop production, but erosion, soil degradation, and depletion of essential plant nutrients reduce our ability to grow the food we need. The net effect of these poor agricultural practices is equivalent to the loss of 15 million hectares (37 million acres), or 1 percent of the world's croplands, each year. Many alternative methods of low-input, sustainable agriculture—some new and some old—offer safer, more healthful, and more ecologically sound farming than our current practices.

Figure 7.1 Even in the United States, a land of plenty, some people are unable to feed themselves. Soup kitchens and relief shelters, such as this one, are often inadequate to meet the need for food.

This chapter will explore both the sources of hunger and what we might do to make agriculture and food distribution systems just, fair, and sustainable.

WORLD HUNGER

The Food and Agricultural Organization (FAO) of the United Nations estimates that at least 750 million people (15 percent of the total world population) have less than the minimum caloric intake necessary to sustain a productive working life. Nearly half of those people have insufficient food to forestall serious health risks or to provide normal growth and development for their children. Some 18 to 20 million people (three quarters of them children) starve to death or die of diseases aggravated by malnourishment each year (fig. 7.2).

Famines

The images that probably first come to your mind when you think about food shortages are the tragic scenes of mass starvation that come from places like Ethiopia, Somalia, Sudan, or Mozambique. These acute shortages, or **famines,** are characterized by large-scale loss of life, social disruption, and economic chaos. Starving people eat their seed grain and slaughter their breeding stock in a desperate attempt to keep themselves and their families alive. Even if better conditions return, they have sacrificed their productive capacity and will take a long time to recover. Famines are characterized by mass migrations as starving people travel to refugee camps in search of food and medical care. Many die on the way or fall prey to robbers.

What causes these terrible tragedies? Environmental conditions are usually the immediate trigger, but poli-

Figure 7.2 Floods, droughts, insects, and other natural disasters are usually the immediate cause of famine; however, politics, economics, and environmental mismanagement are often contributing or underlying causes. Some of the countries with the largest numbers of starving people export cash crops.

tics and economics are often equally important in preventing people from getting the food they need. Adverse weather, insect infestations, and other natural disasters cause crop failures and create food shortages. But these factors have generally been around for a long time, and local people usually have adaptations that get them through hard times if they are allowed to follow traditional patterns of migration and farming. Arbitrary political boundaries, however, along with wars and land seizures by the rich and powerful, block access to areas that once served as refuge during droughts, floods, and other natural disasters. Poor people can neither grow their own food nor find jobs to earn money to buy the food they need.

The African nations that suffered the most widespread and long-lasting famines in recent years are Somalia, Mozambique, Sudan, Ethiopia, and Chad, all of which suffered not only severe drought but also continuing wars. Neighboring countries that experienced comparable weather did not have as severe problems as these unfortunate countries did. The worst famine in the twentieth century and probably the worst famine in human history occurred in China in the late 1960s as a result of adverse weather and poor political decisions on land use and farming (box 7.1).

Feeding a Fifth of the World

*T*wenty years ago, many people were predicting that China would never be able to feed all its people. They said that conditions were hopeless, that "bombs would be kinder than food aid," because the people were going to starve to death anyway. China was one of several countries, including India, Egypt, and Libya, categorized as "can't be helped" by those who believed that population growth had already outstripped the carrying capacity of the land.

In fact, the situation in China did seem impossible. In his 1959 "Great Leap Forward," Chairman Mao Zedong ordered collectivization of farms, clearing of forests, planting of all land to produce food—regardless of its potential for farming—as well as standardization of crops, planting schedules, and agricultural planning. Bureaucratic bungling disrupted planting and harvesting. Peasants who no longer controlled their land nor benefited from working on it resisted the changes. Farm production fell disastrously.

Between 1959 and 1961, China suffered the greatest famine in the world's history. It was kept secret from the western world, and even today we don't know the full extent of the disaster. Bad weather played a part. Droughts in some parts of the country and floods elsewhere ruined crops. Some areas had food but couldn't send it to market because of government policies. Farmers were ordered to plant the wrong crops at the wrong time. Crops were planted on steep hillsides that quickly eroded, smothering good bottomland in mud and silt. Estimates of the number who died range from 10 to 60 million, but most experts agree that a realistic number is 30 million deaths. There surely have never been so many untimely deaths in a single country in so short a time.

Amazingly, just twenty years later, the whole country reversed its course just as suddenly. After Mao's death, Chairman Deng broke up the communes and returned farms to family units. Chinese researchers succeeded in adapting high-yielding rice strains for growth in the cooler climate of the North China Plain. A modified market-incentive system was reestablished, and the results were amazing. Between 1975 and 1985, the total rice yield more than doubled—from 80 to 180 million metric tons per year. Almost half of that increase occurred in just four years, from 1980 to 1984. China is now number one in both rice *and* wheat production.

Nearly all of China's 800 million farmers now cultivate small family plots, getting 50 percent more crops per hectare than did the communes (box fig. 7.1.1). Each crop requires an average of sixty worker days, compared to 250 to 320 commune worker days a year to produce comparable yields. In addition to a diversified mix of crops, farmers are producing commodities such as fish, chickens, pigs, ducks, and rabbits. An ample and varied diet has replaced thin rice porridge as standard fare, and meat consumption has more than quadrupled.

China now has enough food for everyone in the country to have an adequate, if limited, diet. Fewer than 3 percent of the Chinese people are underfed, a lower percentage than in the United States, where about 10 percent of the population goes hungry. China has even begun to export grain to other countries. The country that was thought to be a living example of Malthus's dismal predictions that population would inevitably outstrip food production now seems to have enough food to feed everyone.

Figure 7.1.1 Recent decollectivization and a return to traditional farming practices have more than doubled China's farm yields and improved nutrition.

Chronic Food Shortages

Although we most often think of hunger in terms of catastrophic famine and starvation, more people in the world are affected by **chronic food shortages** than by acute famines. About 1 million people are thought to have died in sub-Saharan Africa in the early 1980s as a result of the drought-related famine. By contrast, approximately 18 to 20 million people—mostly children—die *every year* from the effects of chronic undernutrition and malnutrition. How can this tragic situation persist in the modern world?

Although world food production has increased dramatically over the last forty years and the average per capita daily calorie intake is slightly above the minimum recommended for good health, there are still many hungry people in the world. While North America and Europe have more food than we need or can use, sub-Saharan Africa has only about 80 percent, on average, of the minimum daily caloric requirement. India has the largest number of hungry residents—at least 300 million people, or about one-third of its population. Bangladesh probably has the highest proportion of hungry people; 80 percent of its population has less than the minimum daily caloric intake.

NUTRITIONAL REQUIREMENTS

For many years Americans were advised to eat daily servings of four major food groups: meat, dairy products, grains, and fruits and vegetables. In 1992, after much controversy, the U.S. Department of Agriculture revised these recommendations (fig. 7.3), suggesting that we eat more grains, fruits, and vegetables while decreasing our consumption of meat, milk, fats, and oils. What nutritional needs do these food groups meet that are lacking in the diets of poor people? In this section we will look at some basic nutritional requirements as well as a few serious nutritional deficiencies.

Energy Needs

The amount of energy each of us needs to remain strong and healthy depends on body weight, climate, state of health, stress level, and basic metabolism. A small, sedentary adult living in a warm climate might require less than 2,000 calories per day to remain healthy, whereas a large, muscular person living a vigorous outdoor life in a cold climate might need 6,000 to 7,000 calories per day to stay warm and active. The FAO estimates that the average minimum daily caloric intake over the whole world is about 2,500 calories per day.

People who receive less than 90 percent of their minimum dietary intake on a long-term basis are considered **undernourished.** While not starving to death, they tend not to have enough energy for an active, productive life. Lack of energy and nutrients also tends to make them more susceptible to infectious diseases. Poor diet and poverty create a vicious cycle. People who are weak or sick because of poor diet can't work; without an adequate income, they can't afford good food. Table 7.1 shows the per capita daily caloric intake for the major geographical regions of the world. The *average* world caloric intake is now above 2,600 calories per day, but at least one billion people—one-fifth of the world's population—have less than this amount.

Those who receive less than 80 percent of their minimum daily caloric requirements are considered

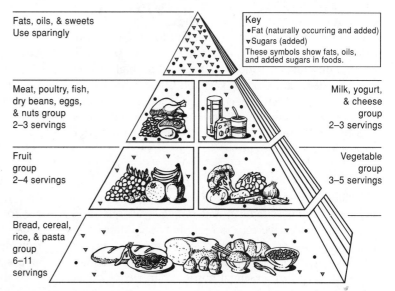

Food Guide Pyramid
A Guide to Daily Food Choices

Fats, oils, & sweets
Use sparingly

Key
• Fat (naturally occurring and added)
▼ Sugars (added)
These symbols show fats, oils, and added sugars in foods.

Meat, poultry, fish, dry beans, eggs, & nuts group 2–3 servings

Milk, yogurt, & cheese group 2–3 servings

Fruit group 2–4 servings

Vegetable group 3–5 servings

Bread, cereal, rice, & pasta group 6–11 servings

Figure 7.3 The food pyramid recommended by the U.S. Department of Agriculture. You should eat two to four times as much bread, cereal, rice, and pasta as milk, meat, eggs, or nuts. Fats, oils, and sweets should be eaten sparingly, if at all.

TABLE 7.1 Food available for human consumption

| | Calories per person per day | |
	1961–1963	1984–1986
World Total	2,300	2,690
Africa (sub-Sahara)	2,040	2,060
North Africa/Near East	2,240	3,050
Asia	1,830	2,430
Latin America	2,380	2,700
North America	3,180	3,620
Western Europe	3,090	3,380
Eastern Europe	3,140	3,410

Source: Data from World Resources 1990–1991.

seriously undernourished. Children who are seriously undernourished are likely to suffer from permanently stunted growth, mental retardation, and other social and developmental disorders. Infectious diseases that are only an inconvenience for well-fed individuals become lethal threats to those who are poorly nourished. Diarrhea rarely kills a well-fed person, but children weakened by nutritional deficiencies are highly susceptible to this and a host of other diseases. One child in four in the developing world—10 million to 12 million children per year—dies of diseases that could be prevented with a better diet, clean water, and simple medicines. One child dies every two seconds from problems caused by lack of food.

In the richer countries of the world, the most common dietary problem is overnutrition from getting too many calories. The average daily caloric intake in North America and Europe is above 3,500 calories, nearly one-third more than is needed for adequate nutrition. At least 20 percent of Americans are seriously overweight. Overnutrition contributes to high blood pressure, heart attacks, strokes, and other cardiovascular diseases that have become the leading causes of death in most developed countries since infectious diseases have been reduced or controlled by better sanitation and health care.

Nutritional Needs

In addition to calories, we need specific nutrients, such as proteins, vitamins, and minerals, in our diet. It is possible to have excess food and still suffer from **malnourishment,** a nutritional imbalance caused by a lack of specific dietary components or an inability to absorb or utilize essential nutrients. In richer countries, people often eat too much meat, salt, and fat and too little fiber, vitamins, trace minerals, and other components lost from highly processed foods. In poorer countries, people often lack specific nutrients because they cannot afford more expensive foods such as meat, fruits, and vegetables that would provide a balanced diet. Let's look now at some essential dietary requirements along with some important types of malnutrition.

Proteins

Proteins make up most of the metabolic machinery and cellular structure essential to life. The average adult human needs about 40 g (1.5 oz) of protein per day. Pregnant women, growing children, and adolescents need up to twice as much to create new cells and build growing tissues. Not only is the total *amount* of protein crucial, but the *quality* of that protein also is of vital importance. We depend on our diet to supply us with the ten essential amino acids we cannot make for ourselves. Furthermore, these are not interchangeable. They have to be present in a balanced ratio in order for us to synthesize functional proteins. Corn, for instance, has a reasonably high protein content but is generally low in lysine and arginine. Adding beans, peas, or milk to a corn-based diet can help balance the amino acid content.

The two most widespread human protein deficiency diseases are kwashiorkor and marasmus. **Marasmus** (from the Greek word *marasmos* meaning "to waste away") is caused by a diet low in both calories and protein. A child suffering from severe marasmus is generally thin and shriveled like a tiny, very old starving person. **Kwashiorkor** is a West African word meaning "displaced child." It occurs when people eat a starchy diet that is low in protein or has

poor-quality protein, even though it may have plenty of calories. Children with kwashiorkor often have reddish-orange hair and puffy, discolored skin and a bloated belly (fig. 7.4). They become anemic and listless and have low resistance to even the mildest diseases and infections. Even if they survive childhood diseases, they are likely to suffer from stunted growth, mental retardation, and other developmental problems. Providing quality dietary protein is one of the world's greatest problems.

Minerals

We generally need only small amounts of most minerals, but deficiencies can have serious health effects. The most common mineral deficiencies worldwide are for calcium, iodine, and iron. Calcium deficiency causes irritability, muscle cramps, and bone defects. Iron deficiency leads to anemia (low levels of hemoglobin in the blood), which more often is caused by an inability to absorb iron from food than from a lack of iron in the diet.

The main symptoms of iodine deficiency are goiter (swollen thyroid glands in the neck) and hypothyroidism (listlessness and other metabolic symptoms due to low thyroid hormone levels). Hypothyroidism in early childhood can cause developmental abnormalities such as mental retardation and

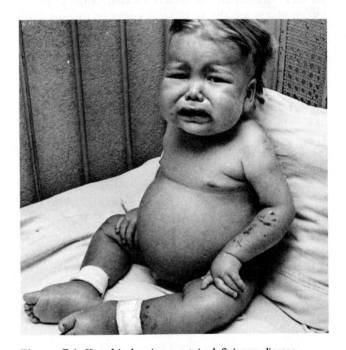

Figure 7.4 Kwashiorkor is a protein-deficiency disease resulting from a diet low in protein or a diet lacking the high-quality protein necessary for normal growth and development. A bloated abdomen, puffy skin, and reddish-orange bleached hair are characteristic of this disease. Children are often listless, apathetic, and very susceptible to infectious diseases. Persistent kwashiorkor can result in mental and developmental retardation or death.

deaf-mutism. It is estimated that 180 million people worldwide have symptoms of iodine deficiency and that 3 million suffer from cretinism (severe mental deficiency), mostly in South and Southeast Asia. In some villages, 15 to 20 percent of the children are brain damaged due to iodine deficiency. Since the human body needs only a teaspoonful of iodine in a whole lifetime, this problem is both technically simple and inexpensive to solve. Adding potassium iodide to salt costs only a few cents per year per person and is highly successful in preventing deficiencies.

Vitamins

Vitamins are organic molecules essential for life (the Latin *vita* = life), but we cannot make them for ourselves and must get them from our diet. We usually require only minute amounts (milligrams per day) of vitamins and get all we need from a varied diet of fruits, vegetables, whole grains, and dairy products. A deficiency of tryptophan and niacin results in **pellagra,** the symptoms of which include lassitude, torpor, dermatitis, diarrhea, dementia, and death. Pellagra used to be very common in the southern United States, where poor people subsisted mainly on corn. It still is tragically common in parts of India and Africa where jowah (a sorghum species) is the only food available to very poor people. In a strict vegetarian diet, vitamin B$_{12}$ is usually lacking. Animal tissues and products are the ordinary sources of this essential dietary ingredient, but it is also present in tempeh, a cultured soybean food from Indonesia.

Vitamin A deficiency causes xerophthalmia (literally, dry eyes) and retinal degeneration, especially in children. More than five hundred thousand children in less-developed countries lose their sight every year because of these diseases. Within a few weeks of becoming blind, 60 to 70 percent of these children die. An additional 6 to 7 million children show signs of moderate vitamin A deficiency and, therefore, are more vulnerable to infectious diseases.

Other vitamin deficiency diseases such as scurvy, beriberi, anemia, and rickets are still a problem for people in countries where diets are limited to starchy foods such as manioc, polished rice, maize, or wheat noodles. Enrichment of flour and milk with vitamins A and D has largely eliminated deficiency problems in developed countries. In fact, in richer countries there are now concerns about excess vitamins in our diets.

WORLD FOOD RESOURCES

Of the thousands of edible plants and animals in the world, only about a dozen types of seeds and grains, a few root crops, twenty or so common fruits and vegetables, six mammals, two domestic fowl, and a few fish and other forms of marine life make up almost all the food humans eat.

TABLE 7.2 Some important food crops

Crop	1986 Yield (million metric tons)
Wheat	522
Rice	469
Maize (corn)	449
Potatoes	290
Barley and oats	178
Cassava and sweet potato	140
Sugar (cane and beet)	100
Pulses (legumes)	96
Sorghum and millet	72
Vegetable oils	47
Vegetables and fruits	347
Meat and milk	140
Fish and seafood	70

Major Crops

The three crops on which humanity depends for the majority of its nutrients and calories are wheat, rice, and maize. Together, about 1,440 million metric tons of these three crops are grown each year, roughly half of all agricultural crops (table 7.2). Wheat and rice are especially important since they are the staple foods for most of the 4 billion people in the developing countries of the world. These two grass species supply around 60 percent of the calories consumed directly by humans.

Potatoes, barley, oats, and rye are staples in mountainous regions and high latitudes (northern Europe, north Asia) because they grow well in cool, moist climates. Cassava, sweet potatoes, and other roots and tubers grow well in warm, wet areas and are staples in Amazonia, Africa, Melanesia, and the South Pacific. Sorghum and millet are drought resistant and are staples in the dry regions of Africa.

Meat and Milk

Meat and milk are highly prized by people nearly everywhere, but their distribution is highly inequitable. Although the industrialized, more highly developed countries of North America, Europe, and Japan make up only 20 percent of the total population, they consume 80 percent of all meat and milk in the world. The 80 percent of the world's people in less-developed countries raise 60 percent of the 3 billion domestic ruminants and 6 billion poultry in the world but consume only 20 percent of all animal products. The grazing lands that support many of these domestic animals are discussed further in chapter 8.

The United States is the largest producer of maize, growing about 8 billion bushels (nearly 200 million metric tons) or slightly less than half of the world total. Most of the grain we produce is not consumed directly by humans. About 90 percent is used to feed dairy and beef cattle, hogs, poultry, and other animals. It is used especially to fatten beef cattle during the last three months before they are slaughtered. As chapter 2 shows, there is a great loss of energy with each step up the food chain. This means we could feed far more people if we ate more grain directly rather than feeding it to livestock. Every 16 kg of grain and soybeans fed to beef cattle in feedlots produces 1 kg of edible meat. The other 15 kg are used by the animal for energy or body parts we do not eat, or it is eliminated. If we were to eat the grain directly, we would get 21 times more calories and 8 times more protein than we get by eating the meat it produces. Hogs and poultry are about 2 and 4 times as efficient, respectively, as cattle in converting feed to edible meat.

Fish and Seafood

Fish and seafood contribute about 70 million metric tons of high-quality protein to the world's diet, about half as much as that from land animals. This is an important source of protein in many countries, contributing up to one-half of the animal protein and one-fourth of the total dietary protein in Japan, for instance.

There are indications that we have already surpassed the sustainable harvest of fish from most of the world's oceans. Our current fishing methods are wasteful, destroying habitat and young fish by indiscriminate trawling. Several tons of unwanted species are discarded for every ton of seafood brought to market. Much of what is discarded could be eaten, but it is the wrong size, too young, out of season, or a variety that we do not like to eat. Changing a name from dogfish or hagfish to ocean perch or sea trout may make an unwanted species acceptable. Tragically, much of the catch that is tossed back into the sea is too badly injured by decompression or being crushed in the net to survive.

What are our alternatives? Obviously, we could use more selective harvest methods and less selective eating preferences. There also is great potential for raising fish and crustaceans in ponds or in estuaries. Under controlled conditions, even environmentally sensitive fish like trout can be raised in high-density ponds (fig. 7.5). Genetic engineering techniques are being used to breed "superfish" that will grow faster and increase yields in the way that the "green revolution" has done for plants.

Increases in World Food Production

Contrary to what many pessimists predicted (see chapter 4), total crop yields have grown faster than

Figure 7.5 Aquaculture—growing fish and other edible species in confined ponds—can be highly productive and add a valuable source of high-quality protein to our diets, but it can also cause environmental problems in natural water bodies.

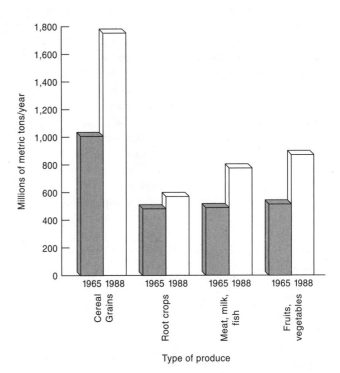

Figure 7.6 World crop production 1965–1988 in millions of metric tons per year. Note that except for root crops, increases have been approximately 70 percent over this twenty-three-year period.
Data from World Resources Institute.

human populations over the past two centuries. Whether we can sustain these advances in food output and tolerate the environmental costs associated with them are difficult questions, however.

Our successes in increased food production are due to a variety of factors including greater use of irrigation, fertilizers, and pesticides; expanded croplands; and new high-yielding crop varieties. Figure 7.6

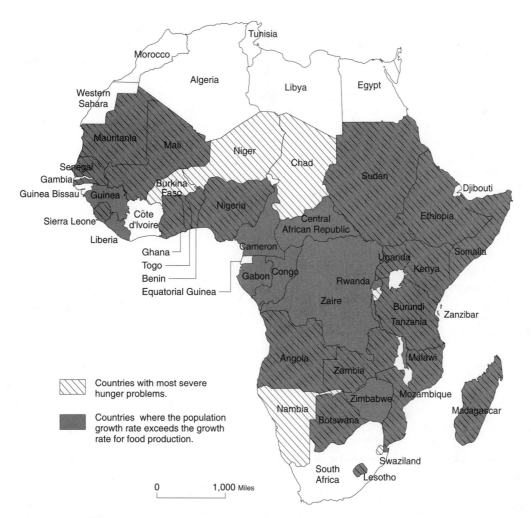

Figure 7.7 Thirty-one of forty-one countries in sub-Saharan Africa have population growth rates that exceed their growth rate of food production. Food shortages have been common in a broad, horseshoe-shaped band, as can be seen on the map.

Source: The World Bank.

Legend:
- Countries with most severe hunger problems.
- Countries where the population growth rate exceeds the growth rate for food production.

0   1,000 Miles

shows gains in major food categories from 1965 to 1988. World grain production grew nearly 70 percent during that time, and the developing countries of the world generally doubled their output. Pulses (legumes), fruits, melons, vegetables, and oil production also increased by about 70 percent, and meat, milk, and fish supplies grew by nearly 60 percent.

These improvements in world food production mean that everyone now could have 2,600 calories per day if food were equally divided. Unfortunately, food supplies are not equitably distributed. Even countries with huge food surpluses such as the United States have large numbers of citizens who are undernourished or malnourished. Some areas of the world have especially severe food shortages. The conditions in sub-Saharan Africa are of great concern. Although the total amount of food produced there has doubled over the past thirty years, population has grown even faster, so food available per person remains below acceptable levels in thirty-one out of forty-one countries in the area (fig. 7.7).

International Food Trade

One way to even out food distribution between areas with excess and those with shortages is through world food trade. International food shipments have risen dramatically in the past fifty years. In 1939, only ten countries imported more than 1 million tons of food. In 1980, this number had risen to forty countries. More importantly, world food trade now accounts for more than 20 percent of all food consumed. The world is becoming more and more interdependent.

World trade in agricultural commodities has brought both benefits and problems for farmers and planners in many countries besides the United States. On one hand, international trade makes up the food deficit in many countries, staving off famine and social disruption. On the other hand, nations that become dependent on imported food supplies may divert so much of their foreign earnings to buying expensive imports that they can

Food and Agriculture   141

never develop indigenous resources. They also may become politically indentured to the countries that sell them food.

Food-producing countries also can become captive to international trade. Often a country finds it expedient to specialize in a few crops that fit its climate and work force. It then can trade some of its surplus for other basic foods that are grown more efficiently elsewhere. The foreign exchange generated by these **cash crops** can provide funds necessary to buy tools, information, and technology needed for development. Too often, however, land for growing these export crops is taken out of staple food production for local consumption. This necessitates importing foreign food and drives up local food prices.

Often, the export crops grown by developing countries are not food supplies for needy neighbor nations but rather luxury foods, drugs, and other nonessential items to be shipped to the rich countries of the world (fig. 7.8). Little of the money generated by these crops trickles down to the underclass of the producing countries. Most of it goes to landowners, money lenders, export brokers, or political leaders. These export-oriented policies contribute significantly to malnutrition problems in many countries. In Guatemala, for example, 97 percent of the citrus crop is exported, while a majority of the local population suffers from vitamin C deficiency. In Central America, beef production and export increased nearly sixfold in the 1960s and 1970s, while per capita meat consumption fell by 50 percent.

International **food aid** by the more developed countries is often selective and highly political. We give aid to governments that support our policies rather than to the countries with the most hungry people. In 1990, for instance, Egypt received $5.6 billion in development aid—more than twice as much as any other country—while Israel and Jordan received by far the highest per capita aid, nearly $300 per person. In contrast, India and China received less than $2 per person, while the average for all poorer countries was less than $10 per capita. Much of the food aid we do give hurts rather than helps, because food prices are driven down when markets are flooded with imported commodities. This makes production by local farmers uneconomical and reduces indigenous food production, trapping some countries in permanent dependence on welfare. It also often helps support unpopular and repressive governments.

SOIL RESOURCES

Maintaining adequate food production to feed the world depends on well-maintained farms and healthy soil (fig. 7.9). Present agricultural practices, however, are depleting soil resources in many parts of the world at unsustainable rates. How can we provide an adequate diet for everyone and also preserve this essential resource? This section will examine the characteristics of soil, as well as how we use and abuse it.

Soil, A Renewable Resource

Soil is a marvelous substance, a resource of astonishing beauty, complexity, and frailty. It is a complex mixture of weathered mineral materials from rocks, partially decomposed organic molecules, and a host of living organisms. It can be considered an ecosystem by itself.

With careful husbandry, soil can be replenished and renewed indefinitely. Many modern farming techniques deplete soil nutrients, however, and expose

Figure 7.8 Cash crops, such as these coffee beans being picked in Colombia, often replace production of staples necessary to feed the local population of developing countries. Government policies encourage export commodities to raise foreign exchange.

Figure 7.9 Good soil is both a gift and a responsibility. With work, care, and understanding, we can use it productively and pass it on to the next generation more fertile than we received it.

Figure 7.10 Soil ecosystems include numerous consumer organisms, as depicted here: (1) snail (2) termite (3) constricting fungus killing nematodes (4) earthworm (5) wood roach (6) centipede (7) carabid beetle (8) slug (9) soil fungus (10) wireworm (11) soil protozoans (12) earthworm (13) sow bug (14) ants (15) mite (16) springtail (17) pseudoscorpion (18) cicada nymph.

the soil to the erosive forces of wind and moving water. As a result, we are essentially mining this resource and using it much faster than it is being replaced.

Building good soil is a slow process. Under the best circumstances, good topsoil accumulates at a rate of about 10 tons per hectare (2.5 acres) per year—enough soil to make a layer about 1 mm deep when spread over a hectare. Under poor conditions, it can take thousands of years to build that much soil. Perhaps one-third to one-half of the world's current croplands are losing topsoil faster than it is being replaced. In some of the worst spots, erosion carries away about 2.5 cm (1 in) of topsoil per year. With losses like that, agricultural production has already begun to fall in many areas.

Soil Organisms

Without soil organisms, the earth would be covered with sterile mineral particles far different from the rich, living soil ecosystems on which we depend for most of our food. The activity of the myriad organisms living in the soil (fig. 7.10) creates its structure, fertility, and tilth (structure suitable for tilling or cultivation). Algae, bacteria, and fungi flourish in the top few centimeters of soil. A single gram of soil (about a teaspoonful) can contain hundreds of millions of these microscopic cells. Bacteria and fungi decompose organic detritus and recycle nutrients that plants can use for additional growth.

The roots of most higher plants form symbiotic relationships with certain specific kinds of fungi called mycorrhizae (from the Greek words for fungus and roots). These fungi aid in the uptake of phosphorus, zinc, copper, and probably other nutrients from the soil and are vitally important for plant growth. Destruction of the mycorrhizal and bacterial flora by poor soil management can inhibit higher-plant growth.

Roundworms, segmented worms, mites, and tiny insects can swarm by the thousands in that same gram of soil. Burrowing animals, such as gophers, moles, insect larvae, and worms, tunnel deeper in the soil, mixing and aerating it. Plant roots also penetrate lower soil levels, drawing up soluble minerals and secreting acids that decompose mineral particles. Fallen plant litter adds new organic material to the soil, returning nutrients to be recycled.

Soil Profiles

Most soils are stratified into horizontal layers called **soil horizons** that reveal much about the history and usefulness of the soil. The thickness, color, texture, and composition of each horizon are used to classify the soil. A cross-sectional view of the horizons in a soil is called a soil profile. Figure 7.11 shows the series of horizons generally seen in a soil profile. Soil scientists give each horizon a letter name. Not all soils have all of these horizons; one or more may be missing, depending on the soil type and history of a specific area.

The upper soil layer, called **topsoil** (or A horizon), ranges from a thickness of 1 meter or more under virgin prairie to zero in some deserts. Topsoil contains most of the living organisms and organic material in the soil, and

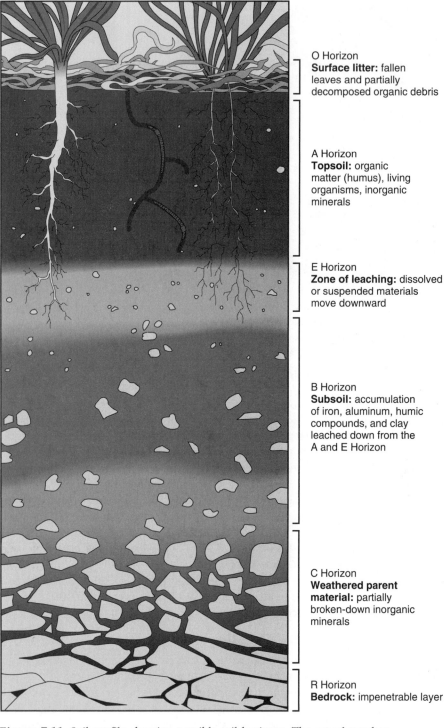

O Horizon
Surface litter: fallen leaves and partially decomposed organic debris

A Horizon
Topsoil: organic matter (humus), living organisms, inorganic minerals

E Horizon
Zone of leaching: dissolved or suspended materials move downward

B Horizon
Subsoil: accumulation of iron, aluminum, humic compounds, and clay leached down from the A and E Horizon

C Horizon
Weathered parent material: partially broken-down inorganic minerals

R Horizon
Bedrock: impenetrable layer

Figure 7.11 Soil profile showing possible soil horizons. The actual number, composition, and thickness of these layers varies in different soil types.

it is in this layer that most plants spread their roots to absorb water and nutrients.

Beneath the topsoil is the **subsoil** (or B horizon), which usually has a lower organic content and higher concentrations of fine mineral particles. Soluble compounds and clay particles carried by water percolating down from the layers above often accumulate in the subsoil. Subsoil particles can become cemented to-gether to form hardpan, a dense, impermeable layer that blocks plant root growth and prevents water from draining properly.

Beneath the subsoil is the parent material (C horizon), made of relatively undecomposed mineral particles and un-weathered rock fragments with very little organic material. Weathering of this layer pro-duces new soil particles for the layers above. About 97 percent of all the parent material in the United States was transported to its present site by geologic forces (glaciers, wind, and water) and is not directly re-lated to the bedrock below it.

Erosion and Land Degradation

The FAO of the United Nations estimates that total world crop-land losses amount to an area equal in size to the United States and Mexico combined (11 mil-lion hectares or 27 million acres) *every year.* Conversion to non-agricultural uses—urbanization, highways, industrial sites, strip-mining, abandonment of mar-ginal farmland—is responsible for one-fourth of that loss. Toxi-fication by hazardous wastes, chemical spills, salinization, alka-linization, and waterlogging of ir-rigated lands, misapplication of pesticides, and deposition of at-mospheric pollutants degrade about 2 million hectares (5 mil-lion acres) each year. Finally, around 6 million hectares of cropland are rendered unproduc-tive each year by **erosion**: soil removal by the abrasive forces of wind, water, or ice.

In some places, erosion occurs so rapidly that any-one can see it happen (fig. 7.12). Often, however, erosion is more subtle. It can be a creeping disaster that occurs in small increments. A thin layer of top-soil is washed off fields year after year until eventually nothing is left but poor-quality subsoil that requires more and more fertilizer and water to produce any crop at all.

Figure 7.12 Erosion caused by water flowing across an unprotected field. Cover crops, crop residue, and terracing are all effective means of reducing this problem.

The net effect, worldwide, of this general, widespread topsoil erosion is a reduction in crop production equivalent to removing about 1 percent of world cropland each year. Many farmers are able to compensate for this loss by applying more fertilizer and by bringing new land into cultivation. Continuation of current erosion rates, however, could reduce agricultural production by 25 percent in Central America and Africa and 20 percent in South America by the year 2000. The total annual soil loss from croplands is thought to be 25 billion metric tons. About twice that much soil is lost from rangelands, forests, and urban construction sites each year.

In addition to reduced land fertility, this erosion results in sediment-loading of rivers and lakes, siltation of reservoirs, smothering of wetlands and coral reefs, and clogging of water intakes and waterpower turbines. It makes rivers unnavigable, increases the destructiveness and frequency of floods, and causes gullying that turns good lands into useless wastelands.

Erosion in the United States

The total amount of soil lost to erosion in the United States appears to be the highest in the world. The United States Department of Agriculture reported in 1990 that about 40 percent of all U.S. farmland is eroding at rates that reduce long-term productivity (fig. 7.13). About 2.5 billion metric tons of soil wash away or are blown from cultivated croplands in this country each year, and 3.3 billion metric tons are lost from forests, pastures, stream banks, and construction sites. Altogether, that's enough soil to fill about 50 million boxcars. Imagine 500,000 trains, each 100 cars long, carrying topsoil away from fields and forests and dumping it into rivers, lakes, and oceans. A frequently quoted estimate is that two bushels of soil are lost from Iowa farmland for every bushel of corn produced. We are essentially mining the soil to produce crops.

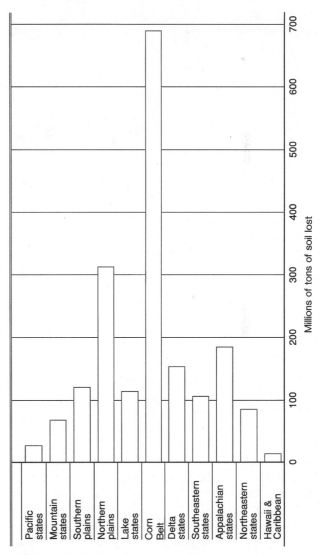

Figure 7.13 Estimated total annual soil erosion on cropland, by region. Although the rate of erosion on farmland is not as high per acre in the Corn Belt as in the border states, there is so much more land under cultivation (95 percent of all land in Iowa is cropland by our estimate) that the total impact is higher there.

Source: U.S. Department of Agriculture.

Figure 7.14 Father and sons walking in the face of a dust storm, Cimarron County, Oklahoma in 1936. This farm has since recovered and is now fertile and productive.

When European settlers moved onto the American prairies in the nineteenth century, the topsoil depth averaged 25 cm (10 in). In a century of farming, we have lost half of the topsoil, on average, and perhaps two-thirds of the soil carbon that had been built up under forests and grasslands in the United States. In some places, there is no longer *any* topsoil left. Farmers plant their crops in rocky subsoil. Even with increasing amounts of expensive fertilizer and water, yields are getting smaller and smaller.

When Hugh Bennett founded the Soil Conservation Service in 1935 during the dust bowl era (fig. 7.14), the total soil loss each year was about 3 billion tons. It is now nearly twice that much. The U.S. General Accounting Office says that 84 percent of U.S. farms have soil losses greater than 5 tons per acre (11 metric tons per hectare). Five tons per acre is generally considered the soil tolerance level, or the rate at which the ecosystem can replace soil.

Intensive farming practices are largely responsible for this situation. Row crops, such as corn and soybeans, leave soil exposed for much of the growing season. Deep plowing and heavy herbicide applications create weed-free fields that look neat but are subject to erosion. Many farmers now plow in the fall to save a few days at spring planting time, letting fields lie bare and exposed to high winds all winter.

Because big machines cannot easily follow contours, they often go straight up and down the hills, creating ready-made gullies for water to follow. Farmers sometimes plow through grass-lined watercourses and have pulled out windbreaks and fencerows to accommodate the large machines and to get every last meter into production. Consequently, wind and water carry away the topsoil.

Pressed by economic conditions, most farmers have abandoned traditional crop rotation patterns and the custom of resting land as pasture or fallow every few years. Continuous monoculture cropping can increase soil loss tenfold over other farming patterns. A soil study in Iowa showed that a three-year rotation of corn, wheat, and clover lost an average of only 2.7 short tons per acre (6 metric tons per hectare). By comparison, continuous wheat production on the same land caused nearly four times as much erosion and continuous corn cropping resulted in seven times as much soil loss as the rotation with wheat and clover.

Erosion in Other Countries

Data on soil condition and soil erosion in other countries are less complete and less easily accessible than U.S. data, but it is evident that many places have serious soil erosion problems. One way of estimating soil loss rates is to look at sediment loads in rivers (table 7.3). The Yellow (Huang) River, for example, drains a large area of loess (wind-blown silt) deposits on the North China Plain that once was covered by forest and grassland. The forests were cut down and the grasslands were converted to cropland (see box 7.1). This plateau is now scarred by gullies up to 40 m (130 ft) deep, and the soil loss is thought to be at least 480 metric tons per hectare per year. This would be equivalent to 3 cm (1.2 in) of topsoil per year. During the spring runoff, when erosion is at its greatest, sediment in the Yellow River often makes up 50 percent of the river volume. Technically, the river is classified as liquid mud at this point.

Perhaps the worst erosion problem in the world, per hectare of farmland, is in Ethiopia. Although Ethiopia has only 1/100 as much cropland in cultivation as the United States, it is thought to lose 2 billion metric tons of soil each year to erosion. This high rate of erosion is both a cause and consequence of famine, poverty, and continued social unrest in that country.

Haiti is another country with severely degraded soil. Once covered with lush tropical forest, the land has been denuded for firewood and cropland. Erosion has been so bad that some experts now say the country has absolutely no topsoil left; poor peasant farmers have difficulty raising any crops at all, and overcrowded boatloads make desperate attempts to migrate to the United States. Economist

TABLE 7.3 *Estimated annual soil erosion in selected river basins*

River name	Principal country	Estimated annual soil erosion (tons/hectare)
Congo	Zaire	3
Nile	Egypt, Sudan	8
Mississippi	United States	10
Amazon	Brazil	13
Mekong	Laos, Kampuchea, Vietnam	43
Irrawaddy	Burma	139
Ganges	India, Nepal	270
Huang (Yellow)	China	479

Source: World Resources Institute, 1987.

Lester Brown of Worldwatch Institute warns that Haiti may never recover from this ecodisaster.

OTHER AGRICULTURAL RESOURCES

Soil is only part of the agricultural resource picture. Agriculture is also dependent upon water, nutrients, and favorable climates to grow crops, and upon mechanical energy to tend and harvest them.

Water

All plants need water to grow. Agriculture accounts for the largest single share of global water use. About three-quarters of all fresh water withdrawn from rivers, lakes, and groundwater supplies is used for irrigation. Although estimates vary widely (as do definitions of irrigated land), about 15 percent of all cropland, worldwide, is irrigated (fig. 7.15).

Some countries are water rich (chapter 11) and can readily afford to irrigate farmland, while other countries are water poor and must use water very carefully. The efficiency of irrigation water use is rather low in most countries. High evaporative and seepage losses from unlined and uncovered canals often mean that as much as 80 percent of water withdrawn for irrigation never reaches its intended destination. Farmers tend to overirrigate because water prices are relatively low and because they lack the technology to meter water and distribute just the amount needed.

Profligate use not only wastes water, it often results in **waterlogging.** Waterlogged soil is saturated with water, and plant roots die from lack of oxygen. **Salinization,** in which mineral salts accumulate in

Figure 7.15 Irrigation allows us to grow crops on otherwise infertile land but carries problems with waterlogging and salinization. Worldwide irrigation accounts for about 70 percent of all human water use. This may become a limiting factor in food production as clean water becomes more scarce.

the soil and kill plants, occurs particularly when soils in dry climates are irrigated profusely. Excessive irrigation accelerates movement of silt and chemicals into surface waters. The largest source of nonpoint chemical water pollution in the United States is runoff from farm fields.

Water conservation techniques can greatly reduce problems arising from excess water use. Conservation also makes more water available for other uses or for expanded crop production where water is in short supply. Drip irrigation, in which a series of small perforated tubes are laid across the field, at or just under the surface of the soil, is the most efficient way to water crops. The exact amount of water each plant needs is delivered directly to the roots where it does the most good with a minimum of evaporative loss or oversoaking of the soil. This technology is expensive to install and operate, however.

Fertilizer

In addition to water, sunshine, and carbon dioxide, plants need small amounts of inorganic nutrients for growth. The major elements required by most plants are nitrogen, potassium, phosphorus, calcium, magnesium, and sulfur. Calcium is usually plentiful in soil, but nitrogen, potassium, and phosphorus availability often limits plant growth. Addition of these elements in fertilizer usually stimulates growth and greatly increases crop yields. A good deal of the doubling in worldwide crop production since 1950 has come from increased inorganic fertilizer use, mostly nitrogen, phosphorus, and potassium. In 1950, the average amount of fertilizer used per acre was 20 kg per hectare. In 1990, this had increased to an average of 91 kg per hectare worldwide. This change represents an

Food and Agriculture 147

increase in total fertilizer use from 30 million metric tons in 1950 to 134 million metric tons in 1990.

Farmers often overfertilize because they are unaware of the specific content of their soils or the needs of their crops. Notice in figure 7.16 that while European farmers use more than twice as much fertilizer per hectare as do North American farmers, their yields are not proportionally higher. Overfertilization wastes money and has negative environmental impacts. Phosphates and nitrates from farm fields and cattle feedlots are a major cause of aquatic ecosystem pollution. Fertilizer runoff percolates through the soil to contaminate groundwater supplies. Nitrate levels in groundwater have risen to dangerous levels in many areas where intense farming is practiced. England, France, Denmark, Germany, the Netherlands, and the United States have reported nitrate concentrations above the safe level of 11.3 mg per liter in drinking water in farming areas. Young children are especially sensitive to the presence of nitrates. Using nitrate-contaminated water to mix infant formula can be fatal for newborns.

What are some alternate ways to fertilize crops? Manure, crop residues, ashes, composted refuse, and green manure (crops grown specifically to add nutrients to the soil) are important natural sources of soil nutrients. Nitrogen-fixing bacteria living symbiotically in root nodules of legumes (a broad group of plants including peas, beans, vetch, alfalfa, and leucaena trees) are valuable for making nitrogen available as a plant nutrient (chapter 3). Interplanting or rotating beans or some other leguminous crop with such crops as corn and wheat are traditional ways of increasing nitrogen availability. In some cases, growing and then plowing under a leguminous green manure crop, such as alfalfa, clover, or vetch, every third or fourth year is necessary to maintain soil fertility.

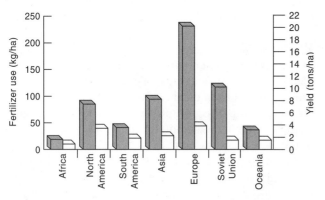

Figure 7.16 Fertilizer use and yield by region. Notice that high fertilizer applications in Europe increase yield, but not in a linear ratio compared to yields in America. The excess fertilizer used in Europe causes serious water pollution problems as well as raises the cost of food. (Data is for the former Soviet Union.)

Source: Data from *World Resources, 1990–1991.*

What is the potential for increasing world food supply by increasing fertilizer use in low-production countries? Africa, for instance, uses an average of only 19 kg of fertilizer per hectare (17 lb per acre), or about one-fourth of the world average. India also has a relatively low average fertilizer level (30 kg per hectare). It has been estimated that both these areas could at least triple their crop production by raising fertilizer use to the world average. Other increases could be achieved by using currently idle land, by introducing new high-yield crop varieties, and by investing in irrigation where water is available. All these steps, however, will provide only illusory relief unless careful thought is given to how agriculture is to be made stable and renewable, how food is to be distributed to those who need it, and how population growth can be brought into line with realistic future planning.

Energy

Farming as it is generally practiced in the industrialized countries is highly energy-intensive. Fossil fuels supply almost all of this energy. The largest energy consumption is for liquid fuels for farm machinery used in planting, cultivating, harvesting, and transporting crops to market. The second largest energy cost is the energy contained in chemical stocks used to synthesize fertilizers, pesticides, and other agricultural chemicals. In the western United States, water pumping for irrigation is a major energy consumer, representing a *triple* mining of resources. We use oil, gas, and coal to pump groundwater that is used to grow crops that deplete the soil!

After crops leave the farm, additional energy is used in food processing, distribution, storage, and cooking. It has been estimated that the average food item in the American diet travels 2,000 km (1,250 mi) between the farm that grew it and the person who consumes it. The energy required for this complex processing and distribution system may be five times as much as is used directly in farming. Altogether the food system in the United States consumes about 16 percent of the total energy we use. Most of our foods require more energy to produce, process, and get to market than they yield when we eat them (table 7.4).

HOW MANY PEOPLE CAN THE WORLD FEED?

A number of studies have estimated the maximum number of people the world could support, given predicted supplies of fertilizers, water, arable cropland, and other factors of agricultural production. The results have ranged from pessimistic warnings that there are already more people than we can feed to optimistic claims that the world could support thirty to fifty times the present population. One set of estimates by

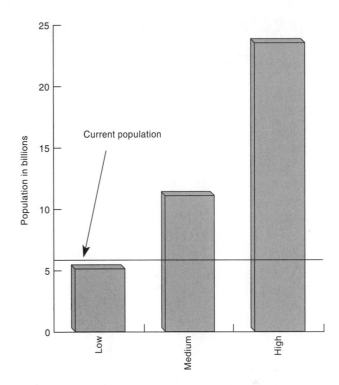

Figure 7.17 Estimates of potential human populations given maximum use of available cropland and varying agricultural inputs such as water, fertilizer, and energy. Notice that low inputs would not feed current populations, while high inputs might feed four times the present populations. High inputs are probably not sustainable, however.

Data from the U.N. Food and Agriculture Organization.

the FAO is shown in figure 7.17. This study predicts that at a low level of agricultural inputs, developing countries could feed only about 2.5 billion people, and the maximum world population would be less than the present 5.5 billion people if everyone is to be fed adequately.

As inputs are increased in this model, the maximum population also increases to around 24 billion people. This says nothing about the quality of life, however, or the effects on other species of conversion of all possible land to crop production. Even with high levels of input, southwest Asia could not support many more people than it does now. Africa and South America, on the other hand, potentially could support far more people if agriculture were modernized and more land were converted to agricultural uses.

Bringing New Land into Cultivation

How much land is available for agriculture? Are we running out of land, or are we about to? Over the whole world, some 3.2 billion hectares (about 25 percent of all ice-free land) is potentially suitable for agriculture. We currently use half of that land for crops. This represents 0.28 hectares (0.69 acres) per person, worldwide. Asia has the most cultivated land in total, but the least land per capita (table 7.5). With 451 million hectares to support 3.1 billion people, each person has only 0.15 hectares (about one-third acre).

The largest increases in cropland over the past thirty years occurred in South America and Oceania where forests and grazing lands are rapidly being converted to farms. Many developing countries are reaching the limit of lands that can be exploited for agriculture without unacceptable social and environmental costs, whereas other areas still have considerable potential for opening new agricultural lands. East Asia, for instance, already uses nearly three-quarters of its potentially arable land. Most of its remaining land has severe restrictions for agricultural use. Further increases in crop production will probably have to come from higher yields per hectare. Latin America, by contrast, uses only about one-fifth of its potential land, and Africa uses only about one-fourth of the land that theoretically could grow crops. However, there would be serious ecological tradeoffs in putting much of this land into agricultural production.

While land surveys tell us that much more land in the world *could* be cultivated, not all of that land necessarily *should* be farmed. Much of it is more valuable in its natural state. The soils over much of tropical Asia, Africa, and South America are old, weathered, and generally infertile. Most of the nutrients are in the

Region	Hectare/person	Percent in use
World	0.28	40
Africa	0.28	25
North and Central America	0.69	58
South America	0.43	19
Asia	0.15	72
Europe	0.28	81
Soviet Union	0.80	65
Oceania	1.82	30

Source: *World Resources,* 1990–1991.

standing plants, not in the soil. In many cases, clearing land for agriculture in the tropics has resulted in tragic losses of biodiversity and the valuable ecological services that it provides. Ultimately, much of this land is turned into useless scrub or semidesert when farmed or grazed intensively.

The FAO estimates that a growth rate of 1 percent per year in total agricultural land is not unreasonable for the immediate future. The world could add 250 million hectares (625 million acres) of new cropland by the end of this century. The major potential expansion areas, according to the FAO, are tropical forests, savannahs, seasonally flooded alluvial zones (flood plains), and hill land. Each of these environmental types has special problems that must be carefully considered if social and biological disasters are to be avoided.

The Green Revolution

Most of the recent growth in world food supply has come from higher yields per hectare of land. The United States, for instance, has had nearly a sixfold increase in corn yields in the past century. This represents a growth from an average of less than 40 bushels per acre fifty years ago to more than 200 bushels per acre today. The two main contributors to this remarkable change have been better-yielding crop varieties and the use of fertilizer. Much of the progress in new hybrid maize varieties, fertilizers, farming procedures, and machines to carry out high-intensity agriculture came from research and training at U.S. land-grant universities.

Starting shortly after the Second World War, the Ford and Rockefeller Foundations (along with a number of governments and other agencies) set up agricultural research stations in less-developed countries to breed tropical wheat and rice varieties that could provide yield gains similar to those obtained in the United States. Collectively, these are known as the Consultative Group on International Agricultural Research (CGIAR). The results have been spectacular for many crops. "Miracle" strains of rice and wheat have made it possible to triple or quadruple yields per hectare. This dramatic yield increase from new crop varieties has been called the **green revolution.**

These new strains are really "high responders" rather than high yielders, however. That is, they respond more efficiently to increases in fertilizer and water and have a higher yield under optimum conditions than do traditional, native species. The average, worldwide grain yields using these new varieties have more than doubled. The benefits of these higher yields, however, are not accessible to everyone. Poor farmers cannot afford the seed, fertilizer, water, pesticides, fuel, and farm equipment necessary to cultivate the new strains. Often only the most prosperous farms are able to participate in the green revolution. The crop surpluses they produce are likely to drive prices down, so the marginal farmers are even worse off than before.

Plant conservators are busy hunting for native crop varieties that might be disease resistant or able to grow in a unique set of environmental conditions. For instance, many irrigated soils in dry climates tend to accumulate surface salt and mineral deposits. Some wild strains are salt tolerant, and there is a possibility that they could be foundation plants for salt-resistant domestic varieties.

Most of our commercial crops originated in the tropics or semiarid lands of the Middle East, Africa, or Central America. Some of the less-developed countries that still have wild ancestors of cultivated crops have raised objections to having scientists from the developed countries collecting samples of their flora and fauna. They feel entitled to a share of the profits that may come from the use of their native species for future crop development. There are threats of "gene wars" over control of this resource. Who is right? Should genetic resources be considered a salable commodity, like mineral and crop resources, to benefit their native country, or are they a globally shared resource? See chapter 9 for further discussion of this topic.

TOWARD A JUST AND SUSTAINABLE AGRICULTURE

How, then, shall we feed the world? Can we make agriculture compatible with sustainable ecological and social systems? Having discussed some of the problems that beset modern agriculture, we will now consider some ways to overcome problems and make farming and food production just and lasting enterprises. This goal is usually termed **sustainable agriculture** or regenerative farming and calls for ecologically sound, socially just, economically viable, and humane methods of food production (box 7.2).

Sustainable Agriculture in America

*L*and stewardship is a concept that has been missing for many years in large-scale American agriculture. Recently, however, a few farmers who are disillusioned with modern chemical farming, the cycle of overproduction and falling prices, and the economic uncertainties of depending on a single large crop every year, are exploring some new and some old alternatives. They are looking for farming methods that protect the land as well as their own health. In the process, they are rediscovering the pride of treating the land well and learning that the land rewards good stewardship.

Dick and Sharon Thompson of Boone, Iowa (box fig. 7.2.1), are two of those farmers seeking alternatives to chemical-dependent farming. After earning a master of science degree in animal husbandry from Iowa State University, Dick Thompson inherited his family's farm in 1957 and began practicing the modern version of farming that he had learned at the university: removing fences, expanding fields, and specializing on a single crop. Soils cannot support the same crops year after year without rest, because nutrients drawn from the soil by plants are lost by harvesting. Farmers were advised to replace those nutrients using chemical fertilizers. Pesticides were to be used to keep all harmful insects away from crops. Liberal amounts of herbicides were to be applied to prevent competition from weeds and keep the fields "clean" between the rows.

These modern farming techniques are practiced by the vast majority of American farmers, and agricultural schools continue to teach them. Unfortunately, some modern farming techniques have brought many farmers to bankruptcy, and they are causing alarming rates of erosion, sending millions of tons of precious, irreplaceable topsoil downriver to the sea every year.

Like their neighbors, Dick and Sharon Thompson practiced high-intensity, single-crop farming with pesticides and fertilizers for ten years, but they felt that something was not right. Their hogs and cattle were sick. Fertilizer, pesticide, and petroleum prices were rising faster than crop prices. They began looking for a better way. In 1967 a friend took them to a program on natural farming. For the Thompsons, it was almost a religious revelation to hear that natural, organic farming methods could be the answer to their problems. They stopped using chemicals and started developing what Dick Thompson calls "regenerative agriculture."

The Thompsons have become experts at ridge-tilling without herbicides, using special machinery that turns the ground into rows of tiny ridges and valleys. Before fall harvest, they plant oats, hay, or hairy vetch (a soil-enriching legume) in the valleys. During the cold, windy winter, these plants protect the soil from erosion, and in spring they inhibit weeds until the new crop is planted

Figure 7.2.1 Dick and Sharon Thompson of Boone, Iowa practice sustainable farming to produce healthy crops that are economically successful.

on the ridges. Any spring weeds in the valleys are left as ground cover until they can be removed easily with a cultivator. Unlike his neighbors, Dick Thompson doesn't leave his fields bare all winter by plowing in the fall, a practice he found encouraged weed growth, thus requiring extra herbicides, and that left the soil exposed to winter wind erosion.

The Thompsons grow a variety of crops and practice crop rotation, allowing the soil to rest and regenerate itself. They pasture their cattle, an unusual idea in Iowa, and feed them hay and grain crops from their own fields. They carefully collect and compost manure to re-enrich the fields from which the cattle feed has been taken. Their annual yields have remained high, and their costs have fallen considerably.

Dick and Sharon Thompson have done a lot of experimentation over the years. They have used new technology, such as ridge-tilling, and ancient ideas, such as regular crop rotation. They have tried many different methods and combinations, but they have never retreated from regenerative agriculture. They actively work to recruit other farmers to their methods. Every September they hold a field day, inviting farmers from across the nation to see their work in action and begin thinking about working out their own methods of regenerative farming and land stewardship.

The Thompsons have found many rewards in their practice of sustainable agriculture. Soil fertility is improving, and the earthworms that can live again in their pesticide-free soil are bringing back birds. They no longer are contaminating their groundwater and surface water with toxic chemicals. They are being energy-efficient and cost-effective, and they are getting recognition and praise from environmental and agricultural organizations around the country. The Thompsons are especially proud that they are fostering an attitude of caring for the land so that it will remain rich and fertile.

Soil Conservation

With careful husbandry, soil is a renewable resource that can be replenished and renewed indefinitely. Some rice paddies in Southeast Asia, for instance, have been farmed continuously for a thousand years without any apparent loss of fertility. The rice-growing cultures that depend on these fields have developed management practices that return organic material to the paddy and carefully nurture the soil's ability to sustain life (fig. 7.18).

Managing Topography

Water runs downhill; the faster it runs, the more soil it carries from the fields. Comparisons of erosion rates in Africa have shown that a 5 percent slope in a plowed field has three times the water runoff volume and eight times the soil erosion rate of a comparable field with a 1 percent slope. Water runoff can be reduced by **contour plowing,** plowing across the hill rather than up and down it. Contour plowing is often combined with strip-farming, that is, planting different kinds of crops in alternating strips along the land contours (fig. 7.19). When one crop is harvested, the other is still present to protect the soil and keep water from running straight downhill. The ridges created by cultivation make little dams that trap water and allow it to seep into the soil rather than running off. In areas where rainfall is very heavy, tied ridges are often useful. This involves a series of ridges running at right angles to each other, so that water runoff is blocked in all directions and is encouraged to soak into the soil.

Terracing involves shaping the land to create level shelves of earth to hold water and soil (see fig. 7.18). The edges of the terrace are planted with soil-anchoring plant species. This is an expensive procedure, requiring either much hand labor or expensive machinery, but making possible the farming of very steep hillsides. The rice terraces in the Chico River Valley in the Philippines, which rise as much as 300 m (1,000 ft) above the valley floor, are considered one of the wonders of the world.

Planting perennial species (plants that grow for more than two years) is the only suitable use for some lands and some soil types. Establishing forest, grassland, or crops such as tea or coffee that do not have to be cultivated every year may be necessary to protect certain unstable soils on sloping sites or watercourses (low areas where water runs off after a rain).

Providing Ground Cover

Often, the easiest way to provide cover that protects soil from erosion is to leave crop residues on the land after harvest. They not only cover the surface to break the erosive effects of wind and water, but they

Figure 7.18 Rice terraces on Java, Indonesia. Some rice paddies have been cultivated for hundreds or even thousands of years without any apparent loss of productivity.

Figure 7.19 Contour plowing and strip-cropping on these Wisconsin farms protect the soil from erosion and help maintain fertility as well as forming a beautiful landscape. With care and good stewardship, we can increase the carrying capacity of the land and create a sustainable environment.

also reduce evaporation and soil temperature in hot climates and protect ground organisms that help aerate and rebuild soil. In some experiments, 1 ton of crop residue per acre (0.4 hectare) has increased water infiltration 99 percent, reduced runoff 99 percent, and reduced erosion 98 percent. Leaving crop residues on the field can increase the potential for disease and pest problems, however, and may require increased use of pesticides and herbicides.

Where crop residues are not adequate to protect the soil or are inappropriate for subsequent crops or

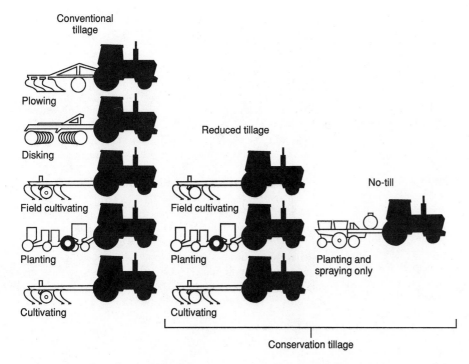

Conventional tillage

Plowing

Disking

Field cultivating

Planting

Cultivating

Reduced tillage

Field cultivating

Planting

Cultivating

No-till

Planting and spraying only

Conservation tillage

Figure 7.20 Reduced tillage and no-till farming protect soil from erosion, save energy, and reduce soil compaction. These methods often depend on increased use of pesticides, however.

farming methods, such cover crops as rye, alfalfa, or clover can be planted immediately after harvest to hold and protect the soil. These cover crops can be plowed under at planting time to provide green manure. Another method is to flatten cover crops with a roller and drill seeds through the residue to provide a continuous protective cover during early stages of crop growth.

Using Reduced Tillage Systems

Farmers have traditionally used a moldboard plow to till the soil, digging a deep trench and turning the topsoil upside down, then disking to break up clods (fig. 7.20). In the 1800s, it was shown that cultivating a field fully—until it was "clean"—increased crop production. It helped control weeds and pests, reducing competition; it brought fresh nutrients to the surface, providing a good seedbed; and it improved surface drainage and aerated the soil. This is still true for many crops and many soil types, but it is not always the best way to grow crops. We are finding that less plowing and cultivation often makes for better water management, preserves soil, saves energy, and increases crop yields.

There are three major **reduced tillage systems.** *Minimum till* uses a disc or chisel plow rather than a traditional moldboard plow. A chisel plow is a curved chisel-like blade that gouges a trench in the soil in which seeds can be planted. It leaves up to 75 per-

cent of plant debris on the surface between the rows, preventing erosion. *Conserv-till* farming uses a coulter, a sharp disc like a pizza cutter. It slices through the soil, opening up a furrow or slot just wide enough to insert seeds. This disturbs the soil very little and leaves almost all plant debris on the surface. *No-till* planting is accomplished by drilling seeds into the ground directly through mulch and ground cover. This allows a cover crop to be interseeded with a subsequent crop.

Farmers who use these conservation tillage techniques often must depend on pesticides (insecticides, fungicides, and herbicides) to control insects and weeds. Increased use of toxic agricultural chemicals is a matter of great concern (box 7.3). Massive use of pesticides is not, however, a necessary corollary of soil conservation. It is possible to combat pests and diseases with crop rotation, trap crops, natural repellents, and biological controls.

Integrated Pest Management

In many cases, improved pest management techniques can cut pesticide use between 50 and 90 percent without reducing crop production or allowing increasing diseases. This approach is called **integrated pest management (IPM),** a flexible, ecologically based pest-control strategy that uses a combination of techniques applied at specific times, aimed at specific crops and pests. IPM often involves such biological controls as predators (wasps, ladybugs, praying mantises) or pathogens (viruses, bacteria, fungi) to control pests safely. Releasing sterile males to interfere with insect pest reproduction, introducing hormones that upset development, or using sex attractants to bait traps also can be effective.

Cultural practices also are a part of IPM. Crop rotation (growing a different crop each year in a four- to six-year cycle) keeps pest populations from building up. Flooding fields before planting or burning crop residues and replanting with a cover crop can suppress both weeds and insect pests. Home gardeners often plant a border of insect-repelling plants, such as garlic or marigolds. Where there is no alternative to using a chemical toxin for pest control, a single heavy dose of a nonpersistent pesticide might be

Focus on Pesticides

What do you do when you discover cockroaches in your cupboard or dandelions in your yard? Do you look for a chemical pesticide to take care of the problem? If so, you are reacting as most Americans do. You are also contributing to a serious environmental problem. The EPA ranks pesticide pollution as the most urgent concern facing the United States. Some medical experts think pesticide contamination of air, water, and food is a major cause of chronic health problems in the United States and around the world. Pesticide residues contaminate groundwater drinking supplies in about half of the United States, and pesticide production is probably the largest source of toxic wastes in this country.

What is a pest, and what are pesticides? A pest is a troublesome animal, plant, microbe, or person. Put another way, a pest is any organism that reduces the availability, quality, or value of resources useful to humans.

Both definitions are highly subjective. The same organism that may be a pest in one case may be considered a friend in another. Only about one hundred species of plants and animals cause 90 percent of all damage to foods and crops. A **pesticide** is any chemical that kills, controls, drives away, or modifies the behavior of a pest so that it is no longer troublesome. Pesticides are identified by their target organism, such as insecticides, rodenticides, fungicides, or herbicides (box table 7.3).

Synthetic chemical pesticides have brought us many benefits. Most agronomists believe that the rapid increases in food productivity in this century would have been impossible without effective pest control. Even with our extensive use of chemicals, pests and diseases reduce or destroy crops amounting to about half the world's food supply. In 1945, crop losses in the United States caused by insects, diseases, and weeds amounted to 32 percent of the harvest. In 1980, these losses had risen to 37 percent of the harvest, despite the use of 450,000 metric

BOX TABLE 7.3 Pesticides

Type	Examples	Characteristics
Insecticides		
Inorganic chemicals	Mercury, lead, arsenic, copper sulfate	Highly toxic to many organisms, persistent, bioaccumulates
Organochlorines	DDT, methoxychlor, heptachlor, HCH, pentachlorophenol, chlordane, toxaphene, aldrin, endrin, dieldrin, lindane	Mostly neurotoxins, cheap, persistent, fast acting, easy to apply, broad spectrum, bioaccumulates, biomagnifies
Organophosphates	Parathion, malathion, diazinon, dichlorvos, phosdrin, disulfoton, TEPP, DDVP	More soluble, extremely toxic nerve poisons, fast acting, quickly degraded, toxic to many organisms, very dangerous to farm workers
Carbamates and urethanes	Carbaryl (Sevin), aldicarb, carbofuran, methomyl, Temik, mancozeb	Quickly degraded, do not bioaccumulate, toxic to broad spectrum of organisms, fast acting, very toxic to honey bees
Formamidines	Amitraz, chlordimeform (Fundal and Galecron)	Neurotoxins specific for certain stages of insect development, act synergistically with other insecticides
Microbes	*Bacillus thuringensis* *Bacillus popillae* Viral diseases	Kills caterpillars Kills beetles Attack a variety of moths and caterpillars
Plant products and synthetic analogs	Nicotine, rotenone, pyrethrum, allethrin, decamethrin, resmethrin, fenvalerate, permethrin, tetramethrin	Natural botanical products and synthetic analogs, fast acting, broad insecticide action, low toxicity to mammals, expensive
Fungicides	Captan, maneb, zeneb, dinocap, folpet, pentachlorophenol, methyl bromide, carbon bisulfide, chlorothalonil (Bravo)	Most prevent fungal spore germination and stop plant diseases; among most widely used pesticides in United States.
Fumigants	Ethylene dibromide, dibromochloropropane, carbon tetrachloride, carbon disulfide, methyl bromide	Used to kill nematodes, fungi, insects, and other pests in soil, grain, fruits; highly toxic, cause nerve damage, sterility, cancer, birth defects
Herbicides	2,4 D; 2,4,5 T; paraquat, dinoseb, diaquat, atrazine, Silvex, linuron	Block photosynthesis, act as hormones to disrupt plant growth and development, or kill soil microorganisms essential for plant growth.

tons (1 billion pounds) of pesticides. The U.S. Department of Agriculture estimates that food in the United States would cost 30 to 50 percent more than it does now if we had no chemical pesticides. The control of disease-carrying insects has spared billions of people from diseases such as malaria, yellow fever, typhus, sleeping sickness, river blindness, and filariasis. Millions of lives have been saved. The reductions in medical expenses, social disruption, and days lost from work probably outweigh the direct costs of pesticide application by a thousand to one.

Problems have been associated with widespread use of pesticides, however. Farmers soon discovered that pests tend to reproduce rapidly and develop pesticide resistance very quickly. Natural predators are killed and ecological balances that controlled pests are upset by broad-spectrum chemicals. Pests then rebound to higher levels than ever before, and secondary pests are created when organisms that were not previously a problem begin to grow out of control. Opponents of pesticide use claim that we are on a pesticide treadmill where ever-increasing amounts of toxins are required just to stay even with the pests. Persistent chemicals (such as DDT) are bioaccumulated and magnified in food chains, eventually reaching toxic levels in such top carnivores as bald eagles, peregrine falcons, brown pelicans, salmon, seals, and humans. One of the main threats faced by many endangered species is pesticide poisoning.

New types of pesticides, such as organophosphates and carbamates, have been synthesized to combat pest resistance and reduce environmental accumulation. These chemicals break down much more quickly than most chlorinated hydrocarbons and, therefore, do not bioaccumulate as much. They are fast acting and extremely toxic to target species. Unfortunately, they also are very toxic to nontarget species, including humans. Factory workers who manufacture these compounds often have severe or acute reactions, including neurological damage, muscle paralysis, skin burns and severe acne, hallucinations, tremors, memory loss, and sterility. Farm workers who apply pesticides or work in fields where they have been used have similar problems. The World Health Organization estimates there are 2 million pesticide poisonings in the world each year, and at least 10,000 people die of immediate pesticide effects.

Many pesticides are known or suspected to be mutagens (cause mutations), teratogens (cause birth defects), carcinogens (cause cancer), and cumulative metabolic poisons in low chronic doses (see chapter 6 for definitions). No one knows how much these chemicals are responsible for long-term human health problems. Of the six hundred active pesticide ingredients on the market, the EPA has completed a preliminary assessment of only 20 percent. These active ingredients are combined with another 900 chemical solvents, thickeners, propellants, stabilizers, absorbents, and other "inert" ingredients to make more than 50,000 commercial products. The EPA estimates that it will take twenty years to test all the products now on the market. Meanwhile, these chemicals are accumulating in the environment and in our bodies, and new ones are being introduced. In a

1985 survey by the National Institute of Environmental Health Sciences, 100 percent of the Americans tested had detectable DDT residues in their body, and 90 percent also had traces of chlordane, heptachlor, aldrin, dieldrin, or hexachlorobenzene.

Herbicides make up nearly two-thirds of all the pesticides used in the United States. They reduce cultivation costs for farmers, increase crop yields, and allow intensive, monoculture farming; but, the clean, bare fields produced by these techniques result in high soil erosion rates. Corn and soybeans account for about one-third of all herbicide use. Kansas farmers who work regularly with herbicides have a sixfold increase in the rate of non-Hodgkin's lymphoma (a cancer of the lymphoid system) compared to those who are not exposed to these chemicals. The 20 to 30 percent of herbicides not applied to crops are spread on golf courses, parks, roadsides, and lawns. Much of our herbicide use could be eliminated by other control methods or by changing our attitudes about what is aesthetically pleasing.

Insecticides account for about one-quarter of our pesticide use. Cotton is one of our most intensely treated crops, consuming about one-quarter of all insecticides and half of all chlorinated hydrocarbons used in the United States. Mosquito and termite control are major civil and domestic uses. Because insecticides are generally more acutely toxic than herbicides, they account for a far larger proportion of immediate illnesses—in spite of being used in far smaller quantities.

Fungicides make up approximately one-sixth of the pesticides we use. They protect crops from fungal diseases, such as wheat rust, mildew, smut, and various blights, and are used in paints and wood preservatives. A major use is to prevent spoilage of fruits and vegetables. Much, perhaps most, of the produce you buy in a grocery store has been treated with a fungicide either before or after harvest. Seven of the ten pesticides listed by the EPA as the greatest health concern in the United States are fungicides (Captan, Maneb, Zeneb, Mancozeb, Captifol, Folpet, and Chlorothalonil). The other three are a herbicide (Linuron) and two insecticides (Permethrin and Chlordimeform).

The U.S. National Academy of Sciences estimates that tighter controls on these compounds would reduce the risk of cancer from pesticides in the United States by 80 percent. The fifteen foods considered the greatest risk to the public in terms of pesticide-related cancer are (in order of importance): tomatoes, beef, potatoes, oranges, lettuce, apples, peaches, pork, wheat, soybeans, beans, carrots, chicken, corn, and grapes.

What can you do to reduce your exposure to these dangerous chemicals? Try growing your own pesticide-free food. By joining a food co-op or shopping at a local farmer's market you can buy "organic" food and support growers who practice pesticide-free farming. Become politically active. Write to your congressional representatives and ask them to support legislation that regulates pesticide residues in food and supports alternative, sustainable farming practices.

applied just at the time insects or weeds are most vulnerable. Trap crops, a small area planted a week or two before the main crop, are also useful. This plot matures before the rest of the field and attracts pests away from other plants. The trap crop then is sprayed heavily with pesticides—enough so that no pests are likely to escape. The trap crop is destroyed so that workers will not be exposed to the pesticide and consumers will not be at risk. The rest of the field should be mostly free of both pests and pesticides.

IPM programs are already in use all over the United States on a variety of crops. Massachusetts apple growers who use IPM, for instance, have cut pesticide use by 43 percent in the past ten years, while maintaining per-acre yields of marketable fruit equal to that of farmers who use conventional techniques. Some of the most dramatic IPM success stories have come in the Third World. In Brazil, pesticide use has been reduced up to 90 percent on soybeans. In Costa Rica, use of IPM on banana trees has eliminated pesticides altogether in one region. In Africa, mealybugs were destroying up to 60 percent of the cassava crop (the staple food for 200 million people) before IPM was introduced in 1982. A tiny wasp that destroys mealybug eggs was discovered and now controls this pest in over 65 million hectares (160 million acres) in thirteen countries.

Acting Locally: Box 7.4

What Can We Do Personally?

Since most of us now live in urban areas we tend to feel removed from the problems of farming and world food supplies. We are intimately involved, however, both as consumers and as citizens, in daily decisions about how food is grown and distributed. The first thing that each of us should do is to become educated about the issues involved. Using that knowledge, we can develop good nutritional habits so that we eat only those foods and amounts we really need. For instance, most of us could benefit both ourselves and the world at large by eating lower on the food chain. We don't really need as much meat and animal fat as we consume. It is not only wasteful, it is not very good for us.

Many of us could grow more of our own food. Even a small yard or boulevard can be used to grow delicious vegetables. Think of how much time, money, and resources are spent growing grass and shrubs that only provide ground cover and decoration. Suburban lawns are often the most heavily fertilized, herbicided, watered, and cared-for lands in the country, using dozens or even hundreds of times as much energy, water, and chemicals per unit area as the most intensely farmed cropland.

Gardening is healthful and enjoyable and can provide valuable, wholesome produce. A well-tended garden also can be attractive. City dwellers have found that fresh vegetables can be grown in pots on a windowsill, a balcony, or a rooftop (box fig. 7.4.1). Cities have many acres of level rooftops that can serve admirably for gardens, as is done in some European countries. A plastic bag full of dirt with a couple of holes punched in it can grow a surprising harvest of tomatoes, lettuce, or squash.

Figure 7.4.1 Growing some of your own food can be healthful and fun. Even a small balcony or patio can be turned into a minigarden.

For the foods that you need to buy, consider shopping at a farmer's market. The produce is fresh, and profits go directly to the person who grows the crop. A local food co-op or owner-operated grocery store also is likely to buy from local farmers and to feature pesticide-free foods.

Accept fruits and vegetables that may not be absolutely perfect. You can trim away a few bad spots or wash off an insect or two. If you overlook a worm or a bug, it may be less harmful to you than the toxic chemicals used to make supermarket foods sterile and cosmetically perfect!

Summary

World food supplies have been rising at an unprecedented rate and have grown faster than populations in every continent except Africa. There is enough food to supply everyone in the world with more than the minimum daily food requirements, but food is inequitably distributed. The FAO estimates that 750 million people are chronically undernourished or malnourished, and 18 to 20 million (mostly children) die each year from diseases related to malnutrition. Additional millions survive on a deficient diet but suffer from stunted growth, mental retardation, and developmental disorders.

The three major crops that are the main source of calories and nutrients for most of the world's people are rice, wheat, and maize. Animal protein from meat, milk, and a variety of seafoods are also important components of the human diet. Protein deficiency diseases and a lack of essential vitamins and minerals tragically stunt the growth and development of many children in poorer countries of the world.

The biggest gains in food production have been in Asia, North America, and Latin America, all of which have nearly tripled their yearly crop yields. The only major region in which food production has failed to keep pace with population growth has been sub-Saharan Africa, where adverse weather, insect infestations, wars, inept governments, social and religious factors, economics, and international politics have intervened.

World food trade and international food aid help transfer food from areas of abundance to areas of shortage, but they also undercut local food supplies by encouraging the conversion of land from production of food for local consumption to production of cash crops for export. They also widen economic and social disparities that increase poverty and powerlessness and make it even more difficult for the poorest people to feed themselves.

Fertile, tillable soil for growing crops is an indispensable resource for human life. Soil is a complex system of inorganic minerals, air, water, dead organic matter, and a myriad of different kinds of living organisms. There are hundreds of thousands of different kinds of soils, each produced by a unique history, climate, topography, bedrock, transported material, and community of living organisms.

It is estimated that 25 billion tons of soil are lost from croplands each year because of wind and water erosion. Perhaps twice as much is lost from rangelands and permanent pastures. This erosion causes pollution and siltation of rivers, reservoirs, estuaries, wetlands, and offshore reefs and banks. The net effect of this loss is worldwide crop reduction equivalent to losing 15 million hectares (37 million acres), or 1 percent of the world's cropland, each year.

More land could be put into agricultural production, especially in Latin America and Africa, but this would mean loss of valuable forests and grasslands and could result in major ecological destruction. It is possible that food production could be expanded considerably, even on existing farmland, given the proper inputs of fertilizer, water, high-yield crops, and technology. This will be essential if human populations continue to grow as they have during this century. Whether it will be possible to supply agricultural inputs and expand crop production remains to be seen.

Many new and alternative methods could be used in farming to reduce soil erosion, avoid dangerous chemicals, improve yields, and make agriculture just and sustainable. Some alternative methods are developed through scientific research; others are discovered in traditional cultures and practices nearly forgotten in our mechanization and industrialization of farming. Some authors advocate returning to low-input, regenerative, "organic" farming that may be more sustainable and more healthful than our current practices.

Review Questions

1. How many calories does the average human adult need to meet minimum daily requirements to carry out a healthy, active life?
2. What is the difference between undernutrition and malnutrition? Describe some of the consequences of poor diet.
3. What are the three major grain crops of the world? What are some of the other major crops?
4. Why are less-developed countries eager to grow cash crops? What are some of the effects of export agriculture on the local food supplies and internal economies of the producing countries?
5. What is soil? Why are soil organisms so important?
6. Describe four kinds of erosion. Why is erosion a problem?
7. Explain some possible effects of overirrigation.
8. What can farmers do to increase agricultural production without increasing land use?
9. How many people could be fed adequately by increasing agricultural inputs?
10. What is low-input, sustainable agriculture?

Questions for Critical or Reflective Thinking

1. Try to imagine what it would be like to be hungry for most of your life. How can those of us who are overnourished understand what it is like to be starving?

2. Paul Ehrlich once wrote: "The carrying capacity of Earth for saints would be much larger than for real people. Real people do not distribute things equitably, and that lowers the carrying capacity." How would you respond?

3. Optimists claim that the fact that food production has kept pace with population growth proves that Malthus was wrong in predicting famine and disaster. Do you agree?

4. How many people do you think the world can feed? What would the optimum human population be?

5. Why do you suppose Ethiopia, Somalia, Chad, and the Sudan have had such terrible famines in recent years? What are the relative effects of natural and human factors in these tragedies?

6. Look at the two opening quotes at the beginning of this chapter. How can two experts come to such different conclusions? What premises or assumptions underlie the positions stated?

7. If Erick Eckholm and Paul Ehrlich were debating the issue of food security and optimum human population size, where would

they agree and disagree? Make a list of points of agreement and disagreement. With which ones do you agree?

8. Soil erosion is limiting our capacity to feed ourselves. Why do we allow it to continue?

9. Given what you have learned about pesticides in this chapter, do you favor their continued use? What restrictions or conditions would you impose on them?

10. Analyze the benefits and disadvantages of low-input, sustainable agriculture. Who is for or against a shift to alternative agricultural systems and why do they take the position they do?

Key Terms

cash crops 142
chronic food shortages 136
contour plowing 152
erosion 144
famines 135
food aid 142
green revolution 150
integrated pest management (IPM) 153
kwashiorkor 138
malnourishment 138
marasmus 138
pellagra 139

pesticide 154
reduced tillage systems 153
salinization 147
soil 142
soil horizons 143
subsoil 144
sustainable agriculture 150
terracing 152
topsoil 143
undernourished 137
vitamins 139
waterlogging 147

Suggested Readings

Altieri, M. *Agroecology: The Scientific Basis of Alternative Agriculture.* 2d ed. Boulder: Westview Press, 1986. Describes the ecological basis of agriculture, soil conservation, nutrient restoration, and biological pest control.

Barbier, E. B. "Sustaining Agriculture on Marginal Land." *Environment* 31, no. 9 (1989): 12. Describes national policies that emphasize low-input technology, land reform, and economic incentives to encourage sustainable systems on marginal land.

Berry, W. *The Gift of Good Land.* San Francisco: North Point Press, 1981. A series of essays on a variety of topics ranging from Amish farms to sheep and scythes by one of America's best environmental writers.

Berry, W. *The Unsettling of America: Culture and Agriculture.* San Francisco: Sierra Club Books, 1977. A thoughtful and passionate argument against the misuse of the land and destruction of rural culture.

Brown, J. L. "Hunger in America." *Scientific American* 256, no. 2 (February 1987): 36. Report of a study by the Physician's Task Force on Hunger in America.

Brown, L. R. "The New World Order." In *State of the World 1991.* New York: Norton, 1991. A good survey of the constraints and potential of world agriculture.

Consultative Group on International Agricultural Research (CGIAR). *1989 Annual Report.* Washington, D.C., 1990. Reports current progress in the green revolution.

Crosson, P. R., and N. J. Rosenberg. "Strategies for Agriculture." *Scientific American* 261, no. 3 (September 1989): 128. Agricultural research will probably yield many new technologies for expanding food production. The challenge will be getting farmers to use them.

Ehrlich, A. H., and P. R. Ehrlich. "Why Do People Starve?" *The Amicus Journal* 9, no. 2 (Spring 1987): 42. A discussion of limits in the world's carrying capacity for humans.

Fukuoka, M. *One Straw Revolution: An Introduction to Natural Farming.* Emmaus, Pa.: Rodale Press, 1978. An account of one man's successful experiments in organic gardening.

Gipps, T. *Breaking the Pesticide Habit.* Minneapolis: International Alliance for Sustainable Agriculture, 1990. Presents alternatives to the "dirty dozen" most dangerous pesticides in pest control.

Hildyard, N., et al. "Declaration of the International Movement for Ecological Agriculture." *The Ecologist* 21, no. 2 (April 1991): 1107. The conclusion of a special issue devoted to a critique of the FAO and modern, high-technology agriculture with strategies for encouraging traditional indigenous forms of food production.

Jackson, W., W. Berry, and B. Colman. *Meeting the Expectations of the Land: Essays in Sustainable Agriculture and Stewardship.* San Francisco: North Point Press, 1984. An eloquent and insightful discussion of sustainable farming.

Lappe, F., and J. Collins. *World Hunger, 12 Myths.* San Francisco: Food First Institute for Food and Development Policy, 1986. An argument that maldistribution of resources is responsible for famine.

National Research Council. *Alternative Agriculture.* Washington, D.C.: National Academy Press, 1989. Success stories in a variety of applications of low-input regenerative agriculture.

Nearing, H., and S. Nearing. *Living the Good Life.* New York: Schocken Books, 1948. A practical description of how to live sanely and simply in a troubled world.

Postel, S. "Saving Water for Agriculture." *State of the World 1990.* Washington, D.C.: Worldwatch Institute, 1990. A good discussion of water use in agriculture and its associated problems.

Power, J., and R. Follet. "Monoculture." *Scientific American.* 256, no. 3 (March 1987): 78. The practice of growing the same crop on the same land repeatedly has some advantages for the farmer but may not always be good agronomy.

Reagnold, J. P., R. I. Papendick, and J. F. Parr. "Sustainable Agriculture." *Scientific American* 262, no. 6 (June 1990): 112. Describes environmental and economic rewards of alternative agriculture.

Shiva, V. "The Failure of the Green Revolution." *The Ecologist* 21, no. 2 (April 1991): 57. Criticizes the green revolution as ecologically destructive and socially inequitable. Argues that the evidence produced in its favor is little more than myth.

World Bank. *Sub-Saharan Africa: From Crisis to Sustainable Growth.* Washington, D.C.: The World Bank, 1989. Policies and practical examples of sustainable agriculture under difficult conditions.

8

*F*ORESTS, RANGELANDS, PARKS, AND WILDERNESS

*We abuse land because we regard it as a commodity belonging to us. When we see land
as a community to which we belong, we may begin to use it with love and respect.*

Aldo Leopold, *A Sand County Almanac*

O b j e c t i v e s

*After studying this chapter, you should be
able to:*

■ Discuss the major world land uses and how human
activities impact these areas.

■ List some forest types and the products we derive from
them.

■ Report on how and why tropical forests are being
disrupted as well as how they might be better used.

■ Understand the major issues concerning forests in more
highly developed countries such as the United States
and Canada.

■ Describe how overgrazing causes desertification of
rangelands.

■ Relate the origins of national parks and wildernesses as
well as what is being done today to preserve areas of
unique biodiversity and natural heritage.

■ Examine land ownership patterns and explain why land
reform is essential for social justice as well as
environmental protection.

INTRODUCTION

The ways we use land clearly
indicate our priorities and values. A series of land-use decisions—often seemingly small
and insignificant, but generally irreversible—has shaped
our history and our future. In
a real way, we create our environment through care or
abuse of the land (fig. 8.1).
This chapter surveys land use
by humans, focusing especially
on forests, rangelands, parks,
and wilderness areas. We will
look at some problems associated with these resources as
well as ways they can be—
and are being—protected.

Throughout history, land
ownership has been the traditional source of human
wealth and power. Although
land no longer is the most
important resource in most industrialized countries,
its use and control still shape societies and economies in many parts of the world. Inequitable patterns of land ownership and poor land-use policies

Figure 8.1 The ways we use and abuse the land clearly indicate our values and
priorities. Ancient, old-growth forests are disappearing rapidly everywhere in the world,
but especially in the tropics, home to the earth's greatest abundance of biodiversity.

lie at the root of many social and environmental
problems. We also will study some of the reform
movements around the world that seek to improve
the distribution of resources and the use of land.

160

WORLD LAND AREA CHARACTERISTICS

The earth's total land area is about 144.8 million sq km (55.9 million sq mi), or about 29 percent of the surface of the globe. Figure 8.2 shows the distribution of four major land-use categories. Notice that the largest single category in this figure is the "other" lands, a residual group including tundra, marsh, desert, scrub forest, bare rock, ice and snow, or other sparsely populated lands. About one-third of the land in this category is so barren that it lacks plant cover altogether. While deserts and other infertile lands are usually unsuitable for intensive human use, they play an important role in biogeochemical cycles and as a refuge for biological diversity. Presently, only about 3 percent of the world's land surface is formally protected in parks, wildlife refuges, and nature preserves. Some land-use planners suggest that at least 10 percent of the land should be set aside to protect natural ecosystems and endangered species.

Notice that only about 10 percent of the earth's landmass is now used for crops. Some agricultural experts claim that as much as half of the 7.2 billion hectares (18 billion acres) of present forests and grazing lands worldwide could be converted to crop production, given the proper inputs of water, fertilizer, erosion control, and mechanical preparation. This land could feed a vastly larger human population (perhaps ten times the present number), but sustained intensive agriculture could result in serious environmental and social problems.

Growing human populations and expanding agricultural and industrial land uses already have brought about major changes in land utilization and land characteristics in recent years. Some 30 million hectares (74 million acres) of former forests and pastures have been converted to croplands in the past decade worldwide, while half again as much has been degraded to wastelands that are useless for most purposes. We will examine some of the causes and consequences of these destructive land-use practices later in this chapter. Since agriculture and biological resources are discussed in detail in chapter 9, we will focus here primarily on the forest and grazing lands themselves.

WORLD FORESTS

Forests play a vital role in regulating climate, controlling water runoff, providing shelter and food for wildlife, and purifying the air. They produce valuable materials, such as wood and paper pulp, on which we all depend. Furthermore, forests have scenic, cultural, and historic values that deserve to be protected. In this section, we will look at forest distribution, use, and management.

Forest Distribution

Before large-scale human disturbances of the world began many thousands of years ago, forests probably covered 6 billion hectares (15 billion acres). Since then, nearly *one-third* of that area has been converted to cropland, pasture, settlements, or unproductive wastelands. The 4 billion hectares still forested covers about 28 percent of the earth's surface, nearly three times as much as all croplands. About two-thirds of the forest is classified as **closed canopy forest** (where tree crowns spread over 20 percent or more of the ground) and has potential for commercial timber harvests. The rest is **open canopy forest** or **woodland,** in which tree crowns cover less than 20 percent of the ground.

The world's principal forest types are shown in figure 8.3. Canada, Northern Europe, and Russia have vast areas of closed canopy temperate deciduous or boreal coniferous forests (see chapter 3 for further description), much of which remains in near-pristine state. The United States once had similar forests, but much has been fragmented by timber harvesting or conversion to other uses. South America and Central Africa have the largest remaining closed canopy, broad-leaved, seasonally deciduous or evergreen moist tropical forests. Africa also has large areas of open woodlands, mainly in the dry savannahs and thorn brush of the sub-Saharan region.

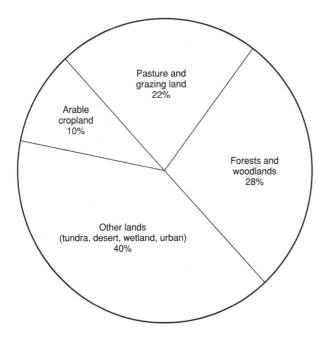

Figure 8.2 World land use, 1992.
Source: Data from *World Resources 1992–1993.*

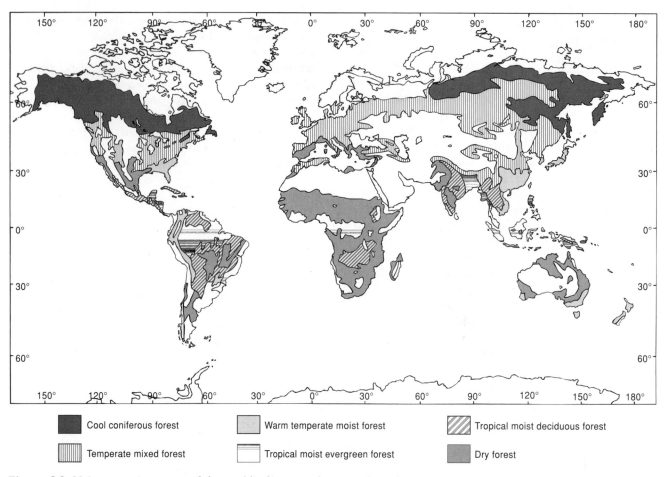

Cool coniferous forest   Warm temperate moist forest   Tropical moist deciduous forest

Temperate mixed forest   Tropical moist evergreen forest   Dry forest

Figure 8.3 Main vegetation zones of the world's forests under natural conditions.

Forest Products

Wood plays a part in more activities of the modern economy than any other commodity. There is hardly any industry that does not use wood or some wood products somewhere in its manufacturing and marketing processes. Furthermore, consider the impact our information explosion has had on use of paper, a wood product. Total wood consumption is about 3.2 billion metric tons or three billion cubic meters annually.

Industrial timber (or roundwood) used for lumber, plywood, veneer, particleboard, and chipboard accounts for slightly less than one-half of worldwide wood consumption (about 1.5 billion tons per year). This exceeds the use of steel and plastics combined. International trade in wood and wood products amounts to more than $100 billion each year. Paper pulp accounts for only about 6 percent of the annual wood harvest. Developed countries produce approximately 60 percent of all industrial wood and account for about 80 percent of its consumption. Less-developed countries, mainly in the tropics, produce the other 40 percent of industrial wood but use only about 20 percent.

The United States, the former Soviet Union, and Canada are the largest producers of both industrial wood (lumber and panels) and paper pulp. Although old-growth, virgin forest with trees large enough to make plywood or clear furniture lumber is diminishing everywhere (fig. 8.4), much of the industrial logging in North America and Europe occurs in managed forests where cut trees are replaced by new seedlings. In contrast, hardwood logging in Southeast Asia, Africa, and Latin America usually leaves devastated forests that have lost both soil fertility and biological capital. Japan is by far the largest net importer of wood in the world, purchasing about 43 million cubic meters per year. China is the second largest net importer, buying some 7 million cubic meters annually. Interestingly, the United States is both a major exporter and importer of wood and wood products since we buy wood and paper pulp from Canada and Latin America at the same time we sell raw logs, rough lumber, and waste paper to countries like Japan and China.

More than half of the people in the world depend on firewood or charcoal as their principal source of heating and cooking fuel (fig. 8.5). Consequently, **fuelwood** accounts for slightly more than one-half of all wood harvested worldwide. Unfortunately, burgeoning populations and dwindling forests are causing wood shortages in many less-developed countries. About 1.5 billion people who depend on fuelwood as their primary energy source have less than they need. At present rates of population growth and wood consumption, the demand probably will be twice the available fuelwood supply by 2025. The average amount of wood used for cooking and heating in less-developed countries averages about one

Figure 8.4 The northern white pine forests of Minnesota and Wisconsin provided a wealth of timber a century ago. Without reforestation, however, the resource was rapidly depleted, and loggers moved to the lush forests of the West.

Figure 8.5 More than half of the world's people rely on firewood or charcoal as their main source of fuel. If present trends continue, the demand for wood will be twice the available supply, and a severe energy shortage for the poorest people will result.

cubic meter per person per year, roughly equal to the amount of wood that each American consumes as paper products alone.

Forest Management

Roughly 25 percent of the world's forests are managed scientifically for wood production. **Forest management** involves planning for sustainable harvests, with particular attention to forest regeneration. Aside from human use, what are some factors that contribute to forest loss? Fires, insects, and diseases damage up to one-quarter of the annual growth in temperate forests. Recently, reduced forest growth and sudden die-off of certain tree species in industrialized countries have caused great concern. It is thought that long-range transport of air pollutants (chapter 10) is contributing to this sudden forest death, but not all the causes and solutions are yet understood.

Most countries replant far less forest than is harvested or converted to other uses each year, but there are some outstanding examples of successful reforestation. China, for instance, has planted an average of 4.5 million hectares per year during the past decade. South Korea, after losing nearly all its trees during the civil war thirty years ago, is now about 70 percent forested again. In spite of being the world's largest net importer of wood, Japan has increased forests to approximately 68 percent of its land area. Strict environmental laws and constraints on the harvesting of local forests encourage imports so that Japan's forests are being preserved while it uses those of its trading partners. It is estimated that two-thirds of all tropical hardwoods cut in Asia are shipped to Japan.

Most reforestation projects involve large plantations of single-species, single-use, intensive cropping called **monoculture agroforestry.** Although this is an efficient approach, high density of a single species encourages pest and disease infestations. High levels of pesticides and herbicides often are required. This type of management lends itself to clear-cut harvesting, which not only saves money and labor, but also tends to leave soil exposed to erosion. Monoculture also often requires higher fertilizer inputs than does a mixed-species forest. Obviously, the biological community is different than it would be in the natural state. Where profits from these agroforest projects go to absentee landlords or government agencies, local people have little incentive to prevent fires or keep grazing animals out of newly planted areas.

Very promising alternative reforestation plans are being promoted by conservation and public service organizations that sponsor planting of community woodlots of fast-growing, multipurpose trees, such as *Leucaena.* Millions of seedlings have been planted in hundreds of self-help projects in Asia, Africa, and Latin America. *Leucaena* is a legume, so it fixes nitrogen and improves the soil. Its nutritious leaves are good livestock fodder. It can grow up to 3 m per year and quickly provides shade, forage for livestock, firewood, and good lumber for building.

TROPICAL FORESTS

The richest and most diverse terrestrial ecosystems on the earth are the tropical forests (chapter 3). Although they now occupy less than 10 percent of the earth's land surface, these forests are thought to contain more than two-thirds of all plant biomass and at least one-half of all plant, animal, and microbial species in the world.

The Diminishing Forests

While temperate forests are expanding slightly due to reforestation and abandonment of marginal farmlands, tropical forests are shrinking rapidly (see fig. 8.1). Biogeographers suggest that tropical forest destruction threatens extinction of millions of species of organisms (see chapter 9 for more detail), a human-caused environmental calamity of unprecedented proportions.

At the beginning of this century, an estimated 20 million square km of the tropics were covered with closed-canopy forest, an area about twice as big as the United States. About 25 percent of that forest has been seriously degraded and another 25 percent has been completely destroyed by human activities, mostly in the past thirty to forty years. Noted tropical

forest ecologist Norman Meyers estimates that current rates of forest clearing are about 200,000 sq km (an area the size of Minnesota) each year. Half of this destruction is the work of farmers or ranchers who move into the jungles in search of agricultural or grazing land (fig. 8.6). Most of the rest is caused by commercial logging or fuelwood gathering.

Brazil has by far both the largest total area and the highest rate of deforestation in the world, estimated by the United Nations to have been 6.8 million hectares of forest cut annually between 1986 and 1991 (fig. 8.7). Brazil disputes these figures and claims the rate is closer to 2 million hectares per year. The United Nations estimates that West Africa has the highest regional forest losses (about 2 percent per year), followed closely by Central America and Mexico (1.8 percent per year) and Southeast Asia (1.6 percent per year). This forest destruction is often accompanied by burning that makes a major contribution of carbon dioxide to the atmosphere (chapter 10).

The coastal forests of Ecuador, Sierra Leone, Ghana, Madagascar, Cameroon, Liberia, and Brazil, for instance, are almost completely gone. Haiti was once 80 percent forested; today, essentially all that forest has been cut and the land lies barren and eroded. India, Burma, Cambodia, Thailand, and Vietnam all have little virgin lowland forest left. Malaysia and Indonesia are liquidating their forests on the Island of Borneo at a disastrous rate.

A Cycle of Destruction

The first step in forest destruction is usually an invasion by loggers seeking valuable hardwoods, such as teak, mahogany, sandalwood, or ebony. Although only one or two trees per hectare might be taken, widespread devastation usually results. Because the canopy

Figure 8.6 Cattle graze on recently cleared tropical rainforest land in Costa Rica. About two-thirds of the forest in Central America has been destroyed, mostly in the past few decades as land is converted to pasture or cropland. Unfortunately, the soil is poorly suited to grazing or farming, and these ventures usually fail in a few years.

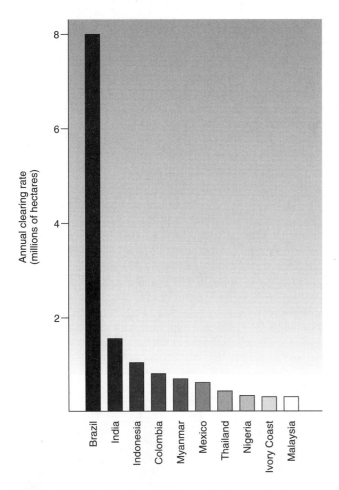

Figure 8.7 Deforestation rates in tropical countries.

of tropical forests is usually tightly linked by vines and interlocking branches, felling one tree can easily bring down a dozen others. Tractors dragging out logs damage more trees, and construction of roads takes large land areas. Insects and infections invade wounded trees. Tropical trees, which usually have shallow root systems, are easily toppled by wind and erosion when they are no longer supported by their neighbors. Up to three-fourths of the canopy may be destroyed for the sake of a few logs (see fig. 8.1). Obviously, the complex biological community of the layered canopy is severely disrupted by this practice.

What happens next? Bulldozed roads make it possible for land-hungry immigrants to move into the forest in search of land to farm. Some governments are encouraging such resettlement to ease population pressures in other areas. People with little experience or understanding of the complex rainforest ecosystem try to turn it into farms. Instead of farming small, temporary clearings of mixed crops as do indigenous tribal people, these newcomers tend to plant large fields of annual row crops. The result is ecological disaster. Rains wash away the topsoil and the tropical sun bakes the exposed subsoil into an impervious hardpan that is nearly useless for crops or grazing, forcing farmers and ranchers further into the forest to start the cycle of destruction all over again.

River degradation is another disastrous result of forest clearing. Tropical rivers carry two-thirds of all freshwater runoff in the world. In an undisturbed forest, rivers are usually clear and clean and flow year-round because of the "sponge effect" of the thick root mat created by the trees. When the forest is disrupted and the thin forest soil is exposed, erosion quickly carries away the soil, silting river bottoms, filling reservoirs, ruining hydroelectric and irrigation projects, filling estuaries, and smothering coral reefs offshore. In Malaysia, sediment yield from an undisturbed primary forest was about 100 cubic m per sq km per year. After forest clearing, the same river carried 2,500 cubic m per sq km per year.

Thinking Globally: Box 8.1

Murder in the Rainforest

*F*rancisco "Chico" Mendes was born in 1944 on the far upper reaches of the Amazon River in Acre Province of northwestern Brazil. Like his father before him, Chico was a *seringueiro,* one of about one hundred thousand independent rubber tappers who live with their families in remote areas of the tropical rainforest. Like the other *seringueiros,* Chico grew a few crops in the small clearing around his house. His main income, however, came from several hundred wild rubber trees in the forest that he visited every week. By carefully making two long V-shaped cuts in the bark of the rubber tree, Chico was able to collect a cup or two of milky latex sap every week from each tree. The latex was dried to make natural rubber (box fig. 8.1.1).

With only two cuts per week, the *seringueiro* doesn't hurt the tree, and production can continue for many years. From his father, Chico inherited trees that had been tapped for fifty years or more. In addition to rubber, the *seringueiros* collect Brazil nuts, wild fruits, and other natural products from the forest. This sustainable harvest allows them to live there without destroying the forest or reducing its natural diversity.

Other people have plans for the forest, however, that don't include *seringueiros* or wild rubber trees. Land speculators and big ranching companies can make quick profits by cutting down the valuable tropical hardwoods and converting the forest to cattle pasture. The first year

Figure 8.1.1 Chico Mendes was a leader of the *seringueiros*, or rubber tappers, in Acre province of the Brazilian Amazon. His murder in 1988 became an international scandal and led directly to establishment of extractive forest reserves that will preserve both the forest and the way of life of his people.

after forest clearing and burning, the grass grows luxuriantly. Soon, however, pounding tropical rains wash away nutrients, so that after a few years it takes 2 to 5 hectares (5 to 10 acres) to support a single cow.

The ranchers need to clear enormous areas each year to maintain their herds. Satellite photos showed that 12 million hectares (30 million acres) of forest—an area the size of South Carolina—were cleared and burned in Brazil in 1988. This destruction was encouraged by low land prices, favorable tax rates, and direct subsidies to ranchers to convert virgin forest to pasture. In addition, roads were built and maintained by the government to open up the forest and make it possible to ship cattle to market.

In the early 1970s, Brazil built highway BR 364 into Rondonia Province, immediately to the south of Acre, with the help of a $200 million loan from the World Bank. A flood of loggers, land speculators, ranchers, and landless peasants poured into the area. Within a decade, more than a third of the Rondonian forest had been destroyed. Native tribal people and *seringueiros* were driven out.

By the 1980s, the developers turned their attention to Acre Province. Brazil asked the Inter-American Development Bank (IDB) for another $200 million to extend BR 364 into the Xapuri region of western Acre where Chico lived. Chico and other *seringueiros* organized to resist this development. They proposed that instead of cutting down the forest it should be declared an "extractive reserve" where traditional patterns of harvest could continue unhindered.

Ecologists and plant biologists supported the idea of sustainable forest yield. They found that logging brought a one-time profit of $3,184 per hectare, on average, followed by only about $150 per year as cattle pasture. By contrast, the native wild fruits, nuts, rubber, and other natural products gathered by the *seringueiros* from that same hectare had a potential value of $6,820 every year.

In 1986, Chico Mendes ran for a seat in the legislature as a Worker's Party candidate on a forest-protection platform. His opponent was heavily financed by loggers, ranchers, and land speculators. Chico won the popular vote but lost in the electoral count. Charges of fraud and vote buying were widespread. Shortly after the election, paving of the Acre road began.

With the help of American conservation organizations, Chico made a trip to Washington, D.C. in 1987 to lobby the United States Congress on behalf of forest protection. His arguments were persuasive. The Senate Appropriations Committee denied $200 million (out of a total of $258 million) requested by the IDB because of the environmental destruction caused by building road BR 364. The IDB immediately suspended payments to Brazil.

When Chico returned home, he was invited to meet with the Interior Ministry. In a major victory for the *seringueiros,* several extractive reserves were established in 1988, ending almost all forest clearing in Acre. The ranchers were furious, especially Darli Alves da Silva, who claimed a large area in the Seringal Cachoeira extractive reserve where Chico had grown up. He vowed that Chico would not live out the year.

On the evening of December 22, 1988, Chico was playing cards at home with his wife, Ilzmar, and the two policemen who were assigned to guard him. He took a towel and went out the back door to the shower in the outdoor bathroom. As he opened the door, Chico was hit in the chest by a shotgun blast from the backyard. He staggered back into the house as the policemen ran out the front door. Although the police station was right across the street, no one responded to Ilzmar's cries for help. Chico died immediately.

The local police said they had no clues or suspects in the case. A storm of protest, both local and international, forced the Brazilian government to intervene. Investigators soon followed a trail of evidence to the ranch of Darli da Silva. In 1990, Darli, his son Darci Pereia da Silva (one of his thirty children), and a ranchhand, Jerdeir Pereia, were convicted of murder. They escaped from prison in 1993, apparently with outside help.

Responsibility for the killing of Chico Mendes may lead beyond Darli, Darci, and Jerdeir. Other prominent ranchers and land brokers have been accused of paying Darli to kill Chico. Even further, it can be argued that politics, economics, and world trade played a role. Certainly international development loans and cash crops that encourage forest destruction are factors. Perhaps each of us bears some responsibility as well. Do you know where the wood veneer on products you buy comes from, or where the beef in your fast-food burger or hot dog was grown? What connection might each of us have to the death of a *seringueiro* deep in the Amazonian forest of Brazil?

Forest Protection

What can be done to stop this destruction and encourage careful management? While there is much discouraging news, there are also some hopeful signs for tropical forest conservation. In 1985, 74 tropical countries representing both forest product consumers and producers worked out a blueprint for forest protection and management called the Tropical Forestry Action Plan. This document calls for (1) reform of national policies that encourage forest destruction, (2) creation of new international agreements on trade, aid, and debt relief, (3) recognition of the rights of indigenous people, and (4) coordination of overlapping jurisdictions that hinder forest protection. Critics of this plan claim, however, that it was engineered by the rich countries to preserve their supply of wood, and that it just channels more aid money into logging rather than really preserving forests. We will have to wait to see how effective this plan will be.

Many tropical countries have realized that their forests represent a valuable resource, and they are taking individual steps to protect them. Indonesia has announced plans to preserve 100,000 sq km, one-tenth of its original forest. Zaire and Brazil each plan to protect 350,000 sq km (about the size of Norway) in parks and forest preserves. Costa Rica has one of the best plans for forest protection in the world. Attempts are being made there to not only **rehabilitate land** (utilitarian program to make an area useful to humans) but also to **restore ecosystems** (reinstate an entire community of organisms to naturally occurring association). One of the best known of these projects is ecologist Dan Janzen's work on restoration of the dry tropical forest of Guanacaste National Park in Costa Rica (box 8.2).

Thinking Globally: Box 8.2

Restoring a Dry Tropical Forest

When the Spanish *conquistadores* arrived in Central America in the sixteenth century, about 5.5 million hectares (21,000 square miles) of dry tropical forest stretched along the Pacific coast from Columbia to Mexico. In contrast to the evergreen rainforests and cloud forests on the Atlantic side of the isthmus, dry forests have distinct seasons. During the wet summer months, the vegetation is dense and lush. In the winter, however, when rains are sparse, trees and bushes lose their leaves and the whole forest becomes open and desertlike.

This dry forest was much easier to turn into farms and ranches than were moist forests. Its climate is healthier, and its soil was more fertile and conducive to agriculture. Today, only about 1 percent of Central America's dry forest remains in anything like its original condition, making it one of the most threatened ecosystems in the world. As the forest has disappeared, many of its unique plant and animal species have become rare and endangered. If much more forest is lost, hundreds or even thousands of species will become extinct.

An exciting project is currently under way in Costa Rica where scientists and local residents have joined together to restore about 700 square kilometers (28,000 acres) of dry tropical forest to approximately its original condition. A new national park called Guanacaste (named after the Costa Rican national tree that once grew in this forest) is being created from private lands, an existing park, and other public land holdings. Under the leadership of entomologist Dan Janzen, attempts are being made to understand the ecosystem and to reintroduce native plants and animals in an effort to restore—rather than just rehabilitate—the forest.

How is this possible after the land has been abused and degraded for centuries? Isn't it long past the point at which it can be rescued? Fortunately, according to Janzen, most of the original flora and fauna have not been completely eliminated, only reduced. Small areas containing most of the indigenous (native) species remain scattered across the countryside. The challenge is to find these species and create habitats where they can thrive and re-create the forest.

Fire is one of the greatest threats to the forest. Every year during the dry season local people accidentally or deliberately start fires that sweep across the land, destroying native species and converting forest to grassland full of non-native invaders. Creating breaks to control the spread of fire and persuading residents to fight fires rather than set them is the first step toward restoring the forest.

Contrary to what you might expect, grazing animals are not excluded from Guanacaste National Park. In fact, they are encouraged because they are efficient seed dispersers. The forest probably coevolved with a fauna that included large, hooved grazing animals before humans arrived, so many plant species actually depend on animals for regeneration. Horses, monkeys, goats, birds, and even turtles eat fruits and pass their seeds through their digestive system days or weeks later. This not only distributes seeds to new locations, it also provides fertilizer for their initial growth. Furthermore, some seeds have tough outer coverings that are weakened by digestive acids and enzymes, facilitating germination. Being able to use the new national park for grazing during restoration makes the whole process much more attractive to its neighbors.

Involving local people in the project and making the park economically beneficial to them is another of the essential keys to successful restoration (box fig. 8.2.1).

When they see how a park will help them, residents will be enthusiastic participants. Native people, with their knowledge of the forest and their skills as land stewards, can be an invaluable resource in the restoration process.

Once Guanacaste National Park is reconstituted, locals can work as guides and rangers or provide services to tourists who come to visit and view wildlife. Providing jobs in the area will help stem the tide of urbanization and also preserve local culture, saving biodiversity and cultural heritage simultaneously. This exciting project may serve as an inspiration and guide to similar efforts in many areas of the world where bad land-use practices threaten both wildlife and indigenous people.

Figure 8.2.1 Former cowboys guard recovering pasturelands in the new Guanacaste National Park in Costa Rica where they once herded cattle. By involving local people in park management, the park service benefits both from their knowledge of the area and their support of ecosystem restoration.

Debt-for-Nature Swaps

Financing nature protection is often a problem in developing countries where the need is greatest. One promising approach is called debt-for-nature swaps. Banks, governments, and lending institutions now hold nearly $1 trillion in loans to developing countries. There is little prospect of ever collecting much of this debt, and banks are often willing to sell bonds at a steep discount—perhaps as little as 10 cents on the dollar. Conservation organizations buy debt obligations on the secondary market at a discount and then offer to cancel the debt if the debtor country will agree to protect or restore an area of biological importance.

The first such swap was made in 1987. Conservation International bought $650,000 of Bolivia's debt for $100,000—an 85 percent discount. In exchange for canceling this debt, Bolivia agreed to protect nearly 1 million hectares (2.47 million acres) around the Beni Biosphere Reserve in the Andean foothills. Ecuador and Costa Rica have had a different kind of debt-for-nature swap. They exchanged debt for local currency bonds that are used to fund activities of local private conservation organizations in their countries. This has the dual advantage of building and supporting indigenous environmental groups while also protecting the land.

Some other countries involved in debt-for-nature swaps include Madagascar, Zambia, Peru, Mexico, and Tanzania. In 1990 the United States Food, Agriculture, and Conservation Act authorized conversion of some or all of the $7 billion in outstanding USAID or Food for Peace loans to conservation programs. This could have a major impact on forest and biodiversity preser-

vation. Critics charge, however, that these swaps compromise national sovereignty and have little security. How can we be sure that countries with high political instability will continue to honor previous agreements for land protection?

NORTHERN FORESTS

The two main issues in timber management in developed countries such as the United States and Canada are (1) cutting of the last remnants of old-growth forest and (2) methods used in timber harvest. Canada is third and the United States is fourth in the world—after Russia and Brazil, which are first and second, respectively—in terms of total forest area. More than 90 percent of Canada's forests are crown lands (managed by federal, provincial, or territorial governments) while only 26 percent of U.S. forests are contained in the 191 national forests administered by the U.S. Forest Service.

Ancient Forests

Only a century ago, most of the coastal ranges of Washington, Oregon, northern California, British Columbia, and southeastern Alaska were clothed in a lush forest of huge, ancient trees (fig. 8.8). The moist, mild climate and rich soil of the lowland valleys nurtured magnificent stands of redwood in California and Douglas fir, western red cedar, hemlock, and Sitka spruce along the rest of the coast. Everyone knows that redwoods can be huge and very old, but did you know that these other species can reach 3 to 4 m (9 to 12 ft) in diameter, 90 m in height (as high as a twenty-story building), and can grow 1,000 or more years old? These temperate rainforests are probably second only to tropical rainforests in terms of terrestrial

Figure 8.8 The Quinauet Rain Forest on the Olympic Peninsula in Washington is an excellent example of the old-growth forests of the Pacific northwest. Huge Sitka spruce, Douglas fir, hemlock, and western cedar form a dense, multilayered, closed canopy that shelters a number of rare and endangered plant and animal species. The forest floor is dark, moist, and quiet, cushioned with a layer of fallen needles, moss, and ferns. Downed tree trunks serve as a nursery bed for new growth. Unfortunately, these temperate rainforests are disappearing at an alarming rate and only scattered remnants remain.

biodiversity, and they accumulate more total biomass in standing vegetation than any other ecosystem on the earth.

These old-growth forests (where at least some trees are more than 200 years old) are extremely complex ecologically. Only in recent years have we begun to realize how many different species live there and how interrelated their life cycles are. Many endemic species such as the Northern Spotted Owl, Vaux's Swift, and the Marbled Murlets are so highly adapted to the unique conditions of these ancient forests that they can live nowhere else.

Before loggers and settlers arrived, there were probably 12.5 million hectares (31 million acres) of virgin temperate rainforest in the Pacific Northwest. Less than 10 percent of that forest in the United States still remains, and 80 percent of what is left is scheduled to be cut down in the near future. British Columbia has felled at least 60 percent of its richest and most productive ancient forests and is now cutting some 240 million hectares (600 million acres) annually, about ten times the rate of old-growth harvest in the United States. At these rates, no ancient forests will remain in North America in fifty years except for a fringe around the base of the mountains in a few national parks.

Wilderness and Wildlife Protection

Many environmentalists would like to save all the remaining virgin forest in the United States as a refuge for endangered wildlife, a laboratory for scientific study, and a place for recreation and spiritual renewal. The economic pressures to harvest the valuable giant trees are considerable, however. The forest products industry employs about 150,000 people in the Pacific Northwest and adds about $7 billion annually to the economy. This is about one-fifth of the gross state product in Oregon. Many small towns depend almost entirely on logging for their economic life.

In 1989, environmentalists sued the U.S. Forest Service over plans to clear-cut most of the remaining old-growth forest, arguing that spotted owls are endangered and must be protected under the Endangered Species Act. A federal judge agreed and ordered some 1 million hectares (2.5 million acres) of ancient forest to be set aside to preserve the last 1,000 pairs of owls. This would be about half the remaining virgin forest in Washington and Oregon. The timber industry claimed that 40,000 jobs would be lost, although environmentalists dispute this number. Outrage in the logging communities was loud and clear. Convoys of logging trucks converged on protest sites while angry crowds burned environmentalists in effigy. Bumper stickers urged "Save a logger; eat an owl" and "I love owls: poached, fried, or stewed."

Environmentalists agree that logging jobs are disappearing but claim the loss is mostly due to mechanization, a naturally dwindling resource base, and the shipping of raw logs to Japan. They argue that the big trees are disappearing anyway. The question is whether to stop cutting now while a few are left, or in a few years when they are all gone. The workers will have to be retrained anyway; why not sooner rather than later?

Harvest Methods

Loggers generally prefer to **clear-cut** the forest—that is, cut every tree in a given stand regardless of species or size (fig. 8.9). This method makes it possible to use large machines to fell, trim, and haul logs very cheaply, but it often wastes many small trees that could have protected the soil and regenerated the forest. This heavily mechanized approach also often drives out wildlife and destroys natural communities. Forests of early successional species, such as aspen, jack pine, and lodgepole pine often respond well to clear-cutting if the harvest blocks are less than 5 hectares (12 acres), but loggers usually find such small areas uneconomical.

The lush Douglas fir and redwood forests of the rainy Pacific Coast ranges also regenerate rapidly after clear-cutting. The problem is that many of these forests are on steep slopes where erosion is a serious problem. Hillsides are stripped of soil that fills streams and smothers aquatic life. British Columbia has felled at least 60 percent of the province's richest and most

Figure 8.9 Clear-cutting and road cutting on steep, unstable slopes expose soil to erosion, damage watersheds, displace wildlife, and make forest regeneration difficult, if not impossible.

productive ancient forests. About 5 percent of the province—but only 2 percent of the temperate rainforest—is preserved in parks, ecological reserves, or recreational areas.

Vancouver Island, 384 km long (240 mi), is being logged faster than any other part of British Columbia. Indigenous people have blocked roads and brought law suits to protest destruction of traditional lands and subsistence ways of life. People concerned about commercial and sport fishing have joined the battle both in British Columbia and in the United States. They argue that salmon spawning depends on the clear cold streams of the native forests. Harvesting timber often destroys this valuable resource. The income from a single year's salmon run can outweigh all the profits from timber harvesting, and the salmon return year after year while 1,000-year-old trees will never be seen again.

RANGELANDS

Pasture (generally enclosed domestic meadows or managed grazing lands) and **open range** (unfenced, natural grazing lands) occupy about 24 percent of the world's land surface, generally where rainfall is too scarce or seasonal to support forests or croplands. Still home to vast herds of wildlife in many places (fig. 8.10), the main commercial use of this land is raising livestock. More than 3 billion domestic grazing animals turn grass and forage into protein-rich meat and milk that make a valuable contribution to human nutrition. Asia has the largest number of domestic grazing animals of any region, because of its vast grassy steppes. Australia and New Zealand are major ranching countries with more sheep than people. The 3 billion hectares (12 million sq mi) of permanent grazing land (both pasture and open range) in the

world is about twice the area of all agricultural crops. When you add to this more than 1 billion hectares of open woodlands and 4 billion hectares of other lands (e.g., desert, tundra, marsh, and scrub) that are used seasonally or in favorable years for raising livestock, nearly one-half of the total landmass of the earth is used at least occasionally as grazing land for domestic animals.

Range Management

By carefully monitoring the numbers of animals and the condition of the range, ranchers and pastoralists (people who live by herding animals) can adjust to variations in rainfall, seasonal plant conditions, and nutritional quality of forage to keep livestock healthy and avoid overusing any particular area. Conscientious management can actually improve the quality of the range.

Some nomadic pastoralists who follow traditional migration routes and animal management practices produce admirable yields from harsh and inhospitable regions. They can be ten times more productive than dryland farmers in the same area and come very close to maintaining the ecological balance, diversity, and productivity of wild ecosystems on their native range. Nomadic herding requires large open areas, however, and wars, political problems, travel restrictions, incursions by agriculturalists, growing populations, and changing climatic conditions on many traditional ranges have combined to disrupt an ancient and effective way of life. The social and environmental consequences often are tragic.

Overgrazing and Desertification

About one-third of the world's range is severely degraded by overgrazing (fig. 8.11). Among the countries with the most damage and the greatest area at

Figure 8.10 Range grazers are a valuable part of grassland ecosystems; they may be the best way to harvest biomass for human consumption. Native species often forage more efficiently, resist harsh climates, and are more pest- and disease-resistant than domestic livestock.

Figure 8.11 Sheep grazing on land in Guatemala that was once tropical forest. Soils in the tropics are often thin and nutrient-poor. When forests are felled, heavy rains carry away the soil and fields quickly degrade to barren scrubland. In some areas, land clearing has resulted in climate changes that have turned once lush forests into desert.

risk are Pakistan, Sudan, Zambia, Somalia, Iraq, and Bolivia. Usually, the first symptom of improper range management is elimination of the most palatable herbs and grasses. Grazing animals tend to select species they prefer and leave the tougher, less tasty plants. When native plant species are removed from the range, weedy invaders move in. Gradually, the nutritional value of the available forage declines. As overgrazing progresses, hungry animals strip the ground bare, and their hooves pulverize the soil, hastening erosion.

The process of denuding and degrading a once fertile land initiates a desert-producing cycle that feeds on itself and is called **desertification.** With nothing to hold back surface runoff, rain drains off quickly before it can soak into the soil to nourish plants or replenish groundwater. Springs and wells dry up. Trees and bushes not killed by browsing animals or humans scavenging for firewood or fodder for their animals die from drought. When the earth is denuded, the microclimate near the ground becomes inhospitable to seed germination. The dry barren surface reflects more of the sun's heat, changing wind patterns, driving away moisture-laden clouds, and leading to further desiccation. The International Soil Information Center in the Netherlands estimates that overgrazing causes some 35 percent of worldwide soil degradation, while deforestation and poor agricultural practices each contribute about 30 percent. Deserts have been called the footprints of civilization.

These processes are ancient, but in recent years they have been accelerated by expanding populations and political conditions that force people to overuse fragile lands (fig. 8.12). Those places that are most severely affected by drought are the desert margins, where rainfall is the single most important determinant in success or failure of both natural and human systems. In good years, herds and farms prosper and the human population grows. When drought comes, there is no reserve of food or water and starvation and suffering are widespread. Can we reverse this process? In some places, people are reclaiming deserts and repairing the effects of neglect and misuse.

Forage Conversion by Domestic Animals

Ruminant animals, such as cows, sheep, goats, buffalo, camels, and llamas, are especially efficient at turning plant material into protein because bacterial digestion in their multiple stomachs allows them to utilize cellulose and other complex carbohydrates that many mammals (including humans) cannot digest. As a result, they can forage on plant material from which we could otherwise extract little food value. Many grazers have very different feeding preferences and habits. Often the most effective use of rangelands is to maintain small mixed-species herds so that all vegetation types are utilized equally and none is overgrazed. Cattle and sheep, for instance, prefer grass and herbaceous plants, goats will browse on low woody shrubs, and camels can thrive on tree leaves and larger woody plants.

Worldwide, 85 percent of the forage for ruminants comes from native rangelands and pasture. In the United States, however, only 15 percent of livestock feed comes from native grasslands. The rest is made up of crops grown specifically for feed, particularly alfalfa, corn, and oats. Grain surpluses and our taste for well-marbled (fatty) meat have shifted livestock growing in the developed countries to feedlot confinement and high-quality diets. In the United States, roughly 90 percent of our total grain crop is used for livestock feed.

Harvesting Wild Animals

A few people in the world still depend on wild animals for a substantial part of their food. About one-half of the meat eaten in Botswana, for instance, is harvested from the wild. There are good reasons to turn even more to native species for a meat source.

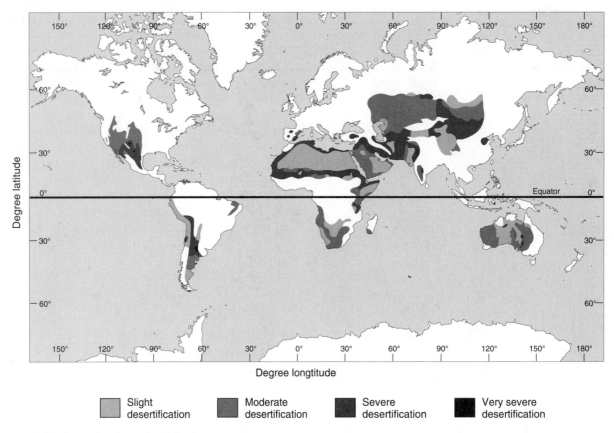

Figure 8.12 Desertification of arid lands.

On the African savannah, researchers are finding that springbok, eland, impala, kudu, gnu, oryx, and other native animals forage more efficiently, resist harsh climates, are more pest- and disease-resistant, and fend off predators better than domestic livestock. Native species also are members of natural biological communities and demonstrate niche diversification, spreading feeding pressure over numerous plant populations in an area. A study by the U.S. National Academy of Sciences concluded that the semiarid lands of the African Sahel can support only about 20 to 28 kg (44 to 62 lb) of cattle per hectare, but can produce nearly three times as much meat from wild ungulates (hooved mammals) in the same area. In the United States, native bison and elk are raised as "novelty" meat sources. Might there be a greater future role for ranching of wild animals on our own rangelands (box 8.3)?

RANGELANDS IN THE UNITED STATES

The United States has approximately 319 million hectares (788 million acres) of grazing lands. Most of this rangeland is in the West, and about 60 percent is privately owned. Of the 120 million cattle and 20 million sheep in the United States, only about 2 percent of the cattle and 10 percent of the sheep graze on public rangelands. Federal lands, thus, are not very important, overall, in livestock production, but they do have important local economic and environmental ramifications. The Bureau of Land Management (BLM) controls 84 million hectares (200 million acres) of grazing lands, and the U.S. Forest Service manages about one-fourth as much.

The BLM manages more land than any other agency in the United States, but the agency is little known outside of the western states where most of its lands are located. It was created in 1946 by a merger of the Grazing Service and the General Land Office. Its formation signaled an end to public land disposal and a commitment to permanent management by the government. The BLM has such a strong inclination toward resource utilization that critics claim the initials really stand for "bureau of livestock and mining." While only 25 percent of BLM land is considered suitable for grazing, its policies have an important effect on the economy and environment of western states.

*I*n 1990, Frank and Debora Popper of Rutgers University stirred up a furious controversy on the Great Plains by suggesting that some 3.6 million sq km (139,000 sq mi) of farms and ranches be turned into a wildlife preserve named the Buffalo Commons. Citing the dwindling water supplies, soil erosion, harsh climate, diminishing human population, and economic decline of the high plains, they argued that attempts to settle and domesticate the area have been the "largest, longest-running social and environmental miscalculation in our history."

The Poppers propose turning about one-quarter of the area between the Rocky Mountains and the one hundredth meridian back into open range stocked with native species such as bison, antelope, elk, and deer. These native grazers have varied feeding patterns that spread the demands for food among grasses and perennial broad-leaved plants and are thus more in balance with the native vegetation than those of cattle. Native wildlife also are better adapted to the harsh conditions of the high plains, requiring less water, winter feed, shelter, and other care than do domestic livestock. The commons over which wild animals would once again wander might be owned and managed by a consortium of public and private agencies. The few remaining residents might earn more money by harvesting wild game and guiding tourists than they do now in conventional farming and ranching.

The proposal for a wilderness like Buffalo Commons is not meant to force out farmers and ranchers who want to stay on their lands. Rather, it is a recognition that the land is emptying spontaneously as residents move away in search of jobs and a better life. The population of many areas has fallen below the critical mass of about six

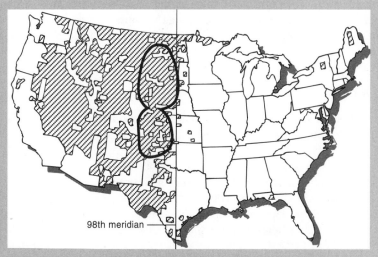

Figure 8.3.1 Shaded areas on this map indicate counties with fewer than six people per square mile. The Buffalo Commons could be established in a part of this depopulated area (circled) on the high plains of Kansas, Nebraska, and North and South Dakota.

Source: Data from F.T. Poppar, *Earth Island Journal*, Spring 1991, p. 34. Copyright © 1991 Earth Island Institute, San Franisco.

people per square mile required to sustain cultural and economic systems (box fig. 8.3.1). It is these relatively empty areas that might be turned back into native prairie.

Restoring something approaching the natural ecosystem also could help the plains adapt to potential climate change and avoid dust bowl conditions when drought comes again. It might also help preserve native American culture by offering an alternative to reservation life. Interestingly, this proposal is very similar to the recommendations of artist George Catlin in 1832 and of explorer and geologist John Wesley Powell in his 1878 *Report on the Arid Lands*. It is no surprise, however, that many ranchers, farmers, and developers strongly oppose the idea of a Buffalo Commons. What do you think? Is this a crazy idea?

State of the Range

The health of most public grazing lands in the United States is not good. Political and economic pressures encourage managers to increase grazing allotments beyond the carrying capacity of the range. Lack of enforcement of existing regulations and limited funds for range improvement have resulted in overgrazing, loss of native forage species, and erosion. A 1986 National Resources Defense Council survey concluded that only 30 percent of public rangelands are in "fair" condition, and 55 percent are "poor" or "very poor."

Overgrazing has allowed populations of unpalatable or inedible species, such as sage, mesquite, cheatgrass, and cactus, to build up on both public and private rangelands. Furthermore, competing herbivores, such as grasshoppers, jackrabbits, prairie dogs, and feral burros and horses, further damage the range and reduce the available forage. (A feral animal is a domestic animal that has taken up a wild existence.) Control programs using poison baits, traps, and hunting to reduce the numbers of these competing native and feral herbivores have been highly controversial because they are seen as both inhumane and dangerous. The case of wild burros and mustangs is especially diffi-cult. They reduce critical winter forage needed by both native wildlife and domestic stock. Most ecologists and land managers believe that the herds of feral animals must be controlled, but attempts to reduce their numbers or remove them from the range meet with vigorous opposition from those for whom they are a symbol of the freedom and romance of the West.

NATIONAL PARKS

The idea of establishing parks where nature is preserved for its own sake and where common people can enjoy wildlife, scenery, and outdoor recreation in a natural setting is a unique American contribution to the world. Parks have existed in other countries, of course, but they were usually menageries of exotic animals (and sometimes people), bizarre natural features, or private playgrounds for the rich and powerful.

The Origin of American Parks

Yosemite Park (established in 1864) and Yellowstone Park (1872) are generally regarded as the first national parks in the world set aside to protect the beauties of nature. Yosemite was authorized by President Abraham Lincoln in the midst of the American Civil War, surely a remarkable gesture of optimism and vision (fig. 8.13).

From the earliest days, the parks needed protection from the people for whom they were established. Army patrols were assigned to guard the park's resources, but hunters slaughtered game and timber thieves cut down the forests. Vandals wantonly de-stroyed natural wonders. In 1890 one group, hoping to witness a spectacular eruption, dumped a thousand pounds of rubbish into Old Faithful geyser. By the 1920s, auto tourism had become the most popular way to visit the parks, and a flood of people over-loaded the limited facilities. Concessionaires, sensing lucrative business possibilities, sold the entertainment value of the parks. Park Service priorities stressed the importance of visitors over natural features or wild-life. The evening entertainment at Camp Curry in Yosemite, for instance, drew two thousand tourists and featured jazz bands, vaudeville acts, a bear-feeding show, and dancing until midnight. One critic said that the honky-tonk atmosphere reminded him of Coney Island.

The National Park System Today

Our national park system has grown to more than 280,000 sq km (108,000 sq mi) in 341 parks, monuments, historic sites, and recreation areas. Each year we spend about 300 million visitor days in this system. The most heavily visited units are the urban recreation areas, parkways, and historic sites. The jewels of the park system, however, and what most people imagine when they think of a national park, are the great wilderness parks of the West: Yellowstone, Yosemite, Glacier, Rocky Mountain, Grand Canyon, Olympic, and Canyonlands. Passage of the Alaska Lands Act of 1980 nearly doubled the national park system.

Current Problems

Although the national parks are a wonderful resource, offering natural beauty and the potential for a peaceful and meaningful experience, they have their problems. Too many people concentrated in an area too small and coming to the parks for the wrong reasons remains a problem in most parks. Roads, trails, buildings, and natural communities have suffered from years of neglect. In 1988, the General Accounting Office reported that the parks needed $1.9 billion merely for repair and restoration. Little money was allocated for land acquisition in the 1980s. In 1978, the budget was $681 million; for 1989, the Reagan administration requested only $17 million.

Figure 8.13 Yosemite National Park, established in 1864, was the world's first public park set aside to protect natural beauty—a unique American contribution to the world.

Yosemite National Park is a good example of park problems: 95 percent of the overnight visitors never leave the 18 sq km (7 sq mi) valley floor, which is less than 1 percent of the total park area. They fill the valley with noise and smoke, trample fragile meadows to dust, and spend hours in traffic jams. Park rangers have become traffic cops and crowd-control specialists rather than naturalists. On Cape Cod National Seashore and in the new California Desert Park, dune buggies, dirt bikes, and off-road vehicles (ORV) run over fragile sand dunes, disturbing vegetation and wildlife and destroying the aesthetic experience of those who come to enjoy nature (fig. 8.14).

Pollution also has come to the parks. The haze over the Blue Ridge Parkway is no longer blue but gray-brown because of air pollution carried in by long-range transport. Visitors to the Grand Canyon could once see mountains 160 km (100 mi) away; now the air is so smoggy that you can't see from one rim to the other during one-third of the year. The main culprits are power plants in Utah and Arizona that supply electricity to Los Angeles, Phoenix, and other urban areas. Acid rain threatens sensitive lakes in the high mountains of the West, as well as in the Great Lakes and eastern states. Ozone is damaging the giant redwoods in California's Sequoia National Park, and unknown agents (probably air pollutants) are killing trees in the Great Smoky Mountains.

Mining and oil interests continue to push for permission to dig and drill in the parks, especially on the

Figure 8.14 Off-road machines such as motorcycles, dune buggies, and four-wheel drive vehicles can cause extensive and long-lasting damage to sensitive ecosystems. Tracks can persist for decades in deserts and wetlands where recovery is slow. Wildlife vegetation, solitude, and the natural beauty for which parks and recreation areas were set aside are negatively affected.

3 million acres of private inholdings in the parks. These forces were successful in excluding mineral lands in the Misty Fjords and Cape Kruzenstern National Monuments in Alaska. Placer mines, which wash sediments out of hillsides with high-pressure hoses in Denali National Park, dump thousands of tons of sediment each day into valuable salmon streams. Uranium mines at the edge of the Grand Canyon threaten to contaminate the Colorado River and the park's water supply with radioactive materials.

In Florida's Everglades National Park, water flow through the "river of grass" has been disrupted and polluted by encroaching farms and urban areas. Wading-bird populations have declined by 90 percent, down from 2.5 million in the 1930s to 250,000 now.

WILDERNESS AREAS

American culture and mythology are strongly influenced by our recent history as a wilderness country. As historian Frederick Jackson Turner pointed out in a series of articles and speeches around the turn of the century, Americans have traditionally believed that the wilderness was not only a source of wealth but also a symbol of strength, self-reliance, wisdom, and character. The frontier was seen as a source of continuous generation of democracy, social progress, economic growth, and national energy.

By 1900, the last frontier was closed, and people began to realize that the wilderness was endangered. The first official wilderness area in the United States was established in 1924 on the Gila National Forest in New Mexico, where noted author Aldo Leopold started his career as a young forest ranger in 1909. By 1939, under the leadership of wilderness advocate Bob Marshall, the U.S. Forest Service (USFS) had designated seventy-five large Wilderness Areas and forty-two smaller Wild Areas totaling nearly 8.4 million hectares (20.7 million acres). These were only administrative classifications, however, that could be changed at any time by the chief of the Forest Service. In 1956, the Wilderness Society (founded by Leopold and Marshall) began to work for legislative protection of these and other primitive areas. This campaign culminated in the passage in 1964 of the Wilderness Act, which legally defined **wilderness** as "an area of undeveloped land which is affected primarily by the forces of nature, where man is a visitor who does not remain; it contains ecological, geological, or other features of scientific or historic value; it possesses outstanding opportunities for solitude or a primitive and unconfined type of recreation; and it is an area large enough so that continued use will not change its unspoiled, natural conditions" (fig. 8.15).

Today, 264 units of the National Wilderness System encompass nearly 36 million hectares (88 million acres). Almost two-thirds of that total was added to the system by passage of the Alaska National Interest Conservation Land Act in 1980. Additional wilderness areas are being evaluated for protected status. The USFS was instructed by the 1964 act to carry out a roadless area review and evaluation (RARE) on all de facto wilderness areas under its jurisdiction. Using a deliberately "pure" interpretation that excluded all lands with any history of roads or development (even if all traces of human impact were long gone), it finally decided in 1979 that only about one-fourth of its 23 million hectares (56 million acres) of roadless areas qualified for protection.

Figure 8.15 The Absaroka-Beartooth Wilderness, north of Yellowstone National Park, preserves a piece of relatively untouched natural ecosystem. Areas such as this serve as a refuge for endangered wildlife, a place for outdoor recreation, a laboratory for scientific research, and a source of wonder, awe, and inspiration.

These proposed wilderness areas now are being considered by Congress on a state-by-state basis. Each bill usually pits environmental groups who want more wilderness against loggers, miners, ranchers, and others who want less wilderness. The arguments for saving wilderness are that it provides (1) a refuge for endangered wildlife, (2) an opportunity for solitude and primitive recreation, (3) a baseline for ecological research, and (4) an area where we have chosen simply to leave things in their natural state. The arguments against more wilderness are that timber, energy resources, and critical minerals contained on these lands are essential for economic development. To people who live in remote areas, jobs, personal freedom, and local control of resources seem more important than abstract values of wilderness. Wilderness proponents point out that 96 percent of the country already is open for resource exploitation; the 4 percent that has been set aside is mostly land that developers didn't want anyway.

The last large areas still being studied for wilderness preservation are on BLM land. The 1976 Federal Land Policy and Management Act (FLPMA) ordered the BLM to review all its roadless areas for wilderness potential. Applying the same "pure" standards used by the Forest Service, the BLM found only one-fourth of its 112 million hectares (276.6 million acres) suitable for wilderness. Preservationists argue that twice that much land should be considered.

WORLD PARKS AND PRESERVES

The idea of setting aside nature preserves has spread quickly as the world has become aware of the rapid disappearance of wildlife and wild places. So far, about 427 million hectares (3 percent of the earth's land) have been preserved in parks and wildlife refuges. Arctic biomes, deserts, and dry scrublands that have no economic value are overrepresented, making up nearly two-thirds of all protected areas. Other important biogeographical regions such as tropical or temperate rainforests are badly underrepresented, however. The International Union for the Conservation of Nature and Natural Resources (IUCN) has identified three thousand areas totaling about 3 billion hectares as worthy of national park or wildlife refuge status. The most significant of these areas are designated World Biosphere Reserves or World Heritage Sites. So far, only 144 of these special refuges have been selected (fig. 8.16).

The tropics are suffering the greatest destruction and loss of species in the world, especially in tropical rainforests. People in some of the affected countries are beginning to realize that the biological richness of their environment may be their most valuable resource and that its preservation is vital for sustainable development. Tourism alone can bring more into a country over the long term than extractive industries such as logging and mining.

Among the countries with the most admirable plans to protect natural values are Costa Rica, Tanzania, Rwanda, Botswana, Benin, Senegal, Central Africa Republic, Zimbabwe, Butan, and Switzerland, each of which has designated 10 percent or more of its land

Figure 8.16 Elephants have disappeared from much of their former range in Africa and Asia. They are endangered or threatened nearly everywhere. Will we be able to preserve this magnificent and interesting species?

as ecological protectorates. This is a larger percentage than the United States has set aside in parks, wilderness, and wildlife refuges combined. Brazil has even more ambitious plans, calling for some 231,600 square km or 18 percent of the country to be protected in nature preserves.

Does park or preserve status effectively protect a wild area? Present levels of protection and management cannot always be adequately administered and are subject to changes in political climate. Many problems threaten natural resources and environmental quality in the parks. Dam building for hydroelectric and irrigation projects, deforestation, oil drilling and pipelines, gold prospecting, timber and wildlife poachers, air pollution, and off-road vehicles all menace natural values in these areas. Many of the countries with the most important biomes lack funds, trained personnel, and experience to manage some of the areas under their control. Management of biological resources and endangered species is discussed further in chapter 15.

The IUCN has developed a **world conservation strategy** for natural resources that includes the following three objectives: "(1) To maintain essential ecological processes and life-support systems (such as soil regeneration and protection, recycling of nutrients, and cleansing of waters) on which human survival and development depend. (2) To preserve genetic diversity, on which depend the breeding programs necessary for protection and improvement of cultivated plants and domesticated animals. (3) To ensure that any utilization of species and ecosystems is sustainable." These goals are further elaborated in the ecological plan of action adopted by the IUCN and shown in table 8.1.

Conservation and Economic Development

Many of the most seriously threatened species and ecosystems of the world are in the developing countries, especially in the tropics. This situation concerns all of us because these countries are the guardians of biological resources that may be vital to all our futures. Unfortunately, where political and economic systems fail to provide people with land, jobs, and food, disenfranchised citizens turn to legally protected lands, plants, and animals for their needs. Immediate human survival always takes precedence over long-term environmental goals. Clearly, the struggle to save species and unique ecosystems cannot be divorced from the broader struggle to achieve a social order or global balance in which the basic needs of all are met.

At the 1982 World Congress on National Parks on the island of Bali, Indonesia, five hundred scientists, managers, and politicians discussed the design and location of biological reserves and the ecological, economic, and social factors that impinge on wildlife preservation. They concluded that conservation and rural development are not necessarily incompatible. In many cases, sustainable production of food, fiber, medicines, and water in rural areas depends on ecosystem services derived from adjacent conservation reserves. Tourism associated with wildlife watching and outdoor recreation can be a welcome source of income for underdeveloped countries. If local people share in the benefits of saving wildlife, they probably will cooperate and programs will be successful. To restate Thoreau's famous dictum: "In broadly shared economic progress is preservation of the world."

LAND OWNERSHIP AND LAND REFORM

Many of the problems discussed in this chapter have their roots in land ownership and public policies concerning resource use. Some land tenure patterns have supported or even created inequality, ignorance, and environmental degradation, while other systems of land use have promoted equality, liberty, and progress. In this final section, we will look at how fair access to the land and its benefits can help bring about better conditions for humans and their environment.

Who Owns How Much?

No place on earth is completely unoccupied. Every place, even Antarctica, is claimed by someone, although many places are very sparsely populated. Indigenous people still own and occupy about 25 percent of the land, but everywhere native lands are under threat of political or economic annexation.

Whatever political and economic system prevails in a given area, the largest landowners usually wield the most power and reap the most benefits, while the landless and near landless suffer at the bottom of the scale. This is a hard fact of life. The World Bank estimates that more than one billion people live in "absolute poverty . . . at the very margin of existence." Three-fourths of these wretched people are rural poor who have too little land to support themselves. Most of the other 250 million absolute poor are urban slum dwellers, many of whom migrated to the city after being forced off the land in their rural homes (chapter 14).

In many countries inequitable land ownership is a legacy of colonial estate systems on which landless peasants still work in virtual serfdom (fig. 8.17). For instance, the United Nations reports that only 7 percent of the landowners in Latin America own or control 93 percent of the productive agricultural land. By far the largest share of landless poor in the world are in South Asia. In Haiti, the poorest and most environmentally degraded country in the Western Hemisphere, 1 percent of the population owns 90 percent of the land.

The 86 million landless households in India comprise about 500 million people; they are two-thirds of both the total population of India and of all the landless rural people in the world. Not surprisingly, the countries with the widest disparity in wealth are often the most troubled by social unrest and political instability. As Adam Smith said in *The Wealth of Nations* in 1776, "No society can surely be flourishing and happy, of which the far greater part of the members are poor and miserable."

Figure 8.17 Landless rural peasants still work most of the land in Asia, Latin America, and Africa while they live in poverty.

Land tenure is not just a question of social justice and human dignity. Political, economic, and ecological side effects of inequitable land distribution affect those of us far from the immediate location of the problem. For example, people forced from their homes by increasing farm mechanization and cash crop production in developing countries often are forced to try to make a living on marginal land that should not be cultivated. The local ecological damage caused by their struggle for land can, collectively, have a global impact. The contribution of concentrated land ownership to these environmental problems, however, receives less attention than do the threatening ecological trends themselves.

Land Reform

Throughout history, some land redistribution movements have been successful and others have not. In many countries, significant land reforms have occurred only after violent revolutionary movements or civil wars. An important factor in revolutions in Cuba, Mexico, China, Peru, the former Soviet Union, and Nicaragua was the breaking of feudal or colonial land-ownership patterns. This struggle continues in many countries. Hundreds of large and small peasant movements demand land redistribution, economic security, and political autonomy in Latin America, Africa, and Asia. Entrenched powers respond with economic reprisals, intimidation, "disappearances," and even open warfare. Millions of people have been killed in such struggles in recent centuries alone.

Consider Brazil: 340 rich landowners possess 46.8 million hectares (117 million acres) of cropland, while 9.7 million families (70 percent of all rural people) are landless or nearly landless. Studies have shown that 13 percent of the land on the biggest estates is completely idle, while another 76 percent is unimproved pasture. When President José Sarney took over the government in 1985 after twenty-one years of right-wing military rule, he ordered sweeping land reform to redistribute land to the rural poor. Although the government has offered to pay for expropriated land, the large landowners are resisting and have spent millions of dollars to raise and arm private armies to protect the status quo. These inequities and the tensions they create are important factors in the invasion of the rainforest described in box 8.1.

In some countries, land reform has been relatively peaceful and also successful. In Taiwan in 1949, only 33 percent of the farm families owned the land they worked, whereas 59 percent of rural families farm their own land today. South Korea also has carried out sweeping land redistribution. In the 1950s, more than half of all farmers were landless, but now 90 percent of all South Korean farmers own at least part of the land they till. In general, those countries that have had successful land reform also have stabilized population growth and have begun industrialization and development.

What are the ecological implications of land ownership? Absentee landlords have little personal contact with the land, may not know or care about what happens to it, and often won't let sharecroppers cultivate the same land from year to year for fear that they may lay claim to it. This gives tenant farmers little incentive to protect or improve the land because they can't count on reaping any long-term benefits. In fact, if tenants do improve the land or increase their yields, their rents may be raised. Also, where land is aggregated into collective farms or large estates worked by landless peasants, productivity and health of the soil tend to suffer.

Far from being a costly concession to the idea of equality, land reform often can provide a key to agricultural modernization. In many countries, the economic case for land reform rivals the social case for redistributive policies. Many studies have shown that the productivity of owner-operated farms is significantly higher than that of corporate or absentee landowner farms. Independent farmers, especially those who have only a few hectares to grow crops, tend to lavish a great deal of effort and attention on their small plots, and their yields per hectare often are twice those of larger farms.

Figure 8.18 Landless peasants have been moved from Java to the jungles of Sumatra, but land in the area is not good for farming.

Resettlement Projects

Mass movements of people into new settlement areas often have serious adverse environmental consequences like those described in box 8.1. A number of current transmigration projects are continuing to cause large-scale tropical forest destruction and disruption of indigenous cultures (fig. 8.18). Indonesia, for example, announced plans in the early 1980s to move some 60 million people from the crowded islands of Java and Bali to the relatively sparsely settled outer islands of Sumatra, Borneo, and Irian Jaya (New Guinea). The rainforests to which these new settlers were sent were generally unsuitable for continuous agriculture for all the reasons we have already discussed in this chapter. The cycle of land degradation they set in motion left both the people and the land impoverished. Indigenous people also lost their traditional hunting and gardening plots, and many traditional cultures have been seriously disrupted or lost altogether.

After years of disastrous crops and incredible hardships, most of the Indonesian transmigrants ended up as ecological refugees in the slums of Jakarta or other burgeoning cities. This project has been scaled back in recent years, but other governments still see population transfer programs as a convenient way to claim territory, solve population pressures, and move subsistence farmers out of areas where they get in the way of large-scale, export-oriented, industrialized farming.

Summary

About 60 percent of the earth's land is used by humans in one way or another. Three major uses of this land are croplands, forests, and rangelands. The remaining 40 percent of the land is too steep, rocky, inhospitable in climate, or otherwise unsuitable for human use. Much of this land is valuable, however, as wilderness, refuge, wildlife habitat, or simply open space.

Forests cover a little more than one-quarter of the earth's land area, providing a variety of useful products such as lumber, pulpwood, and firewood. Tropical rainforests are in danger of destruction by overuse and exploitation, threatening extinction of millions of species, exposing land to erosion, and affecting global atmospheric and climatic patterns.

Range and pasture lands occupy a little more than one-fifth of the earth's land area. When not managed properly, these lands can be degraded and turned permanently into desert. Croplands occupy about 10 percent of the earth's land area, much of it also suffering the effects of poor land management.

Some lands should be preserved in their natural state, safe from human exploitation. We have made considerable progress in recent years creating parks, wildlife refuges, and wilderness areas. Debate continues, however, over management and expansion of these protected areas.

Land ownership plays a central role in determining how we use and care for the land. Inequitable access to land is common in many countries and forces the poor to use land unsuited to agriculture, while good land is monopolized by the rich. Redistributing the land more fairly could have beneficial results both socially and environmentally.

Review Questions

1. Which land category constitutes the greatest land area?
2. What are the major forest products and who uses them?
3. What are the advantages and disadvantages of monoculture agroforestry?
4. What is clear-cutting and how does it damage forest ecosystems?
5. What are some results of tropical deforestation?
6. How does overgrazing encourage undesirable forage species to flourish?
7. Why can grazing animals generate food on land that would otherwise be unusable by humans?
8. What are the three objectives of the conservation strategy proposed by the International Union for the Conservation of Nature?
9. List four current problems that threaten National Parks and suggest some things that could be done about them.
10. Describe how patterns of land ownership affect social equity and environmental management.

Questions for Critical or Reflective Thinking

1. On average, each American uses about 600 lbs of paper per year while a person in a less-developed country might use about 10 lbs per year. Is this wrong, considering that most of our paper pulp now comes from second-growth forests or commercial agroforestry plantations?
2. Landless poor people in tropical countries clear and burn rainforests to grow crops. If you had it in your power, would you force them out of the forest?
3. Japan is by far the largest net importer of tropical wood in the world while protecting and expanding its own forests. What additional information would you need to know to decide whether this is fair and equitable?
4. What messages are conveyed—explicitly or implicitly—by the first photograph in this chapter? What assumptions or attitudes on your part influence your reaction to it? If you were designing this chapter, what photograph would you choose as the introductory illustration?
5. Try some role-playing with your family, friends, or classmates. Imagine that one of you is a rancher whose family has lived on the high plains for several generations but now finds it difficult to make ends meet. Another might be a wildlife manager who favors a Buffalo Commons wildlife refuge. What issues would you want to discuss?

6. Loggers in the Pacific Northwest depend on jobs cutting old-growth timber. Wildlife needs intact forests to survive. How can we reconcile these competing interests? Make a list of pros and cons for cutting or preserving the forest. How would you weigh one against the other?

7. Surveys of National Park visitors show that what they really want in the parks are fast-food restaurants, shopping opportunities, parking spaces, and entertainment such as video arcades and amusement park rides. Put yourself in the place of a park superintendent. If you allow these activities you degrade some of the values of the park, but you raise funds and support for necessary maintenance. What would you do?

8. Wilderness means different things to different people. Describe what you think it would mean to a homeless urban person, a Third World farmer, an ecologist, and a young, upwardly mobile professional person. Where would they agree and disagree?

Key Terms

clear-cut 169
closed canopy forest 161
desertification 171
forest management 163
fuelwood 163
industrial timber 162
monoculture agroforestry 164
open canopy forest 161

open range 170
pasture 170
rehabilitate land 167
restore ecosystems 167
wilderness 175
woodland 161
world conservation strategy 177

Suggested Readings

Beasley, Conger. "The Forest for the Trees: Old Growth Logging on the Olympic Peninsula." *Buzzworm* 2, no. 1 (1992): 24-36. A good primer on ancient forests and their future.

Connelly, Joel. "Big Cut." *Sierra* 76, no. 3 (1992): 42. Loggers threaten magnificent ancient forests on British Columbia's Vancouver Island and the traditional way of life that depends on them.

Cooper, Marc. "Alerce Dreams." *Sierra* 77, no. 1 (1992): 122. A north/south coalition hopes to buy Chile's majestic *alerce* forests before all are cut for lumber.

Douglas, J., and R. A. Hart. *Forest Farming: Towards a Solution to Problems of World Hunger and Conservation.* Boulder: Westview Press, 1985. A description of agrosilvaculture, or tree farming, based on ecological principles, appropriate technology, and local community involvement.

Finkelstein, Max. "National Park Dreams." *Borealis,* Spring 1992, 32-42. Canada's Green Plan to protect natural areas is discussed in the quarterly journal of the Canadian Parks and Wilderness Society.

Hecht, S., and A. Cockburn. *Fate of the Forest.* New York: HarperCollins, 1991. Traces European exploitation of the Amazon starting with the rubber boom of the last century and continuing to the present. Argues that only "socialist ecology" can save the forest.

Marshall, G. "The Political Economy of Logging: A Case Study in Corruption." *The Ecologist* 20, no. 5 (1990): 174. Examines logging practices and corruption in Papau, New Guinea.

Mitchell, John G. "Love and War in the Big Woods." *Wilderness* 55, no. 196 (1992): 11-23. Describes the largest wilderness area in the Eastern United States, most of which is privately owned by logging and paper companies that want to liquidate their value.

Nations, J. D., and D. I. Komer. "Rainforests and the Hamburger Society." *Environment.* 25, no. 3 (1983): 12. A classic study of the destructive farming and land management practices in Central America.

Repetto, R. "Deforestation in the Tropics." *Scientific American* 262, no. 4 (1990): 36. Discusses policies to encourage forest preservation and to stop destruction of an irreplaceable resource.

Revkin, A. *The Burning Season: The Murder of Chico Mendes and the Fight for the Amazon Rain Forest.* New York: Houghton Mifflin, 1991. An excellent account of the complex fight to save the rainforest and of the rise to prominence and martyrdom of Chico Mendes.

Rifkin, Jeremy. "Beyond Beef." *The Utne Reader* 50 (March/April 1992): 96. Eating meat is bad for our health and bad for the environment according to vegetarians. Is it time to reassess our diet?

Royte, Elizabeth. "Imagining Paseo Pantera." *Audubon* 94, no. 6 (1992): 74–80. The Carr brothers dream of creating a 1,500-mile-long wilderness corridor to protect wildlife from Mexico to Panama. Can they succeed?

Shane, D. R. *Hoofprints on the Forest; Cattle Ranching and the Destruction of Latin America's Tropical Forest.* Philadelphia: Institute for the Study of Human Issues, 1986. A vivid account of the role that cattle ranching has played in the destruction of tropical forests in Latin America.

Shoumatoff, A. *The World Is Burning: Murder in the Rain Forest.* New York: Little, Brown, 1991. A frenzied and rather self-centered account of the reporter's personal travels in Brazil. Captures some of the flavor of the rainforest and its inhabitants.

Solbrig, Otto T., and Michael D. Young. "Toward a Sustainable and Equitable Future for Savannas." *Environment* 34, no. 3 (1992): 6. Shows how policies that preserve both the ecological base and the welfare of the nomadic people who traditionally inhabit drylands can bring about long-lasting protection of these lands.

Spears, J., and E. S. Ayensu. "Resources, Development and the New Century: Forestry." In *Global Possible,* edited by R. Repetto. New Haven: Yale University Press, 1985.

Stone, R. D. *Dreams of Amazonia.* New York: Viking-Penguin, 1985. A travelogue of local color, sights, sounds, and smells of the world's largest rainforest written by a former reporter and present officer of the World Wildlife Fund.

Sutton, S. L., T. C. Whitmore, and A. C. Chadwick, eds. *Tropical Rain Forest: Ecology and Management.* London: Blackwell Scientific Publications, 1983. Research papers and reviews from a symposium on the biology and politics of tropical rainforests.

Tropical Forests: A Call for Action. Part I. Washington, D.C.: World Resources Institute, 1985. Report of an International Task Force of the World Resources Institute, the World Bank, and the United Nations Development Program.

Weisman, Alan, and Sandy Tolan. "Out of Time." *Audubon* 94, no. 6 (1992): 68–73. As "civilization" intrudes on Latin America, traditional cultures are dying.

Williams, Ted. "Incite: He's Going to Have an Accident." *Audubon* 93, no. 2 (March 1991): 30. How Idaho cattle barons threatened a tough ranger who criticized their misuse of public lands.

Witte, John. "Deforestation in Zaire." *The Ecologist* 22, no. 2 (1992): 58. With the timber resources of West Africa nearly exhausted, logging companies are now moving into the forests of Zaire followed by impoverished, landless settlers.

Wuerther, G. "How the West Was Eaten." *Wilderness* 54, no. 192 (Spring 1991): 28. Grazing leases controlled by a few cattle and sheep barons are degrading the range while paying only a fraction of their true costs.

9
*B*IOLOGICAL *RESOURCES*

He prayeth well who loveth well both mankind, bird, and beast.

Samuel Taylor Coleridge

Objectives

After studying this chapter you should be able to:

- Report on the total number, relative distribution, and abundance of biological species.

- Describe the major categories of benefits that we derive from other organisms and give an example of each category.

- Compare and contrast human and natural causes of depletion or extinction of species as well as disruption of entire ecosystems.

- Sum up the direct and indirect ways that humans cause losses of biological resources.

- Evaluate the effectiveness of the Endangered Species Act and CITES in protecting endangered species.

- Understand how gap analysis, ecosystem management, and habitat protection can contribute to preserving biological resources.

- Appreciate the importance of social and economic factors in the design and function of nature reserves.

INTRODUCTION

As far as we know, our planet is the only place in the universe that supports life, yet there are few places on Earth that are not home to some kind of organism. From the most arid desert to the dripping rainforest, from the highest mountain peak to the deepest ocean trench, life occurs in a marvelous spectrum of sizes, colors, shapes, life cycles, and interrelationships. Think, for a moment, how remarkable, varied, abundant, and important the other living creatures are with whom we share the earth (fig. 9.1).

Although our understanding of the earth's organisms—its **biological resources**—is still imperfect, there is no doubt that the abundance and diversity of living organisms provide many benefits and make our world a beautiful and interesting place to live. Unfortunately, rapidly expanding human populations and resource consumption now threaten to deplete three types of biodiversity: the number of species, the genetic richness within species, and the variety of biological communities in the world.

Extinction, the irrevocable eradication of species, is a normal process of the natural world. Species die out and are replaced by others, often their own descendants, as part of evolutionary change. In undisturbed ecosystems, the rate of extinction appears to be about one species lost every ten years. In this century, however, human impacts on populations and ecosystems have accelerated that rate, causing hundreds or perhaps even thousands of species, subspe-

Figure 9.1 Living organisms can be both incredibly abundant and wonderfully diverse. The species with which we share the planet provide us with many benefits and make our world a beautiful and interesting place to live. To our shame, humans have often squandered these biological resources through carelessness, greed, and ignorance. Because living organisms are self-replenishing, however, populations can be rebuilt through careful management.

cies, and varieties to become extinct every year. If present trends continue, we may destroy *millions* of kinds of plants and animals in the next few decades.

In this chapter, we will look at some ways humans are causing depletion of biological abundance and loss of biodiversity, both directly and indirectly. First, we will consider some benefits we derive from other organisms to illustrate why we should care about the fate of the species threatened by our activities. Then we will examine some ways that biological resources can be—and are being—protected.

BIOLOGICAL RESOURCES

Living organisms can be both incredibly abundant and fascinatingly diverse. This section of the chapter discusses the nature and extent of the biological resources that the earth offers.

How Many Species Are There?

At the end of the great exploration era of the nineteenth century, some scientists confidently declared that every important kind of living thing on the earth would soon be found and named. Since then, however, further studies suggest that millions of new species and varieties remain to be studied scientifically. Ironically, we spend far more money to catalog the stars in the sky than we do to investigate the living organisms with which we share this planet, even though the stars are much more likely to be around for a long time and are less likely to be of value to us than are biological resources.

We now believe that the approximately 1.7 million species presently known (table 9.1) are only a small fraction of the total number that exist. Taxonomists (scientists who classify species) estimate that there are somewhere between 3 million and 30 million different species. In fact, there may be 30 million species of tropical insects alone. Most of the organisms yet to be discovered and classified will probably be invertebrates (animals without backbones), bacteria, and fungi.

Of all the world's species, only 10 to 15 percent live in North America and Europe. We're pretty certain that most of the larger and more conspicuous organisms here have been studied and classified. It is a rare occurrence to find a new species of higher plant or animal in the developed countries of the world. By contrast, tropical countries have an incredible diversity of organisms, most of which have never even been seen by scientists.

The Malay Peninsula, for instance, has at least 8,000 species of flowering plants, while Britain, with an area twice as large, has only 1,400 species. There are probably more botanists in Britain than there are species of higher plants. South America, on the other hand, has fewer than one hundred botanists to study perhaps 200,000 species of plants. The greatest diversity of organisms in the world are tropical arthropods (insects, spiders, and their kin), which probably make up 90 percent or more of all species on the earth.

Biological Abundance

If it is difficult to imagine the number of species alive today, it is staggering to try to comprehend the total number of living organisms in the world. If you have ever witnessed the spectacle of huge flocks of migrating waterfowl or the profuse life of a coral reef community, you have experienced some of the abundance of life on the earth. Many of the most plentiful species, however, are invisible to the unaided eye or are in remote places that most of us never visit.

The most numerous multicellular organism is thought to be krill: small, shrimplike crustaceans (fig. 9.2) that

Groups	Identified species	Estimated species
Mammals	4,170	4,300
Birds	8,715	9,000
Reptiles	5,115	6,000
Amphibians	3,125	3,500
Fishes	21,000	23,000
Invertebrates	1,300,000	4,400,000*
Vascular Plants	250,000	280,000
Nonvascular Plants	150,000	200,000
Total†	1,742,000	4,926,000

TABLE 9.1 *Number of living species by taxonomic group*

*This figure is a minimum. Recent research suggests there could be as many as 30 million insect species in tropical forests alone.

†Totals are rounded.

Source: Data from the World Resources Institute, 1986.

Leggitt

Figure 9.2 Krill are thought to be the most numerous multicellular organisms on Earth. These small (1–5 cm), shrimplike crustaceans (perhaps one quadrillion individuals) inhabit the Antarctic oceans where they are eaten by whales, penguins, fish, and birds in teeming multitudes.

live in Antarctic oceans and form the base of a rich food web that includes whales, seals, penguins, and fish. Summer krill swarms are estimated to contain 650 million metric tons of biomass and close to 10^{15} (one million billion) individual organisms. Schemes to harvest some of this vast resource for fertilizer or animal food threaten the whole antarctic food web. Humans probably make up the next largest biomass for a single species (about 250 million metric tons), even though we are not nearly as numerous as some others.

BENEFITS FROM BIOLOGICAL RESOURCES

We benefit from other organisms in many ways, some of which we may not appreciate until the species that provides that service is gone. Even seemingly obscure and insignificant organisms can play irreplaceable roles in the life-support systems on which we all depend. Although it seems that the earth teems with an inexhaustible diversity of life, that may not be the case.

Food

All of our food comes from other organisms. Although most of what we in developed countries eat comes from domesticated plants and animals, we still get at least part of our diet from wild species. Most seafood, for instance, is harvested from free-roaming wild organisms. Seafood isn't a very large percentage of the total caloric intake for most people; however, it constitutes a valuable, easily digested protein source that makes the difference between adequate nutrition and malnourishment in many developing countries—especially for women and children, who need more protein. Wild amphibians, reptiles, birds, and mammals also are used as food, to varying degrees, around the world.

Many wild plant species could make important contributions to human food supplies either as they are, or as a source of genetic material to improve domestic crops. Noted tropical ecologist Norman Meyers estimates that as many as eighty thousand edible wild plant species could be utilized by humans. Villagers in Indonesia, for instance, are thought to use some four thousand native plant and animal species for food. Few of these species have been explored for possible domestication or more widespread cultivation. A 1975 study by the U.S. National Academy of Science found that New Guinea has two hundred and fifty-one edible fruits, only forty-three of which have been cultivated widely.

Unfortunately, potentially valuable food species and the wild ancestors of our domestic crops are being destroyed by forest clearing, grazing, and conversion of wild lands to domestic crops before they can be identified and their genes can be preserved. Later in this chapter we will look at some of the programs underway to find useful wild species and preserve them in gene banks, botanical gardens, zoos, and nature preserves.

Industrial and Commercial Products

In addition to lumber, firewood, paper pulp, and other wood products, we get many valuable commercial products from forests and other natural communities. Rattan, cane, sisal, rubber, pectins, resins, gums, tannins, vegetable oils, waxes, and essential oils are among the products gathered in the wild. Many wild species may have useful characteristics. Guayule, for instance, is a shrub native to the deserts of Texas and New Mexico. It produces latex that is essentially identical to that harvested from rubber (*Hevea*) trees. Species in the Euphorbia family (milkweeds, etc.) are also being investigated as a source of rubber, alkaloids, and other valuable organic chemicals.

Medicine

Wild plants and animals are sources of drugs, analgesics (pain killers), pharmaceuticals, laxatives, antibiotics, heart regulators, anticancer and antiparasite drugs, blood pressure regulators, anticoagulants, enzymes, and hormones (table 9.2). More than half of all prescriptions in the United States contain some natural products. The total value of drugs from natural sources amounts to nearly $3 billion per year in the United States alone. Patent medicines, such as laxatives and cough and cold remedies from natural sources, add an equal amount to the economy. There

TABLE 9.2 Some natural medicinal products

Product	Source	Use
Penicillin	Fungus	Antibiotic
Bacitracin	Bacterium	Antibiotic
Tetracycline	Bacterium	Antibiotic
Erythromycin	Bacterium	Antibiotic
Digitalis	Foxglove	Heart stimulant
Quinine	Chincona bark	Malaria treatment
Diosgenin	Mexican yam	Birth control drug
Cortisone	Mexican yam	Anti-inflammation treatment
Cytarabine	Sponge	Leukemia cure
Vinblastine, vincristine	Periwinkle plant	Anticancer drugs
Reserpine	Rauwolfia	Hypertension drug
Bee venom	Bee	Arthritis relief
Allantoin	Blowfly larva	Wound healer
Morphine	Poppy	Analgesic

may be many unrecognized medicinal products yet to be discovered in wild populations.

Consider the success story of vinblastine and vincristine. These anticancer alkaloids are derived from the Madagascar periwinkle (*Catharanthus roseus*) (fig. 9.3). They inhibit the growth of cancer cells and are very effective in treating certain kinds of cancer. Twenty years ago, before these drugs were introduced, childhood leukemias were invariably fatal. Now the remission rate for some childhood leukemias is 99 percent. Hodgkin's disease was 98 percent fatal a few years ago, but now only 40 percent die, thanks to these compounds. The total value of the periwinkle crop is roughly $15 million per year, although Madagascar gets little of those profits.

In 1992, Merck, the world's largest pharmaceutical company, agreed to pay $1 million to the Instituto Nacional de Biodiversidad (INBIO) of Costa Rica for plant, insect, and microbe samples to be screened for useful medicinal products. INBIO, a public-private collaboration, is using local people trained as practical "parataxonimists" to locate and catalog all the native flora and fauna—between 500,000 and 1 million species—in Costa Rica. Selling information and specimens will help with scientific work as well as provide funds to protect nature. This may be a powerful model both for scientific information gathering and as a way for developing countries to share in the profits from their native resources.

Ecological Benefits

Natural biological communities play important ecological roles in producing and sustaining habitable environments. Soil formation, waste disposal, air and water purification, nutrient cycling, solar energy absorption, and management of biogeochemical and hydrological cycles all depend, to a significant extent, on nondomestic plants, animals, and microbes (chapter 3). The earth's ecosystems represent the culmination of historic evolutionary processes of immense antiquity and majesty. They have resulted from billions of years of evolution under conditions that may never have occurred anywhere else in the universe. Wild species maintain ecological processes at no cost to us, and they represent a genetic library of information we could never reproduce.

Wild species also provide a valuable but often unrecognized service in suppressing pests and disease-carrying organisms. It is estimated that 95 percent of the potential pests and disease-carrying organisms in the world are controlled by other species that prey upon them or compete with them in some way. In most cases, we are not even aware of those interactions. We find out how valuable natural predators are when we try to control systems with synthetic chemicals, because broad-spectrum biocides kill both pests and natural predators. As a result, pest populations often surge to higher levels than before (chapter 7). By preserving natural areas and conserving wild species, we utilize the stabilizing diversity of nature that keeps pest organisms in balance.

Aesthetic and Cultural Benefits

Wild species enhance our appreciation and enjoyment of the environment in many ways. All of our familiar domestic plants are derived originally from wild ancestors, many of which are now endangered in their native habitat. African violets, for instance, perhaps our most familiar flowering houseplant, have almost totally disappeared from Tanzania, the only place they grow in the wild.

Gathering wild food is no longer a necessity for most people in developed countries, but many millions enjoy hunting, fishing, mushroom picking, nut gathering, and other outdoor activities that involve wild species. These activities provide healthful physical exercise, a chance to renew pioneer skills, and an opportunity to enjoy a wild environment. Other people find pleasure in simply observing and photographing wild species. There are some 8 million birdwatchers and 42 million hunters and anglers in the United States. The total amount spent each year in America on these activities is probably several billion dollars. Inexpensive intercontinental air flights now make it possible to visit remote and exotic places to enjoy nature (box 9.1).

Leggitt

Figure 9.3 The rosy periwinkle from Madagascar provides recently discovered anticancer drugs that now make childhood leukemias and Hodgkin's disease highly remissible.

Ecotourism

*T*ravel is now the largest industry in the world, generating around $3 trillion per year in total revenues. A growing segment of this market is **ecotourism,** a combination of adventure travel, nature appreciation, and cultural exploration in wild settings that appeals to those who want something more exciting and intellectually stimulating than simply basking on a beach. Trekking, hiking, bird watching, nature photography, wildlife safaris, camping, mountain climbing, fishing, botanical study, and river rafting, canoeing, or kayaking are some of the favorite forms of ecotourism (box fig. 9.1.1). Experiencing other cultures, especially those of rural or native people who have traditional relationships to the land, is usually an important aspect of such travel.

Although only about 10 percent of the total vacation market, ecotourism currently generates some $20 billion per year in domestic and international receipts. This can provide both funding and an incentive for developing countries to preserve endangered wildlife and threatened habitats. Creating jobs for local people gives them alternatives to destructive harvesting practices that may previously have been their only source of income. Our interest in traditional customs and crafts serves to validate them in the eyes of young people who might be tempted to abandon their history. Dancers and artisans are paid to practice their art and keep it alive where it might otherwise be lost.

Hunting was the first form of tourism in many remote places. Kenya, for instance, developed an extensive infrastructure to serve big-game safaris in the early part of this century. By 1978, however, game was becoming scarce and a hunting ban eliminated the jobs of many safari guides and bearers. Happily, ecotourists—who shoot with cameras rather than rifles—have more than replaced big-game hunters. Wildlife biologist Michael Soulé estimates that one maned lion in Kenya's Amboseli National Park is worth $515,000 for tourist viewing but is worth only $8,500 for hunting. Another interesting comparison is that the economic yield from tourists who come to see one lion in Amboseli is equal to the income from a herd of 30,000 cows, making ecotourism more profitable than farming in this region.

Not all ecotourism is benign and beneficial, however. We who visit other countries must be careful to avoid alienating or humiliating local people by flaunting our wealth or treating them as subhuman curiosities. In many places, insensitive tourists consume resources at an exorbitant rate, drive up prices, defile holy places, and offend local sensibilities with unseemly behavior. Nepal, for instance, has suffered from the sudden popularity of its beautiful mountains. The number of visitors increased from 45,000 in 1970 to 250,000 in 1990. Large numbers

Figure 9.1.1 American students learn about tropical flowers in the Monte Verde cloud forest in Costa Rica. One of the star attractions of this lush forest is the Resplendent Quetzal, perhaps the world's most beautiful bird.

of trees are felled to supply firewood for cooking, hot showers, and campfires for trekkers. A typical climbing expedition may use as much wood in two months as a local family would use in two years. The popular route from Namche Bazaar to Everest Base Camp has become so littered that it is now called the "garbage trail." Unfortunately, only about twenty cents of the three dollars per day spent by the average trekker remains in the local economy.

With care, however, these problems can be avoided and ecotourism can be helpful to you, your local hosts, and the environment. The following list of suggestions will help you plan a responsible, positive trip.

1. *Pretrip preparation:* Learn about the history, geography, ecology, and culture of the area you will visit. Understand the do's and don'ts that will keep you from violating local customs and sensibilities.
2. *Environmental impact:* Stay on designated trails and camp in established sites, if available. Take only photographs and memories and leave only goodwill wherever you go.
3. *Resource impact:* Minimize your use of scarce fuels, food, and water resources. Do you know where your wastes and garbage go?
4. *Cultural impact:* Respect the privacy and dignity of those you meet and try to understand how you would feel in their place. Don't take photos without asking first. Be considerate of religious and cultural sites and practices. Be as aware of cultural pollution as you are of environmental pollution.

For many people, the value of wildlife goes beyond the opportunity to shoot or photograph or even see a particular species. They argue that **existence value,** simply knowing that a species exists, is reason enough to protect and preserve it. We contribute to programs to save bald eagles, redwood trees, whales, whooping cranes, giant pandas (fig. 9.4), and a host of other rare and endangered organisms because we like to know they still exist somewhere, even if we may never have an opportunity to see them. A particular species or community of organisms may have emotional value for a group of people because they feel that their identity is inextricably linked to the natural components of the environment that shaped their culture. This may be expressed as a religious value, or it may be a psychological need for access to wildlife. In either case, we often place a high value on the preservation of certain wild species.

DESTRUCTION OF BIOLOGICAL RESOURCES

Biological resources are diminished or destroyed in a number of ways. Natural changes in the environment eliminate once successful species or reduce their numbers to mere remnant populations. Humans disrupt ecosystems and extirpate species, both deliberately and accidentally.

Kinds of Losses

Three kinds of losses are of major concern: (1) *depletion of abundance* of a once plentiful species, (2) *species extinction,* and (3) *ecosystem disruption.* There is an increasing gradient of seriousness in these losses. If a once abundant species is depleted, it probably can be restored if its proper habitat still exists and if its ecological niche hasn't been usurped. The herds of deer, elk, antelope, and buffalo that once roamed the American plains have been replaced by even larger numbers of domestic cattle and sheep. Since the wild species are still preserved in wildlife refuges, however, they could be reintroduced if we chose to do so (*see* box 8.3).

Extinction, on the other hand, is a permanent loss. Every species represents a unique set of characteristics resulting from long interaction between the genetic system and its environment. Because past evolutionary conditions can never be reproduced exactly, each species represents an irreplaceable resource. When a species is destroyed, not only is that particular set of characteristics lost, but also all of the potential adaptations and developments that might have appeared in its offspring are lost.

Ecosystem disruption is even more serious. In a healthy ecosystem there is usually enough diversity so that a single species, if lost, can be replaced by other organisms that use the same resources. Beyond a

Figure 9.4 Fewer than a thousand giant pandas remain in the wild in China. Human encroachment has reduced their habitat, while hunters kill them for their fur. Since they eat only bamboo shoots and leaves, they are totally dependent on a specific forest type. In the 1970s, huge areas of bamboo flowered and died. Many pandas starved.

certain point, however, losses can leave an ecosystem so impoverished that major ecological processes are disrupted and some niches cannot be filled. A catastrophic **decline spiral** can be set in motion that not only destroys a particular community but may even affect the whole biosphere. There is a worry that human disruptions of natural systems, such as by deforestation and marine pollution, could have worldwide effects in some cases.

Natural Causes of Extinction

Extinction is neither a new phenomenon nor a process caused only by humans. Studies of the fossil record suggest that more than 99 percent of all species that ever existed are now extinct. Most of those species disappeared long before humans came on the scene. Species arise through processes of mutation, isolation, and natural selection (*see* box 3.2). Evolution can proceed gradually over millions of years or may occur in large jumps when new organisms migrate into an area or when environmental conditions change rapidly. New forms replace their own parents or drive out less adapted competitors. In a sense, species that are replaced by their descendants are not completely lost. The tiny *Hypohippus,* for instance, has been replaced by the much larger modern horse, but most of its genes probably still survive in its distant offspring.

Mass Extinctions

The geological record shows that a number of mass extinctions occurred in geological history. The best known of these cycles occurred at the end of the Cretaceous period when dinosaurs disappeared, along with at least 50 percent of existing genera and 15 percent of marine animal families. An even greater disaster occurred at the end of the Permian period about 250 million years ago when two-thirds of all marine species and nearly half of all plant and animal families died out over a period of about 10,000 years—a short time by geological standards. Current theories suggest that these catastrophes were caused by climate changes, perhaps triggered when large asteroids struck the earth. It is possible that global climate change caused by accumulation of "greenhouse" gases in the atmosphere might do something similar (chapter 10).

Although extinction is obviously disastrous for the organisms to which it occurs, it may provide an opening for the next stages of evolution. We can speculate that the end of the dinosaurs made possible the age of mammals, of which we are the beneficiaries. An important lesson learned from fossil studies is that evolutionary change is based on the species and characteristics that happen to be available in a particular environment at a particular time. Neither the individual organisms present in a given ecosystem nor the ecosystem itself are necessarily perfectly designed or ideally adapted to existing conditions. Evolutionary history suggests that this may not be the best of all possible worlds. We shouldn't assume that every ecosystem is a perfect and fragile balance of irreplaceable species. On the other hand, we can't assume that if we exterminate existing life-forms, they will be replaced with equal or better ones. As Aldo Leopold said, "The first rule of intelligent tinkering is to save all the pieces."

Human-Caused Extinction

Humans have a long history of biological resource depletion. Stone-age human hunters, for example, may have been responsible for the extermination of the "megafauna" of both America and Eurasia during the Pleistocene era, some 20,000 years ago. Mastodons, mammoths, giant bison, ground sloths, early horses, and cameloids all disappeared from North America about the time humans first appeared here. Climatic change may have been partially or primarily responsible, but vast "boneyards" in Europe and villages constructed entirely of mammoth bones in Siberia suggest the hunting prowess of our ancestors.

Ancient civilizations were probably responsible for the extinction of many species. Misuse of the land and destruction of biological resources have played a role in the demise of every major civilization from the Babylonian, Sumerian, Indus, and Mayan cultures to ancient Greece, Rome, and China.

Current Extinction Rates

The rate at which species are lost appears to have increased dramatically over the past one hundred years. Before humans became a major factor, extinction rates from natural causes appear to have been one species lost every five to ten years. Between A.D. 1600 and 1900, human activities seem to have been responsible for the extermination of about one species per year. Harvard entomologist E. O. Wilson estimates that we now are pushing 20,000 species a year into extinction. We cannot be absolutely sure of these rates because many parts of the world haven't been thoroughly explored, and many species may have disappeared before they were studied and classified by biologists. In North America, sixty-one species of flowering plants and six bird species are known to have been eliminated since Europeans arrived.

The biggest reason for the current increase in extinctions is habitat loss. Destruction of tropical forests, coral reefs, estuaries, marshes, and other biologically rich ecosystems threatens to eliminate thousands or even millions of species in a human-caused mass extinction that could rival those of geologic history. By destroying habitat, we eliminate not only prominent species but also many obscure ones of which we may not even be aware. It has been suggested that half of all existing species could be endangered or extinct by the middle of the next century if this destruction continues.

WAYS HUMANS CAUSE BIOLOGICAL LOSSES

People rarely intend to completely destroy other species, but through overhunting, overfishing, habitat destruction, and introduction of diseases and exotic competitors, we have reduced biological abundance or driven species into extinction (fig. 9.5). In this section, we will look at some of the ways we deplete or destroy biological resources.

Direct Impacts

Some of our actions are deliberately aimed at a particular target species. In this section we will look at some examples of direct impacts on other species.

Hunting and Food Gathering

Overharvesting of food species is probably the most obvious direct way in which humans destroy biological resources. The dodo, for example, was a large, flightless bird from the island of Mauritius in the Indian Ocean (fig. 9.6). Because the birds had evolved in an ecosystem with no predators, they were slow, trusting, and easy to catch, providing a welcome source of fresh meat for visiting sailing ships. By the seventeenth century, dodos had been completely exterminated.

Some other well-known overhunting cases include the near extermination of both the great whales and

L. O'Keefe

Figure 9.6 Dodos were large flightless birds once on the island of Mauritius in the Indian Ocean. Because they had no fear of humans, they were easy to kill. Sailors and early settlers exterminated all of them several hundred years ago.

the American bison (or buffalo). In 1850, some 60 million bison roamed the western plains. Forty years later, only about 150 bison survived in the wild and 250 survived in captivity. Many had been killed only for their hides or tongues, leaving as much as a ton of meat in each carcass to rot on the plains. Much of the destruction was carried out by the U.S. Army to vanquish native peoples who depended on bison for food, clothing, and shelter and thereby force them onto reservations.

Before commercial whaling began in the early nineteenth century, something like 2.5 million great whales inhabited the world's oceans. Great fortunes were made by catching these magnificent creatures and rendering them down into oil, but populations were quickly depleted by unrestrained exploitation. The introduction of steamships and explosive harpoons made it possible to catch and kill even the fastest whales, leading to a pattern of successive devastation of new species as old stocks crashed (fig. 9.7). The largest species—the blue and the humpback—were the first to go. As these grew rare, the fin whale and the smaller sei were hunted to near extinction. Now the little minke is the target of whalers.

In 1985, the International Whaling Commission called for a moratorium on the taking of all great whales. Iceland and Japan found a loophole, however, each killing hundreds of whales per year for "scientific

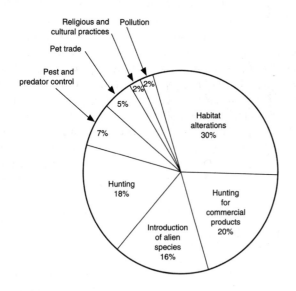

Figure 9.5 Estimated contribution of various human activities in wildlife extinction. More than one activity may be involved in extermination of a particular species.

Source: Data from World Wildlife Fund.

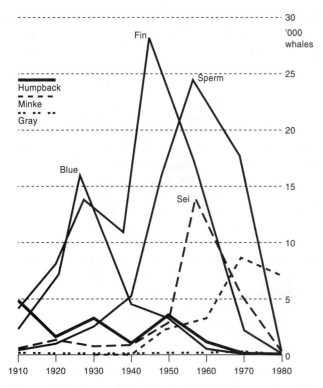

Figure 9.7 This world catch of whales shows a series of sharp peaks as each species in turn is hunted to exhaustion. Now only the minke, the smallest of the great whales, is present in a significant population.

Source: Data from Geoffrey Lean, et al., *World Wildlife Fund Atlas of the Environment.* Copyright © 1990 Prentice-Hall, Inc.

research." The meat of animals taken in these programs is sold for human consumption. In 1992, Norway threatened to resume whaling in spite of protests from the world community. It may be too late for some species anyway. Population numbers of the largest whales are so low that individuals may not be able to find mates to propagate the species.

Fishing

Fish stocks have been seriously depleted by overharvesting in many parts of the world. A classic case of the collapse of a fishery is the California sardine. Commercial fishing for this once abundant species began in the 1920s. In 1936, 750,000 tons were harvested. Twenty years later, in 1956, the total catch was 1 ton. Similarly, cod, haddock, and other groundfish once swarmed in such immense schools around the fertile shoals off eastern Canada that early fishermen simply dipped them up in wicker baskets (fig. 9.8). The 1990 Canadian cod catch, however, was down 75 percent from the historic high of 1.5 million tons in 1968. Overexploitation has caused a decrease in both the total number and the size of the fish caught. In 1977 the average fish taken was 2.4 kg; ten years later the average was only 1 kg (2.2 lbs). In 1992, a two-

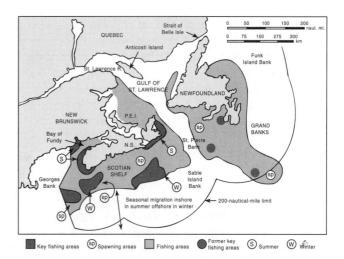

Figure 9.8 Key fishing areas for haddock and cod in Eastern Canada. Both the average size and the total number of fish caught annually have declined precipitously causing the government to declare a two-year moratorium on fishing that has idled 20,000 workers and shut down a $700 million per year industry.

year ban on cod fishing was imposed along the entire northeastern coast of Newfoundland and Labrador, idling 20,000 workers and closing down a $700 million per year industry.

Highly efficient but terribly destructive modern fishing techniques play a large role in these catastrophic declines. Huge trawlers churn up spawning beds and feeding banks, sucking up everything on the bottom, including baby fish, invertebrates, and non-economic species. Only large fish are kept, but after the trauma of sudden decompression and entrainment in the pumps, most of what goes back into the ocean is dead or dying.

In the 1970s and 1980s, huge drift nets caused tremendous damage to many pelagic species. These nets, which could be as large as 48 km (30 mi) long and 10 m (32.5 ft) deep, were simply dropped into the ocean and allowed to drift with the currents. As they swept across the oceans, these "walls of death" could trap and drown millions of sea birds and marine mammals and kill uncounted tons of unwanted fish species. In the mid-1980s, the Asian squid-fishing fleets laid 3 million km (2 million mi) of drift nets annually in the Pacific. Thousands of miles of drift nets were lost or abandoned every year. These ghost nets will continue to catch and kill fish and diving birds for years as they drift across the oceans. In 1992, the U.N. General Assembly called for a worldwide moratorium on drift-net fishing. Whether this ban will be effective remains to be seen.

Shrimp trawlers catch several tons of "trash fish" for every ton of shrimp they keep. They also catch and drown an estimated 10,000 endangered sea turtles in their nets each year. The National Marine

Fisheries Service invented a metal and net box called a turtle excluder device (TED) to save turtles from drowning in shrimp nets. The TED has slats that let shrimp into the net but push turtles up and out a trap door. The U.S. Fish and Wildlife Service ordered shrimpers to install and use TEDs, but many refused to comply because they claim shrimp catches are reduced.

Cashing in on Animal Products

Undoubtedly, the most valuable nonfood products from wild animals are furs and skins (fig. 9.9). Fur coat sales amounted to some $1.5 billion worldwide in 1985. The fur industry says that only 10 percent of the furs sold in the United States each year come from wild animals, but trapping opponents say the total is closer to 50 percent. The U.S. Department of the Interior reports that 100 million steel-jaw, leg-hold traps are used in the United States each year and that approximately 25 million animals are trapped. Two-thirds of those animals are discarded, however, because they are the wrong species or are too badly mangled or decomposed to be useful. Publicity and protests by environmental and animal rights groups have had a dramatic impact on the killing of baby harp and hooded seals. The annual kill dropped from some 200,000 in 1983 to about 5,000 in 1985. Nevertheless, there is still a very damaging traffic in skins and furs of endangered species that we will discuss later in this chapter.

In spite of progress in wildlife protection and international conventions against trade in endangered species, overharvesting of many species continues today because the economic pressures are great. Tiger or leopard fur coats can bring $100,000 in Japan or Europe. The population of African black rhinos dropped from approximately 65,000 in 1973 to about 3,000 in 1990 because of a demand for rhino horn handles for jambiyya (traditional curved daggers) of the Middle East or for folk medicine in Africa and Asia.

Perhaps no animal symbolizes wasteful exploitation as much as the African elephant. In 1979, there were an estimated 1.3 million elephants in Africa. Ten years later, only 625,000 remained. An estimated 100,000 elephants are killed by poachers each year for the ivory trade (fig. 9.10). It is little wonder that people are tempted by this opportunity for wealth. In the 1970s, when wholesale prices for ivory jumped from $10 to over $100 per kg (2.2 lbs), a medium-sized set of tusks was equal to ten years' income for a subsistence farmer. The Convention on International Trade in Endangered Species (CITES) voted in 1989 to ban all ivory trade. The ban seems to be working, and the world ivory market has collapsed.

Harvesting Wild Plants

Plants also are threatened by overharvesting. Wild ginseng has been nearly eliminated in many areas because of the Asian demand for roots that are used as an aphrodisiac and folk medicine. Cactus "rustlers" steal cacti by the ton from the American southwest and Mexico. In 1977, some 10 million cactus plants were shipped from Texas to both domestic and foreign markets. With prices ranging as high as $1,000 for rare specimens, it's not surprising that many are now endangered. TRAFFIC (a program of the World Wildlife Fund) reported that some twenty thousand illegal orchids were exported by smugglers in 1980.

Figure 9.9 Endangered wildlife products. These leopard and cheetah skins were seized by United States Fish and Wildlife agents. Smugglers of products from endangered species are subject to heavy fines and/or jail terms. Unfortunately, however, as long as there are consumers for these products, some hunters and dealers will be encouraged to take the risks.

Figure 9.10 Soldiers load illegal elephant tusks seized from poachers. In 1989, Kenya burned 2,400 confiscated tusks worth more than $3 million in an attempt to end the global ivory trade that threatens elephants with extinction.

Pet and Scientific Trade

The trade in wild species for pets is an enormous business. Worldwide, some 5 million live birds are sold each year for pets, mostly in Europe and North America. In 1980, pet traders imported (often illegally) into the United States some 2 million reptiles, one million amphibians and mammals, 500,000 birds, and 128 million tropical fish. Keeping an aquarium is one of the most popular hobbies in America. About 75 percent of all saltwater tropical aquarium fish sold come from the marvelously rich coral reefs of the Philippines and Indonesia (fig. 9.11). These fish are caught by divers who use plastic squeeze bottles of cyanide to stun their prey. Far more fish die with this technique than are caught. Worst of all, it kills the coral animals that create the reef. A single irresponsible diver can destroy all of the life on 200 square meters of reef in a day. Altogether, thousands of divers now working destroy about 50 sq km of reefs each year. Net fishing would prevent almost all this destruction. It could be enforced if pet owners would insist on net-caught fish.

Smuggling of rare and endangered species is particularly lucrative. Bird collectors will pay $10,000 for a hyacinth macaw from Brazil or up to $12,000 for a pair of golden-shouldered parakeets from Australia. A rare albino python might bring $20,000 in Germany. The destruction in this live animal trade is enormous. It is generally estimated that fifty animals are caught or killed for every live animal that gets to market. It is the buyers, many of whom say they love animals, who keep this trade going.

Every year, approximately 100 million animals are used in scientific research. Most of those animals are lab-reared rats and mice, but some experiments require chimpanzees, monkeys, and other species obtained from the wild where they are becoming rare.

Figure 9.11 A diver uses cyanide to stun tropical fish being caught for the aquarium trade. Many fish are killed by the method, while others die later during shipment. Even worse is the fact that cyanide kills the coral reef itself.

Unfortunately, one of the main techniques for capture is to kill a mother and take her baby. Many young animals don't survive capture and shipping. We might be able to spare many of these deaths if we develop alternate techniques for medical research, such as human cell cultures and organ cultures.

Predator and Pest Control

Some animal populations have been greatly reduced or even deliberately exterminated because they are regarded as dangerous to humans or livestock or because they compete with our use of resources. The first recorded case of human-caused extinction was extermination of the European lion in A.D. 80. Between 1937 and 1970, U.S. government predator control agents trapped, poisoned, or shot 23,800 bears, 7,255 mountain lions, 1,574 gray wolves, 50,283 red wolves, 477,104 bobcats, and 2,823,056 coyotes. Uncounted millions of animals were killed unintentionally by poisoned bait or misplaced traps, or intentionally by private individuals for bounty or sport.

Predator control programs demonstrate the paradox that can exist when conflicts between humans and wildlife are not examined in a holistic manner. In 1990, predator-related livestock losses (both confirmed and unconfirmed) in the United States were claimed to be $27.4 million. The cost of predator control that year by the Office of Animal Damage Control (as it is now called) amounted to $38 million, mainly to kill 86,500 coyotes. It would have been much cheaper to leave the wild animals alone and pay farmers and ranchers for their losses.

Many poisons used in pest and predator control are wide-spectrum, long-lasting biocides. Once they contaminate soil or water or get into the food chain, they persist, moving from organism to organism and affecting many nontarget species. One of the most controversial of these agents is 1080. First introduced in 1945, this powerful, persistent poison is capable of killing not only the primary target animal but also a whole series of scavengers and decomposers that eat the poisoned carrion as it spreads through the food chain. More than 150,000 baits laced with 1080 were distributed before its use was banned in 1972. Recently, ranchers have been seeking permission to use 1080 collars on livestock. They believe that wolves and coyotes will grab their prey by the neck, puncturing the collar and getting a lethal dose of poison. Each collar, however, contains enough poison to kill 185 coyotes or six men. A much better solution would be to employ more herders and guard dogs to watch the sheep and cattle.

Indirect Damage

Although hunting and some other human pursuits that kill wildlife directly often have disastrous effects on specific species, processes that destroy habitat and disrupt ecosystems can endanger far more species than any other human actions. We now turn to some indirect ways humans affect plants and animals.

Habitat Destruction

There are many historic examples of human disturbances of natural systems. In much of the Middle East, Asia, and the Americas, once-fertile areas have become deserts because of unsound forestry, grazing, and agricultural practices (chapter 10). The whole Mediterranean Basin, in fact, has been described as a "goatscape" because of the destruction caused by domestic animals. Technology now makes it possible for us to destroy vast areas even faster than in the past. Native forests are cut down, marginal land is plowed for crops, and plant cover that protects the land from erosion is destroyed with herbicides or tractors dragging huge chains.

Undoubtedly the greatest current losses in terms of biological diversity and unique species occur when tropical moist forests are disrupted. Of some 10 million sq km of original tropical forests, more than a third has been disturbed to some extent by humans (see chapter 8). It is thought that more than half of all the species on earth live in the tropics. Upsetting these complex ecosystems could endanger millions of species.

If the global climate changes as rapidly and to the extent predicted by the worst-case scenarios for the "greenhouse effect" (see chapter 10), there could be massive losses of species unable to adapt to new conditions in the next century. This could rival the catastrophic die-offs at the ends of geologic epochs in the past.

Exotic Species Introductions

Exotics—organisms introduced into habitats where they are not native—are considered to be among the most damaging agents of habitat alteration and degradation in the world. They can be thought of as biological pollutants. Introducing species accidentally or intentionally from one habitat into another where they have never been before is risky business. Freed from the predators, parasites, pathogens, and competitors that normally keep their numbers in check, exotics often exhibit explosive population growth that crowds out native species. Their aggressive invasion might be considered a kind of ecological cancer.

Species such as carp and purple loosestrife are sometimes released intentionally because they are expected to be beneficial or attractive. Others such as zebra mussels and spiny water fleas are introduced accidentally, brought in to new habitats on vehicles, animals, commercial goods, or even clothing (box 9.2). Once established, exotics are rarely eliminated.

Many nuisance birds of our cities, such as starlings, English sparrows, and common pigeons, are aliens that were introduced by "acclimatizers" who spent considerable sums of money to import these species from Europe. These birds have prospered because they are opportunists that do well in urban areas, but in many places they have driven out more desirable native species.

Plants also can multiply explosively when introduced into a new environment. Kudzu, a cultivated legume in Asia, has run amok in the southern United States where it blankets trees, houses, and whole hillsides, smothering slower-growing native species. Water hyacinths are floating water plants with waxy green leaves and beautiful blue flower spikes. They were brought to the United States from Argentina in 1880 for an exhibition in New Orleans. People took plants home and either deliberately put them in streams and lakes or just tossed them out. Within ten years, this prolific weed had spread across the South from Florida to Texas, completely covering the surface of many waterways, blocking boat traffic, and smothering fish and native plants. Mechanical harvesters and poisons were tried, but the plant population doubled every two weeks. Ironically, there now appear to be some benefits to this pest that we have fought for nearly a century. The water hyacinth has been proposed as a source of biomass for energy production because it grows so fast. It is also useful in water purification, absorbing large amounts of contaminants, such as heavy metals and toxic organic compounds.

Diseases

Disease organisms, or pathogens, may be considered predators. To be successful over the long term, a pathogen must establish a balance in which it is vigorous enough to reproduce but not so lethal that it completely destroys its host. When a disease is introduced into a new environment, however, this balance may be lacking and an epidemic may sweep through the area.

The American chestnut was once the heart of many Eastern hardwood forests. In the Appalachian Mountains, at least one of every four trees was a chestnut. Often over 45 m (150 ft) tall, 3 m (10 ft) in diameter, fast growing, and able to sprout quickly from a cut stump, it was a forester's dream. Its nutritious nuts were important food for birds (like the passenger pigeon), forest mammals, and humans. The wood was straight-grained, light, rot-resistant, and used for everything from fence posts to fine furniture. Its bark was used to tan leather. In 1904, a shipment of nursery stock from China brought a fungal blight to the United States, and within about forty years the American chestnut had all but disappeared from its native range. Efforts are now underway to transfer blight-resistant genes into the few remaining American

Alien Invaders in the Great Lakes

*T*he opening of the Saint Lawrence Seaway and the Welland Canal early in this century were a financial boon to cities and industries around the Great Lakes. With access to world markets, producers ship millions of dollars worth of agricultural commodities, raw materials, and finished products each year to and from Canadian and American ports thousands of miles from the open ocean. Biological organisms have been transported as well, however, some with disastrous environmental consequences. A number of alien invaders have made the transoceanic crossing and have become established here, to the detriment of native organisms (box figure 9.2.1). The following list describes some of the greatest problems.

1. **Sea lamprey** (*Petromyzon marinus*) is a predatory, eel-like fish native to coastal regions along both sides of the Atlantic Ocean. They clamp onto the side of another fish by means of a suckerlike mouth and then rasp through the skin with concentric rings of small, sharp teeth to feed on blood and tissue fluids. Since entering the Great Lakes through the Welland Canal about 1921, they have contributed to the catastrophic collapse of native whitefish and lake trout populations.

2. **Zebra mussel** (*Dreissena polymorpha*) is a fingernail-sized mussel native to the Caspian Sea. First discovered in 1986 in the St. Clair River near Detroit where larvae were probably discharged along with freshwater ballast, these incredibly prolific bivalves have spread explosively throughout the Great Lakes. Female zebra mussels can produce up to 1 million eggs per year, which develop into microscopic, free-swimming larvae that form shells and attach to any surface.

    Unchecked by natural predators—in Europe they are eaten by fish and diving ducks—the mussels can reach population densities of 23,000 animals per square meter. They clog pipes and shut down water intake systems. They pile up on boats, buoys, and piers, crowding out native species. One utility estimated it will cost $50 million to $100 million per year to scrape mussels from cooling pipes at their power plants. One beneficial effect is that the mussels have made Lake Erie noticeably clearer by filtering its water through feeding apparatus. Native drum, a bottom feeding fish, eat these mussels and may help control their spread.

3. **Ruffe** (*Gymnocephalus cernuus*) is a small perch native to central and eastern Europe. It was introduced into the Duluth harbor, probably

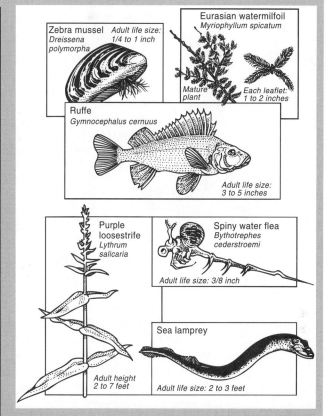

Figure 9.2.1 Some of the exotic organisms that have invaded the Great Lakes with human help. These aliens are driving out native species and irreparably damaging the ecosystem.

in tanker ballast water, around 1985 and is now spreading through Lake Superior where it competes with native perch and other forage fish. A single female can lay 90,000 eggs per year. They rarely grow larger than 12.5 cm (5 in), but sharp spines in their fins make them difficult for larger fish to eat.

4. **Spiny water flea** (*Bythotrephes cederstroemi*) is not an insect at all but a tiny (less than half an inch long) crustacean with a long, sharp, barbed tail spine. A native of Great Britain and northern Europe, these animals were first noticed in Lake Huron in 1984—probably imported in ballast water of a transoceanic freighter. Since then, populations have exploded, and the animal can now be found throughout the Great Lakes and in some inland lakes. These carnivorous crustaceans compete with young perch and other small fish for zooplankton. The flea's sharp spines make them hard to eat so they may overpopulate lakes and deplete food sources of game fish and other desirable species.

5. **Eurasian water milfoil** (*Myriophyllum spicatum*) is a feathery aquatic plant that is often

used in aquaria. It was probably dumped into lakes by someone cleaning out an aquarium. In nutrient-rich lakes it can form thick mats of floating vegetation that interferes with water recreation and crowds out native water plants. Dead, rotting milfoil mats rob water of its oxygen and smother fish. A key factor in the plant's rapid spread is its ability to reproduce through stem fragmentation and underground runners. A single segment of stem and leaves can infect a lake and start a new colony. Mechanical clearing of weed beds creates thousands of new fragments that cling to boats and trailers, spreading the plants from lake to lake.

6. **Purple loosestrife** (*Lythrum salicaria*) is a common ornamental plant that grows well in both wet and dry habitats and produces an attractive, long-lasting spike of purple flowers.

First grown in gardens, this fast-growing weed has spread across forty states and all Canadian border provinces. Purple loosestrife grows especially well in wetlands, crowding out native species and creating dense, impenetrable stands that are unsuitable as cover, food, or nesting sites for most animals. Eradicating an established stand is difficult because of the enormous number of seeds stored in the soil, some of which remain viable for years. Several European insects that attack only purple loosestrife are being tested as a possible long-term biological control.

Text and watermilfoil adapted with permission from Minnesota Department of Natural Resources, A Field Guide to Aquatic, Exotic Plants and Animals, *1992; Ruffe and Purple loosestrife, courtesy of Wisconsin Fish and Wildlife; Sea Lamprey and Spiny water flea, courtesy of Don Luce; Zebra mussel after U.S. Fish & Wildlife Service.*

chestnuts that weren't reached by the fungus or to find biological controls for the fungus that causes the disease.

A similar disaster has all but exterminated the American elm. A shipment of European elm logs introduced a fungal disease fatal to the American elm. The loss is most noticeable in prairie towns of the Midwest where elms formed arching colonnades over the streets and provided an oasis of shade from the summer sun. As Dutch elm disease swept across the country, some towns lost all of their trees in just a few years. Des Moines, Iowa, had to remove some 250,000 trees in five years at a cost of about five million dollars.

Pollution

We have known for a long time that toxic pollutants can have disastrous effects on local populations of organisms. The publication of *Silent Spring* by Rachel Carson in 1962 alerted the public to a much more insidious and widespread effect of pollution. Carson pointed out that pesticides and other pollutants were causing declines in many nontarget organisms and warned that they might even drive certain species into extinction. The book and its warnings were attacked by many as being unfounded, but subsequent studies have shown that many of Carson's concerns are valid.

Persistent pesticides like DDT (dichloro-diphenyl-trichloroethane) accumulate in food chains and are especially dangerous to top carnivores, such as hawks, eagles, and game fish. Largely because of DDT accumulated from their prey, peregrine falcons and brown pelicans disappeared from large parts of their former ranges in the United States. DDT disrupts hormone regulation of calcium levels and causes thin eggshells that break when nesting parents try to incubate them,

resulting in reproductive failure. Since DDT was phased out in the early 1970s, its levels in the environment have been falling and both falcons and pelicans have been making a comeback.

Lead poisoning is another major cause of mortality for many species of wildlife. Bottom-feeding waterfowl, such as ducks, swans, and cranes, ingest spent shotgun pellets that fall into lakes and marshes. They store the pellets, instead of stones, in their gizzards and the lead slowly accumulates in their blood and other tissues. The U.S. Fish and Wildlife Service (USFWS) estimates that 3,000 metric tons of lead shot are deposited annually in wetlands and that between 2 and 3 million waterfowl die each year from lead poisoning. Scavengers, such as condors and bald eagles, eat birds and mammals that have lead shot or fragments of bullets in their bodies. In 1987, the last five wild California condors were trapped and put into zoos. Lead poisoning was thought to be a major cause of their high mortality in the wild. In 1976, the USFWS banned the use of lead shot in certain areas with high accumulations of spent shot. Hunters are required to use nontoxic steel shot, but they complain that it causes excessive wear in gun barrels and has poor ballistic characteristics.

Genetic Assimilation

Some rare and endangered species are threatened by **genetic assimilation** because they crossbreed with closely related species that are more numerous or more vigorous. Opportunistic plants or animals that are introduced into a habitat or displaced from their normal ranges by human actions may genetically overwhelm local populations. For example, the rare southern red wolf crossbreeds with the more prolific coyote, whose range has been expanding because of its successful adaptation to human presence. The red wolf was nearly extinct by 1970 but is now being re-introduced through captive breeding programs. Black ducks have declined severely in the eastern United States and Canada in recent years. Hunting pressures and habitat loss are factors in this decline, but so is interbreeding with mallards forced into black duck habitat by destruction of prairie potholes in the west.

BIOLOGICAL RESOURCES MANAGEMENT

Over the years, we have gradually become aware of the harm we have done—and continue to do—to wildlife and biological resources. Slowly, we are adopting national legislation and international treaties to protect these irreplaceable assets. Parks, wildlife refuges, nature preserves, zoos, and restoration programs have been established to protect nature and rebuild depleted populations. There has been encouraging progress in this area, but much remains to be done.

Hunting and Fishing Laws

In 1874, a bill was introduced in the U.S. Congress to protect the American bison, whose numbers were already falling to dangerous levels. This initiative failed, however, because most legislators believed that all wildlife—and nature in general—was so abundant and prolific that it could never be depleted by human activity. As we discussed earlier in this chapter, however, by the end of the nineteenth century bisons had plunged from some 60 million to only a few hundred animals.

By the 1890s, most states had enacted some hunting and fishing restrictions. The general idea behind these laws was to conserve the resource for future human use rather than to preserve wildlife for its own sake. The wildlife regulations and refuges established since that time have been remarkably successful for many species. At the turn of the century, there were an estimated half million white-tailed deer in the United States; now there are some 14 million. Wild turkeys and wood ducks were nearly all gone fifty years ago. By restoring habitat, planting food crops, transplanting breeding stock, building shelters or houses, protecting these birds during breeding season, and employing other conservation measures, conservationists have restored populations of these beautiful and interesting birds to several million each. Snowy egrets, which were almost wiped out by plume hunters eighty years ago, are now common again.

The Endangered Species Act

Passage of the U.S. Endangered Species Act of 1973 and the Committee on the Status of Endangered Wildlife in Canada (COSEWIC) in 1976 represented a pow-

erful new approach to wildlife protection. Where earlier regulations had been focused almost exclusively on "game" animals, these programs seek to identify all endangered species and populations and to save as much biodiversity as possible, regardless of its usefulness to humans. **Endangered species** are those considered in imminent danger of extinction, while **threatened species** are those that have declined significantly in total numbers and may be on the verge of extinction in certain localities. **Vulnerable species** are those that are naturally rare or have been depleted by human activities to a level that puts them at risk. Bald eagles, gray wolves, brown (or grizzly) bears, sea otters, and a number of native orchids and other rare plants are considered either vulnerable or threatened in many places even though they remain locally abundant over parts of their former range.

The United States currently has some 1,200 species on its endangered and threatened species lists, and 4,000 candidate species waiting to be considered. The number of species listed in different taxonomic groups reflects much more on which organisms we consider interesting and desirable than the actual number in each category. Compare figure 9.12 with table 9.1, for instance.

Canada, which generally has less diversity because of its boreal location (chapter 3), has designated a total of forty-six endangered and fifty threatened

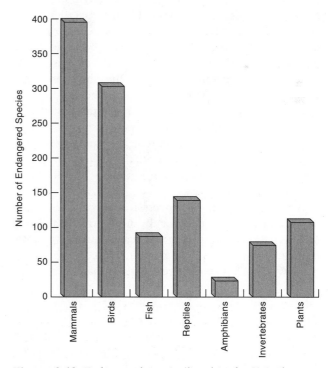

Figure 9.12 Endangered species listed in the United States in 1990 by taxonomic group. Note that the number listed in each group bears no relation to the total number of species in that group. It reflects, instead, our interest in different kinds of organisms.

species. During the 1980s, listing of species in the United States slowed to a virtual stop. Hundreds of species were classified as "warranted (deserving of protection) but precluded" for lack of funds or political will. At least eighteen species have gone extinct since being listed for protection and at least twice that number disappeared while waiting for consideration.

Some of the species protected by the Endangered Species Act are naturally rare and exotic, causing some people to grumble about spending money or disrupting commercial interests to save obscure organisms such as the Coachella Valley fringe-toed lizard, Furbisher's lousewort, or Higgins' eye pearly mussel. This raises some interesting ethical questions about the values of seemingly minor species and our place in nature.

Even if you believe that we are entitled to do whatever we please with other organisms, however, there is still an argument that these species are indicators of the health of the environment. Their loss may have implications that we don't yet understand and may portend changes that are vitally important to us. Endangered species have often been surrogates or champions for protecting whole communities threatened by our activities (fig. 9.13).

Recovery Programs

The Endangered Species Act not only prohibits projects or actions that jeopardize endangered species or critical habitats, it also requires the U.S. Fish and Wildlife Service to prepare a **species recovery plan** that spells out how the species can be restored to numbers that permit its delisting. A few of these plans have been gratifyingly successful. The American alligator was listed as endangered in 1967 because hunting (for meat, skins, and sport) and habitat destruction had reduced populations to precarious levels. Protection has been so effective that the species is now plentiful and has been delisted throughout its entire southern range. Florida alone estimates that it has at least 1 million alligators.

Among the more controversial recovery plans now being considered is that for the northern spotted owl (chapter 8), whose protection may require a moratorium on harvesting of old-growth forest in the Pacific Northwest. An even more costly program may be needed to save native populations of Columbia River salmon and steelhead endangered by hydropower dams and water storage reservoirs that block their migration routes to the sea. Fish ladders were built at many dams to facilitate migration, but they have not been very effective.

Opening up floodgates to allow young fish to run downriver and adults to return to spawning grounds could have grave consequences for barge traffic, farmers, electric rate payers, and others who have come

Figure 9.13 Endangered species sometimes serve as surrogates for an entire biological community.

Copyright 1990 by Herblock in *The Washington Post*.

to depend on abundant water and cheap power. Electric rates in Washington and Oregon could increase 10 to 30 percent, costing somewhere around $200 million per year. Some economists would say that we haven't been paying the true cost of using the river; we have been discounting the value of salmon and wild nature. This just represents an internalization of formerly external costs (chapter 5).

CONVENTION ON INTERNATIONAL TRADE IN ENDANGERED SPECIES

The 1975 Convention on International Trade in Endangered Species (CITES) was a significant step toward worldwide protection of endangered flora and fauna. It regulated trade in living specimens and products derived from listed species, but it has not been foolproof. Species are smuggled out of countries where they are threatened or endangered, and documents are falsified to make it appear they have come from areas where the species are still common (see fig. 9.9). Investigations and enforcement are especially difficult in developing countries where wildlife is disappearing most rapidly. Still, eliminating markets for endangered wildlife is an effective way of stopping poaching. In 1988, the World Conservation Centre reported that 430 metric tons of ivory were sold on the world markets. The 1990 world ban on ivory trade dried up almost all of that traffic. Environmentalists hope that bans passed in 1992 on trade in endangered tropical birds and sea turtles will have similarly beneficial effects.

Habitat Protection

Over the past decade, growing numbers of scientists, land managers, policy makers, and developers have been making the case that it is time to focus on a rational, continent-wide preservation of ecosystems that support maximum biological diversity rather than a species-by-species battle for the rarest or most popular organisms. Our earlier focus on individual species can lead to a situation in which we spend millions of dollars to breed plants or animals in captivity but then have no natural habitat where they can be released. While flagship species such as California condors or Indian tigers may be reproducing well in zoos and wild animal parks, the ecosystems that they formerly inhabited may be disappearing.

A leader of this new form of conservation is J. Michael Scott, who was project leader of the California condor recovery program in the mid-1980s and before that spent ten years working on endangered species in Hawaii. In making maps of endangered species, Scott discovered that even Hawaii, where more than 50 percent of the land is federally owned, has many vegetation types completely outside of natural preserves (fig. 9.14). The gaps between protected areas may contain more endangered species than are preserved within them.

This observation has led to an approach called **gap analysis** in which conservationists and wildlife managers look for unprotected landscapes that are rich in species. Computers and geographical information systems (GIS) make it possible to store, manage, retrieve, and analyze vast amounts of data and create detailed, high-resolution maps relatively easily (fig. 9.15). This seems likely to save more species and whole communities than would a piecemeal approach.

Ecosystem Management

One of the biggest problems with managing wildlife populations and natural areas is that park and nature preserve boundaries are often determined by political jurisdictions rather than self-contained ecosystems. Airsheds, watersheds, and animal territories or migration routes often extend far beyond official boundaries and yet profoundly affect communities that we are attempting to preserve. Yellowstone and Grand Teton parks in northwestern Wyoming are examples of this principle. Although about 1 million hectares (2.5 million acres) in total size, these parks probably cannot preserve viable populations of large predators such as wolves and grizzly bears. Management policies in the surrounding national forests and private lands seriously affect conditions in the park (fig. 9.16). The natural **biogeographical area** (an entire ecosystem and its associated land, water, air, and wildlife resources) must be managed as a unit if we are to preserve all its values.

Size and Location of Nature Preserves

What is the optimum size and shape of a wildlife preserve? Ideally, a reserve should be large enough to keep an ecosystem intact and should be designed to isolate critical core areas from damaging external forces. Numerous studies have shown that small preserves can't support large species such as elephants or tigers that need a large territory. Small areas also rapidly lose species diversity through genetic

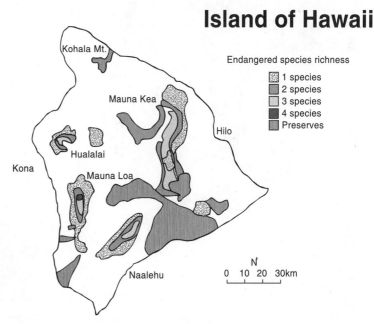

Figure 9.14 An example of the biodiversity maps produced by J. Michael Scott and the Fish and Wildlife Service. Notice that few of the areas of endangered species richness are protected in preserves, which were selected more for scenery or recreation than for biology.

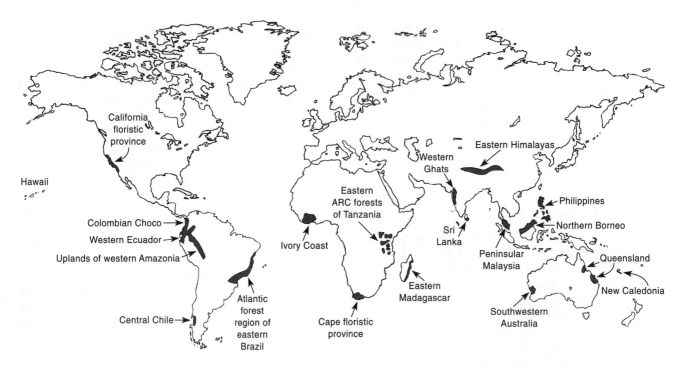

Figure 9.15 'Hotspots' that are especially rich in plant biodiversity, according to tropical ecologist Norman Meyers.

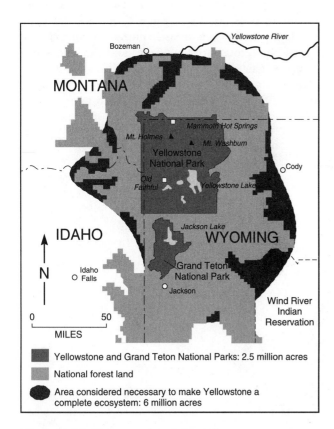

Figure 9.16 This map shows the Yellowstone ecosystem complex or biogeographical region, which extends far beyond the park boundaries. Ecologists believe that it is necessary to manage the entire region if the park itself is to remain viable.

problems that arise in limited populations. When inevitable adverse environmental conditions occur, there is neither room to seek out alternative habitats nor a sufficient supply of organisms to survive bad times.

Given human needs and pressures, however, large preserves aren't always possible. The relative positioning of several small reserves and the use of **corridors** of natural habitat that allow movement of species from one area to another can help maintain genetic exchange and prevent the high extinction rates characteristic of isolated and badly fragmented areas (fig. 9.17).

Figure 9.17 Corridors of natural vegetation allow endangered species to migrate between adjacent protected areas. This helps prevent stress due to changing conditions and effectively increases the genetic pool.

An interesting experiment funded by the World Wildlife Fund and the Smithsonian is being carried out in the rainforest of Brazil to determine the effects of shape and size on biological reserves. Some twenty-three test sites, ranging in size from one hectare (2.47 acres) to 10,000 hectares have been established. Some areas are surrounded by clear-cuts and newly created pastures, while others remain connected to the surrounding forest (fig. 9.18). Selected species are regularly inventoried to follow their dynamics after disturbance. As was expected, some species disappear very quickly, especially from small areas. Sun-loving species flourish in the newly created forest edges, but deep-forest, shade-loving species move out, particularly when size or shape reduce the distance from the edge to the center below a certain minimum. This demonstrates the importance of surrounding some reserves with buffer zones that maintain the balance of edge and shade species.

Even when preserves contain adequate area and habitat conditions to perpetuate a given species, there may be serious long-term genetic problems if only a small number of individuals are available to originate a population. We call this a **founder bottleneck.** It can lead to fertility and infant survival problems, and can greatly reduce both the diversity and the adaptability of a species. African cheetahs, for instance, appear to have undergone a catastrophic population reduction sometime in the past. Genetic tests suggest that all existing cheetahs descended fairly recently from a single female ancestor. The species has almost no genetic diversity. Infertility is high (60 to 70 percent) and the species may be doomed, no matter how large and well managed their habitat may be.

Social and Economic Factors

Nature preservation is closely tied to appropriate rural development in areas surrounding parks and preserves. Traditional seasonal wildlife migration routes often extend into or through areas claimed for human occupation. Conflicts between humans and wildlife can be minimized by careful planning and accommodation. The establishment of buffer zones to separate intensive human activities from central core regions of the preserve can reduce both predation by wildlife on gardens and flocks, as well as protecting wild species from predation by humans (fig. 9.19).

As we discussed earlier, providing jobs and income for local people through ecotourism and sustainable harvesting of natural products gives residents an incentive to protect wildlife and to cooperate with reserve management plans. Some promising programs attempt to integrate human needs and biological resource preservation. In 1986, UNESCO (United Nations Educational, Scientific, and Cultural Organization) initiated its **Man and Biosphere (MAB) program** that encourages traditional native cultures to live in and manage peripheral zones around protected areas. Mexico's 545,000-hectare (2,100 sq mi) Sian Ka'an Reserve on the Caribbean coast of the Yucatan Peninsula, for example, is an exemplary MAB reserve. The core area includes 112 km of coral reef and adjacent bays and marshes. Sustainable resource use in the rest

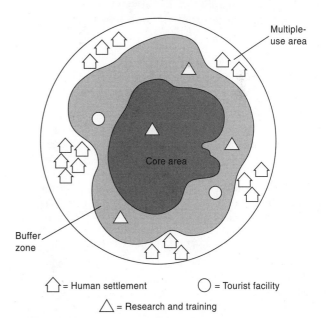

Figure 9.19 A model biosphere reserve. Traditional parks and wildlife refuges have well-defined boundaries to keep wildlife in and people out. Biosphere reserves, by contrast, recognize the need for people to have access to resources. Critical ecosystem is preserved in the core. Research and tourism are allowed in the buffer zone, while sustainable resource harvesting and permanent habitations are situated in the multiple use area around the perimeter.

Figure 9.18 How small can a nature preserve be? In an ambitious research project, scientists in the Brazilian rainforest are carefully tracking wildlife in plots of various sizes either connected to existing forests or surrounded by clear-cuts. As you might expect, the largest and most highly specialized species are the first to disappear.

of the reserve includes lobster fishing, small-scale farming, and coconut cultivation. The Amigos de Sian Ka'an, a local community organization, played a central role in establishing the reserve and is working to protect the resource base while it improves living standards for local people.

An even grander plan called Passeo Pantera (path of the panther) envisions a thousand-mile-long series of reserves and protected areas interconnected by natural corridors and managed buffer zones that would preserve both wildlife and native cultures along the entire Caribbean coast of Central America from the Yucatan to Panama. It is questionable whether the politically unstable and financially troubled countries involved could get together to establish such a plan. Competing goals and objectives of conservation groups, international development banks, and powerful neighbors such as the United States also complicate the picture, but the idea is a noble one whose time may have come.

Summary

In this chapter, we have briefly surveyed world biological resources and the ways humans benefit from wildlife. Natural causes of wildlife destruction include evolutionary replacement and mass extinction. Human actions also threaten our wildlife resources. Among the direct threats are overharvesting of animals and plants for food and various industrial and commercial products. Millions of live wild plants and animals are collected for pets, houseplants, and medical research. We also exterminate many plants and animals deliberately or inadvertently when we use pesticides to destroy "pest" species. Examples of indirect threats to biological resources are habitat destruction, the introduction of exotic species and diseases, pollution of the environment, and genetic assimilation.

The potential value of the species that may be lost if environmental destruction continues is enormous. It is also possible that the changes we are causing could disrupt vital ecological services on which we all depend for life.

The first hunting and fishing laws in the United States were introduced more than a century ago to restrict overexploitation and preserve species for future uses. The Endangered Species Act and CITES represent a new attitude toward wildlife in which we protect organisms just because they are rare and endangered. Now we are expanding our concern from protecting individual species to protecting habitat, threatened landscapes, and entire biogeographical regions. Social, cultural, and economic factors must also be considered if we want to protect biological resources on a long-term, sustainable basis.

Review Questions

1. What is the range of estimates of the total number of species on the earth? Why is the range so great?
2. What group of organisms has the largest number of species? What is the most abundant single species? Where do they live?
3. Define extinction. What is the natural rate of extinction in an undisturbed ecosystem?
4. What are rosy periwinkles and what products do we derive from them?
5. Describe some foods we obtain from wild plant species.
6. List three categories of damage to biological resources caused by humans.
7. What is the current rate of extinction and how does this compare to historic rates?
8. Compare the scope and effects of the Endangered Species Act and CITES.
9. Describe ten ways that humans directly or indirectly cause biological losses.
10. What is gap analysis and how is it related to ecosystem management and design of nature preserves?

Questions for Critical or Reflective Thinking

1. One reviewer said that this chapter is the most biased in this book. Do you agree? Is the level of moral outrage expressed here appropriate? Does it interfere with objectivity or effective communication? What is the proper balance between emotion and objectivity in a subject such as this?
2. One reason that we don't know how many species exist is that scientists don't agree on what constitutes a species. One taxonomist studying a group of

plant or animal specimens might divide them into a hundred species. Another taxonomist examining the same specimens might group them into ten or twenty species. What might lead to such a great discrepancy?

3. Put yourself in the place of a Canadian fishing industry worker. If you continue to catch cod and haddock, the species will quickly become economically extinct if not completely exterminated. On the other hand, there are no other jobs in your village and the dole (welfare) will barely keep you alive. What would you do?

4. According to our understanding of evolution, stronger species dominate and flourish at the expense of weaker species. Since we humans are now strong enough to exploit our environment, do we have any practical or ethical reasons to restrain ourselves?

5. How could people have believed a century ago that nature is so vast and fertile that human actions could never have a lasting impact on wildlife populations? Are similar examples of denial or misjudgment occurring now?

6. In the past, mass extinction has allowed for new growth, including the evolution of our own species. Should we assume that another mass extinction would be a bad thing? Could it possibly be beneficial to us? To the world?

7. Try designing a Man and Biosphere mixed-use preserve using a real-world example of some endangered species, a vulnerable landscape, and a threatened indigenous culture. How large would you make the buffer zone compared to the core area? What resource harvesting or other activities would you allow?

8. Suppose that preserving healthy populations of grizzly bears requires that we set aside some large fraction of Yellowstone Park as a zone into which no humans will ever again be allowed to enter for any reason. Would you support protecting bears even if no one ever sees them or could even be sure that they still exist?

9. Suppose that you had trespassed into the bear sanctuary and were attacked by a bear. Should park rangers shoot the bear or let it eat you?

Key Terms

biogeographical area 199
biological resources 183
corridors 200
decline spiral 189
ecotourism 187
endangered species 197
existence value 188
extinction 183

founder bottleneck 201
gap analysis 199
genetic assimilation 196
Man and Biosphere (MAB) program 201
species recovery plan 198
threatened species 197
vulnerable species 197

Suggested Readings

Abramovitz, J. N. *Investing in Biological Diversity.* Washington D.C.: World Resources Institute, 1991. Analyzes projects in 100 developing countries designed to protect and restore biodiversity.

Allen, L. "Plugging the Gaps." *Nature Conservancy News,* September/October 1992, 8. A good example of how gap analysis helps in preserving endangered species.

Baker, B. "Aliens!" *The Amicus Journal,* Winter 1992, 8. Kudzu, zebra mussels, and killer bees—give them a niche, they take a mile.

Campbell, T. "Net Losses." *Sierra* 76, no. 2 (1991): 48. A chilling investigative report on the destructive practice of drift-net fishing in the North Pacific.

Derr, M. "Raiders of the Reef." *Audubon* 94, no. 2 (March/April 1992): 48. Aquarists bring pieces of the world's coral reefs into their homes but are destroying the resource.

Durant, M., and M. Saito. "The Hazardous Life of our Rarest Plants." *Audubon* 87, no. 4 (July 1985): 50. Far more endemic races, varieties, and subspecies of higher plants are threatened with extinction than are higher animals. This is a good description of some examples of plant losses with lovely paintings of endangered species.

Durell, L. *State of the Ark: An Atlas of Conservation in Action.* Garden City, N.Y.: Doubleday, 1986. A popular and colorful overview of wildlife conservation.

Gardner, F. "Who Benefits from Ecotourism?" *Earth Island Journal* 6, no. 2 (Spring 1991): 30. A hard look at the effects of ecotourism in Guatemala. See also article by Michael Passoff in same issue.

Halpern, S. "Losing Ground." *Audubon* 94, no. 4 (July/August 1992): 70. Whooping cranes have made a comeback. It's their Texas refuge that's eroding.

Laycock, G. "The Importation of Animals." *Sierra* 63, no. 3 (April 1978): 20. A description of deliberate introduction of exotic species by "acclimatizers" and their effects.

Lean, G., et al. *Atlas of the Environment.* New York: Prentice Hall, 1990. Good source of information and graphics about many environmental issues sponsored by the World Wildlife Fund.

Leopold, A. *Game Management.* New York: Scribner's, 1933. The first textbook of wildlife management published in America. A classic by one of America's greatest naturalists.

Luoma, J. R. "Born to Be Wild." *Audubon* 94, no. 1 (January/February 1992): 50. Breeding programs may be able to save endangered animals, but as habitat disappears, will these creatures be condemned to captivity?

Mann, C. C., and M. L. Plummer. "The Butterfly Problem." *Atlantic Monthly,* January 1992, 47. A comprehensive discussion of endangered species protection from an anthropocentric perspective.

May, R. M. "How Many Species Inhabit the Earth?" *Scientific American* 267, no. 4 (October 1992): 42. A thoughtful discussion of biodiversity and the problem of species identification.

McNeely, J. A., et al. *Conserving the World's Biodiversity.* New York: International Union for the Conservation of Nature, 1990. A valuable overview of the effects of human actions on biodiversity.

Peters, R. L., ed. *Consequences of the Greenhouse Effect for Biological Diversity.* New Haven: Yale University Press, 1991. A comprehensive series of articles linking two of the most important environmental problems of our day.

Rauber, P. "Last Refuge." *Sierra* 77, no. 1 (January/February 1992): 36. Environmentalists campaign to save the Arctic coastal plain from oil drilling—while on site the reporter considers what to do when the bear comes but the plane doesn't.

Soulé, M. *Conservation Biology: The Science of Scarcity and Diversity.* Sunderland, Mass.: Sinauer Associates, 1986. A conservation biology textbook with contributions by many authors on theory as well as applications in specific case studies.

Tattersall, I. "Madagascar's Lemurs." *Scientific American* 268, no. 1 (January 1993): 110. The lemur's diverse Madagascaran habitats are disappearing fast and so are they. Unless hunting and deforestation cease, all may be lost.

Tennesen, M. "Poaching: Ancient Traditions vs the Law." *Audubon* 93, no. 4 (July/August 1991): 90. Traditional folk medicines contribute to wildlife destruction. Which will have to go?

Terborgh, J. "Why American Songbirds are Vanishing." *Scientific American* 266, no. 5 (May 1992): 98. Pesticides and tropical forest destruction contribute to songbird declines, but so do forest fragmentation and parasitism in summer ranges.

Whelan, T. *Nature Tourism.* Washington, D.C.: Island Press, 1991. A compendium of case studies from many countries on the benefits and disadvantages of adventure travel.

Wille, C. "Riches from the Rainforest." *Nature Conservancy Annual Report,* January/February 1993, 10. Describes how "barefoot parataxonomists" are searching for "green gold" in Costa Rica.

Wilson, E. O. *The Diversity of Life.* Cambridge, Mass.: Harvard University Press, 1992. A fascinating description of biodiversity by one of the world's leading field biologists.

World Wildlife Fund. *The Official World Wildlife Guide to Endangered Species of North America.* New York: Beacham Publishers, 1990. A two-volume set describing each of the officially listed endangered species in the United States. Volume I treats plants and mammals. Volume II describes other species.

10
AIR RESOURCES

Welcome carbon monoxide; hello sulfur dioxide. The air, the air is everywhere.

Gerome Ragni and James Rado, *Hair*

Objectives

After studying this chapter, you should be able to:

- Understand how solar energy warms the atmosphere and how humans are perturbing that system.
- Identify the major categories of pollution and their sources.
- Report on the causes and consequences of stratospheric ozone depletion.
- Describe the risks of indoor air pollutants, including radon.
- Explain how air pollutants damage human health, natural ecosystems, and the built environment.
- Compare different techniques for air pollution control.
- Sum up the current state of air pollution and prospects for the future at home as well as around the world.

INTRODUCTION

How does the air taste, feel, smell, and look in your home or your neighborhood? Chances are that wherever you live, the air is contaminated to some degree. Smoke, haze, dust, odors, corrosive gases, noise, and toxic compounds are present nearly everywhere, even in the most remote, pristine wilderness. Air pollution is generally the most widespread and obvious kind of environmental damage. According to the Environmental Protection Agency (EPA), some 147 million metric tons of air pollution (not counting carbon dioxide or wind-blown soil) are released into the atmosphere each year in the United States by human activities. Total worldwide emissions of these pollutants are around 2 billion metric tons per year. The air in a typical industrial city may contain unhealthy concentrations of hundreds of different toxic substances; indoor air can be even worse.

Over the past twenty years, air quality has improved appreciably in most cities in Western Europe, North America, and Japan. At the same time, however, we have discovered dangers from air pollutants that did not exist or were not recognized in the past. The manufacturing, shipping, use, and disposal of thousands of new supertoxic chemicals have introduced a great variety of new hazardous materials into the air we breathe. Exposure to parts per billion or even parts per trillion of these chemicals may be dangerous to sensitive members of the community.

In this chapter, we will examine the major types and sources of air pollution. We will study how they enter and move through the atmosphere and how they are changed into new forms, concentrated or dis-persed, and removed from the air by physical and chemical processes. We also will look at some of the major effects of air pollution on human health, ecosystems, and materials. Finally, we will survey some of the control methods available to reduce air pollution or mitigate its effects, and the results of air pollution control efforts on ambient air quality in North America and elsewhere.

ATMOSPHERE AND CLIMATE

We live at the bottom of a virtual ocean of air composed primarily of nitrogen (78 percent) and oxygen (21 percent) with traces of other gases and variable amounts of water vapor. Extending upward about 1,600 km (1,000 mi), this vast, restless envelope of gases is far more turbulent and mobile than the oceans of water (fig. 10.1). The current composition of the earth's atmosphere is unique in our solar system in having significant amounts of free oxygen and water vapor. We believe that virtually all the molecular oxygen in the air was produced by photosynthesis. If oxygen were not present, heterotrophs (like us) who oxidize organic compounds as an energy source could not exist.

Solar Energy Warms the Earth

The sun supplies the earth with an enormous amount of energy. Although it fluctuates from time to time, incoming solar energy at the top of the atmosphere averages about 340 watts per sq m. About half of this energy is reflected or absorbed by the atmosphere, and half the earth faces away from the sun at any given time. The amount reaching the earth's surface is

Figure 10.1 The atmospheric processes that purify and redistribute water, moderate temperatures, and balance the chemical composition of the air are essential in making life possible.

still about 178,000 terawatts (trillion watts) per year. This is 15,000 times the commercial energy used by humans each year. Energy flows through space as photons (light units) of electromagnetic energy that seem simultaneously to be waves of pure energy and infinitesimal particles of matter. Figure 2.6 shows the frequencies and wavelengths of some members of this family.

On average, clouds and the atmosphere absorb or reflect about half the **insolation** (*in*coming *sol*ar ra*diation*) that reaches the earth (fig. 10.2). Eventually, all the energy absorbed at the earth's surface is reradiated back into space. However, most of the solar energy reaching the earth is visible light, to which the atmosphere is relatively transparent, while the energy reemitted by the earth is mainly infrared radiation (heat energy). These longer wavelengths are absorbed by carbon dioxide and water vapor, trapping much of the heat close to the earth's surface. If the atmosphere were as transparent to infrared radiation as it is to visible light, the earth's surface temperature would be about 35° C (63° F) colder than it is now.

The warming caused by infrared absorption is called the **greenhouse effect** because the atmosphere, like the glass of a greenhouse, transmits sunlight while trapping heat inside. (The analogy is not totally correct, however, because glass is much more transparent to infrared radiation than is air; greenhouses stay warm mainly because the glass blocks air movement.) Increasing atmospheric carbon dioxide due to human activities could cause major climatic changes. This may be *the* environmental concern of the next century (box 10.1).

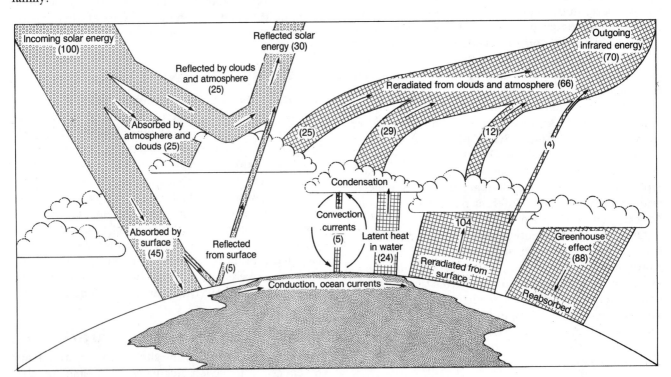

Figure 10.2 Energy balance between incoming and outgoing radiation. The atmosphere absorbs or reflects about half of the solar energy reaching the Earth. Most of the energy reemitted from the earth's surface is long-wave, infrared energy. Most of this infrared energy is absorbed by aerosols and gases in the atmosphere and is reradiated toward the planet, keeping the surface much warmer than it would otherwise be. This is known as the greenhouse effect. The numbers shown are arbitrary units. Note that for 100 units of incoming solar energy, 100 units are reradiated to space, but more than 100 units are radiated from the earth's surface because of the greenhouse effect.

The Greenhouse Effect and Global Climate Change

*A*re we altering the atmosphere in ways that could lead to disastrous, worldwide climate change? Some climatologists think so, forecasting an apocalyptic future in which summer heat is unbearable, farms are turned to deserts, famines sweep the globe, melting polar ice caps raise sea levels and flood coastal regions, and entire forests die along with thousands or even millions of species that can't migrate or adapt to sudden climatic changes.

About half of this predicted warming would be due to carbon dioxide (CO_2) released by burning fossil fuels, making cement, and cutting and burning forests. Together, these activities release about 8.5 billion metric tons of CO_2 annually, causing atmospheric levels to rise about 0.4 percent each year. Two hundred years ago, at the beginning of the Industrial Revolution, the atmosphere contained about 280 parts per million (ppm) CO_2; now it contains about 350 ppm.

If current trends continue, preindustrial CO_2 concentrations will have doubled by the year 2075. Computer models predict that doubling atmospheric CO_2 will cause global temperatures to rise 1.5 to 4.5° C (3° to 9° F). This may not sound like much change, but the difference between the temperature now and the last ice age about ten thousand years ago when glaciers covered much of North America was only about 5° C.

If a greenhouse warming occurs, it probably will not be distributed evenly around the globe. Additional ocean evaporation and cloud cover will likely keep tropical coastal areas about as they are now. The greatest temperature changes are predicted to be at high latitudes and in the middle of continents. Siberia and the Canadian arctic might experience increases of 10° to 12° C (18° to 22° F). Chicago might go from an average of fifteen to forty-eight days each year above 32° C (90° F). Calcutta, however, where the temperature is always hot, will not get much hotter.

Changing precipitation patterns might be one of the most serious consequences of the greenhouse effect. Some models predict that the midcontinents of North America, South America, and Asia will be significantly drier than they are now. This could have calamitous effects on world food supplies. The models don't agree on this point, however. While some show the American corn belt to be drier, others predict that it will be wetter than now. Model building is not yet an exact science.

Another disastrous effect could be rising sea levels. Some oceanographers calculate that thermal expansion alone could raise the sea level by one meter or more in the next century. If polar ice caps melt, sea level could rise by 30 meters (90 feet). Most of the world's major cities are on coasts only a few meters above sea level. About half the world's population would be threatened

and a large amount of valuable farmland would be lost. The United States Gulf Coast, for instance, would lose about 5,000 sq km, an area the size of Delaware.

Carbon dioxide is not the only gas that could cause climate warming. Methane, chlorofluorocarbons (CFCs), nitrous oxide, and other trace gases also absorb infrared radiation and warm the atmosphere (box fig. 10.1.1). Although these gases are more rare than CO_2, they absorb heat much more effectively. Methane, for instance, absorbs twenty to thirty times as much—molecule for molecule—as CO_2, and CFCs absorb approximately twenty thousand times as much. Collectively, these minor greenhouse gases would have warming effects comparable to doubling CO_2 concentrations. The biggest share of greenhouse gases are produced by the developed countries of the world (box fig. 10.1.2). Together, the United States, the former Soviet Union, Europe, and Japan are responsible for about two-thirds of all potential global warming.

Most scientists are convinced that global climate change is inevitable if CO_2 levels continue to rise. The only uncertainty is when the change will occur and how severe it will be. One of my friends says: "We know the climate will warm 1 to 4 degrees; we just don't know whether it will be Fahrenheit or Centigrade." Some climate modelers claim that warming has already started. They point to the fact that four out of the five warmest years in history were in the late 1980s and early 1990s. In 1992, however, the warming trend was deflected by sulfate aerosols from Mt. Pinatubo in the Philippines, which cooled the whole world by about one degree Celcius.

A few skeptics argue that the data are so variable and climatic processes so poorly known that we cannot yet be sure what will happen to the climate. They remind us that current climate models do not adequately represent the effects of water vapor, clouds, air currents, ocean

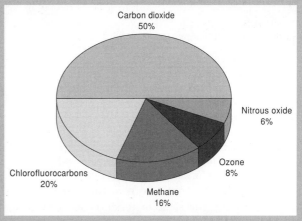

Figure 10.1.1 Contributions to global warming by greenhouse gases in 1990.

Source: Data from *Student Study Guide,* World Resources Institute.

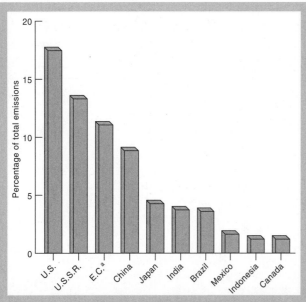

Figure 10.1.2 Countries with the highest greenhouse gas emissions in 1989. These countries account for about two-thirds of all global warming. [a]The European Community (EC) comprises 12 countries in Western Europe.

Source: *Student Study Guide,* World Resources Institute.

circulation, sulfate aerosols, decreasing solar radiation, biogeochemical ocean processes, or the possible growth stimulation of green plants and ocean plankton by higher CO_2. If the models' assumptions are changed slightly, the predictions change from positive (warming) to negative (cooling) trends.

Some people even claim that rising CO_2 levels could be beneficial. Sherwood B. Idso of the Department of Agriculture Water Conservation laboratory in Phoenix says that higher CO_2 will stimulate growth of many plants and result in more efficient water utilization. Lands now unsuitable for agriculture might come into production, and crop yields might increase rather than decrease. Growth of phytoplankton (single-celled photosynthetic organisms) in the ocean might absorb most of the additional CO_2 and balance world temperatures. This could increase productivity of marine ecosystems and make more seafood available for human consumption.

What should we do about greenhouse gases in the face of such uncertainty? Should we take steps now to reduce them or wait to see if their effects are as negative as some models suggest? Unfortunately, by the time we have evidence that a disaster is underway, it may be too late to make changes. Some authorities say it is too late already. Others argue that it would be wasteful to spend hundreds of billions of dollars and cause substantial disruptions to society to reduce these gases if they pose no threat.

There are, however, actions that could help reduce the greenhouse effect if it is real and that would be useful in their own right if it is not. For instance, the industrialized countries have agreed to phase out chlorofluorocarbons because they damage stratospheric ozone as well as absorb heat. We could cut CO_2 emissions and save money by increasing energy efficiency in our homes, automobiles, factories, and offices. Increasing your automobile mileage from 25 to 50 mpg, for instance, would save about $3,000 and reduce CO_2 production by 20 metric tons over the life of the vehicle.

We could (and should) switch to alternative, renewable energy sources, such as solar, wind, tidal, and hydropower. This would reduce pollution, save fossil fuels, and reduce warming. We could stop destruction of tropical forests and plant more trees, which would save biological resources, shelter and feed wildlife, and provide a useful source of environmentally responsible wood products. Trapping and burning methane from landfills and manure would provide useful energy while also reducing greenhouse effects. All of these actions are virtuous in their own right and deserve to be pursued, regardless of potential climate change. We might be able to reduce greenhouse emissions by 20 percent (enough to stabilize effects) over the next few decades without much sacrifice.

In many ways, your response to this issue comes down to a matter of philosophy and outlook. Are you pessimistic or optimistic? Do you believe it is more prudent to assume the worst-case scenario, or is it better to look on the bright side? Are draconian measures justified now, or should we be cautious until the science becomes clearer? This may well be the most difficult and the most important question in environmental science in the next few decades.

TYPES AND SOURCES OF AIR POLLUTION

Pollution is generally defined as unwanted environmental change caused by humans. This definition is problematic in the case of many types of air pollutants for which there are both natural and human sources of exactly the same material. Is smoke considered pollution if it comes from a human-started fire but merely a normal part of the environment if it is caused by wildfires? If the environmental or health effects are identical, should we care about the source? Since we can do little about natural origins of pollutants, most regulations focus on anthropogenic (human caused) emissions. This section surveys the sources and characteristics of some major air pollutants.

Primary, Secondary, and Fugitive Emissions

Primary pollutants are those released directly into the air in a harmful form. **Secondary pollutants,** by contrast, are modified to a hazardous form by chemical or physical reactions that occur as atmospheric components mix and interact. The most dangerous components of urban smog are often secondary products of photochemical reactions as many pollutants mix and "cook" in a noxious chemical soup. We will study some of these reactions in detail later in this chapter.

Fugitive emissions are those that do not go through a smokestack. By far the most massive example of this category is dust from soil erosion, strip mining, rock crushing, and building construction (and destruction). In the United States, natural and anthropogenic sources of fugitive dust add up to some 100 million metric tons per year. This can be a serious health hazard to those with respiratory problems.

Conventional or "Criteria" Pollutants

The Clean Air Act of 1970 designated seven major pollutants (sulfur dioxide, carbon monoxide, particulates, hydrocarbons, nitrogen oxides, photochemical oxidants, and lead) for which maximum **ambient air** (the air all around us) levels are mandated. These seven conventional or **criteria pollutants** contribute the largest volume of air quality degradation and also are considered the most serious threat of all air pollutants to human health and welfare. Figure 10.3 summarizes the major human sources of the first five criteria pollutants.

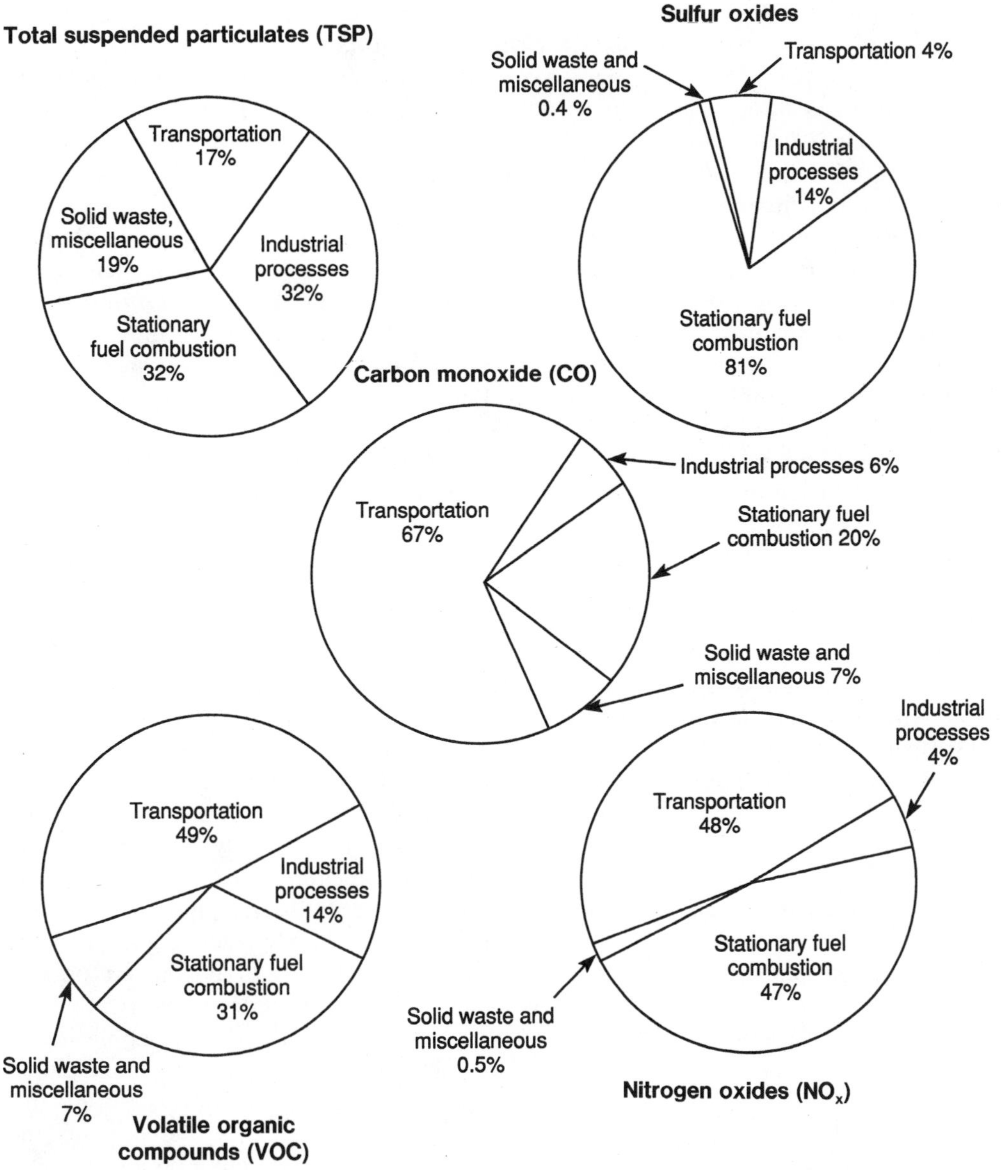

Sulfur Compounds

Natural sulfur sources such as sea spray evaporation, wind-blown sulfate-containing dust from arid soils, volcanic fumes, and biogenic emissions (from living organisms) of organic sulfur-containing compounds release about three times as much sulfur into the atmosphere as do human sources. In most urban areas, however, anthropogenic sources contribute as much as 90 percent of the sulfur in the air. The predominant form of anthropogenic sulfur is sulfur dioxide (SO_2) from combustion of sulfur-containing fuel (coal and oil), purification of sour (sulfur-containing) natural gas or oil, and industrial processes,

Figure 10.3 Sources of five major pollutants in the United States. Yearly emissions are 7.5 megatons TSP, 21.5 megatons SO_2 + SO_4, 79 megatons CO, 20.3 megatons NO_x, and 18.6 megatons VOC from anthropogenic sources. Totals may not be 100 percent due to rounding of figures.

such as smelting of sulfide ores. China and the United States are the largest sources of anthropogenic sulfur, primarily from coal burning.

Sulfur dioxide is a colorless corrosive gas that is directly damaging to both plants and animals. Once in the atmosphere, it can be further oxidized to sulfur trioxide (SO_3), which reacts with water vapor or dissolves in water droplets to form sulfuric acid (H_2SO_4). Very small solid particles or liquid droplets can transport the acidic sulfate ion (SO_4^-) long distances through the air or deep into the lungs where it is very damaging. Sulfur dioxide and sulfate ions are probably second only to smoking as causes of air-pollution-related health damage. Sulfate particles and droplets reduce air visibility as much as 80 percent over much of the United States.

Nitrogen Compounds
Nitrogen oxides are highly reactive gases formed when nitrogen in fuel or combustion air is heated to high temperatures in the presence of oxygen, or when bacteria in soil or water oxidize nitrogen-containing compounds. The initial product, nitric oxide (NO), oxidizes further in the atmosphere to nitrogen dioxide (NO_2), a reddish brown gas that gives photochemical smog its distinctive color. Because of their interconvertibility, the general term NO_x is used to describe these gases. Nitrogen oxides combine with water to make nitric acid (HNO_3), which is also a major component of atmospheric acidification.

Worldwide, anthropogenic sources account for around half of all nitrogen oxide emissions. In the United States about 95 percent of all human-caused NO_x is produced by fuel combustion in transportation and electric power generation (fig. 10.3). Ammonia from fertilizer and decaying organic material is oxidized to NO_x and is an important source of nitrogen loading in rural areas. Nitrous oxide (N_2O) is an intermediate in soil denitrification that absorbs ultraviolet light and plays an important role in climate modification.

Carbon Oxides
The predominant form of carbon in the air is carbon dioxide (CO_2). It is nontoxic and innocuous in the concentrations normally found in the atmosphere, but we are adding carbon dioxide to the air so rapidly (about 0.4 percent per year) that we may cause disastrous global climate changes (*see* box 10.1). Some 90 percent of the CO_2 emitted each year is from respiration (oxidation of organic compounds by plant and animal cells). These releases are usually balanced by an equal uptake by photosynthesis in green plants.

Burning of fossil fuels and biomass each contribute about 5 billion metric tons per year to the air.

Carbon monoxide (CO) is a colorless, odorless, highly toxic gas produced by incomplete fuel combustion, biomass or solid waste incineration, or partially anaerobic decomposition of organic material. CO inhibits respiration in animals by binding irreversibly to hemoglobin. About 1 billion metric tons of CO are released to the atmosphere each year, half of that from human activities. In the United States, two-thirds of the CO emissions are created by internal combustion engines in transportation. Land-clearing fires and cooking fires also are major sources. About 90 percent of the CO in the air is consumed in photochemical reactions that produce ozone.

Metals and Halogens
Many toxic metals are mined and used in manufacturing processes or occur as trace elements in fuels, especially in coal. These metals are released to the air in the form of metal fumes or suspended particulates by fuel combustion, ore smelting, and disposal of wastes. Worldwide lead emissions amount to about 2 million metric tons per year, or two-thirds of all metallic air pollution. Most of this lead is from leaded gasoline, still the only kind available in most developing countries. Lead is a metabolic poison and a neurotoxin that binds to essential enzymes and cellular components and inactivates them. An estimated 20 percent of all inner-city children in the United States suffer some degree of mental retardation from high environmental lead levels.

Mercury is another dangerous neurotoxin that is widespread in the environment. The two largest sources of atmospheric mercury appear to be coal-burning power plants and mercuric fungicides in house paint. Mercury batteries discarded in domestic waste make garbage incinerators another dangerous source of mercury vapors.

Other toxic metals of concern are nickel, beryllium, cadmium, thallium, uranium, cesium, and plutonium. Some 780,000 tons of highly toxic arsenic are released from metal smelters, coal combustion, and pesticide use each year. Halogens (fluorine, chlorine, bromine, and iodine) are highly reactive and generally toxic in their elemental form. Halogenated organic molecules such as chlorofluorocarbons (CFCs) are cheap, stable, and useful in a variety of applications. Hundreds of millions of tons are used annually in spray propellants, refrigeration compressors, and for foam blowing. Unfortunately, they diffuse into the stratosphere where they release chlorine and fluorine ions that destroy the ozone shield that protects the earth from ultraviolet radiation (box 10.2).

A Hole in the Ozone Shield

*I*n 1985, the British Antarctic Atmospheric Survey announced a startling and disturbing discovery: ozone levels in the stratosphere over the South Pole drop precipitously during September and October as the sun returns at the end of the long polar spring. This ozone depletion has been occurring since the late 1960s (box fig. 10.2.1) but was not recognized because researchers programmed their computers to ignore changes in ozone levels that were presumed to be erroneous. In 1992, stratospheric ozone was depleted over an area of 23 million square kilometers (about the size of the continental United States). Also in 1992, sulfate aerosols released by Mt. Pinatubo in the Philippines allowed chlorine radicals to attack ozone lower in the atmosphere than ever before. Between 14 and 18 kilometers in altitude, one hundred percent of the ozone was destroyed, but the polar vortex was less stable than usual and the hole quickly collapsed. Chlorine monoxide levels were reported to be high in the Arctic as well, but unusually warm weather melted the ice crystals necessary for ozone-destroying reactions.

Ozone is a harmful pollutant in the lower atmosphere, damaging plants, building materials, and human health, but it is an irreplaceable resource in the upper atmosphere where it screens out more than 99 percent of the dangerous ultraviolet (UV) rays from the sun. Without this shield, organisms on the earth's surface would be subjected to life-threatening radiation burns and genetic damage; thus, it is urgent that we learn what is depleting the ozone in Antarctic skies and whether it is a general phenomenon or limited to the unique conditions of the polar atmosphere.

The exceptionally cold temperatures (−85 to −90° C) of the Antarctic winter play a role in ozone destruction. During the long, dark months of the polar winter, isolation of the air mass within the polar vortex allows stratospheric temperatures to drop low enough for water vapor to freeze, creating a dense haze of ice particles at high altitudes. Molecules containing ozone and chlorine are adsorbed on the surface of the ice crystals. When the sun returns in the spring and provides UV energy to liberate active chlorine atoms from precursor molecules, they already are in close contact with ozone, and destructive chemical reactions proceed quickly (box fig.10.2.2).

The most likely sources of chlorine in the stratosphere appear to be chlorine monoxide (ClO) and hydrochloric acid (HCl) derived from chlorofluorocarbons (CFCs) and halon gases. CFCs were first synthesized in 1928 by scientists at General Motors as propellants for spray paint. Known commonly by the trade name, Freon, these nontoxic, nonflammable, chemically inert, cheaply produced, and versatile compounds soon found a wide variety of applications.

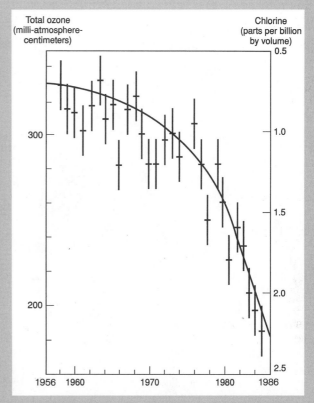

Figure 10.2.1 Monthly means of total ozone at Halley Bay, Antarctica, October 1956–1986. Bars represent standard deviation above and below mean.

Source: F. Sherwood Rowland. University of California at Irvine (unpublished October 1986).

Until 1978, aerosol spray cans used more CFCs than any other application, but concerns about a possible role in ozone destruction prompted the United States, Canada, and the Scandinavian countries to ban nonessential propellant uses. Other countries continue to use CFCs as spray propellants, however, and this use still represents nearly one-third of the 320,000 metric tons of CFC consumed annually worldwide. CFCs are mainly used now as refrigerants and solvents for cleaning printed circuits and electronic parts. They also are used extensively as blowing agents for making styrofoam cups, fast-food containers, and other foam products of modern society. Halons are used mainly as fire extinguishers.

Recent studies suggest that total ozone levels in the atmosphere have declined 4 to 5 percent over the past decade, but sunspot activity and other sources of normal fluctuation could account for most of that amount. Calculations based on atmospheric chemistry modeling suggest that continuing emissions of CFCs at 1987 levels could reduce ozone levels another 4 to 5 percent by the middle of the next century.

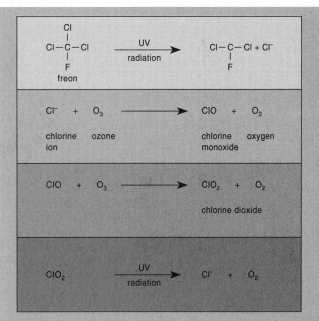

Figure 10.2.2 Photodissociation of Freon II. When struck by UV radiation, chloroflurocarbons release chlorine ions. These chlorine ions combine with ozone to make chlorine monoxide plus oxygen. The chlorine monoxide reacts with another ozone molecule to make chlorine dioxide, which dissociates into oxygen and a reactive chlorine ion that starts the cycle over again.

The extra UV light allowed to reach the earth's surface by this depletion of the ozone shield could cause a million extra cases of skin cancer (about 30,000 would probably be fatal) each year, as well as cataracts, suppression of the immune system, severe sunburns, and ac-

celerated skin aging. It also could damage plants (including crops) and degrade plastics and other polymers, causing losses amounting to billions of dollars each year.

Some of this cancer increase seems already to have begun. Southern Australia is experiencing a skin cancer epidemic that may be related to ozone loss. Some oncologists predict that two out of three Australians will have skin cancer before age 70. While much of this is due to fair skin and a love of sun bathing, ozone-depleted air spreading from polar regions has increased UV irradiation over Australia.

In response to this threat, eighty-one nations signed a treaty in 1989 in Helsinki, Finland, agreeing to phase out production of CFCs by the end of the century. This treaty does not cover carbon tetrachloride or methyl chloroform. In addition, China, India, and a group of developing nations asked for help in acquiring new technology so they can obtain refrigeration. Fortunately, an alternative to CFCs already exists. The DuPont Company, which developed CFCs thirty years ago, announced that it has developed a family of hydrofluorocarbons that can replace CFCs with only a fraction of the damage to the atmosphere. However, these hydrofluorocarbons are more expensive and their health safety has not yet been established.

Many observers are amazed and encouraged that nations could respond so quickly to the threat of ozone depletion. Others doubt that the reductions in CFC use will be quick enough to protect us from the danger of UV irradiation. This issue, like so many in environmental studies, raises difficult questions about how much disruption and expense is justified when scientific evidence is equivocal but the potential risks are very great.

Particulate Material

An **aerosol** is any system of solid particles or liquid droplets suspended in a gaseous medium. For convenience, we generally describe all atmospheric aerosols, whether solid or liquid, as **particulate material.** This includes dust, ash, soot, lint, smoke, pollen, spores, algal cells, and many other suspended materials. Anthropogenic particulate emissions amount to about 100 million metric tons per year worldwide. Windblown dust, volcanic ash, and other natural materials may contribute one hundred times that much.

Particulates often are the most apparent form of air pollution since they reduce visibility and leave dirty deposits on windows, painted surfaces, and textiles. Respirable particles smaller than 2.5 micrometers are among the most dangerous of this group because they can be drawn into the lungs where they damage delicate respiratory tissues. Asbestos fibers and cigarette smoke are among the most dangerous respirable particles in urban and indoor air because they are carcinogenic. The World Health Organization estimates that

70 percent of the global urban population, primarily in developing countries, breathes air that has unhealthy particulate concentrations.

Volatile Organic Compounds

Volatile organic compounds (VOC) are organic chemicals that exist as gases in the air. Plants are the largest source of VOC, releasing an estimated 800 million tons of isoprene and terpene hydrocarbons each year. About 400 million tons of methane (CH_4) are produced by natural wetlands and rice paddies and by bacteria in the guts of termites and ruminant animals. These volatile hydrocarbons are generally oxidized to CO and CO_2 in the atmosphere.

In addition to these natural VOCs, a large number of other synthetic organic chemicals, such as benzene, toluene, formaldehyde, vinyl chloride, phenols, chloroform, and trichloroethylene, are released into the air by human activities. About 28 million tons of these compounds are emitted each year in the United States, mainly unburned or partially burned hydrocarbons from transportation, power plants, chemical plants, and petroleum refineries. These chemicals play an important role in the formation of photochemical oxidants.

In 1987, the EPA began requiring industries to report releases of some 332 toxic organic chemicals into the air. The expectation was that these emissions would amount to about 36,000 metric tons per year. To everyone's surprise, the reports totaled 2 *million* metric tons or 5 billion pounds. The largest carcinogen emission was 52,000 tons (115 million lbs) of dichloromethane, which is used as an industrial solvent and paint stripper. This startling revelation was instrumental in the regulation of these hazardous organic compounds in the 1990 Clean Air Act revisions.

Photochemical Oxidants

Photochemical oxidants are products of secondary atmospheric reactions (fig. 10.4) driven by solar energy in the form of ultraviolet light (UV). One of the most important of these reactions involves formation of singlet (atomic) oxygen by splitting either molecular oxygen (O_2) or nitrogen dioxide (NO_2). This singlet oxygen then reacts with another molecule of O_2 to make **ozone** (O_3). Ozone formed in the stratosphere provides a valuable shield for the biosphere by absorbing incoming ultraviolet radiation. In ambient air, however, O_3 is a strong oxidizing reagent and damages vegetation, building materials (such as paint, rubber, and plastics), and sensitive tissues (such as eyes and lungs). Ozone has an acrid, biting odor that is a distinctive characteristic of photochemical smog. Hydrocarbons in the air contribute to accumulation of ozone by removing NO in the formation of compounds, such as peroxyacetyl nitrate (PAN), which is another damaging photochemical oxidant.

Unconventional or "Noncriteria" Pollutants

The EPA has authority under the Clean Air Act to set emission standards (regulating the amount released) for certain unconventional or noncriteria pollutants that are considered especially toxic or hazardous. Among the materials regulated by emission standards are asbestos, benzene, beryllium, mercury, polychlorinated biphenyls, PCBs, and vinyl chloride. The 1990 amendments to the Clean Air Act require the EPA to establish regulations for another 250 toxic air pollutants.

Some other unconventional forms of air pollution deserve mention as well. Aesthetic degradation includes any undesirable changes in the physical characteristics or chemistry of the atmosphere. Noise, odors, and light pollution are examples of atmospheric degradation that may not be life-threatening but reduce the quality of our lives. This is a very subjective category. Odors and noise (such as loud music) that are offensive to some may be attractive to others. Often the most sensitive device for odor detection is the human nose. We can smell styrene, for example, at 44 parts per billion (ppb). Trained panels of odor testers often are used to evaluate air samples. Factories that emit noxious chemicals sometimes spray "maskants" or perfumes into smokestacks to cover objectionable odors.

Indoor Air Pollution

We have spent a considerable amount of effort and money to control air pollution, but we have only recently become aware that most of us are exposed to far higher concentrations of toxic air pollutants indoors than outside. These pollutants can be trapped and concentrated indoors, especially when houses are closed tightly during heating or air conditioning seasons. Furthermore, people generally spend more time indoors than outside and are therefore exposed to higher doses of these pollutants.

Smoking is without doubt the most important air pollutant in the United States in terms of human health. The Surgeon General estimates that 350,000 smokers die each year from emphysema, heart attacks, strokes, lung cancer, or other diseases caused by smoking. The EPA estimates that 3,000 people die each year from lung cancer caused by secondhand smoke. This is twice as many cancers as from all toxic emissions from all industries.

Other major indoor air pollution health hazards include asbestos, formaldehyde, vinyl chloride, radon, and combustion gases. Asbestos was once widely used in floor and ceiling tiles, plaster, cement, insulation, and soundproofing. It is a serious concern in indoor air because of its carcinogenicity. Formaldehyde is used in more than three thousand products, including such building materials as particle board, waferboard, and urea-formaldehyde foam insulation. Vinyl chloride is used in plastic plumbing pipe, floor and wall coverings, and countertops. New carpets and drapes typically contain two dozen chemicals designed to kill bacteria and molds, resist stains, bind fibers, and retain colors.

```
1. O₂ + UV ⟶ O* + O* (oxygen ion or free radical)
2. NO₂ + UV ⟶ NO + O*
3. O* + O₂ ⟶ O₃
4. O₃ + NO ⟶ O₂ + NO₂
5. NO + VOC ⟶ NO₂ + PAN (peroxyacetylnitrate) + aldehydesᵃ

Net results:

NO₂ + UV + VOC + O₂ ⟶ NO₂ + O₃ + PAN + aldehydes

ᵃExamples of aldehydes are formaldehyde, acetylaldehyde, and
benzaldehyde.
```

Figure 10.4 Some photochemical atmospheric reactions that contribute to smog formation.

Acting Locally: Box 10.3

Radon in Indoor Air

The Environmental Protection Agency (EPA) estimates that one home in ten in the United States has excessive indoor radon levels, and that 5,000 to 20,000 of the 136,000 deaths from lung cancer can be attributed to radiation from indoor radon. This makes indoor radon second only to smoking as an air pollution health hazard. It appears that radon causes ten times more deaths each year than all other regulated toxic air pollutants combined.

Radon is a colorless, odorless, and tasteless gas that is produced naturally from uranium in rocks and soil. It is found almost everywhere, although some areas have geological structures that are especially high in radon. Radon gas is chemically inert, but it gives rise through a series of radioactive decay reactions to "daughters" such as polonium that bind to biological tissues and expose them to damaging radioactivity. Most of the health risk from radon actually comes from polonium.

In surveys of areas suspected to be at high risk for radon, one home in four exceeds the "action level" of 4 picocuries per liter (pCi/l). The highest individual concentrations were over 200 pCi/l in Pennsylvania, Ohio, and on a South Dakota Indian reservation. Iowa has the highest percentage of homes with excess radon. Nearly three quarters of Iowa homes exceeded 4 pCi/l.

Continuously breathing air with 4 pCi/l of radon and its daughters gives a radiation dose of about 2,000 millirem per year. This is equivalent to two hundred chest X rays each year and carries about a 1-in-100 lifetime risk of dying from lung cancer. In other words, breathing 4 pCi/l continuously is equivalent to smoking about half a pack of cigarettes per day. About 1 percent of the homes surveyed had more than 20 pCi/l, which would produce an annual exposure equal to or exceeding that received by underground uranium miners.

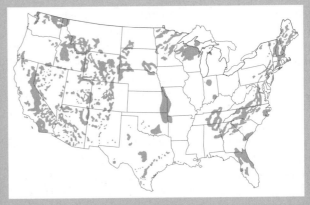

Figure 10.3.1 Areas with potentially high levels of uranium in soil and rocks. Radon levels in air and water may be high in these areas.
Source: U.S. Geological Survey.

The amount of radon entering a house depends not only on the underlying soil and rock type (box fig. 10.3.1), but also on how the house is built and how much ventilation it has. Radon filters up through the ground and seeps into the house through porous basement walls, cracks in the foundation, loose pipe fittings, sump pumps, and enclosed crawl spaces (box fig. 10.3.2). The more cracks and other openings in the basement, the more radon can enter. Radon concentrations in basements are generally about twice those on the first floor. Except for some basement apartments, multistory apartment buildings rarely have a radon problem, and most concern is focused on single-family houses.

If houses are well ventilated, radon is quickly diluted to harmless levels. Outdoor air typically contains 0.1 to 0.2 pCi/l of radon. Houses that are tightly sealed to prevent energy losses tend to trap radon inside, along with formaldehyde and other toxic air pollutants that may have synergistic effects. There often is a conflict between

---

In many cases, indoor air in homes has concentrations of chemicals that would be illegal outside or in the workplace. The EPA has found that concentrations of such compounds as chloroform, benzene, carbon tetrachloride, formaldehyde, and styrene can be seventy times higher in indoor air than in outdoor air. Many people are highly sensitive to these chemicals, and it is not uncommon to trace illness to a "house syndrome" caused by polluted indoor air. Next to smoking, the most serious indoor air pollutant in the United States is probably radon gas that leaks into houses from surrounding soil and rock (box 10.3).

In the less-developed countries of Africa, Asia, and Latin America where such organic fuels as firewood, charcoal, dried dung, and agricultural wastes make up the majority of household energy, smoky, poorly ven-

tilated heating and cooking fires represent the greatest source of indoor air pollution (fig. 10.5). The World Health Organization (WHO) estimates that some 400 million people, mostly women, are adversely affected by pollution from this source. Many women spend long hours each day cooking over open fires or unventilated stoves in enclosed spaces. The levels of carbon monoxide, particulates, aldehydes, and other toxic chemicals can be one hundred times higher than would be legal for outdoor ambient concentrations in the United States. Designing and building cheap, efficient, nonpolluting energy sources for the developing countries would not only save shrinking forests but would make a major impact on health as well.

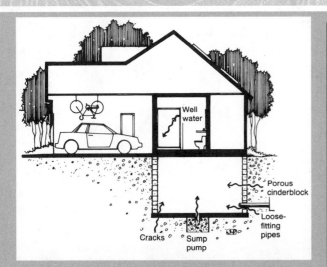

Figure 10.3.2 Five common ways radon can enter your house.

the need to conserve energy and the dangers of indoor air pollution. In cold climates, winter levels of indoor radon are usually twice as high as summer levels when windows are open and air flow is greater. In the south, where air conditioning is common, the opposite may be true.

Radon also accumulates in groundwater. A survey of groundwater-based public drinking water supplies revealed that 20,000 (roughly 40 percent) of such systems in the United States contain more than the 500 pCi/l, considered to be a health risk. The danger is not so much from drinking the water as it is from inhaling radon evaporated during showers, bathing, cooking, and toilet flushing. At 500 pCi/l, the maximum limit for drinking water proposed by the EPA, radon would carry a 1–in–10,000 risk of causing a fatal cancer. This is one

hundred to one thousand times higher than the risk considered acceptable for any other toxic substance in water, and makes radon the leading water pollutant, as well as the leading air pollutant, in terms of health effects.

How can you find out if your house exceeds safe levels of radon? Your state public health department or local branch of the American Lung Association probably has radon detectors or can tell you where to obtain one. There are two main types of detectors: charcoal canisters and alpha-track. The charcoal canister is slightly cheaper but measures radon levels over only three to seven days. The alpha-track gives an average over one to three months and is usually a better representation of actual levels.

What can you do if you find that your house has excess radon? The first step is to seal cracks in the foundation and openings around pipes. Sealing walls with a plastic lining or epoxy paint may be desirable. Gravel or other porous fill around foundations can allow radon to escape outside rather than seep into the house. Drilling holes in the basement floor and installing sealed pipes (perhaps with a small fan to improve air flow) to vent radon from subfloor spaces to the outside can reduce radon infiltration. Finally, you can improve ventilation of indoor spaces to remove radon. Where heat loss is a problem, it may be worthwhile to install an air-to-air heat exchanger. Various combinations of these techniques, depending on particular house design and local conditions, have been able to reduce indoor radon levels by 95 percent or more. Carrying out all of these steps might cost thousands of dollars, but, in many cases, a few hundred dollars invested wisely is sufficient to make most houses relatively safe.

EFFECTS OF AIR POLLUTION

So far we have looked at the major types and sources of air pollutants. Now we will turn our attention to the effects of those pollutants on human health, physical materials, ecosystems, and global climate.

Human Health

The primary human health effects of most air pollutants seems to be injury of delicate tissues that sets in motion an inflammatory response similar to the series of reactions that accompany an infection. **Bronchitis** is a persistent inflammation of airways in the lung that causes a painful cough and involuntary muscle spasms that constrict airways. Acute bronchitis can

obstruct airways so severely that death results. Smoking is undoubtedly the largest cause of chronic bronchitis in most countries. Persistent smog and acid aerosols also can cause this disease.

Severe bronchitis can lead to **emphysema**, an irreversible obstructive disease in which airways become permanently constricted and lung tissues are damaged or even destroyed (fig. 10.6). Breathing becomes difficult, and victims of emphysema make a characteristic whistling sound when they breathe. Cardiovascular stress from lack of oxygen in the blood is a common complication of all obstructive lung diseases. About twice as many people die of heart failure associated with smoking as die of lung cancer.

Irritants in the air are so widespread that about half of all lungs examined at autopsy in the United States show some damage and about 10 percent of the population suffers from pollution-related bronchitis and emphysema. Some 70,000 excess deaths per year are attributable to these diseases, which make them third—behind heart attacks and strokes—as sources of disease mortality in the United States.

Asthma is a distressing disease characterized by unpredictable and disabling shortness of breath caused by sudden episodes of muscle spasms in the bronchial walls. These attacks are often triggered by inhaling allergens, such as dust, pollen, animal dander, or corrosive gases. In some cases, there is no apparent external factor, and internal release of triggering agents is suspected. It isn't known whether asthma is genetic, environmental, or some combination of the two.

Fibrosis is the general name for accumulation of scar tissue in the lung. Among the materials that cause fibrosis are silica or coal dust, asbestos, glass fibers, beryllium and aluminum whiskers, metal fumes, cotton lint, and irritating chemicals, such as the herbicide paraquat. We give each of these diseases an individual name such as silicosis (quartz crystals), black lung (coal), or brown lung (cotton), but they really are very similar in development and effect. As the lung fills up with fibrotic tissue, respiration is blocked and the person slowly suffocates. In some cases, cell growth stimulated by the presence of foreign material in the lung results in tumor formation. Lung cancers are often lethal.

Plant Pathology

In the early days of industrialization, fumes from furnaces, smelters, refineries, and chemical plants often destroyed vegetation and created desolate, barren landscapes around mining and manufacturing centers. The copper-nickel smelter at Sudbury, Ontario, is a notorious example of air pollution effects on vegetation and ecosystems. Early in this century, sulfur dioxide and sulfuric acid released by smelting of sulfide ores caused massive destruction of the plant community over a wide area around the smelter. Rains washed away the exposed soil, leaving a grim moonscape of blackened bedrock (fig. 10.7).

Figure 10.5 Smoky cooking and heating fires may cause more ill health effects than any other source of air pollution except tobacco smoking. Levels of carbon monoxide, particulates, and cancer-causing hydrocarbons can be one thousand times higher indoors than outdoors. Some 400 million people, mostly women and children, spend hours each day in poorly ventilated kitchens and living spaces.

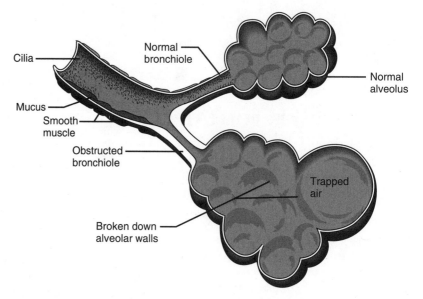

Figure 10.6 Emphysema results from chronic irritation and obstruction of airways and alveoli.

Figure 10.7 Sulfur dioxide emissions and acid precipitation from the International Nickel Company copper smelter (*background*) killed all vegetation over a large area near Sudbury, Ontario. Even the pink granite bedrock has burned black. Recently installed scrubbers have dramatically reduced sulfur emissions. The ecosystem farther away from the smelter is slowly beginning to recover.

"Supertall," 400-meter smokestacks were installed in the 1950s and sulfur scrubbers were added 20 years later, reducing emissions by 90 percent. Liming (spreading limestone on the ground) has been remarkably effective in restoring vegetation, especially in the town of Sudbury, but complete restoration of the ecosystem will take many years.

Similar destruction occurred at many other industrial sites. Copperhill, Tennessee; Butte, Montana; and the Ruhr Valley in Germany are some other well-known examples, but these areas also are showing signs of recovery since corrective measures have been taken. This can be considered either as discouraging testimony to the carelessness of humans or encouraging evidence of the regenerative powers of nature.

Pollutant levels too low to produce visible symptoms of damage may still have important effects on plants. Field studies show that yields in some sensitive crops, such as soybeans and alfalfa, may be reduced as much as 50 percent by currently existing levels of oxidants such as sulfur dioxide and ozone in ambient air. Some plant pathologists suggest that these pollutants are responsible for losses of $10 billion per year to agricultural, ornamental, and forest products.

Acid Deposition

Most people in the United States became aware of problems associated with **acid precipitation** (the deposition of wet acidic solutions or dry acidic particles from the air) within the past decade or so, but English scientist Robert Angus Smith coined the term "acid rain" while studying air chemistry in Manches-

ter, England, in the 1850s. By the 1940s, it was known that pollutants, including atmospheric acids, could be transported long distances by wind currents. This was thought to be only an academic curiosity until it was shown that precipitation of these acids can have far-reaching ecological effects.

Atmospheric Acidity

Normal, unpolluted rain generally has a pH of about 5.6 due to carbonic acid created by CO_2 in air (see box fig. 10.2.1). Volcanic emissions, biological decomposition, and chlorine and sulfates from ocean spray can reduce the pH of rain well below 5.6, while alkaline dust can raise it above 7. In industrialized areas, anthropogenic acids in the air usually far outweigh those from natural sources. Acid rain is only one form in which acid deposition occurs. Fog, snow, mist, and dew also trap and deposit atmospheric contaminants. Furthermore, fallout of dry sulfate, nitrate, and chloride particles can account for as much as half of the acidic deposition in some areas. These particles are converted to acids when they dissolve in surface water or contact moist tissues (e.g., in the lungs). We have considerable evidence that acid aerosols are a human health hazard.

Aquatic Effects

It has been known for about thirty years that acids—principally H_2SO_4 and HNO_3—generated by industrial and automobile emissions in northwestern Europe are carried by prevailing winds to Scandinavia where they are deposited in rain, snow, and dry precipitation. The thin, acidic soils and oligotrophic lakes and streams in the mountains of southern Norway and Sweden have been severely affected by this acid deposition. Some 18,000 lakes in Sweden are now so acidic that they will no longer support game fish or other sensitive aquatic organisms.

There has been a great deal of research on the mechanisms of damage by acidification. Generally, reproduction is the most sensitive stage in the life cycle. Eggs and fry of many fish species are killed when the pH drops to about 5.0. This level of acidification also can disrupt the food chain by killing aquatic plants, insects, and invertebrates on which fish depend for food. At pH levels below 5.0, adult fish die as well. Trout, salmon, and other game fish are usually the most sensitive. Carp, gar, suckers, and other less desirable fish are more resistant. Acids may kill fish in several ways. Acidity alters body chemistry, destroys gills and prevents oxygen uptake, causes bone decalcification, and disrupts muscle contraction. Another dangerous effect (for us as well as fish) is that acid water leaches toxic metals, such as aluminum, out of soil and rocks. Which of these mechanisms is the most important is open to debate, but it is clear that acid deposition has had disastrous effects on sensitive aquatic ecosystems.

In the early 1970s, evidence began to accumulate suggesting that air pollutants are acidifying many lakes in North America. Studies in the Adirondack Mountains of New York revealed that about half of the high-altitude lakes (above 1,000 m or 3,300 ft) are acidified and have no fish. Precipitation records show that the average pH of rain and snow has dropped significantly over a large area of northeastern United States and Canada in the past two decades. Some 48,000 lakes in Ontario are endangered and nearly all of Quebec's surface waters, including about 1 million lakes, are believed to be highly sensitive to acid deposition. Figure 10.8 shows the location of acid-sensitive regions of North America and levels of acidity in precipitation. About 50 percent of the acid deposition in Canada comes from the United States, while only 10 percent of U.S. pollution comes from Canada. Canadians are understandably upset about this imbalance.

Sulfates account for about two-thirds of the acid deposition in eastern North America and most of Europe, while nitrates contribute most of the remaining one-third. In urban areas, where transportation is the major source of pollution, nitric acid is equal to or slightly greater than sulfuric acids in the air. A vigorous program of pollution control has been undertaken by Canada, with promises of 50 percent reduction of SO_2 emissions and significant lowering of NO_x production.

Much of the western United States and Canada has relatively alkaline bedrock and carbonate-rich soil that counterbalance acids from the atmosphere. Recent surveys of the Rocky Mountains, the Sierra Nevadas in California, and the Cascades in Washington, however, have shown that many high mountain lakes and streams have very low buffering capacity (ability to resist pH change) and are susceptible to acidification.

Forest Damage

In the early 1980s, disturbing reports appeared of rapid forest declines in both Europe and North America. One of the earliest was a detailed ecosystem inventory on Camel's Hump Mountain in Vermont. A 1980 survey showed that seedling production, tree density, and viability of spruce-fir forests at high elevations had declined about 50 percent in fifteen years. By 1990, almost all the red spruce, once the dominant species on the upper part of the mountain, were dead or dying. A similar situation was found on Mount Mitchell in North Carolina, where almost all red spruce and Frasier fir above two thousand meters are in a severe decline. In Canada, millions of sugar maples—the national symbol and an important income source for many people— are dying of unknown causes.

Similar damage is reported in Germany, Czechoslovakia, Poland, Austria, and Switzerland (fig. 10.9). Again, high-elevation forests are most severely affected. This is a disaster for mountain villages in the Alps that

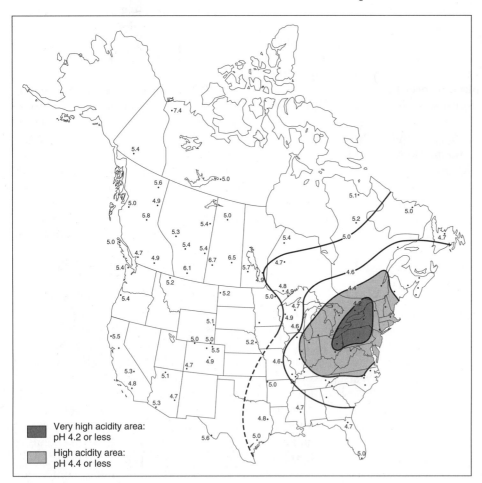

Figure 10.8 The acidity of precipitation over Canada and the United States. Numbers are average annual pH in 1982.

Source: Data from the National Atmospheric Deposition Program and the Canadian Network for Sampling of Precipitation.

Figure 10.9 A Czech forest killed by pollution in the Kakonoski National Park in southwest Poland. A toxic soup of many different air pollutants may combine to cause this damage.

depend on forests to prevent avalanches in the winter. Sweden, Norway, the Netherlands, Romania, China, and the former Soviet Union also have evidence of growth reduction, defoliation, root necrosis, lack of seedling growth, and premature tree death. The species afflicted vary from place to place, but the overall picture is of widespread forest destruction.

This complex phenomenon probably has many contributing factors, but air pollution and deposition of atmospheric acids are thought to be leading causes of forest destruction in many areas. Considerable research has shown that acids are directly toxic to tender shoots and roots. High-altitude forests are subjected to especially intense doses of these acids because clouds saturated with pollutants tend to hang on mountaintops, bathing forests in a toxic soup for days or even weeks at a time.

Scientists have suggested that other mechanisms also may play a role in forest decline. Overfertilization by nitrogen compounds may make trees sensitive to early frost. Essential minerals, such as magnesium, may be washed out of foliage or soil by acidic precipitation. Toxic metals, such as aluminum, may be solubilized by acidic groundwater. Plant pathogens and insect pests may damage trees or attack trees debilitated by air pollution. Microrhizzae (fungi) that form essential mutual associations with trees may be damaged by acid rain.

In addition to acid precipitation, other air pollutants such as sulfur dioxide, ozone, or toxic organic compounds may damage trees. Furthermore, repeated harvesting cycles in commercial forests may remove nutrients and damage ecological relationships essential for healthy tree growth. Perhaps the most likely sce-

nario is that all these environmental factors act cumulatively but in different combinations in the deteriorating health of individual trees and entire forests.

Damage to Buildings and Monuments

In cities throughout the world, some of the oldest and most glorious buildings and works of art are being destroyed by air pollution. Smoke and soot coat buildings, paintings, sculptures, and textiles. Limestone and marble are being destroyed by atmospheric acids at an alarming rate. The Parthenon in Athens, the Taj Mahal in Agra, the Colosseum in Rome, frescoes and statues in Florence, medieval cathedrals in Europe, and the Lincoln Memorial and Washington Monument in Washington, D.C., are slowly dissolving and flaking away because of acidic fumes in the air (fig. 10.10). Medieval stained glass windows in Cologne's Gothic cathedral are so porous from etching by atmospheric acids that pigments disappear and the glass literally crumbles away. Restoration costs for this one building alone are estimated at three to four billion German marks ($1.5 to $2 billion).

On a more mundane level, air pollution also damages ordinary buildings and structures. Corroding steel in reinforced concrete weakens buildings, roads, and bridges. Paint and rubber deteriorate due to oxidization. Limestone, marble, and some kinds of sandstone flake and crumble. The Council on Environmental Quality estimates that U.S. economic losses from architectural damage caused by air pollution amount to about $4.8 billion in direct costs and $5.2 billion in property value losses each year.

AIR POLLUTION CONTROL

What can we do about air pollution? In this section we will look at some of the techniques that can be used to avoid creating pollutants or to clean up efflu-

Figure 10.10 Atmospheric acids, especially sulfuric and nitric acids, have almost completely eaten away the face of this medieval statue. Each year, the total losses from air pollution damage to buildings and materials amounts to billions of dollars.

ents before they are released. We also will look at some legislation that regulates pollutant emissions and ambient air quality.

Particulate Removal Techniques

Filters remove particles physically by trapping them in a porous mesh of cotton cloth, spun glass fibers, or asbestos-cellulose, which allows air to pass through but holds back solids. Collection efficiency is relatively insensitive to fuel type, fly ash composition, particle size, or electrical properties. Filters are generally shaped into giant bags 10 to 15 meters long and 2 or 3 meters wide. Effluent gas is blown into the bottom of the bag and escapes through the sides much like the bag on a vacuum cleaner (fig. 10.11*a*). Every few days or weeks, the bags are opened to remove the dust cake. Thousands of these bags may be lined up in a "baghouse." These filters are usually much cheaper to install and operate than electrostatic filters (described in the following paragraphs).

Electrostatic precipitators are the most common particulate controls in power plants (fig. 10.11*b*). Fly ash particles pick up an electrostatic surface charge as they pass between large electrodes in the effluent stream. This causes the particle to migrate to and accumulate on a collecting plate (the oppositely charged electrode). These precipitators consume a large amount of electricity, but maintenance is relatively simple and collection efficiency can be as high as 99 percent. Performance depends on particle size and chemistry, strength of the electric field, and flue gas velocity.

Sulfur Removal

As we have seen earlier in this chapter, sulfur oxides are among the most damaging of all air pollutants in terms of human health and ecosystem damage. It is important to reduce sulfur loading. This can be done either by using low-sulfur fuel or by removing sulfur from effluents.

Fuel Switching and Fuel Cleaning

Switching from soft coal with a high sulfur content to low-sulfur coal can greatly reduce sulfur emissions. This may eliminate jobs, however, in such areas as Appalachia where the economy is already depressed. Changing to another fuel, such as natural gas or nuclear energy, can eliminate all sulfur emissions as well as those of particulates and heavy metals. Natural gas requires expensive pipelines for delivery, however, and many people prefer the sure dangers of coal pollution to the uncertain dangers of nuclear power (chapter 12). Alternative energy sources, such as solar power, would be preferable to either fossil fuel or nuclear power, but are not yet economically competitive in most areas. In the interim, coal can be crushed and washed to remove sulfur and metals before combus-

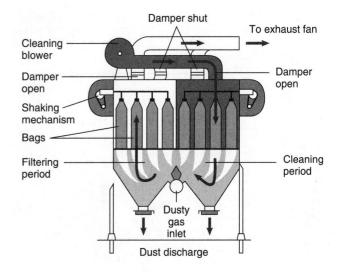

(a) Typical bag filter

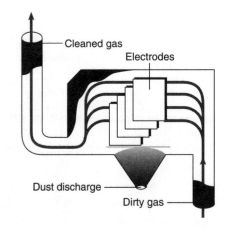

(b) Electrostatic precipitator

Figure 10.11 Typical emission-control devices: (*a*) bag filter and (*b*) electrostatic precipitator. Note that two stages in the operational cycle are shown in (*a*): the filtering period on the left and the period of cleaning the filter bag on the right.

tion. This improves heat content and firing properties but may simply transform air pollution into solid waste and water pollution problems.

Fluidized Bed Combustion

A relatively new technique for burning called fluidized bed combustion offers several advantages over conventional boilers. In this procedure, a mixture of crushed coal and limestone particles about a meter deep (3 ft) is spread on a perforated distribution grid in the combustion chamber. When high-pressure air is forced through the grid, the surface of the fuel rises as much as one meter and resembles a boiling fluid as particles hop up and down. Fresh coal and limestone are fed continuously into the top of the burning bed,

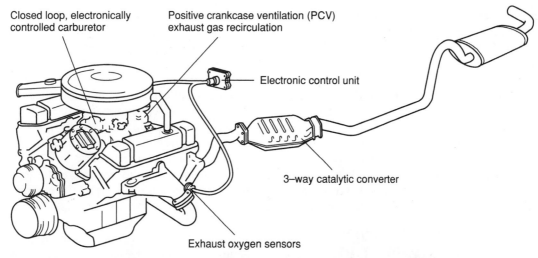

Figure 10.12 Elements of a modern automobile emission-control system. A closed-loop, electronically controlled carburetor or fuel-injector carefully meters fuel/air ratios to optimize combustion. Exhaust oxygen sensors measure completeness of fuel burning. Positive crankcase ventilation captures oil "blow-by" and unburned exhaust gases and recycles them to the cylinder.

while ash and slag are drawn off from below. The rich air supply and constant motion make burning efficient and prevent buildup of large slag clinkers. Steam generator pipes are submerged directly into the fluidized bed, and heat exchange is more efficient than in the water walls of a conventional boiler. More than 90 percent of SO_2 is captured by the limestone particles, and NO_x formation is reduced by holding temperatures below those of other boilers. The efficient burning of this process makes it possible to use cheaper fuel, such as lignite or unwashed subbituminous coal, rather than higher priced hard coal.

Flue Gas Desulfurization

Crushed limestone, lime slurry, or alkali (sodium carbonate or bicarbonate) can be injected into a stack gas stream to remove sulfur after combustion. These processes are often called flue gas scrubbing. Spraying wet alkali solutions or limestone slurry is relatively inexpensive and effective, but maintenance can be difficult. Rock-hard plaster and ash layers coat the spray chamber and have to be chipped off regularly. Corrosive solutions of sulfates, chlorides, and fluorides erode metal surfaces. Electrostatic precipitators don't work well because of fouling and shorting of electrodes after wet scrubbing.

Nitrogen Oxide Control

Undoubtedly the best way to prevent nitrogen oxide pollution is to avoid creating it. A substantial portion of the emissions associated with mining, manufacturing, and energy production could be eliminated through conservation (chapter 12).

Staged burners, in which the flow of air and fuel are carefully controlled, can reduce nitrogen oxide formation by as much as 50 percent. This is true for both internal combustion engines and industrial boilers. Fuel is first burned at high temperatures in an oxygen-poor environment where NO_x cannot form. The residual gases then pass into an afterburner where more air is added and final combustion takes place in an air-rich, fuel-poor, low-temperature environment that also reduces NO_x formation. Stratified-charge engines and new orbital automobile engines use this principle to meet emission standards without catalytic converters.

The approach adopted by U.S. auto makers for NO_x reductions has been to use selective catalysts to change pollutants to harmless substances. Three-way catalytic converters use platinum-palladium and rhodium catalysts to remove up to 90 percent of NO_x, hydrocarbons, and carbon monoxide at the same time (fig. 10.12). Unfortunately, this approach doesn't work on diesel engines, power plants, smelters, and other pollution sources because of problems with back pressure, catalyst life, corrosion, and production of unwanted by-products such as ammonium sulfate that foul the system.

Hydrocarbon Emission Controls

Hydrocarbons and volatile organic compounds are produced by incomplete combustion of fuels or solvent evaporation from chemical factories, painting, dry cleaning, plastic manufacturing, printing, and other industrial processes that use a variety of volatile organic chemicals. Closed systems that prevent escape of fugitive gases can reduce many of these emissions. In automobiles, for instance, positive crankcase ventilation (PCV) systems collect oil that escapes from around the pistons and unburned fuel and channel it back to the engine for combustion. Modification of carburetor and fuel systems prevents evaporation of gasoline (fig. 10.12).

In the same way, controls on fugitive losses from valves, pipes, and storage tanks in industry can have a significant impact on air quality. Afterburners are often the best method for destroying volatile organic chemicals in industrial exhaust stacks. High air–fuel ratios in automobile engines and other burners minimize hydrocarbon and carbon monoxide emissions but also cause excess nitrogen oxide production. Careful monitoring of air–fuel inputs and oxygen levels in exhaust gases can minimize all these pollutants.

CURRENT CONDITIONS AND PROSPECTS FOR THE FUTURE

Although we have not yet achieved the Clean Air Act goals in many parts of the United States, air quality has improved dramatically in the past decade in terms of the major large-volume pollutants. For twenty-three of the largest U.S. cities, the number of days in which air quality reached the hazardous level (PSI greater than 300) is down 93 percent from an average of 1.8 days/year a decade ago to 0.13 days/year now. In those same cities, the average number of days above PSI 200 (very unhealthful) was 11.3 days in 1982, down 65 percent from the average of 32.2 days/year ten years earlier. Unhealthful days (above PSI 100) still average about thirty-three days per year, but that is only one-half the level of the 1970s.

The EPA estimates that emissions of particulate materials are down 60 percent, lead is down 80 percent, SO_2 and CO are down 30 percent, and O_3 is down 18 percent over the past two decades (fig. 10.13). Industrial cities, such as Chicago, Pittsburgh, and Philadelphia, that suffered "smokestack" pollution have had

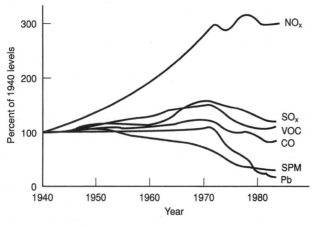

Figure 10.13 Air pollution trends in the United States, 1940–1984. Emissions of six "criteria" pollutants as percent of 1940 levels.
Source: Data from the Environmental Protection Agency (EPA).

90 percent reductions in number of days exceeding National Air Quality standards. Filters, scrubbers, and precipitators on power plants and other large stationary sources are responsible for most of the particulate and SO_2 reductions. Catalytic converters on automobiles are responsible for most of the CO and O_3 reductions. The only conventional "criteria" pollutant that has not dropped significantly is NO_x, which has risen 300 percent since 1940 and dropped only slightly over the past ten years.

Because automobiles are the main source of NO_x, cities where pollution is largely from traffic have had increased PSI levels in recent years. Los Angeles, Anaheim, and Riverside, California, are the only cities in the country in the extreme urban smog category, exceeding the ozone standards an average of 137.5 days per year between 1987 and 1989. Baltimore, New York City, Chicago, Gary, Houston, Milwaukee, Muskegon, Philadelphia, and San Diego are all in the severe category. Eighty-five other urban areas are still considered nonattainment regions. In spite of these local failures, however, 80 percent of the United States now meets the NAAQS goals. This improvement in air quality is perhaps the greatest environmental success story in our history.

The outlook is not so encouraging in other parts of the world, however. The major metropolitan areas of many developing countries are growing at explosive rates to incredible sizes (chapter 14), and environmental quality is abysmal in many of them. The composite average annual levels of SO_2 in São Paulo, Brazil, for instance, are more than 100 $\mu g/m^3$, and peak levels can be up to ten times higher. Mexico City remains notorious for bad air. Its 131,000 industries and 2 million cars and buses spew out more than 5,500 tons of air pollutants daily. Santiago, Chile, averages 299 days per year on which suspended particulates exceed WHO standards of 150 $\mu g/m^3$. Teheran, Iran, is nearly as bad.

While there are few statistics on China's pollution situation, it is known that many of China's 400,000 factories have no air pollution controls. Experts estimate that home coal burners and factories emit 10 million tons of soot and 15 million tons of sulfur dioxide annually. Sheyang, an industrial city in northern China, is thought to have the world's worst particulate problem, with peak winter concentrations of 690 $\mu g/m^3$ (nine times U.S. maximum standards). Airborne particulates in Sheyang exceed WHO standards on 347 days per year. Beijing, Xian, and Guangzhou are nearly as bad. The high incidence of cancer in Shanghai is thought to be linked to air pollution.

As political walls came down across Eastern Europe and the Soviet Union at the end of the 1980s, horrifying environmental conditions in these centrally planned economies were revealed. Inept industrial managers, a rigid bureaucracy, and lack of democracy have created ecological disasters. Where government owns, operates, and regulates industry, there are few checks and balances or incentives to clean up pollution. Much of the Eastern Bloc depends heavily on soft brown coal for its energy. Pollution controls are absent or highly inadequate (fig. 10.14).

Southern Poland and the adjacent Czech Republic are covered most of the time by a permanent cloud of smog from factories and power plants. Acid rain is eating away historic buildings and damaging already inadequate infrastructures. The haze is so dark that drivers must turn on their headlights during the day. Residents complain that washed clothes turn dirty before they can dry. Zabrze, near Katowice in southern Poland, has particulate emissions of 3,600 metric tons per square kilometer. This is more than seven times the emissions in Baltimore, Maryland, or Birmingham, Alabama, the dirtiest cities (for particulates) in the United States. Home gardening in Katowice has been banned because vegetables raised there have unsafe levels of lead and cadmium.

For miles around the infamous Romanian "black town" of Copsa Mica, the countryside is so stained by soot that it looks as if someone had poured black ink over everything. Birth defects afflict 10 percent of infants in northern Bohemia. Workers in factories there get extra hazard pay: burial money, they call it. Life expectancy in these industrial towns is as much as ten years less than the national average. Espenhain, in the industrial belt of former East Germany, has one of the world's highest rates of sulfur dioxide pollution. One of every two children has lung problems; one of every three has heart problems. Brass doorknobs and name plates have been eaten away by the acidic air in a few months.

Not all is pessimistic, however. There have been some spectacular successes in air pollution control. Sweden and West Germany (countries affected by forest losses due to acid precipitation) cut their sulfur emissions by two-thirds between 1970 and 1985. Austria and Switzerland have gone even further. They even regulate motorcycle emissions. The Global Environmental Monitoring System (GEMS) reports declines in particulate levels in twenty-six of thirty-seven cities worldwide. Sulfur dioxide and sulfate particles, which

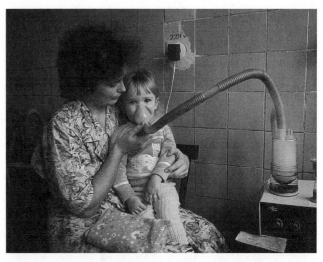

Figure 10.14 A mother gives an oxygen treatment to her child who suffers from respiratory disease triggered by air pollution. Rapid industrialization and lack of pollution controls have left much of the former U.S.S.R. and its allies toxic wastelands. In some villages three-fourths of all children suffer from pollution-related diseases.

cause acid rain and respiratory disease, have declined in twenty of these cities.

Ten years ago Cubatao, Brazil, was described as the "Valley of Death," one of the most dangerously polluted places in the world. In the mid-1980s, a steel plant, a huge oil refinery, and fertilizer and chemical factories churned out thousands of tons of air pollutants every year. Trees died on the surrounding hills. Birth defects and respiratory diseases were alarmingly high.

Since that time, however, the citizens of Cubatao have made remarkable progress in cleaning up their environment. The end of military rule and restoration of democracy allowed residents to publicize their complaints. The environment became an important political issue. The state of São Paulo invested about $100 million, and the private sector spent twice as much, to clean up most pollution sources in the valley. Particulate pollution was reduced 92 percent. Ammonia emissions were reduced 97 percent, other hydrocarbons that cause ozone and smog were cut 78 percent, and sulfur dioxide production fell 84 percent. Fish are returning to the rivers, and forests are regrowing on the mountains. Progress is possible! We hope that similar success stories will be obtainable elsewhere.

Summary

In this chapter, we have studied the atmosphere and how it is warmed by solar energy as well as how human activities are disturbing the natural balance of this system. We also have looked at major categories, types, and sources of air pollution. Burning fossil fuels, biomass, and wastes continues to be the largest source of anthropogenic (human-caused) air pollution.

The six "conventional" large-volume pollutants are nitrogen oxides, sulfur dioxide, carbon monoxide, lead, particulates, and volatile organic compounds. The major sources of air pollution are transportation, industrial processes, stationary fuel combustion, and solid waste disposal.

We also looked at some unconventional pollutants. Indoor air pollutants, including formaldehyde, asbestos, toxic organic chemicals, radon, and tobacco smoke may pose a greater hazard to human health than all of the conventional pollutants combined. Odors, visibility losses, and noise generally are not life-threatening but serve as indicators of our treatment of the environment. Some atmospheric processes play a role in distribution, concentration, chemical modification, and elimination of pollutants. Among the most important of these processes are long-range transport of pollutants and photochemical reactions that produce ozone and other oxidizing chemicals.

Encouraging improvements have been made in ambient outdoor air quality over most of the United States in the past decade. We have made considerable progress in designing and installing pollution-control equipment to reduce the major conventional pollutants. There are many types of scrubbers, filters, catalysts, fuel modification processes, and new burning techniques for controlling pollution. The Clean Air Act regulates air quality in the United States, and its 1990 amendments promise a dramatic improvement in our atmosphere. There is much yet to be done, especially in developing countries and in Eastern Europe, but air-pollution control is, perhaps, our greatest success in environmental protection and an encouraging example of what can be accomplished in this field.

Review Questions

1. What is the difference between bronchitis, emphysema, and fibrosis? What causes these diseases?
2. What are the most important human health threats from air pollution?
3. What is acid deposition? What causes it?
4. What have been the effects of acid deposition on aquatic and terrestrial ecosystems?
5. How do electrostatic precipitators, baghouse filters, flue gas scrubbers, and catalytic converters work?
6. What is the difference between primary and secondary standards in air quality?
7. What will the proposed ban on fluorocarbon production do to your life? How much will you sacrifice to save the ozone layer?
8. What are some of the major toxic air pollutants, and what are their sources?
9. Which of the conventional pollutants has decreased most in the recent past and which has decreased least?
10. Describe some of the current air pollution problems in Central Europe.

Questions for Critical or Reflective Thinking

1. There is considerable uncertainty and disagreement about many issues in this chapter. For instance, some scientists argue that global warming has already begun, while others argue that we don't know enough to say when, how much, or even whether our actions will affect the climate. If you could interview individuals from each of these camps, what would you ask them?
2. Can you think of professional or personal interests that might lead scientists to either minimize or exaggerate the dangers from these environmental problems?
3. Suppose that you were writing an article or preparing a speech about global climate change, radon in homes, or ozone depletion. How sensational or apocalyptic would you be?
4. If you were the environmental czar, how much sacrifice would you ask the public to endure for environmental protection? As a citizen, how much sacrifice would you accept?
5. The EPA has been criticized for frightening the public unnecessarily by warning about radon levels in homes, since most people don't spend much time in their basements. Smoking probably causes hundreds of times as many deaths as radon, but we don't force people to stop smoking. How would you compare the risks of smoking and radon?
6. There are many competing theories about why forests are dying. Most scientists claim that the pollutant on which they work is the main culprit. Given this uncertainty, what should public policy makers do?
7. Originally we thought that "dilution is the solution to pollution." Are there any cases where this still makes sense?

8. It will cost hundreds of billions of dollars to clean up the environmental mess in the former Soviet Union and its client states.

How much can or should we help with this?

9. Overall, how would you summarize the current air pollution situation? Are conditions getting better or worse? How would you rank the importance of the issues discussed in this chapter?

Key Terms

acid precipitation 217
aerosol 212
ambient air 209
asthma 215
bronchitis 215
carbon monoxide 210
criteria pollutants 209
electrostatic precipitators 220
emphysema 215
fibrosis 215
filters 220

fugitive emissions 209
greenhouse effect 206
insolation 206
nitrogen oxides 210
ozone 213
particulate material 212
photochemical oxidants 213
primary pollutants 209
secondary pollutants 209
sulfur dioxide 210
volatile organic compounds (VOC) 212

Suggested Readings

Boutron, C. F., et al. "Decrease in Anthropogenic Lead, Cadmium and Zinc in Greenland Snows Since the Late 1960s." *Nature* 353 (September 1991): 153. Reduced air pollution in industrialized countries is reflected in lower deposition rates in arctic snow.

Bureau of National Affairs. "World Health Unit Says Pollution Remains a Serious Health Threat in Europe." *International Environmental Reporter,* September 1990, 12. A good summary of urban air quality in Europe.

Cicerone, R. J. "Changes in Stratospheric Ozone." *Science* 237, no. 4810 (July 1987): 35. A useful overview of atmospheric chemistry and human effects on stratospheric ozone over Antarctica.

Committee on Global Change. *Toward an Understanding of Global Change.* Washington, D.C.: National Academy Press, 1988. A report of the United States' committee for the International Geosphere-Biosphere Program.

Fischhoff, B. "Report from Poland: Science and Politics in the Midst of Environmental Disaster." *Environment* 33, no. 2 (1991): 21. Decades of neglect and oppression have created an environmental disaster in Eastern Europe. The new democratic governments have little money to clean up the mess they inherited.

French, H. F. "Restoring Eastern European and Soviet Environments." In *State of the World 1991.* Washington, D.C.: World Watch Institute, 1991. Environmental reconstruction is imperative in Eastern Europe and the former Soviet Union. The Green ecological movement has been instrumental in both political change and environmental restoration.

Frenzel, G. "The Restoration of Medieval Stained Glass." *Scientific American* 252, no. 5 (May 1985): 126. Discusses air pollution's effect on building materials and artworks.

Graedel, T. E., and P. J. Crutzen. "The Changing Atmosphere." *Scientific American* 261, no. 3 (September 1989): 58. Describes how human activities are polluting the atmosphere.

Gribbin, J. "Climate and Ozone: The Stratospheric Link." *The Ecologist* 21, no. 3 (May/June 1991): 133. Ozone losses in the stratosphere could have far-reaching consequences both for cancer rates and the global climate.

Jones, P. D., and R. M. Wigley. "Global Warming Trends." *Scientific American* 263, no. 2 (August 1990): 84. A 300-year weather history suggests that climate is warming.

Krahl-Urban, B., et al. *Forest Decline.* Corvallis, Oreg.: U.S. Environmental Protection Agency, 1988. A thorough and beautifully illustrated report on the effects of air pollution on forests in North America and Germany. Presents data and theories on several mechanisms of forest ecosystem damage.

Llyman, F. "As the Ozone Thins, the Plot Thickens." *The Amicus Journal* 13, no. 3 (Summer 1991): 20. Will we join together to protect stratospheric ozone? A thoughtful discussion.

Monastersky, R. "A Star in the Greenhouse: Can the Sun Dampen the Predicted Global Warming?" *Science News* 142: (October 1992): 282. An interesting discussion of the solar "constant," sun spot activity, and climate.

Reilly, W. K. "The New Clean Air Act: An Environmental Milestone." *EPA Journal* 17, no. 1 (1991): 2. In a special volume entirely devoted to this issue, the former head of the EPA explains the significance of the new act not only as a tool for management but also as an environmental precedent.

Sooroos, M. S. "The Odyssey of Arctic Haze: Toward a Global Atmospheric Regime." *Environment* 34, no. 10 (December 1992): 6. Polar air is loaded with contaminants from faraway industrial areas.

Svitil, K. "Holey War." *Discover,* January 1993, 75. A current report on Antarctic ozone and the scientists who study it.

Torrens, I. M. "Developing Clean Coal Technologies." *Environment* 32, no. 6 (July/August 1990): 10. Coal combustion is the single largest source of air pollution in North America. Billions of dollars are being spent to find ways to reduce this pollution.

White, R. M. "Our Climatic Future: Science, Technology, and World Climate Negotiations." *Environment* 33, no. 2 (March 1991): 18. An address from the Second World Climate Conference on international cooperation in environmental protection.

World Resources 1992-93 Washington, D.C.: World Resources Institute, 1992. An excellent, comprehensive summary of global environmental conditions with an extensive bibliography.

11
W*ATER RESOURCES*

The water won't ever clear up till you get the hogs out of the creek.

Jim Hightower, former Texas Commissioner of Agriculture

O b j e c t i v e s

After studying this chapter you should be able to:

- Characterize the major water compartments and how they are replenished by the hydrologic cycle.
- Describe the types of human water use and how this varies by region and national wealth.
- Fathom the causes, consequences, and extent of freshwater shortages around the world.

- Discuss the advantages and disadvantages of schemes for increasing or stabilizing our water supplies.
- Understand the basis, results, and significance of water pollution.
- Give examples of both progress in water pollution control and problems that remain to be solved.

INTRODUCTION

Water is a marvelous substance—flowing, rippling, swirling around obstacles in its path, seeping, dripping, trickling—constantly moving from sea to land and back again (fig. 11.1). Water can be clear, crystalline, and icy green in a mountain stream or black and opaque in a cypress swamp. Water bugs skitter across the surface of a quiet lake; a stream cascades down a stair-step ledge of rock; waves roll endlessly up a sand beach, crash in a welter of foam, and recede. Rain falls in a gentle mist, refreshing plants and animals. A violent thunderstorm floods a meadow, washing away stream banks. Why do we find water so fascinating, and why is it so important in natural and human systems?

The unique properties of water (see box 2.2) make it essential for life. Most of us think of water as an infinitely available, renewable resource because it is constantly purified and redistributed by the action of the sun, wind, and gravity. But in many parts of the world, water supply is increasingly limited. More people making demands on the resource, natural variations in rainfall, and wasteful or extravagant uses create shortages in many areas. To make matters worse, pollution makes whatever water is available unfit for many uses, further exacerbating supply problems. Eminent hydrologist Luna B. Leopold warns that water shortages might be the environmental crisis of the 1990s and that water conservation might be as much a national priority in a few years as energy conservation was in the 1970s.

In this chapter, we will look at the processes that supply fresh water to the land and how humans access and use it. We will survey major water compart-

Figure 11.1 Water is a magical substance. It is the most unique aspect of our world and is essential for life. But will it always be available in pure form and abundant quantities as it has been in the past? Some authors predict that water shortages will be the environmental crisis of the next decade.

ments of the environment and see how they are depleted and polluted by humans and what we can do about these problems.

WATER RESOURCES

The earth is the only place in the universe, as far as we know, where liquid water exists in substantial quantities. Oceans, lakes, rivers, glaciers, and other bodies of liquid or solid water cover about 70 percent of our world's surface. Although water is our most abundant resource, a vast majority of the water on earth is too salty, remote, unreliable, or otherwise unavailable for human use. In this section, we will investigate the compartments in which water resides as well as the hydrologic cycle that replenishes our freshwater supplies.

Major Water Compartments

The total amount of water on our planet is more than 1,404 million cu km (370 billion-billion gal). If the earth's surface were perfectly smooth, an ocean about 3 km (1.9 mi) deep would cover everything. Fortunately for us, continents rise above the general surface level, creating dry land over about 30 percent of the planet. More than 97 percent of all the *liquid* water in the world is in oceans and salt lakes (table 11.1). Oceans play a crucial role in moderating earth's temperature, but they are generally too salty for most human uses.

Of slightly less than 3 percent of all water that is fresh (nonsalty), about three-quarters is tied up in glaciers, ice caps, and snowfields. The next largest reservoir of fresh water is **groundwater,** held in the soil and in porous, water-bearing layers of sand, gravel, and rock called aquifers. Water seeps into aquifers through **recharge zones** (fig. 11.2). Most aquifers refill very slowly, however, and groundwater presently is being removed faster than it can be replenished in many areas.

Precipitation that does not evaporate or seep into the ground runs off over the surface, drawn by the force of gravity back toward the sea in streams and

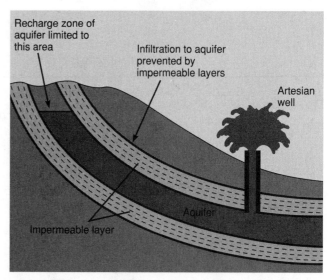

Figure 11.2 An aquifer is a porous, water-bearing layer of sand, gravel, or rock. This aquifer is confined between impermeable layers of rock or clay and bent by geological forces, creating hydrostatic pressure. A break in the overlying layer creates an artesian well or spring.

rivers, or is temporarily stored in lakes and ponds. Although these surface waters represent only a small fraction of the total water supply, they are vitally important to humans and most other organisms.

The atmosphere is among the smallest of the major water reservoirs of the earth in terms of water volume, containing less than 0.001 percent of the total water supply. It also has the most rapid turnover rate. An individual water molecule resides in the atmosphere for about ten days, on average. Transport of water vapor through the air shapes our climate and replenishes our freshwater supplies, as we will discuss in the next section.

Replenishing Freshwater Supplies

The hydrologic cycle (water cycle) describes the circulation of water as it evaporates from land, water, and organisms; enters the atmosphere; condenses and is precipitated to the earth's surfaces; and moves underground by infiltration or overland by runoff into rivers, lakes, and seas (see fig. 2.14). This process supplies fresh water to the land masses while also redistributing heat from warm areas to cooler ones. Plants play an important role in the hydrologic cycle, absorbing groundwater and pumping it into the atmosphere by transpiration (transport plus evaporation). In tropical forests, as much as 75 percent of annual precipitation is returned to the atmosphere by plants.

Everything about global hydrological processes is awesome in scale. Each year the sun evaporates approximately 496,000 cu km of water from the earth's surface. More water evaporates in the tropics than at

	% total water	Average residence** time
Total	100	2,800 years
Ocean	97.6	3,000 years to 30,000 years*
Ice and snow	2.07	1 to 16,000 years*
Groundwater down to 1 km	0.28	From days to thousands of years*
Lakes and reservoirs	0.009	1 to 100 years*
Saline lakes	0.007	10 to 1,000 years*
Soil moisture	0.005	2 weeks to a year
Biological moisture in plants and animals	0.005	1 week
Atmosphere	0.001	8 to 10 days
Swamps and marshes	0.003	From months to years
Rivers and streams	0.0001	10 to 30 days

TABLE 11.1 Earth's water compartments and average residence time

*Depends on depth and other factors.
**Residence is the average time a water molecule would spend in a compartment. It is the recycle time for a reservoir.
Source: Data from U.S. Geological Survey.

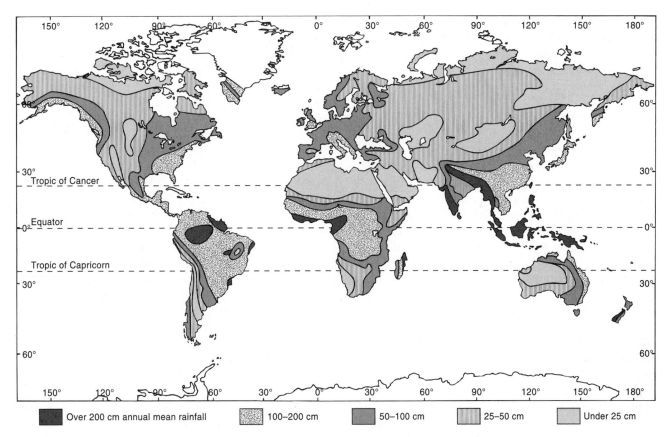

Figure 11.3 Patterns of world precipitation. Notice wet areas that support tropical rainforests along the equator and dry areas where dry, warm air currents create deserts along the Tropics of Cancer and Capricorn.

Source: Data from W. G. Kendrew, *Climate*, 1930. Oxford: Clavendon Press.

Legend: Over 200 cm annual mean rainfall | 100–200 cm | 50–100 cm | 25–50 cm | Under 25 cm

higher altitudes, and more water evaporates over the oceans than over land. Ninety percent of the water evaporated from the oceans falls back on the oceans as rain. The remaining 10 percent is carried by prevailing winds over the continents where it combines with water evaporated from soil, plant surfaces, lakes, streams, and wetlands to provide a total continental precipitation of about 111,000 cu km.

What happens to the surplus water on land, the difference between what falls as precipitation and what evaporates? Some of it is incorporated by plants and animals into biological tissues. A large share of what falls on land seeps into the ground to be stored for a while (from a few days to many thousands of years) as soil moisture or groundwater. Eventually, all the water makes its way back downhill to the oceans. The 41,000 cu km carried back to the ocean each year by surface runoff or underground flow represents the renewable supply available for human uses and for sustaining freshwater-dependent ecosystems.

WATER AVAILABILITY AND USE

Clean, fresh water is essential for nearly every human endeavor. Perhaps more than any other environmental factor, the availability of water determines the location and activities of humans on the earth. Figure 11.3

shows rainfall distribution around the world. Runoff is the excess of precipitation over evaporation and infiltration and represents, in broad terms, the water *available* for human use.

Regional Water Supplies

As you can see in figure 11.3, there is a wide range of precipitation around the world. The richest continents in terms of total water supply are South America and Asia, each of which receives about one-fourth of the total global runoff. In terms of water available per person, South America has the most abundant supply. Its enormous water supply is shared by only 6 percent of the world population. However, most of the rainfall and runoff in South America occurs in areas where infertile soil and inhospitable conditions limit human habitation. Much of the runoff in Asia does occur in regions suitable for agriculture; this is one reason that Asia has nearly 60 percent of all humans on earth.

Brazil has by far the largest renewable water supply of any country, but Iceland is the wealthiest in terms of per capita water supply with some 672,000 cu m of water available per person each year. The most water-poor countries are Kuwait and Bahrain, which have essentially no renewable water supply. China is fourth in the world in terms of total water supply, but its large population makes it below average in terms of per capita availability. It is interesting to consider how water supplies have shaped the history and will frame the future of each of these countries.

Seasonal Variability

Stable runoff, the fraction that is available year-round, is usually more important than total runoff in determining human uses. In most parts of the world, a majority of the precipitation falls during a limited "wet" season. Water is abundant—perhaps unpleasantly so—during this season, but much of that water quickly drains away to oceans and isn't available during the succeeding dry season. In India, for instance, 90 percent of annual precipitation falls between June and September. The rest of the year can be extremely dry and dusty.

Every continent has regions where rainfall is scarce because of topographic effects or wind currents. In addition, cycles of wet and dry years create temporary droughts. Water shortages have their most severe effect in semiarid zones where moisture availability is the critical factor in determining plant and animal distribution. Undisturbed ecosystems often survive extended droughts with little damage, but introduction of domestic animals and agriculture disrupts native vegetation and undermines natural adaptations to low moisture levels.

In the United States, the cycle of drought seems to be about twenty to thirty years. There were severe dry years in the 1870s, 1900s, 1930s, 1950s, and 1970s. The worst of these in economic and social terms were the 1930s. Wasteful farming practices and a series of dry years in the Great Plains combined to create the dust bowl. Wind stripped topsoil from millions of hectares of land and billowing dust clouds turned day into night. Thousands of families were forced to leave farms and migrate to cities. Many people worry that the global climate change (see box 10.1) will make droughts both more frequent and more severe than in the past.

Types of Water Use

In contrast to energy resources, which are consumed when used, water has the potential for being reused many times. In discussing water appropriations, we need to distinguish between different kinds of uses and how they will affect the water being appropriated.

Withdrawal is the total amount of water taken from a lake, river, or aquifer for any purpose. Much of this water is employed in nondestructive ways and is returned to circulation in a form that can be used again. **Consumption** is the fraction of withdrawn water that is lost in transmission, evaporation, absorption, chemical transformation, or otherwise made un-

available for desirable purposes as a result of human use. **Degradation** is a change in water quality due to contamination or pollution so that the water is unsuitable for other desirable service. The total quantity available may remain constant after some uses, but the quality is degraded so the water is no longer as valuable as it was.

Worldwide, humans withdraw about 10 percent of the total annual runoff and about 25 percent of the stable runoff. The remaining three-quarters of the stable supply is generally either uneconomical to tap (it would cost too much to store, ship, purify, or distribute), or there are ecological constraints on its use. Consumption and degradation together account for about half the water withdrawn in most industrial societies. The other half of the water we withdraw would still be valuable for further uses if we could protect it from contamination and make it available to potential consumers.

As you might expect, those countries with a plentiful water supply and a small population withdraw a very small percentage of the water available to them. Canada, Brazil, and the Congo, for instance, withdraw less than 1 percent of their annual runoff. By contrast, in countries such as Kuwait, Libya, and Qatar, where water is one of the most crucial environmental resources, groundwater and surface water withdrawal together amount to more than 100 percent of the renewable supply.

Use by Sector

Worldwide, agriculture is the largest human water use, claiming about 69 percent of total water withdrawal. As you can see in figure 11.4, agriculture's share of the water supply depends strongly on national wealth, ranging from more than 90 percent in low-income nations to less than half of the water used

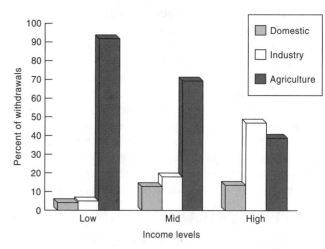

Figure 11.4 Water withdrawals by sector in low-, middle-, and high-income countries.

Source: Data from World Bank, 1992.

Figure 11.5 Center-pivot sprinklers allow farmers to irrigate crops on rolling terrain. Wheels driven by electric or gasoline engines circle around a central deep well and can water an area up to 2.6 km² (640 acres). Agriculture now uses 41 percent of all water withdrawn in the United States. In some areas, irrigation consumes 90 percent of all available water.

TABLE 11.2 Examples of water use*

	Liters	Gallons
Agriculture and food processing:		
1 egg	150	40
1 ear corn	300	80
1 loaf bread	600	160
1 pound beef	9,500	2,500
Industrial and commercial products:		
1 Sunday paper	1,000	280
1 pound steel	110	32
1 pound synthetic rubber	1,100	300
1 pound aluminum	3,800	1,000
1 automobile	380,000	100,000

*These totals are based both on the production of the item itself and its precursors. For example, a pound of beef includes the grain fed to cattle.

in richer countries. The range is even greater in individual countries. For example, 93 percent of all water used in India is agricultural, while Kuwait, which cannot afford to spend its limited water on crops, devotes only 4 percent to agriculture. Canada, where the fields are well watered by natural precipitation, uses only 12 percent of its water for agriculture.

Agricultural water use is notoriously inefficient and highly consumptive. Typically, from 70 to 90 percent of the water withdrawn for agriculture never reaches the crops for which it is intended. The most common type of irrigation is to simply flood the whole field or run water in rows between crops. As much as half is lost through evaporation or seepage from unlined irrigation canals bringing water to fields (fig. 11.5). Most of the rest runs off, evaporates, or infiltrates into the field before it can be used. The water that evaporates or seeps into the ground is generally lost for other purposes, and the runoff from fields is often contaminated with soil, fertilizer, pesticides, and crop residues, making it low quality.

Industry accounts for about one-fourth of all human water use, ranging from 70 percent of withdrawal in wealthy countries such as Germany to 5 percent in less industrialized countries such as Egypt and India. Cooling water for power plants is by far the largest single industrial use of water, typically accounting for about half of all industrial withdrawal. Unlike agriculture, however, only a small fraction of this water is consumed or degraded. Most power plants have a ''once-through'' cooling system that returns water to its source after it passes through the plant. Typically, only 2 to 5 percent of the cooling water is lost through leaks or evaporation. If care is taken to avoid contamination, this water can be used for other purposes.

A few other industries account for the majority of the remaining industrial water use. In the United States, primary metal smelting and fabrication, petroleum refining, pulp and paper manufacturing, and food processing use about two-thirds of the industrial water not used by power plants. You might be surprised to learn how much water is used to manufacture some of the ordinary products that we consume. Table 11.2 shows a sample of products and the water used in their production. Much of the water used by these industries could be recycled and used over again in the factory. This would have benefits both in extending water supplies and in protecting water quality. Although Third World countries typically allocate only about 10 percent of their water withdrawal to industry, this could change rapidly as they industrialize. Water may be as important as energy in determining which countries develop and which remain underdeveloped.

Domestic water for cooking, bathing, and washing dishes and clothes is generally the smallest water use. Although the United States does not have the most abundant water supply in the world, we use more water per person than any other country because we can afford the facilities to extract and use large amounts of water. Still, only 12 percent of total water withdrawal in the United States is for municipal supplies that provide water for direct personal consumption.

FRESH WATER: A SCARCE RESOURCE

Water is a major limiting factor of the environment, both for biological systems (chapter 2) and human societies. Our growing world population is placing great demands upon natural freshwater sources. The world is faced with increasing pressure on water resources and widespread, long-lasting water shortages in many areas for three reasons: (1) rising demand, (2) unequal distribution of usable fresh water, and (3) increasing pollution of existing water supplies. The Russian hydrologist G. P. Kalinin predicts that by the year 2000 humans will be using about half of all the earth's renewable water. Arnon Sofer of Israel's Haifa University warns that Middle Eastern water wars might erupt over competition for this scarce resource. Every continent has areas of water shortage. In this section, we will look at some causes and effects of those water shortages.

Water Shortages

About 2 billion people, nearly half the world's population, already lack an adequate supply of safe drinking water. In many countries, women and children spend a large part of each day walking to and from the nearest water supply or standing in line at the village well (fig. 11.6). An estimated 80 percent of the disease affecting people in the poorest countries of the world is caused by contaminated water supplies and lack of sanitation. In some cases, these shortages are caused by natural forces: the rains fail, rivers change their courses and take water elsewhere, or dry winds evaporate the reserve moisture on which humans and wildlife depend for survival.

In other cases, shortages are human in origin: too many people compete for the resource, overgrazing and inappropriate agricultural practices allow water to run off before it can be captured, or sewage contamination makes existing water supplies unusable. Without wells, storage reservoirs, and delivery systems, people don't have access to the resources available to them. The United Nations declared the 1980s as the decade for solving problems of clean water supplies and adequate sanitation for all. It is estimated that providing clean water and adequate sanitation for everyone in the world would cost about $300 billion, approximately what the United States spends on its military each year.

Depleting Groundwater Supplies

Groundwater provides about 40 percent of the fresh water for agricultural and domestic use in the United States. Nearly half of all Americans and about 95 percent of the rural population depend on groundwater for drinking and other domestic purposes. Overuse of these supplies causes several kinds of problems, including depletion, subsidence, and saltwater infiltration.

In many areas, groundwater is being withdrawn from aquifers faster than natural recharge can replace it. On a local level, this causes a cone of depression in the water table (fig. 11.7). A heavily pumped well can lower the local water table so that shallower wells go dry. On a broader scale, heavy pumping can deplete a whole aquifer (box 11.1). When we pump water out of a reservoir that cannot be refilled in our lifetime, we essentially are mining a nonrenewable resource. Covering aquifer recharge zones with urban development or diverting runoff that once replenished reservoirs ensures that they will not refill.

Figure 11.6 Many people in developing countries have inadequate water supplies. In India, 8,000 villages have no water at all. Women and children, who do almost all the family domestic chores, must walk long distances to the nearest well or river. Often, the water is contaminated and unsafe to drink.

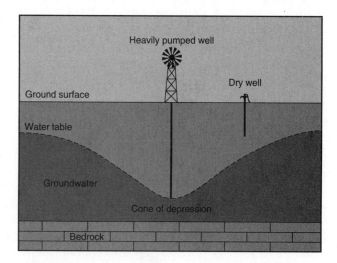

Figure 11.7 A cone of depression forms in the water table under a heavily pumped well. This may dry up nearby shallow wells or make pumping so expensive that it becomes impractical.

The Ogallala Aquifer: An Endangered Resource

*T*he Ogallala Aquifer System is the largest known underground freshwater reservoir in the world. Lying under parts of eight states in the arid, high plains region (box fig. 11.1.1), this reservoir has supplied water to one of the most important agricultural areas in the United States and has been a significant factor in the high rates of productivity we have enjoyed in recent years. However, the enormous amounts of water now being pumped out of the aquifer are depleting it much faster than it can be recharged from surface infiltration. In some areas, water tables are falling as much as 1 m (3 ft) per year. Many farmers are having to abandon irrigation, either because the aquifer has been exhausted under their land or because the costs of pumping from greater and greater depths are no longer justified by the crops produced.

Ranging in thickness from a few meters around the periphery of the formation to more than 400 m (1,200 ft) under the sand hills of Nebraska, this vast aquifer is estimated to have once contained about 2,000 cu km (2 billion acre-feet) of water—more than all surface freshwater on the earth. Most of the water in the aquifer is thought to have been left by the glaciers that melted 15,000 years ago. More than 85 percent of that huge reservoir has already been pumped out to the point at which it is economically unavailable.

Exploitation of the high plains groundwater began about one hundred years ago when pioneer farmers and ranchers set up windmills for domestic supplies and to water crops and livestock. Although this early technology played a vital role in the settling of the American West, it had relatively little effect on the enormous amount of water contained in the aquifer. The real change came in the 1960s with the invention of center-pivot irrigation systems in which large sprinklers ride on a pipe up to 1 km (0.6 mi) long carried on motor-driven wheels that circle around a central well. Because the pipes have flexible joints and the wheels are driven independently, the whole apparatus can traverse hilly land that could not be irrigated by conventional gravity-fed methods.

The amount of irrigated land on the high plains jumped from about 1 million ha (2.5 million acres) in 1950, to 6.5 million ha (16 million acres) in 1980. More than 150,000 wells supplied the water for about 33 percent of all cotton, 50 percent of all grain, and 40 percent of all beef produced in the United States in the early 1970s. More water was being drawn out of the aquifer each year at the peak of irrigation than the entire annual flow of the Colorado River.

In contrast to surface waters, which are tightly controlled by water laws and appropriation rights, there are no limits on how much water a person can pump from the ground, even though it is a shared resource that underlies neighboring land as well. There also are no regulations about how and for what purposes the water can

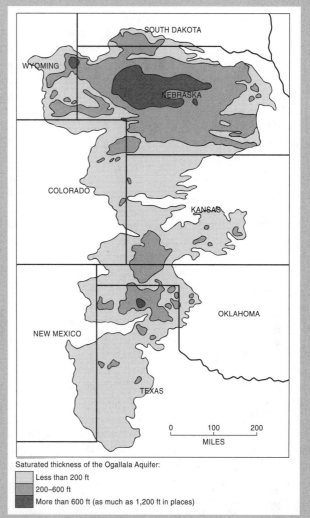

Saturated thickness of the Ogallala Aquifer:

☐ Less than 200 ft
▨ 200–600 ft
▦ More than 600 ft (as much as 1,200 ft in places)

Figure 11.1.1 The Ogallala aquifer, underlying parts of eight plains states, is the largest known body of fresh water in the world. Its water-bearing rock and gravel formations range from 10 m (30 ft) to 400 m (1,200 ft) in thickness. Yearly withdrawal from the aquifer (mostly for agriculture) is equal to the total annual flow of the Colorado River. The water table has dropped as much as 50 m (150 ft) in a few decades in some places.

Source: U.S. Geological Survey.

be used. The rule was, and still is, "Those who pump fastest get most."

What will happen when the aquifer runs dry? Farmers may have to return to dry land farming that typically yields only one-third as much per unit of land as irrigated fields. Alternative sources for municipal and industrial water supplies often do not exist. Pressure undoubtedly will rise for water transfer projects that can ship water from the Great Lakes states or the Mississippi River Valley. The costs of those projects would be billions of dollars and the price of water delivered might be ten times what farmers comfortably can pay. Should the government bear some or all of those costs? And will those states with plentiful water be willing to share it?

Subsidence and Saltwater Infiltration

Withdrawal of large amounts of groundwater causes porous formations to collapse, resulting in **subsidence** or settling of the surface above. The United States Geological Survey estimates that the San Joaquin Valley in California has sunk more than 10 m in the past fifty years because of excessive groundwater pumping. Many of the world's great cities are sinking because of the removal of groundwater or oil and gas beneath them. Unstable city locations include unconsolidated sediments on river floodplains (New Orleans, London, Bangkok), in deltaic coastal marshes (Venice, Houston, Tokyo, Shanghai), or lake beds (Mexico City).

Sinkholes form when the roof of an underground channel or cavern collapses, creating a large surface crater (fig. 11.8). Drawing water from caverns and aquifers accelerates the process of collapse. Sinkholes can form suddenly, dropping cars, houses, and trees without warning into a gaping crater hundreds of meters across. Subsidence and sinkhole formation generally represent permanent loss of an aquifer. When caverns collapse or the pores between rock particles are crushed as water is removed, it is usually impossible to reinflate these formations and refill them with water.

Another consequence of aquifer depletion is saltwater infiltration. Along coastlines and in areas where saltwater deposits are left from ancient oceans, overuse of freshwater reservoirs often allows saltwater to intrude into aquifers used for domestic and agricultural purposes (fig. 11.9).

Figure 11.8 Home lost to a sinkhole in Bartow, Florida. This sinkhole was over 150 m (nearly 500 ft) long, 40 m wide, and 20 m deep. It formed without warning in a matter of minutes.

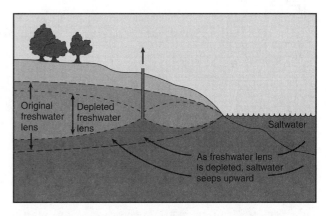

Figure 11.9 Saltwater intrusion into a coastal aquifer as the result of groundwater depletion. Many coastal regions of the United States are losing water sources due to saltwater intrusion.

INCREASING SUPPLIES: PROPOSALS AND PROJECTS

Where do present and impending freshwater shortages leave us now? On a human time scale, the amount of water on the earth is fixed, for all practical purposes, and there is little we can do to make more water. There are, however, several ways to increase local supplies.

Desalination

A technology that might have great potential for increasing freshwater supplies is desalination of ocean water. The most common methods of desalination are distillation (evaporation and recondensation) or reverse osmosis (forcing water under pressure through a semipermeable membrane whose tiny pores allow water to pass but exclude most salts and minerals). Already over one hundred desalination plants are in operation in the United States, and several hundred more are at work in other countries. Total annual freshwater production from these plants is more than 10 billion l (2.6 billion gal). The cost of this water is prohibitively high for most purposes. Only in such places as Oman and Bahrain, where there is no other access to fresh water, does it pay to desalt ocean water with present technology. If a cheap, inexhaustible source of energy were available, however, the oceans could supply all the water we would ever need.

Water Transfer Projects

It is possible to trap runoff with dams and storage reservoirs and transfer water from areas of excess to areas of deficit using canals, tunnels, and underground pipes. Some water transfer projects are truly titanic in scale. California, for instance, transports water from the Colorado River on the eastern border and the Sacramento and Feather rivers in the north to Los Angeles and the

San Joaquin Valley in the south. In 1990, the U.S. Bureau of Reclamation completed the Central Arizona Project, a $4 billion dollar system of huge pumps, canals, and reservoirs that carries water from the Colorado River 300 miles across the desert to Phoenix and Tucson.

There has been much controversy about these massive projects. Some of the water now being delivered to southern California and Arizona is claimed by native people, neighboring states, and even Mexico. Environmentalists claim that this water transfer upsets natural balances of streams, lakes, estuaries, and terrestrial ecosystems. Fishing enthusiasts, whitewater boaters, and others who enjoy the scenic beauty of free-running rivers mourn the loss of rivers drowned in reservoirs or dried up by diversion projects. These projects also have been criticized for using public funds to increase the value of privately held farmland and for encouraging agricultural development and urban growth in arid lands where other uses might be more appropriate.

In China, a huge water transfer project is underway that will carry water from the Chang Jiang (Yangtze) River in the center of the country north to the dry plains around Beijing. The 1,000 km (625 mi) main trunk canal in this system would move about 30 cu km of water per year (three times the annual flow of the Colorado River in the United States). Among the engineering problems to be surmounted are the crossing of several mountain ranges and nearly 150 rivers in its path. The costs of this project include moving earth equivalent to that of about a dozen Panama Canals and an investment equal to 10 million working years. A big question about this project is whether the enormous distribution system required to deliver water to hundreds of thousands of farms and communes can be made to work. An even more disastrous situation has developed in the former Soviet Union where river diversions have dried up the Aral Sea to an alarming extent (box 11.2).

Thinking Globally: Box 11.2

The Aral Sea Is Dying

*T*he Aral Sea is a huge, shallow, saline lake hidden in the remote deserts of Uzbekistan and Kazakhstan in Central Asia (box fig. 11.2.1). Once the world's fourth largest lake in area, its only water sources are two large rivers, the Amu Dar'ya and the Syr Dar'ya. Flowing northward from the Pamir mountains on the Afghan border, these rivers pick up salts as they cross the Kyzyl Kum and Kara Kum deserts. Evaporation from the landlocked sea's surface (it has no outlet) makes the water even saltier.

The Aral Sea's destruction began in 1918 when plans were initiated to draw off water to grow cotton, a badly needed cash crop for the newly formed Soviet Union. The amount of irrigated cropland was more than doubled in the 1950s and 1960s with the completion of the Kara Kum canal. Annual water flows in the Amu Dar'ya and Syr Dar'ya dropped more than 90 percent. In some years no water at all reaches the lake.

Scientists warned Soviet authorities that the sea would die without replenishment, but sacrificing a remote desert lake

for the sake of economic development seemed an acceptable trade. Inefficient irrigation practices drained away the lifeblood of the lake. Dry years in the early 1970s and mid-1980s accelerated water shortages in the region. Now, in a disaster of unprecedented magnitude and rapidity, the Aral Sea is disappearing as we watch.

By 1990, the Aral Sea had lost 40 percent of its surface area and two-thirds of its volume. Surface levels

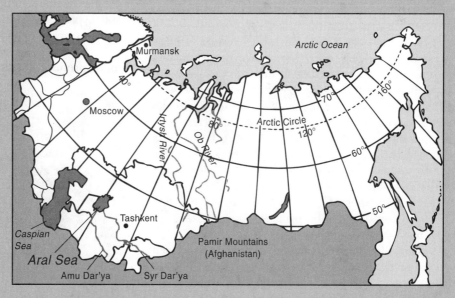

Figure 11.2.1 The Aral Sea in Uzbekistan and Kazakhstan.

Figure 11.2.2 Fishing boats lie marooned on the dry lake bed of the Aral Sea, which has lost two-thirds of its volume in 30 years due to excessive water withdrawals from its tributary rivers.

dropped 14 m (42 ft), turning an area of former seabed about the size of Maryland into a salty, dusty desert. Fishing villages that were once at the sea's edge are now 40 km (25 mi) from water. Boats trapped by falling water levels now lie abandoned in the sand (box fig. 11.2.2). Salinity of the remaining water has tripled and almost no aquatic life remains. Commercial fishing, which brought in 48,000 metric tons in 1957, was completely gone in 1990.

Winds whipping across the dried-up seabed pick up salty dust, poisoning crops and causing innumerable health problems for residents. An estimated 43 million metric tons of salt are blown onto nearby fields and cities each year. Eye irritations, intestinal diseases, skin infections, asthma, bronchitis, and a variety of other health problems have risen sharply in the past twenty years, especially among children. Infant mortality in the Kara-Kalpak Autonomous Republic adjacent to the Aral Sea is 60 per 1,000, twice as high as the rest of the former Soviet Union.

Among adults, throat cancers have increased fivefold in thirty years. Many physicians believe that heavy doses of pesticides used on the cotton fields and transported by runoff water to the lake sediments are now becoming airborne in dust storms. Although officially banned, DDT and other persistent pesticides have been widely used in the area and are now found in mothers' milk. More than 35 million people are threatened by this disaster.

A report by the Institute of Geography of the Soviet Union Academy of Sciences predicts that, without immediate action, the Aral Sea will vanish by 2010. What can be done to avert this calamity? Clearly, one solution would be to stop withdrawing water for irrigation, but that would have disastrous social and economic effects. Making irrigation more efficient might save as much as half the water now lost without reducing crop yields. Restoring river flows to half their former amounts would probably stabilize the sea at present levels.

Perhaps more than technological fixes, what we all need most is a little foresight and humility in dealing with nature. The words painted on the rusting hull of an abandoned fishing boat lying in the desert might express it best: "Forgive us Aral. Please come back."

Dam Difficulties

Dams have been useful over the centuries for ensuring a year-round water supply, but they are far from perfect. One of the main problems with dams is inefficiency. Some dams built in the western United States lose so much water through evaporation and seepage into porous rock beds that they waste more water than they make available. The evaporative loss from Lake Mead and Lake Powell on the Colorado River is about 10 percent of the annual river flow. This amounts to nearly 4,500 l (1,200 gal) for each person in the United States per year. The salts left behind by evaporation nearly double the salinity of the river and make its water unusable when it reaches Mexico. To compensate, the United States is building a $350 million desalination plant at Yuma, Arizona, to try to restore water quality.

As rivers slow in the reservoirs, silt and suspended material settle out. The Colorado drops more than 10 million metric tons of silt per year, which collect behind Boulder and Glen Canyon dams. Imagine twenty thousand dump trucks backed up to Lake Mead and Lake Powell every day, dumping dirt into the water. Within as little as one hundred years, these reservoirs will be full of silt and useless for either water storage or hydroelectric generation (fig. 11.10).

The accumulating sediments that clog reservoirs and make dams useless also represent a loss of valuable nutrients. The Aswan High Dam in Egypt was built to supply irrigation water to make agriculture more productive. Although thousands of hectares are

Figure 11.10 This dam is now useless because its reservoir has filled with silt and sediment. The structure in the center of the photo is the former flood-control drain.

being irrigated, the water available is only about half that anticipated because of evaporation in Lake Nasser behind the dam and seepage losses in unlined canals that deliver the water. Controlling the annual floods of the Nile also has stopped the deposition of nutrient-rich silt on which farmers depended for fertility of their fields. This silt is being replaced with commercial fertilizer costing more than $100 million each year. Furthermore, the nutrients carried by the river once supported a rich fishery in the Mediterranean that was a valuable food source for Egypt. After the dam was installed, sardine fishing declined 97 percent. To make matters worse, growth of snail popula-tions in the shallow permanent canals that distribute water to fields has led to an epidemic of schistosomia-sis. This debilitating disease is caused by blood flukes (parasitic flatworms) spread by snails living in perma-nent ponds and irrigation canals. In some areas, 80 percent of the residents are infected (chapter 6).

WATER CONSERVATION

We could save as much as half of the water we now use without great sacrifice or serious lifestyle changes. Many relatively simple conservation meas-ures in agriculture and industry, or by individuals (box 11.3), could go a long way toward forestalling

Acting Locally: Box 11.3

You Can Make a Difference

Most of us aren't even aware of how much water we use. We turn on the tap and all the clean, clear water we could want comes gushing out. Even though domestic supplies account for only 12 percent of all water with-drawn in the United States, this still amounts to about 650 l (170 gal) per person per day. By contrast, people in less-developed countries average less than 45 l per per-son per day. Do we really need to use so much?

As you can see in box figure 11.3.1, toilet flushing is the largest single domestic water consumption. A typical toilet uses 20 to 25 l (about 5 to 6 gal) per flush. Low-flush toilets are available that use as little as 1 l (about 1 quart) per cycle. If you can't afford new appliances, you can put a couple of bricks or a plastic jug full of water in the tank to reduce the amount that flows out each time. An even cheaper solution is simply to not flush with every use.

Filling a bathtub generally requires 100 to 150 l (30 to 40 gal). A shower that uses 20 l (5 gal) per minute is more efficient than a bath if you limit yourself to less than 5 minutes. Low-flow shower heads are cheap and easy to install. Super-efficient ones give a very satisfactory shower with only 4 l (1 gal) per minute. A low-tech alter-native is to take a "navy" shower: run the water long enough to get wet all over, turn it off to lather up, then turn it on again to rinse off. Some low-flow shower heads have an on/off button so you don't have to adjust the temperature between rinses. If you don't like the cold air that seeps in while the water is off, install a plastic cap or tent over the top of the shower stall.

If you wash your laundry or dishes by machine, make sure you have a full load before starting the cycle. Many dishwashers have a rinse-and-hold feature that enables you to accumulate dishes from several meals to make a full load. If you wash by hand, fill one basin with wash water and another with rinse water. If you are really fru-gal, you should be able to wash a whole meal's worth of dishes in only a couple of liters of water. You will save

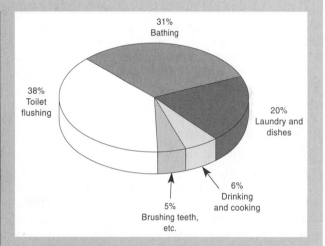

Figure 11.3.1 Typical household water use in the United States.

energy as well as water. Heating water is usually the sec-ond largest household energy use.

Similarly, when you brush your teeth, wash your hands, or shave, don't let the faucet run. You can save 30 to 40 gallons of water each time by drawing a basin or sinkfull. Developing good habits raises our conscious-ness of our environmental impacts in many areas.

How many times have you seen your neighbor washing a car or watering a lawn with so much water that the ex-cess runs down the street like a small river? When you wash your car, use a bucket and a sponge; you can even rinse the same way. Water your yard or garden with only the amount that the plants need. Don't run sprinklers on hot, windy days because most of the water goes into the air rather than the ground. If you live in an arid part of the country, you might consider whether you really need a lush green lawn that requires constant watering, feeding, and care. Planting native ground cover in a "natural lawn" or developing a rock garden or landscape in harmony with the surrounding ecosystem can be both ecologically sound and aesthetically pleasing. We are accustomed to having abun-dant water supplies at low prices, but we may be forced to rethink how we use water in the future.

water shortages that many authorities predict. Isn't it better to adapt to more conservative uses now while we have a choice than to wait until water scarcity forces us to do so?

Industrial and Agricultural Conservation

Nearly half of all water use in the United States is for cooling of electric power plants and other industrial facilities. Some of this water use could be avoided by installing dry cooling systems similar to the radiator of your car. In many cases, cooling water could be reused for irrigation or other purposes in which water does not have to be drinking quality. The waste heat carried by this water could be a valuable resource if techniques were developed for using it.

Of all the agricultural water used, about half is lost to leaks in irrigation canals, application in areas where plants don't grow, runoff, and evaporation. Better farming techniques, such as minimum tillage, leaving crop residue on fields and ground cover on drainage ways, intercropping, use of mulches, and trickle irrigation, could reduce these water losses dramatically.

Price Mechanisms

We have traditionally treated water as if there were an endless supply and the water itself had no intrinsic value. Federal water supply projects charge customers only for the immediate costs of delivery, benefiting large users and discouraging conservation. Farmers in Arizona, for instance, pay less than 1¢ per cu m—less than one-tenth the cost of production—while residents of Phoenix pay about twenty-five times as much for the same water. This allows farmers to grow water-intensive crops in the desert and distorts our whole economy. Paradoxically, there is some good news in this situation. Agriculture accounts for 90 percent of the water consumed in western states. If farmers there could save only 10 percent of the water they use, the amount available for all other uses would double.

Numerous municipalities also have unreasonably low prices for water. In New York City, for example, water was supplied to homes and businesses for many years at a flat rate. There were no meters because it was considered more expensive to install meters and read them than the water was worth. With no incentive to restrict water use or repair leaks, 750,000 cu m (200 million gal) of water were wasted each year from leaky faucets, toilets, and water pipes. The drought of 1988 convinced the city to begin a ten-year, $290 million program to install meters and reduce waste.

If water users were charged the real cost for environmental damage, future use, and public subsidies,

conservation would be more attractive. One way to establish true water cost is to allow it to be marketed in interstate commerce. Laws intended to protect agriculture often prevent municipalities and industries from bidding on water supplies. This policy also protects inefficient and wasteful uses. Allowing the market to determine a price for water can encourage efficiency that makes more water available as if a new supply were being created. In 1982, the U.S. Supreme Court ruled that water is subject to the Interstate Commerce Clause of the Constitution so that state and local laws cannot interfere with its marketing. It will be important, however, as water markets develop, to be sure that environmental, recreational, and wildlife values are not sacrificed to the lure of high-bidding industrial and domestic uses.

WATER POLLUTION

Among the many environmental problems that offend and concern people, perhaps none is as powerful and dramatic as water pollution. Ugly, scummy water full of debris, sludge, and evil-colored foam is surely one of the strongest and most easily recognized symbols of our misuse of the environment.

What Is Water Pollution?

Any physical, biological, or chemical change in water quality that adversely affects living organisms or makes water unsuitable for desired uses can be considered **water pollution** (fig. 11.11). Often, however, a change that adversely affects one organism may be advantageous to another. Nutrients that stimulate oxygen consumption by bacteria and other decomposers in a river or lake, for instance, may be lethal to fish but will stimulate a flourishing community of decomposers. Whether the quality of the water has suffered depends on your perspective. There are natural

Figure 11.11 Our national goal of making all surface waters in the United States "fishable and swimmable" has not been fully met, but scenes like this have been reduced by pollution control efforts.

sources of water contamination, such as poison springs, oil seeps, and sedimentation from erosion, but in this chapter we will focus primarily on human-caused changes that affect water quality or usability.

Pollution control standards and regulations usually distinguish between point and nonpoint pollution sources. Factories, power plants, sewage treatment plants, underground coal mines, and oil wells are classified as **point sources** because they discharge pollution from specific locations, such as drain pipes, ditches, or sewer outfalls. These sources are discrete and identifiable, so they are relatively easy to monitor and regulate. It is generally possible to divert effluent from the waste streams of these sources and treat it before it enters the environment.

In contrast, **nonpoint sources** of water pollution are scattered or diffuse, having no specific location where they discharge into a particular body of water. Nonpoint sources include runoff from farm fields, golf courses, lawns and gardens, construction sites, logging areas, roads, streets, and parking lots. Multiple origins and scattered locations make this pollution more difficult to monitor, regulate, and treat than point sources.

Ocean Pollution

Although the oceans are vast, unmistakable signs of human abuse can be seen even in the most remote places. Garbage and human wastes from coastal cities are dumped into the ocean. Silt, fertilizers, and pesticides wash off farm fields to smother coral reefs and coastal spawning beds and overfertilize estuaries. Every year millions of tons of plastic litter and discarded fishing nets entangle aquatic organisms, dooming them to a slow death (fig. 11.12).

Generally, coastal areas, where the highest concentrations of sea life are found and human activities take

Figure 11.12 Deadly necklace. Marine biologists estimate that castoff nets, plastic beverage yokes, and other packing residue kill hundreds of thousands of birds, mammals, and fish each year.

place, are most critically affected. As I write, news is coming in about yet another supertanker that has gone aground—this one off the coast of the Shetland Islands north of Britain—and is hemorrhaging its load of 600,000 barrels (25 million gallons) of crude oil into the ocean. This is one of the consequences of our addiction to imported oil. The oceans are vitally important to the ecological welfare of the planet, but there isn't space here to discuss them in more detail. Instead, we will focus on freshwater pollution, which is a greater immediate problem for most people in the world.

Waterborne Diseases

The most serious water pollutants in terms of human health worldwide are pathogenic organisms (chapter 6). Altogether, at least 25 million deaths each year are blamed on these water-related diseases, including nearly two-thirds of the mortalities of children under five years old. The main source of these pathogens is from untreated or improperly treated human wastes. Animal wastes from feedlots or fields near waterways and food processing factories with inadequate waste treatment facilities also are sources of disease-causing organisms.

In the more-developed countries, sewage treatment plants and other pollution-control techniques have reduced or eliminated most of the worst sources of pathogens in inland surface waters. Furthermore, drinking water is generally disinfected by chlorination so epidemics of waterborne diseases are rare in these countries. The United Nations estimates that 90 percent of the people in high-income countries have adequate sewage disposal, and 95 percent have clean drinking water.

For poor people, the situation is quite different. The United Nations estimates that three-quarters of the people in less-developed countries have inadequate sanitation, and that fewer than half have access to clean drinking water. Some countries fare much worse than this average; in Ethiopia, for example, 94 percent of the population has insufficient or unsafe water. Conditions are generally worse in remote, rural areas where sewage treatment is usually primitive or nonexistent, and purified water is either unavailable or too expensive to obtain. In the thirty-three poorest countries, 60 percent of the urban population has access to clean drinking water but only 20 percent of rural people do.

Oxygen-demanding Wastes

The amount of oxygen dissolved in water is a good indicator of water quality and of the kinds of life it will support. Water with an oxygen content above 8 parts per million (ppm) will support game fish and other desirable forms of aquatic life (fig. 11.13).

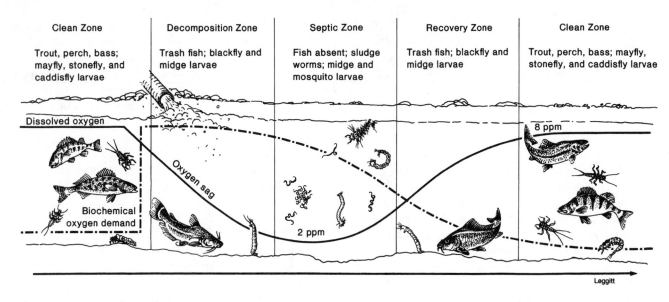

Clean Zone	Decomposition Zone	Septic Zone	Recovery Zone	Clean Zone
Trout, perch, bass; mayfly, stonefly, and caddisfly larvae	Trash fish; blackfly and midge larvae	Fish absent; sludge worms; midge and mosquito larvae	Trash fish; blackfly and midge larvae	Trout, perch, bass; mayfly, stonefly, and caddisfly larvae

Figure 11.13 Oxygen sag downstream of an organic source. A great deal of time and distance may be required for the stream and its inhabitants to recover.

Water with less than 2 ppm oxygen will support only worms, bacteria, fungi, and other decomposers. Oxygen is added to water by diffusion from the air, especially when turbulence and mixing rates are high, and by photosynthesis of green plants, algae, and cyanobacteria. Oxygen is removed from water by respiration and chemical processes that consume oxygen.

The addition of certain organic materials, such as sewage, paper pulp, or food processing wastes, to water stimulates oxygen consumption by decomposers. The impact of these materials on water quality can be expressed in terms of **biological oxygen demand (BOD),** a standard test of the amount of dissolved oxygen utilized by aquatic microorganisms.

Toxic, Inorganic Water Pollutants

In many areas, toxic, inorganic chemicals introduced into water as a result of human activities have become the most serious form of water pollution. Among the chemicals of greatest concern are heavy metals, such as mercury, lead, tin, and cadmium (box 11.4). Supertoxic elements, such as selenium and arsenic, also have reached hazardous levels in some waters. Other inorganic materials, such as acids, salts, nitrates, and chlorine, that normally are not toxic in low concentrations may become concentrated enough to lower water quality or adversely affect biological communities.

Since Roman times, people have known lead to be dangerous to human health. Lead pipes are a serious source of drinking water pollution, especially in older homes or in areas where water is acidic and, therefore, leaches more lead from pipes. Even lead solder in pipe joints and metal containers can be hazardous.

In 1988, the EPA set the maximum limit for lead in public drinking water at 50 parts per billion (ppb). Suppliers of water with 20 to 50 ppb lead must notify customers of possible hazards. Some public health officials argue that lead is neurotoxic at any level and that limits should be less than 10 ppb.

Salts, Acids, and other Nonmetallic Pollutants

Desert soils often contain high salt concentrations that can be mobilized by irrigation and concentrated by evaporation, reaching levels that are toxic for plants and animals. Salt levels in the San Joaquin River in central California rose about 50 percent between 1930 to 1970 as a result of agricultural runoff. Salinity levels in the Colorado River and surrounding farm fields have become so high in recent years that millions of hectares of valuable croplands have had to be abandoned. The United States is building a huge desalination plant at Yuma, Arizona, to reduce salinity in the river. In northern states, millions of tons of sodium chloride and calcium chloride are used to melt road ice in the winter. The corrosive damage to highways and automobiles and the toxic effects on vegetation are enormous. Leaching of road salts into surface waters has a similarly devastating effect on aquatic ecosystems.

Acids are released by mining and as by-products of industrial processes, such as leather tanning, metal smelting and plating, petroleum distillation, and organic chemical synthesis. Coal mining is an especially important source of acid water pollution. Sulfides in coal are solubilized to make sulfuric acid. Thousands of kilometers of streams in the

Thinking Globally: Box 11.4

Minamata Disease

*I*n the early 1950s, people in the small coastal village of Minamata, Japan, noticed strange behavior that they called dancing cats. Inexplicably, cats would begin twitching, stumbling, and jerking about as if they were drunk. Many became "suicidal" and staggered off docks into the ocean. The residents didn't realize it at the time, but they were witnessing an ominous warning of an environmental health crisis that would make the name of their village synonymous with a deadly disease. Their cats were suffering from brain damage that we now know was caused by methyl mercury poisoning.

In 1956, the first human case of neurological damage was reported. A five-year-old girl who had suddenly lapsed into a convulsive delirium was brought into the local clinic. Within a few weeks there seemed to be an epidemic of nervous problems in the village, including numbness, tingling sensations, headaches, blurred vision, slurred speech, and loss of muscle control. For an unlucky few, these milder symptoms were followed by violent trembling, paralysis, and even death. An abnormally high rate of birth defects also occurred. Children were born with tragic deformities, paralysis, and permanent mental retardation (box fig. 11.4.1). Lengthy investigations showed that these symptoms were caused by mercury from fish and seafood that formed a major part of the diet of both humans and their cats.

For years the Chisso Chemical Plant had been releasing residues containing mercury into Minamata Bay. Since elemental mercury is not water soluble, it was as-

sumed that it would sink into the bottom sediments and remain inert. Scientists discovered, however, that bacteria living in the sediments were able to convert metallic mercury into soluble methyl mercury, which was absorbed from the water and concentrated in the tissues of aquatic organisms. People who ate fish and shellfish from the bay were exposed to dangerously high levels of this toxic chemical. Altogether, more than 3,500 people were affected and about 50 died of what became known as Minamata Disease. After nearly twenty years of rancorous protests and litigation, the Chisso Company finally admitted that it was guilty of dumping the mercury and agreed to pay reparations to the victims.

Dumping of mercury into Minimata Bay was stopped twenty years ago. Mud containing mercury was dredged up and buried elsewhere so the bay is now considered safe for fishing. The minds and bodies of those people who ate the mercury-poisoned fish, however, can never be repaired. Have we learned from this tragedy how to anticipate and prevent future environmental disasters?

Figure 11.4.1 A mother from Minamata, Japan, bathes her daughter, who suffered permanent brain damage and birth defects from mercury-contaminated seafood the mother ate while pregnant. This kind of poisoning is now known as Minamata Disease.

United States have been poisoned by acids and metals, some so severely that they are essentially lifeless.

Toxic Organic Chemicals

Thousands of different natural and synthetic organic chemicals are used in the chemical industry to make pesticides, plastics, pharmaceuticals, pigments, and other products that we use in everyday life. Many of these chemicals are highly toxic (chapter 6). Exposure to very low concentrations can cause birth defects, genetic disorders, and cancer. Some synthetic

chemicals are resistant to degradation, allowing them to persist in the environment for many years. Contamination of surface waters and groundwater by these chemicals is a serious threat to human health.

The two most important sources of toxic organic chemicals in water are improper disposal of industrial and household wastes and runoff of pesticides from farm fields, forests, roadsides, golf courses, and other places where they are used in large quantities. The EPA estimates that nearly half a million tons of pesticides are used in the United States each year. Much of

Water Resources 241

this material washes into the nearest waterway, where it passes through ecosystems and may accumulate in high levels in certain nontarget organisms. The bioaccumulation of DDT in aquatic ecosystems was one of the first of these pathways that scientists understood. Polychlorinated biphenyls, dioxins, and other chlorinated hydrocarbons (hydrocarbon molecules that contain chlorine atoms) also have been shown to accumulate to dangerous levels in the fat of salmon, fish-eating birds, and humans.

Hundreds of millions of tons of hazardous organic wastes are thought to be stored in dumps, landfills, lagoons, and underground tanks in the United States (chapter 13). Many, perhaps most, of these sites are leaking toxic chemicals into surface waters or groundwater or both. The EPA estimates that about 26,000 hazardous waste sites will require cleanup because they pose an imminent threat to public health, mostly through water pollution.

CURRENT WATER QUALITY CONDITIONS

In 1989, the EPA announced that 17,365 segments of surface water in the United States and its territories were contaminated by toxic chemicals, sewage, or other pollutants. This contamination affects about 10 percent of the river, stream, coastal water, lake, and estuary mileage in the country. In addition, between 1 to 2 percent of the groundwater near the surface is also polluted. How does this situation compare to past pollution levels? How does the United States compare to other countries? In the next section, we will look at areas of progress and at remaining problems in water pollution control.

Areas of Progress

Like most developed countries, the United States has made encouraging progress in protecting and restoring water quality in rivers and lakes over the past forty years. In 1948, only about one-third of Americans were served by municipal sewage systems, and most of those systems discharged sewage without any treatment or with only primary treatment (the bigger lumps of waste are removed). Most people depended on cesspools and septic systems to dispose of domestic wastes.

Since the passage of the Clean Water Act in 1972, the United States has spent more than $100 billion to build or upgrade thousands of municipal sewage treatment plants. As a result, by 1990, 70 percent of the U.S. population was served by municipal sewage systems. No major city now discharges raw sewage into a river or lake except as overflow during heavy rainstorms. If we avoid polluting water in the first place, we will have more clean water for uses we desire as well as save cleanup costs for those who want to use the resource later. Remember that not everyone can live upstream.

The national goal of making all U.S. surface waters "fishable and swimmable" by 1985 was not met, but 75 percent of the streams, rivers, and lakes monitored by the EPA fully supported their designated uses by that date. Thirteen percent of the monitored streams and rivers on which the EPA has information have improved in recent years in terms of reduced BOD, increased dissolved oxygen content, and greater diversity of aquatic life. Only 3 percent of the monitored stations have recorded a decrease in water quality, while 84 percent have remained about the same since 1972.

Remaining Problems

The greatest impediments to achieving our national goals in water quality are nonpoint discharges of pollutants. These sources are harder to identify and reduce or treat than are specific point sources. The EPA estimates that about three-fourths of the water pollution in the United States comes from soil erosion, fallout of air pollutants, and surface runoff from urban areas, farm fields, and feedlots.

Loading of both nitrates and phosphates in surface water has decreased from point sources but has increased about fourfold since 1972 from nonpoint sources. "Toxic fallout" from the atmosphere is a major nonpoint source of water pollutants. Fossil fuel combustion has become a major source of nitrates, sulfates, arsenic, cadmium, mercury, and other toxic pollutants that find their way into water. Carried to remote areas by atmospheric transport, these combustion products now are found nearly everywhere in the world.

Since DDT and PCBs were banned in the 1970s, levels in birds and fish living in many parts of the Great Lakes have declined significantly (fig. 11.14). Still, large amounts are released slowly by sediments or remain stored in living organisms. Furthermore, toxic compounds banned here but still used elsewhere are transported long distances by wind currents and have become a major source of contamination of the Great Lakes.

Surface Waters in Other Countries

Japan, Australia, and most of Western Europe also have improved surface water quality in recent years. Sewage treatment in the wealthier countries of Europe generally equals or surpasses that in the United States and Canada. Every city dweller in Sweden, for instance, has access to municipal sewage treatment. Denmark and Germany have nearly as good a record. Poorer countries, however, have much less to spend on sanitation. Spain serves only 18 percent

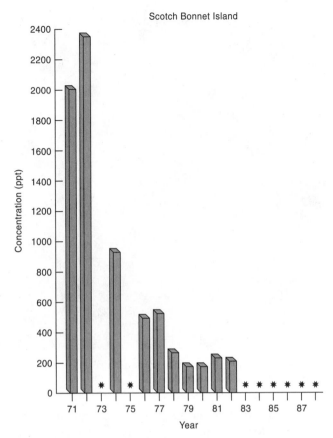

Scotch Bonnet Island

Figure 11.14 Dioxin concentrations in herring gull eggs on Scotch Bonnet Island in Lake Ontario have fallen sharply since 1971. Discontinued production and use of the herbicide 2,4,5-T is a major reason for this decline.
*data not available

Source: Data from *State of Canada's Environment*, Minister of the Environment.

cleanup of the Thames. More than $250 million in public funds plus millions more from industry were spent to curb pollution. By the early 1980s, the river was showing remarkable signs of rejuvenation. Some ninety-five species of fish had returned, including pollution-sensitive salmon, which had not been seen in London for three hundred years.

The less-developed countries of South America, Africa, and Asia have even worse water quality than do the poorer countries of Europe. Sewage treatment is usually either totally lacking or woefully inadequate. Low technological capabilities and little money for pollution control are problems made even worse by burgeoning populations, rapid urbanization, and the shift of much heavy industry (especially the dirtier ones) from developed countries where pollution laws are strict to less-developed countries where regulations are more lenient.

Appalling environmental conditions often result from these combined factors (fig. 11.15). Two-thirds of India's surface waters are contaminated sufficiently to be considered dangerous to human health. The Yamuna River in New Delhi has 7,500 coliform bacteria (from human feces) per 100 ml (thirty-seven times the level considered safe for swimming in the United States) *before* entering the city. The coliform count increases to an incredible 24 *million* organisms per 100 ml as the river leaves the city! At the same time, the river picks up some 20 million l of industrial effluents every day from New Delhi. No wonder disease rates are high and life expectancy is low in this area. Less than 1 percent of India's towns and cities have any sewage treatment, and only eight cities have anything beyond primary treatment.

In Malaysia, forty-two of fifty major rivers are reported to be "ecological disasters." Residues from

Figure 11.15 Severe contamination of this tidal canal in Jakarta exposes nearby residents to a variety of health hazards. It also contributes to the pollution of Jakarta Bay and the Java Sea.

of its population with even primary sewage treatment; Ireland serves only 11 percent. In Greece, less than 1 percent of the population has sewage treatment. Most wastes, both domestic and industrial, are dumped directly into the ocean.

This lack of pollution control is reflected in inland water quality as well. In Poland, 95 percent of all surface water is unfit to drink. The Vistula River, which winds through the country's most heavily industrialized region, is so badly polluted that more than half the river is utterly devoid of life and unsuited even for industrial use. In Russia, the lower Volga River is reported to be on the brink of disaster due to the 300 million tons of solid waste and 20 trillion liters (5 trillion gal) of liquid effluent dumped into it annually.

Some encouraging pollution control stories are also emerging. One of the most outstanding examples is the Thames River in London. Since the beginning of the Industrial Revolution, the Thames had been little more than an open sewer, full of vile and toxic waste products from domestic and industrial sewers. In the 1950s, however, England undertook a massive

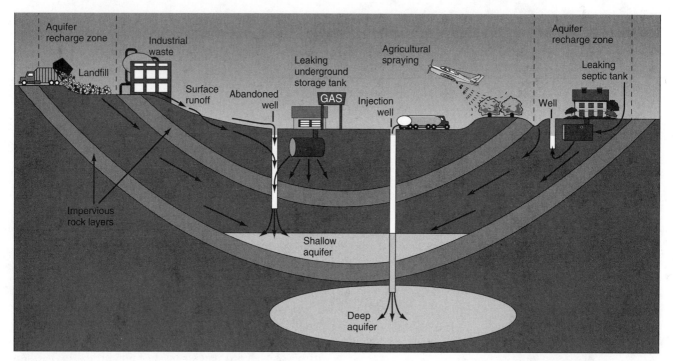

Figure 11.16 Sources of groundwater pollution. Septic systems, landfills, and industrial activities on aquifer recharge zones leach contaminants into aquifers. Wells provide a direct route for injection of pollutants into aquifers.

palm oil and rubber manufacturing, along with heavy erosion from logging of tropical rainforests, have destroyed all higher forms of life in most of these rivers. In the Philippines, domestic sewage makes up 60 to 70 percent of the total volume of Manila's Pasig River. Thousands of people use the river not only for bathing and washing clothes but also as their source of drinking and cooking water. China treats only 2 percent of its sewage. Of seventy-eight monitored rivers in China, fifty-four are reported to be seriously polluted. Of forty-four major cities in China, forty-one use "contaminated" water supplies, and few do more than rudimentary treatment before it is delivered to the public.

Groundwater

About half the people in the United States, including 95 percent of those in rural areas, depend on underground aquifers for their drinking water. This vital resource is threatened in many areas by overuse and pollution. In even more places, the precious remaining reserves are being made unfit for use by a wide variety of industrial, agricultural, and domestic contaminants. For decades it was widely assumed that groundwater was impervious to pollution because soil would bind chemicals and cleanse water as it percolated through. Springwater or artesian well water was considered to be the definitive standard of water purity, but that is no longer true in many areas.

The Office of Technology Assessment estimates that every day some 4.5 trillion liters (1.2 trillion gal) of contaminated water seep into the ground in the United States from septic tanks, cesspools, municipal and industrial landfills and waste disposal sites, surface impoundments, agricultural fields, forests, and wells (fig. 11.16). It is possible, but expensive, to pump water out of aquifers, clean it, and then pump it back. For very large aquifers, pollution may be essentially irreversible.

In a 1982 survey of large public water systems served by groundwater, the EPA found that 45 percent were contaminated with industrial solvents, agricultural fertilizers or pesticides, or other toxic synthetic chemicals. In a 1986 survey in Iowa, pesticides and other synthetic chemicals were detected in half of all wells tested. One-fifth of these wells had nitrate levels from fertilizer infiltration that exceeded federal standards. These high nitrate levels are dangerous to infants (nitrate combines with hemoglobin and results in "blue-baby" syndrome, a potentially deadly lack of oxygen in the blood). They also are transformed into cancer-causing nitrosamines in the human intestine.

The United States has at least 2.5 million underground chemical storage tanks. Most of the tanks are reaching the end of their life expectancy, and one-third of them are believed to be leaking. Many of these tanks were left behind at abandoned gasoline stations or industrial sites, their exact whereabouts and contents unknown. Abandoned wells represent another major source of groundwater contamination. Most domestic wells have no casings to prevent surface contaminants from leaking directly into aquifers that they penetrate. When these wells are no longer in use, they are rarely capped adequately, and people forget where they are. They become direct routes for drainage of surface contaminants into aquifers.

Aquifer recharge zones are the normal route for replenishing groundwater (fig. 11.16). When surface waters that percolate through the recharge zone are contaminated by runoff from farm fields, feedlots, city streets, or industrial sites, the underlying aquifers are threatened. If pollution levels are low and the filtering capacity of surface sediments is high, the aquifer may remain clean; but when pollution levels are high, aquifers become contaminated. Sound land-use practices exclude polluting activities from aquifer recharge zones. When well built and properly maintained, septic systems are appropriate for domestic wastes in rural areas—but they should not be built on recharge zones.

WATER LEGISLATION

The Clean Water Act of 1972 regulates release of "conventional pollutants" (dirt, organic wastes, and sewage), toxic chemicals (from the Toxic Substances Control Act), and "nonconventional pollutants" (those whose toxicity is suspected but not yet determined). For such point sources as industries, municipal facilities, and other discrete sources of conventional pollutants, the act requires discharge permits and **best practical control technology (BPT)** to reduce pollution. It sets national goals of **best available, economically achievable technology (BAT)** for toxic substances and zero discharge for 129 priority toxic pollutants. An important provision of the act provided $52 billion in federal grants to help states and communities construct sewage treatment plants.

Nonpoint sources of pollution (runoff from city streets, construction sites, farmland, mines, and waste dumps) are more difficult to control. When the U.S. Senate voted in 1985 to renew and extend the Clean Water Act, one of the more significant provisions of this bill was a program to require and help states to control nonpoint sources. Another important section of the act protects wetlands and aquifer recharge zones. A powerful clause of the act allows environmental groups to sue polluters for exceeding pollution permits. Using data that polluters themselves are required to submit, groups like the Natural Resources Defense Council and the Sierra Club have won million-dollar settlements under this act. This helps clean up the environment while deterring pollution.

Is it safe to assume that we are well on our way to solving our water pollution problems? We are better off in terms of legislation, policy, and practice than we were in 1960. Laws, however, are only as good as (1) the degree to which they are not weakened by subsequent amendments and exceptions and (2) the degree to which they are funded for research and enforcement. Economic interests cause continued pressure on both of these points, so that the importance of an overriding national attitude to maintain the intent of protective legislation must continually be stressed and retaught.

S u m m a r y

The hydrologic cycle constantly purifies and redistributes fresh water, providing an endlessly renewable resource. Less than 3 percent of all water on the earth is fresh, and about three-quarters of that is locked up in ice or snow or buried in groundwater aquifers. Lakes, rivers, and other surface freshwater bodies make up only about 0.01 percent of all the water in the world, but they provide habitat and nourishment for aquatic ecosystems that play a vital role in the chain of life.

Water is essential for nearly every human endeavor. In poorer countries, agriculture consumes some 70 percent of all water withdrawn for human use. In industrialized countries such as the United States, industrial cooling is the single largest use. Nearly everywhere domestic uses account for only a small fraction of our water supply uses. Only about half the water we withdraw is consumed or degraded so that it is unsuitable for other purposes; much could be reused or recycled. Water conservation and recycling would have both economic and environmental benefits.

Water shortages in many parts of the world result from rising demand, unequal distribution, and increased contamination. Arid zones are especially vulnerable to the effects of natural droughts and land abuse by humans and domestic animals. Lakes, rivers, and groundwater reservoirs are being depleted at an alarming rate, leading not only to water shortages but also to subsidence, sinkhole formation, saltwater intrusion, and permanent loss of aquifers.

Water storage and transfer projects are a response to flooding and water shortages. Giant dams and diversion projects can have environmental and social costs that are not justified by the benefits they provide. Among the problems they pose are evaporation and infiltration losses, siltation of reservoirs, and loss of recreation and wildlife habitat. Watershed management and small dams are preferred by many conservationists as means of flood control and water storage.

There is much we can do to save water. Charging users the true cost of water is a good start toward conservation. We can each use less water in our personal lives, and society can encourage development of water-saving appliances, natural yards, recycling, and efficient water use. Perhaps the most important change we can make is to treat wastes at their sources rather than use precious water resources for waste disposal. Not everyone can live upstream.

Worldwide, the most serious water pollutants, in terms of human health, are pathogenic organisms from human and animal wastes. In industrialized nations, toxic chemical wastes have become an increasing problem. Agricultural and industrial chemicals have been released or spilled into surface waters and are seeping into groundwater supplies. The extent of this problem is probably not yet fully appreciated. In spite of persistent problems, however, progress has been made in more-developed countries in eliminating water pollution. People in poorer countries, however, suffer both from water shortages and availability of clean drinking water and sanitation.

Review Questions

1. What is the difference between withdrawal, consumption, and degradation of water?
2. How do patterns of water use in more-developed countries compare to those of less-developed countries?
3. What is subsidence? What are its results?
4. Describe some problems associated with dam building and water diversion projects.
5. Describe the path a molecule of water might follow through the hydrologic cycle from the ocean to land and back again.
6. What are the major water reservoirs of the world?
7. How much water is fresh and where is it found?
8. Define aquifer. How does water get into an aquifer?
9. Define water pollution. What are the most important types of pollutants?
10. What is the practical difference between best available, economically achievable technology and best practical control technology?

Questions for Critical or Reflective Thinking

1. As is the case in most of this book, I have presented many facts and numbers in this chapter because I want the book to be a useful reference to many readers. You surely don't need to know all the details presented here, but how will you decide which numbers are important to you and which ones can be ignored? (Hint: check with your instructor about this.)
2. Even among the numbers that you need to remember, what level of precision is necessary? When I report that "billions of tons" of sediment are transported by rivers each year, does it matter whether that is metric tons or short tons? How precise do you suppose the estimate is

that 2 billion people don't have access to clean water? How would you define clean water and a sufficient supply?
3. Will it make any difference to people in water-poor countries if we conserve water in countries where there is an abundance? How important is frugality as a good example or a consciousness-raising exercise?
4. Some people in water-rich states in the United States and Canada resist any attempt to transfer water to desert states such as Arizona. They claim that people who live in the desert will have to learn to live on available local supplies. Is this a reasonable position? Could you say the same thing about energy resources and

cold cities such as Minneapolis or Toronto?
5. I have defined water pollution as anything that decreases water quality in ways that adversely affect organisms or make it unsuitable for desired human uses. Suppose that part of the silt in a river is from natural sources and part is human-caused. Is one pollution but the other not?
6. Suppose that a farmer likes having silt in the river because it fertilizes the fields, but a fishing enthusiast dislikes it because it kills game fish. How can we decide whether it is pollution or not?

Key Terms

best available, economically achievable technology (BAT) 245
best practical control technology (BPT) 245
biological oxygen demand (BOD) 240
consumption 230
degradation 230
groundwater 228
nonpoint sources 238

point sources 238
recharge zones 228
sinkholes 234
stable runoff 230
subsidence 234
water pollution 238
withdrawal 230

Suggested Readings

Batisse, M. "Probing the Future of the Mediterranean Basin." *Environment* 32, no. 5 (June 1990): 4. Describes the Mediterranean Blue Plan for international cooperation in cleaning up this sea.

Carrier, J. "Water and the West: The Colorado River." *National Geographic*, June 1991, 2. There isn't enough water in the Colorado River to meet all demands. How will this resource be managed? Great photos.

Colburn, T., and R. Liroff. "Toxics in the Great Lakes." *EPA Journal* 16, no. 6 (November/December 1990): 5. Increasing levels of toxins in the Great Lakes make cleanup more urgent and more complicated.

Fischhoff, B. "Report from Poland: Science and Politics in the Midst of Environmental Disaster." *Environment* 33, no. 2 (March 1991): 12. Air and water pollution were ignored and information was suppressed during forty years of communist rule. Now an enormous task of environmental cleanup faces the country.

Glomb, S. "Measuring Environmental Success." *EPA Journal* 16, no. 6 (November/December 1990): 57. How do we know whether efforts to clean up the major bodies of water are really succeeding?

Leopold, L. B. "Ethos, Equity and Water Resource." *Environment* 32, no. 2 (1990): 16. A useful review of water resources and management by a distinguished hydrologist.

Lewis, J. "The Ogallala Aquifer, an Underground Sea." *EPA Journal* 16, no. 6 (November/December 1991): 42. A good description of the history of and possible solutions to overuse of this imperiled resource.

Maurits la Riviere, J. W. "Threats to the World's Water." *Scientific American* 261, no. 3 (1989): 80. Unless appropriate steps are taken soon, severe water shortages will occur.

Micklin, P. P. "Desiccation of the Aral Sea: A Water Management Disaster in the Soviet Union." *Science* 241 (1988): 1170. An extensive and authoritative account of causes and effects of Aral Sea drying.

Postel, S. "Saving Water for Agriculture." In *State of the World 1990*. Washington, D.C.: Worldwatch Institute, 1990. Strategies for efficient water use in agriculture.

Quammen, D. "A Long River with a Long History." *Audubon*, March 1990, 68. A grand description of the Rio Grande and how it shapes the ecology and culture of the southwest.

Reisner, M., and S. F. Bates. *Overtapped Oasis: Reform or Revolution for Western Water*. Covello, Calif.: Island Press, 1990. A stinging rebuke of business as usual in western water use.

Sears, P. B. *Deserts on the March*. Covello, Calif.: Island Press, 1991. A reissue of the classic 1935 study of human-caused desertification around the world.

Sierra Club Legal Defense Fund. *The Poisoned Well: New Strategies for Groundwater Protection*. San Francisco: Sierra Club Press, 1990. An excellent source of information about sources of and protection against groundwater pollution.

Stegner, W. *Beyond the Hundredth Meridian: John Wesley Powell and the Second Opening of the West*. Cambridge, Mass.: The Riverside Press, 1954. A classic history by the dean of western writers about the role of water in western settlement and politics.

Tuan, Yi-Fu. *The Hydrologic Cycle and The Wisdom of God: A Theme in Geoteleology*. Toronto: University of Toronto Press, 1968. A historic and cultural review of attitudes toward nature with special emphasis on the hydrologic cycle.

Worster, D. *Rivers of Empire*. New York: Pantheon Press, 1985. A fascinating history of water use and abuse in the western United States.

12
ENERGY RESOURCES

Each year the United States wastes more fuel than most of mankind uses. . . . We could lead lives as rich, healthy, and fulfilling—with as much comfort, and with more employment—using less than half the energy now used.

Denis Hayes, president, Green Seal

Objectives

After studying this chapter, you should be able to:

- Recapitulate the major energy sources as well as how energy is used in richer and poorer countries.
- Describe the available coal, oil, and natural gas resources and some problems that our uses of these fossil fuels create.
- Understand how nuclear power reactors work and why they are dangerous.

- Recognize the difficulties in storing nuclear wastes.
- Appreciate the benefits of energy conservation, especially what you might do personally in this area.
- Relate the methods for capturing solar energy, including how photovoltaic cells work.
- Grasp the potential of biomass, hydropower, wind, and other sustainable sources of energy.

INTRODUCTION

Using an external energy source to do useful work is one of the main features that distinguishes humans from most other animals. Our civilization is dependent upon—some would say addicted to—this energy, most of which now comes from nonrenewable sources. The extraction and use of those resources are causing intolerable environmental problems. We desperately need to develop benign, sustainable energy sources.

In this chapter, we will examine the current primary sources of energy in the world as well as explore some alternative sources of energy for the future. The relationship between economic development and energy use is an important aspect of energy policy. We will look briefly, therefore, at disparities in energy use between the more-developed and less-developed countries of the world and ways in which economics and energy access are linked to environmental conditions, quality of life, and future world development.

Major Energy Sources

Fossil fuels (petroleum, natural gas, and coal) now provide about 95 percent of all commercial energy (fig. 12.1). This is a rather recent development. Petroleum and natural gas were not used in large

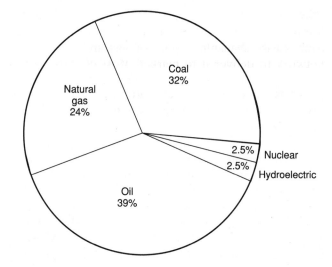

Figure 12.1 Worldwide commercial energy consumption, 1989.
Source: Data from World Resources Institute, 1992.

quantities until this century, but supplies are already running low. Coal makes up about one-third of all the energy we consume. There are large amounts of coal in the world, but the environmental costs of mining and burning it are so great that we may have to curtail our use. The fossil fuel age may turn out to be a rather short episode in the total history of humans.

Figure 12.2 Fossil fuels provided the energy source for the Industrial Revolution and built the world in which we now live. We have used up much of the easily accessible supply of these traditional fuels, however, often through wasteful practices such as the flaring shown here. Adverse environmental effects of burning coal, oil, and natural gas may preclude further use of these fuels. We desperately need to find renewable, environmentally benign sources of energy.

Natural gas is a clean-burning, convenient fuel, but it is difficult to ship across oceans or to store. We in North America are fortunate to have an abundant, easily available supply of gas and a pipeline network to deliver it to market. Most countries cannot afford a pipeline network, and much of the natural gas produced in conjunction with oil pumping is simply burned (flared off), a terrible waste of a valuable resource (fig. 12.2).

Sustainable or renewable energy resources—solar, biomass, hydropower, and wind—currently provide a total of only about 2.5 percent of world power needs, but they have potential to do much more. Worldwide, nuclear power provides about as much energy as renewable sources. We have enough uranium to fuel many more nuclear plants, but, as we will discuss later in this chapter, safety concerns make this option unacceptable to most people.

You may have noticed a conspicuous hole in the energy consumption data given so far: use of biomass, especially fuelwood. The preceding figures apply only to *commercial* energy sources, whereas most biomass burning occurs for residential use, primarily by individuals in the Third World. The importance of biomass to billions of humans can't be neglected, however. Many countries use fuelwood (including charcoal) for more than 75 percent of their nonmus-

cle energy source. The poorest countries, such as Haiti, Mali, Malawi, and Burkina Faso, depend on biomass supplies for more than 90 percent of their total energy for heating and cooking. This is a serious cause of forest destruction and soil loss (chapters 7 and 8).

Per Capita Consumption

Perhaps the most important facts about fossil fuel consumption are that we in the richer countries consume about 78 percent of the natural gas, 65 percent of the oil, and 50 percent of the coal produced worldwide each year. Although we make up less than *one-fifth* of the world's population, we use more than *two-thirds* of the commercial energy supply. North America, for instance, constitutes only 5 percent of the world's population but consumes about one-quarter of the available energy. To be fair, however, some of that energy goes to manufacture goods that are later shipped to developing countries.

Stated another way, each person in the United States uses about 300 GJ (equivalent to 7 *tons* of oil; see table 12.1) each year on average. By contrast, the poorest countries of the world, such as Ethiopia, Kampuchea, Nepal, and Bhutan, generally consume less than one GJ per person per year. This means that each of us in the United States consumes, on average, almost as much energy in a single day as a person in one of these countries would consume in a year.

Clearly, there is a link between energy consumption and the comfort and convenience of our lives. Those of us in the richer countries enjoy many amenities not available to most people in the world. The linkage is not absolute, however. Several European countries, including Sweden, Denmark, and Switzerland, have higher standards of living than the United States by almost any measure but use only about half as much energy per person as we do. This shows how much potential we have for energy savings.

Table 12.1 Some energy units
1 joule (J) = the work done when 1 kg is lifted 1 meter
1 gigajoule (GJ) = 1 billion (10^9) J
1 petajoule (PJ) = 1 million GJ or 947.10^9 Btu
1 British thermal unit (Btu) = energy to heat 1 lb of water 1° F.
1 watt (w) = 1 J per second
1 megawatt (MW) = 1 million watts
1 metric ton of oil = 6.66 barrels of 42 gal each

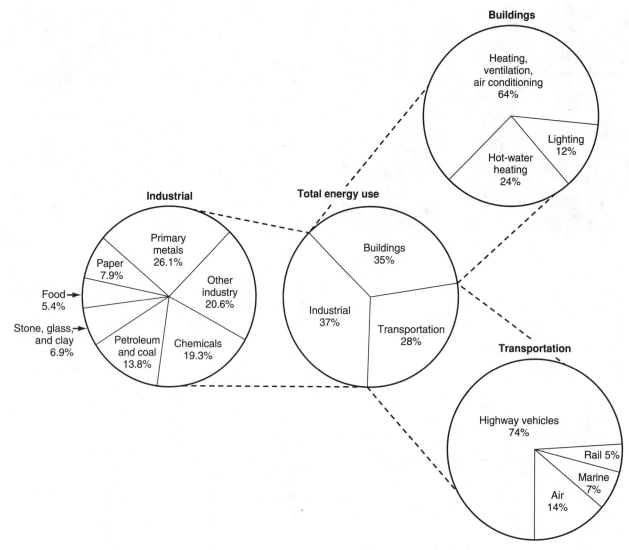

Figure 12.3 How energy is used in the United States.
Source: Data from U.S. Energy Agency.

How Energy Is Used

The largest share (37 percent) of the energy used in the United States is consumed by industry (fig. 12.3). Mining, milling, smelting, and forging of primary metals consume about one-quarter of the industrial energy share. The chemical industry is the second largest industrial user of fossil fuels, but only half of its use is for energy generation. The remainder is raw material for plastics, fertilizers, solvents, lubricants, and hundreds of thousands of organic chemicals in commercial use. The manufacture of cement, glass, bricks, tile, paper, and processed foods also consumes large amounts of energy. Coal, oil, and natural gas each contribute about one-third of the energy used by industry. Residential and commercial buildings use some 35 percent of the primary energy consumed in the United States, mostly for space heating, air conditioning, lighting, and water heating.

Transportation consumes 28 percent of all energy used in the United States each year. Nearly 98 percent of that energy comes from petroleum products refined into liquid fuels. The convenience, comfort, and flexibility of having cheap gasoline and private automobiles to take us wherever we want to go makes our lives much easier than if we had to walk or take public transportation. The environmental, social, political, and economic costs of maintaining this system, however, are a major concern. If we clearly understood all these costs, we might reconsider our priorities.

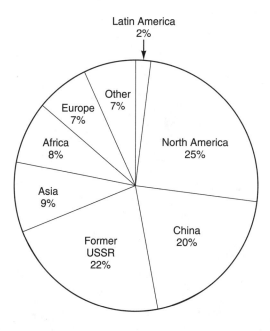

Figure 12.4 Proven-in-place coal reserves.

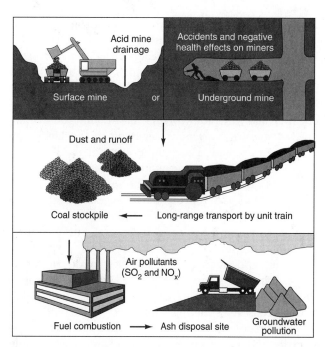

Figure 12.5 The coal story, from surface or underground mine to the power plant. Some associated environmental problems are also indicated.

COAL

Coal is fossilized plant material preserved by burial in sediments and altered by geological forces that compact and condense it into a carbon-rich fuel. Most coal was formed during the Carboniferous period (286 to 360 million years ago) when the earth's climate was much warmer and wetter than it is now. Because coal takes so long to form, it is considered a nonrenewable resource.

Coal Resources and Reserves

World coal deposits are vast: ten times greater than conventional oil and gas resources combined. The total resource is estimated to be 10 trillion metric tons. If all this coal could be extracted and coal consumption continued at present levels, it would last several thousand years. At present rates of consumption, the proven-in-place reserves—those explored and mapped but not necessarily economically recoverable—will last about 200 years.

Where are these coal deposits located? They are not evenly distributed throughout the world (fig. 12.4). Notice that the countries with the largest land area have the most coal. North America has one-fourth of all proven reserves (mostly in the United States). China and the former Soviet Union have nearly as much. Eastern and Western Europe have fairly large coal supplies, in spite of their relatively small size. Latin America has very little coal. Antarctica is thought to have large coal deposits, but they would be difficult, expensive, and ecologically damaging to mine.

Mining

Coal mining is a dirty, dangerous business (fig. 12.5). Underground mines are subject to cave-ins, fires, accidents, and accumulation of poisonous or explosive gases (carbon monoxide, carbon dioxide, methane, hydrogen sulfide) or both. Between 1870 and 1950 more than 30,000 coal miners died of accidents and injuries in Pennsylvania alone, equivalent to one man per day for eighty years. Untold thousands have died of respiratory diseases. In some mines, nearly every miner who did not die early from some other cause was eventually disabled by **black lung disease,** inflammation and fibrosis caused by accumulation of coal dust in the lungs or airways (chapter 6).

Coal mining also contributes to both air and water pollution. Mine drainage and water leaching from coal piles and mine tailings are generally acidic and contaminated with highly toxic chemicals. Thousands of miles of streams in the United States have been poisoned by coal-mining operations. The 900 million tons of coal burned every year in the United States (83 percent for electric power generation) produces about three-quarters of the sulfur dioxide, one-third of the nitrogen oxides, and about half of the industrial carbon dioxide released in the United States each year.

Mining also destroys landscapes. **Strip-mining** or **surface-mining** uses giant machines to remove overburden (overlying layers of dirt and rock). This is cheaper and safer for workers than underground mining but can be extremely damaging to the landscape if proper reclamation techniques are not used. On relatively level land, it is possible to set aside topsoil when cutting down to the coal seam and to recontour the land and replant native vegetation after the coal has been removed. In the United States, reclamation is now required by the Federal Surface Mine Reclamation Act. These procedures are expensive, however, and mining companies often do an inadequate job. On steep slopes, the overburden sometimes slides downhill, smothering forests, filling streambeds, and burying farmland, roads, and even houses. Erosion on steep slopes makes reclamation especially difficult, if not impossible. Some areas simply should not be strip-mined.

OIL

Like coal, petroleum is derived from organic molecules formed by living organisms millions of years ago. Similar conditions can be created artificially, but only at great cost. Pools of oil and gas often accumulate under layers of shale or other impermeable sediments, especially where folding and deformation of rock layers create pockets that will trap upward-moving hydrocarbons (fig. 12.6). Contrary to the image implied by its name, however, an

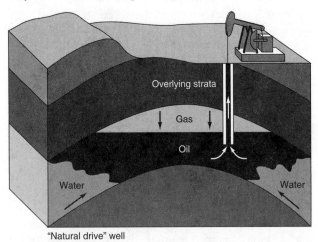

Figure 12.6 Pumping oil from an oil pool trapped between impermeable strata. Natural gas often occurs with petroleum but may be flared off if no pipeline exists to ship the gas to market.

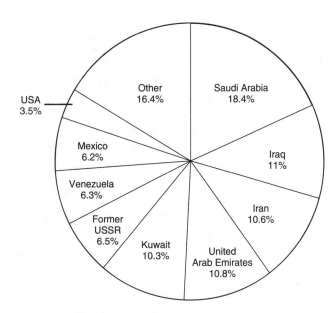

Figure 12.7 World proven recoverable oil reserves by country in 1991. Annual world consumption is about 20 billion barrels or 2.4% of the 823.6 billion barrel proven recoverable reserve.
Source: Data from World Resources Institute, 1991.

oil pool is not usually a reservoir of liquid in an open cavern but rather individual droplets or a thin film of liquid permeating spaces in a porous sandstone or limestone, much like water saturating a sponge.

Oil Resources and Reserves

The total amount of oil in the world is estimated to have been about 4 trillion barrels (600 billion metric tons), half of which is thought to be ultimately recoverable. About one-quarter of the recoverable supply already has been consumed. In 1990 the proven reserves were roughly 1 trillion barrels, enough to last only fifty years at the current consumption rate of 20 billion barrels per year. It is estimated that another 800 billion barrels either remain to be discovered or are not recoverable at current prices with present technology. As oil resources become depleted and prices rise, it probably will become economical to find and bring this oil to the market unless alternative energy sources are developed.

By far the largest supply of proven-in-place oil is in Saudi Arabia, which has around 168 billion barrels, one-fourth of the total world reserve (fig. 12.7). Kuwait had about 10 percent of the proven world oil reserves before Iraq invaded in 1990. Some six hundred wells were blown up and set on fire by the retreating Iraqis. The fires were put out much faster than anyone expected, however, and the environmental damage was less than people feared. Together, the Persian Gulf countries in the Middle East contain nearly two-thirds of the world's proven petroleum supplies. With our insatiable appetite for oil, it is not difficult to see why this volatile region plays such an important role in world affairs.

Originally, the United States is estimated to have had about 10 percent of the total world oil resource. Until 1947, the United States was the leading oil export country in the world. Now, however, domestic production has fallen below 8 million barrels per day, while imports have risen to more than half of our oil supply, making the United States even more vulnerable to events in the Middle East.

Altogether, the United States has already used nearly half of its original recoverable resource of 200 billion barrels. Of the oil thought to remain, about 50 billion barrels are proven in place. If we stopped importing oil and depended exclusively on indigenous supplies, our proven reserve would last *eight* years at current rates of consumption.

The regions of North America with the greatest potential for substantial new oil discoveries are portions of the continental shelf along the California coast, the Beufort Sea, and the Grand Banks (fig. 12.8), all of which are prime wildlife areas. Proposals to do exploratory drilling in these areas have been strongly opposed by many people concerned about the potential for long-term environmental damage from drilling activities and oil spills like the *Exxon Valdez* in Prince William Sound in 1989 or the *Braer* in the Shetland Islands in 1993. Other people argue that energy independence, even if only for a few decades, is worth whatever environmental problems are encountered.

Unconventional Oil

Estimates of our total oil supply usually do not take into account the very large potential from **unconventional oil** resources, such as shale oil and tar sands, which might double the total reserve if they can be extracted with reasonable social, economic, and environmental costs. Canada, for example, has an estimated 270 billion cubic meters of bituminous tar sands, mostly in northern Alberta (fig. 12.8). Liquid petroleum can be extracted from these sands with hot water, chemicals, or other stripping processes. If even part of this resource is recoverable, it would be the world's largest known oil deposit. There are severe environmental constraints, however. A typical plant producing 125,000 barrels of oil per day creates about 15 million cubic meters of toxic sludge and

consumes or contaminates billions of liters of water each year. Similarly vast deposits of oil shale occur in the western United States. Extraction of this oil has comparable problems to those of tar sands.

NATURAL GAS

Natural gas is the world's third largest commercial fuel (after oil and coal), making up 24 percent of global energy consumption. It is the most rapidly growing energy source because it is convenient, cheap, and clean burning. Because natural gas produces only half as much carbon dioxide as an equivalent amount of coal, substitution could help reduce global warming (chapter 10).

Natural Gas Reserves

The republics of the former Soviet Union have 44 percent of known natural gas reserves (mostly in Siberia and the central Asian republics) and account for 36.5 percent of all production. Both Eastern and Western Europe buy substantial quantities of gas from these wells. Figure 12.9 shows the distribution of proven natural gas reserves in the major producing countries of the world.

The total ultimately recoverable natural gas resources are estimated to be 10,000 trillion cubic feet, corresponding to about 80 percent as much energy as the recoverable reserves of crude oil. The proven world reserves are about one-third of the total resource. Because gas consumption rates are approximately half of those for oil, current gas reserves represent roughly a sixty-year supply at present usage rates. Proven reserves in the United States are about 6 percent of the world total, a ten-year supply at current rates of consumption. Known reserves are more than twice as large.

Unconventional Gas Sources

Natural gas resources have not been as extensively investigated in most places as have oil resources. Until recently, gas has been a by-product of crude oil drilling and production. There may be many sources of "unconventional" gas, such as Devonian shales, "tight-sand" gas, and coal-seam methane, that have not been extensively explored but could provide large and valuable fuel sources. Some geologists theorize that the true resources of methane within the earth's crust may be vastly larger than current estimates predict.

Unlike coal and oil, natural gas is presently being formed as a by-product of human activities. Some municipalities heat their waste treatment plants with methane produced during sewage digestion. Methane

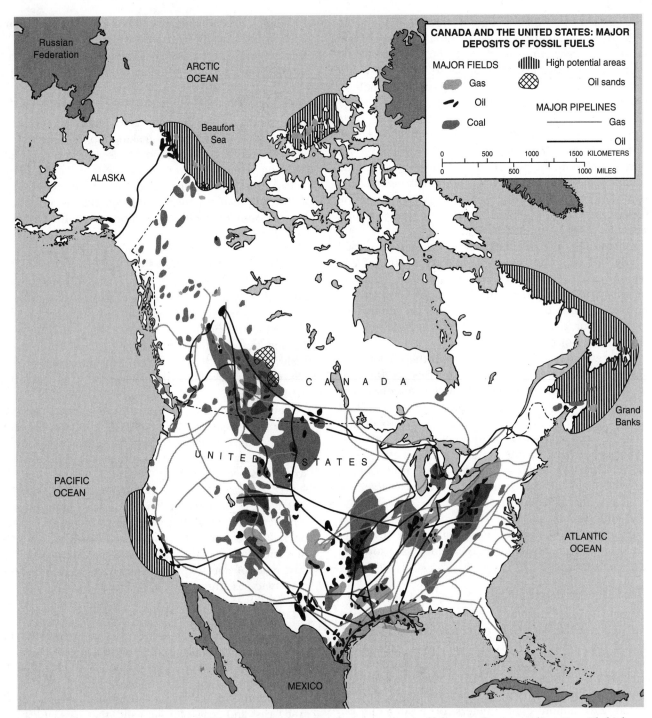

CANADA AND THE UNITED STATES: MAJOR DEPOSITS OF FOSSIL FUELS

MAJOR FIELDS
- Gas
- Oil
- Coal

High potential areas
Oil sands

MAJOR PIPELINES
- Gas
- Oil

Figure 12.8 Canadian and U.S. fossil fuel deposits. Cross-hatched areas show off-shore sedimentary deposits with high potential for oil or gas. The ecological risks of drilling in these areas are high; many environmentalists argue that drilling should be prohibited.

From *Geography: Regions and Concepts,* 6th ed. by H. J. DeBlij and P. O. Muller. Copyright © 1992 John Wiley & Sons, Inc. Reprinted by permission of John Wiley & Sons, Inc.

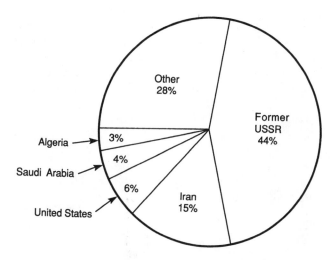

Figure 12.9 Percent of natural gas reserves by region in 1990. The total proven reserve is 3,200 trillion cubic feet. World annual production is approximately 54 trillion cubic feet or 1,409 million tons of oil equivalent.
Source: Data from World Resources Institute, 1990.

is also collected from landfills. Some farmers in both China and the United States generate methane for heating, lighting, and cooking with animal waste digesters. Such domestic methane sources have potential for small-scale, low-cost, local applications.

NUCLEAR POWER

In 1953, President Dwight Eisenhower announced an "Atoms for Peace" program to provide clean, abundant energy from nuclear-powered electrical generators. He predicted that nuclear energy would fill the deficit caused by anticipated oil and natural gas shortages. It would deliver power "too cheap to meter" for continued industrial expansion of both the developed and the developing world. Technology and engineering would tame the evil genie of atomic energy and use its enormous power to do useful work.

Glowing predictions about the future of nuclear energy continued into the early 1970s. Between 1970 and 1974, American utilities ordered 140 new reactors for power plants. Some advocates predicted that by the end of the century there would be 1,500 reactors in the United States alone. The International Atomic Energy Agency (IAEA) projected worldwide electric generating capacities of at least 4.5 million megawatts (MW) by the year 2000, eighteen times more than current nuclear capacity and twice as much as the total world electrical supply from all sources in 1990.

Rapidly increasing construction costs, declining demand for electric power, and public opposition held down the growth of the nuclear industry. Electricity from nuclear power was about half the price of coal in 1970, but twice as much in 1990. If costs double again by 2000, nuclear power will probably become more expensive than solar power.

The United States led the world into the nuclear age, and it appears that we are leading the world out of it. After 1975, only thirteen orders were placed for new nuclear reactors in the United States, and all of those orders subsequently were canceled. It is beginning to look as if the much-acclaimed nuclear power industry may have been a very expensive wild goose chase that may never produce enough energy to compensate for the amount invested in research, development, mining, fuel preparation, and waste storage.

Recently, the nuclear industry has begun an advertising campaign claiming to be environmentally friendly because they do not contribute carbon dioxide to the greenhouse effect (chapter 10). Whether this risky technology will play a substantial role in our energy future remains to be seen.

How do nuclear reactors produce energy? In this section, we will look at the nuclear fuel cycle, reactor designs, and some of the concerns about accidents and security risks associated with nuclear power.

Reactor Fuel

The most commonly used fuel in nuclear power plants is U^{235}, a naturally occurring radioactive isotope of uranium. Ordinarily, U^{235} makes up only about 0.7 percent of uranium ore, too little to sustain a chain reaction. Uranium-bearing ore is extracted from surface or underground mines and the lighter U^{235} isotope is separated from the heavier, nonradioactive uranium by mechanical or chemical procedures. When the concentration of U^{235} reaches about 3 percent, it is packed in hollow steel rods approximately 4 m long. Thousands of these rods containing about 100 tons of uranium are bundled together in the reactor core.

When packed tightly together in the reactor core, the radioactive uranium atoms bombard each other with neutrons produced by fission of the atomic nuclei. As each atom splits, it releases both heat and more neutrons, each of which can cause another atom to split. Thus a self-sustaining chain reaction is set in motion. Rather than continuing to amplify exponentially, as is the case in nuclear bombs, the reaction is moderated by a neutron-absorbing cooling solution that circulates around and between the fuel rods and by **control rods** made of carbon or boron that slow the reaction when they are inserted into spaces between fuel assemblies or that allow it to proceed when they are withdrawn.

Kinds of Reactors in Use

Seventy percent of the nuclear plants in the United States and in the world are **pressurized water reactors (PWR)** in which high-pressure water is circulated through the core to absorb heat from the nuclear fission (fig. 12.10). This primary cooling water is then pumped to a heat exchanger where it generates steam that drives a high-speed turbine generator and produces electricity. Both the reactor vessel and the steam generator are contained in a thick-walled concrete and steel containment building that is designed to withstand accidents. Engineers operate the plant from a complex control room. Overlapping layers of safety mechanisms are designed to prevent accidents, but these fail-safe controls make reactors very expensive and very complex. A typical nuclear power plant has 40,000 valves, compared to only 4,000 in a similar size fossil fuel-fired plant. In some cases, nuclear safety systems are so complex that they confuse operators and cause accidents rather than prevent them. Under normal operating conditions, a PWR releases very little radioactivity and is probably less dangerous for nearby residents than a coal-fired power plant.

Canadian nuclear reactors use deuterium oxide—"heavy" water with the stable isotope of hydrogen—as both a cooling agent and a moderator. These Canadian Deuterium (CANDU) reactors operate with natural, unconcentrated uranium for fuel, eliminating expensive enrichment processes. However, if cooling pumps fail, core temperatures rise very quickly in these reactors. Unless steps are taken immediately, there is risk of a core meltdown.

In Britain, France, and the former Soviet Union, a common reactor design uses graphite, both as a moderator and as the structural material for the reactor core. In the British MAGNOX design (named after the magnesium alloy used for its fuel rods), gaseous carbon dioxide is blown through the core to cool the fuel assemblies and carry heat to the steam generators. In the Soviet design, called RBMK (the Russian initials for a graphite-moderated, water-cooled reactor), low-pressure cooling water circulates through the core in thousands of small metal tubes.

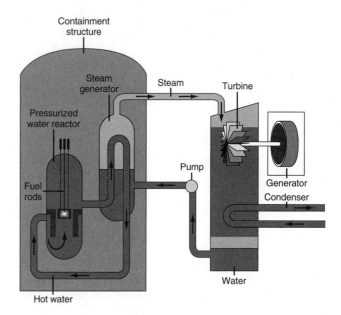

Figure 12.10 Pressurized water nuclear reactor. Water is superheated and pressurized as it flows through the reactor core. Heat is transferred to nonpressurized water in the steam generator. The steam drives the turbogenerator to produce electricity.

These designs were originally thought to be very safe because graphite has high capacity both for capturing neutrons and for dissipating heat. Designers claimed that these reactors could not possibly run out of control. The fuel rods are dispersed in small groups rather than packed in large bundles as in pressurized water designs. However, the small cooling tubes in the Soviet RBMK also have a positive void coefficient. That means if the pumps are turned off, the water turns to steam quickly and the reactor overheats.

Furthermore, graphite burns when heated and exposed to air. The two most disastrous reactor accidents in the world, so far, involved fires in graphite cores that allowed the nuclear fuel to melt and escape into the environment. In 1956, a fire at the Windscale Plutonium Reactor in England spewed out several tons of radioactive material and contaminated hundreds of square kilometers of countryside. In 1986, a fire and explosion at the Chernobyl nuclear plant in the Ukraine released a roughly similar amount of radioactivity (box 12.1).

Alternative Reactor Designs

Several other reactor designs are inherently safer than the ones we now use. Among these are the modular High-Temperature, Gas-Cooled Reactor (HTGCR) and the Process-Inherent Ultimate-Safety (PIUS) reactor.

HTGCR is similar to the British MAGNOX design. Gaseous helium is the coolant, and graphite blocks form the core structure and moderate the reaction. If the reactor core is kept small enough and the fuel is dispersed throughout the core, heat will not build up above 1,600°C, even if all coolant is lost; thus, a meltdown is impossible. These reactors are fueled by graphite or ceramic-coated pellets of uranium oxide or uranium carbide. The pellets are loaded into the reactor from the top, shuffle through the core as the uranium is consumed, and emerge from the bottom as spent fuel. Since the reactors are small, they can be added to a system a few at a time, avoiding the costs, construction time, and long-range commitment of large reactors.

The PIUS design features a huge pressure vessel of concrete and steel, within which the reactor core is submerged in a deep pool of boron-containing water (fig. 12.11). As long as the primary cooling water is flowing, it keeps the borated water away from the core. If the primary coolant pressure is lost, however, the surrounding water floods the core, and the boron poisons the fission reaction. There is enough secondary water in the pool to keep the core cool for at least a week without any external power or cooling. This should be enough time to resolve the problem. If not, operators can add more water and evaluate conditions further.

The Canadian "slow-poke" is a small-scale version of the PIUS design that doesn't produce electricity but might be a useful substitute for coal, oil, or gas burners in district heating plants. The core of this "mini-nuke" is half the size of a kitchen stove and generates only one-tenth to one-one hundredth as much power as a conventional reactor. The fuel sits in a large pool of ordinary water, which it heats to just below boiling and sends to a heat exchanger. A secondary flow of hot water from the exchanger is pumped to nearby buildings for space heating. Promoters claim that a runaway reaction is impossible in this design and that it makes an attractive, cost-efficient alternative to fossil fuels. Despite a widespread aversion to anything nuclear, Switzerland and West Germany are developing similar small nuclear heating plants.

Breeder Reactors

For more than thirty years, nuclear engineers have been proposing **breeder reactors:** high-density,

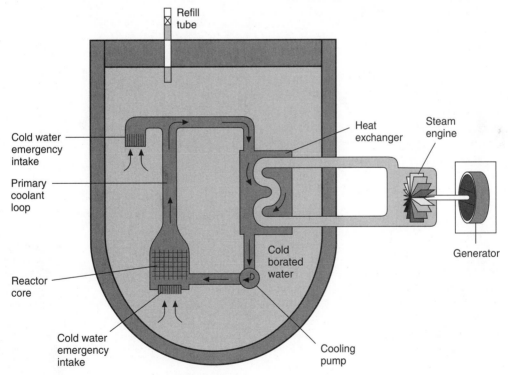

Figure 12.11 A PIUS reactor consists of a core and primary cooling system immersed in a very large pool of cold, borated water. As long as the reactor is operating, the cool water is excluded from the coling circuit by the temperature and pressure differential. Any failure of the primary cooling system would allow cold water to flood the core and shut down the nuclear reaction. The large volume of the pool would cool the reactor for several days without any replenishment.

Thinking Globally: Box 12.1

Chernobyl: The Worst Possible Accident?

*I*n the early morning hours of April 26, 1986, residents of the Ukrainian village of Pripyat saw a spectacular and terrifying sight. A glowing fountain of molten nuclear fuel and burning graphite was spewing into the dark sky through a gaping hole in the roof of the Chernobyl Nuclear Power Plant only a few kilometers away. Although officials assured them that there was nothing to worry about in this "rapid fuel relocation," the villagers knew that something was terribly wrong. They were witnessing the worst possible nuclear power accident, a "meltdown" of the nuclear fuel and rupture of the containment facilities, releasing enormous amounts of radioactivity into the environment (box fig. 12.1.1).

The accident was a result of a risky experiment undertaken by the plant engineers in violation of a number of safety rules and operational procedures. They were testing whether the residual energy of a spinning turbine could provide enough power to run the plant in an emergency shutdown if off-site power were lost. Reactor number four had been slowed down to only 6 percent of its normal operating level. To conserve the small amount of electricity being generated, they then disconnected the emergency core-cooling pumps and other safety devices, unaware that the reactor was dangerously unstable under these conditions.

The heat level in the core began to rise, slowly at first, and then faster and faster. The operators tried to push the control rods into the core to slow the reaction, but the graphite pile had been deformed by the heat so

Figure 12.1.1 The explosion and fire at the Chernobyl Nuclear Power Plant in the Soviet Ukraine in 1986 was the world's worst civilian nuclear accident so far. It marked a major turning point in the history of nuclear power.

high-pressure fission reactors that produce fuel rather than consume it. These machines create fissionable plutonium and thorium isotopes from the abundant, but nonradioactive, forms of uranium. The starting

material for this reaction is plutonium reclaimed from spent fuel from conventional fission reactors. After about ten years of operation, a breeder reactor would produce enough plutonium to start another reactor.

that the rods wouldn't go in. In 4.5 seconds, the power level rose two thousandfold, far above the rated capacity of the cooling system. Chemical explosions (probably hydrogen gas released from the expanding core) ripped open the fuel rods and cooling tubes. Cooling water flashed into steam and blew off the 1,000-ton concrete cap on top of the reactor. Molten uranium fuel puddled in the bottom of the reactor, accelerating the nuclear fission reactions. The metal superstructure of the containment building was blown apart and a column of burning graphite, molten uranium, and radioactive ashes billowed 1,000 m (3,000 ft) into the air.

Panic and confusion ensued. Officials first denied that anything was wrong. The village of Pripyat was not evacuated for thirty-six hours. There was no public announcement for three days. The first international warning came not from Soviet authorities but from Swedish scientists 2,000 km away who detected unusually high levels of radioactive fallout and traced air flows back to the southern Soviet Union.

There were many acts of heroism during this emergency. Firefighters climbed to the roof of the burning reactor building to pour water into the blazing inferno. Engineers dived into the suppression pool beneath the burning core to open a drain to prevent another steam explosion. Bus drivers made repeated trips into the contaminated area to evacuate nearby residents. Helicopter pilots hovered over the gaping maw of the ruined building to drop more than 7,000 tons of lead shot, sand, clay, limestone, and boron carbide onto the burning nuclear core to smother the fire and suppress the nuclear fission reactions. The main fire was put out within a few hours, but the graphite core continued to smolder for weeks. It wasn't finally extinguished until tunnels were dug beneath the reactor building and liquid nitrogen was injected under the core to cool it.

Altogether, more than 135,000 people were evacuated from seventy-one villages in the local area, and 250,000 children from Kiev, 80 km (50 mi) to the south, were sent on an "early summer holiday" after the accident. Several hundred people were hospitalized for radiation sickness. Thirty-one people are officially reported to have died from direct effects of radiation. Critics claimed that the total was ten times higher.

The amount of radioactive fallout varied from area to area, depending on wind patterns and rainfall. Some places had heavy doses while neighboring regions had very little. One band of fallout spread across Yugoslavia, France, and Italy. Another crossed Germany and Scandinavia. Small amounts of radiation even reached North America. Altogether, about 7 tons of fuel containing 50 to 100 million curies were released, roughly 5 percent of the reactor fuel.

Assessments of the long-term effects of this accident differ widely. The U.S. Department of Energy estimates that between 400 and 28,000 cancer deaths will occur in the Northern Hemisphere in the next fifty years as a result of Chernobyl. Critics claim that at least 100,000 deaths from cancer and genetic defects could result worldwide from Chernobyl in the next fifty years. Some areas already appear to be showing effects from Chernobyl. Leukemia rates in Minsk have doubled from 41 per million in 1985 to 93 per million in 1990.

For the present, the damaged reactor has been entombed in a giant, concrete sarcophagus that is showing signs of deterioration. Forests have been cleared and topsoil removed over wide areas to clean up radioactivity. In the worst areas, thousands of square meters of plastic film have been laid on the ground to contain radioactive dust. Both Chernobyl and Pripyat have been declared unsafe for habitation. Along with dozens of smaller towns and hundreds of farms, they have been abandoned and will be completely destroyed.

More than 600,000 people worked on decontamination and reconstruction. Those who worked in the most radioactive areas could stay there for only a few minutes before being rotated out. What health effects those workers will experience is not known. The immediate direct costs in the Soviet Union were roughly $3 billion; total costs might be one hundred times that much.

In 1990, Soviet officials revealed that contamination was much higher and more widespread than previously admitted. A program is underway to resettle 250,000 people from the Ukraine and Belarus; another 3 million people might eventually have to be moved. Other countries have filed claims in international court seeking compensation for damages from this accident. Public opinion has swung against nuclear power in many countries since the Chernobyl accident, which may have been the death knell for the nuclear power industry.

Sufficient uranium currently is stockpiled in the United States to produce electricity for one hundred years at present rates of consumption, if breeder reactors can be made to work safely and dependably.

Several problems have held back the breeder reactor program in the United States. One concern is safety. The reactor core of the breeder must be at a very high density for the breeding reaction to occur.

Water doesn't have enough heat capacity to carry away the high heat in the core, so liquid sodium is used as a coolant. Liquid sodium is very corrosive and difficult to handle. It burns with an intense flame if exposed to oxygen and explodes if it comes into contact with water. Because of its intense heat, a breeder reactor will melt down and self-destruct within a few seconds if the primary coolant is lost, as opposed to a few minutes for a normal fission reactor.

Another very serious concern about breeder reactors is that they produce excess plutonium, which is among the most toxic substances known and can be used to make bombs. It is essential to have a spent-fuel reprocessing industry if breeders are used, but the existence of large amounts of plutonium in the world would surely be a dangerous and destabilizing development. The chances of accidental spills during shipping, or of some material falling into the hands of terrorists or unstable politicians are very high. In 1993, 1.7 tons of plutonium (enough to make about 125 nuclear bombs) was shipped from a reprocessing plant in France to Japan, where it is intended to be used in a large new breeder-reactor program. Japan plans to purchase and ship 30 additional tons of this dangerous material. Worldwide protests have been launched in opposition to this plan.

RADIOACTIVE WASTE MANAGEMENT

One of the most difficult problems associated with nuclear power is the disposal of wastes produced during mining, fuel production, and reactor operation. Eventually, even the old reactors will have to be dismantled and stored somewhere. How these wastes are managed may ultimately be the overriding obstacle to nuclear power.

Enormous piles of mine wastes and mill tailings that have accumulated and now are blowing in the wind and washing into streams in mining and reprocessing areas represent another serious waste disposal problem. Production of 1,000 tons of uranium fuel typically generates 100,000 tons of tailings and 3.5 million liters of liquid waste. There now are approximately 200 million tons of radioactive waste in piles around mines and processing plants in the United States. Canada has even more radioactive mine waste on the surface than does the United States.

In addition to the leftovers from fuel production, there are about 100,000 tons of low-level waste (contaminated tools, clothing, building materials, etc.) and about 15,000 tons of high-level (very radioactive)

wastes in the United States. The high-level wastes consist mainly of spent fuel rods from commercial nuclear power plants and assorted wastes from nuclear weapons production. For the past twenty years, spent fuel assemblies from commercial reactors have been stored in deep water-filled pools at the power plants. These pools were originally intended only as temporary storage until the wastes were shipped to reprocessing centers or permanent disposal sites.

Since neither of these options now exists, utilities have been stuck holding large amounts of dangerous wastes. Space in these pools is running out and fuel rods are being packed closer together than was intended, a problem that is quickly approaching crisis proportions. High-level waste is especially dangerous since it contains high concentrations of very toxic radioactive elements, such as cesium, iodine, plutonium, and strontium, in addition to natural uranium, thorium, and radon. Since the pools are full, utilities are asking for permission to store these wastes outside their plants in large metal casks. This request is meeting with fierce opposition from people who fear that these casks will leak. Since most nuclear power plants are built near rivers, lakes, or seacoasts, a radioactive leak could quickly spread over a very large area if materials get into surface waters.

In 1987, the Department of Energy announced plans to build the first high-level waste repository at a cost of $6 billion to $10 billion on a barren desert ridge near Yucca Mountain, Nevada. Wastes are to be buried deep in the ground where it is hoped that they will remain unexposed to groundwater and earthquakes for the tens of thousands of years required for the radioactive materials to decay to a safe level (fig. 12.12). Although the area is very dry now, we can't be sure that it will always remain that way. We also don't know how long the metal canisters will contain the wastes, or where they might migrate over thousands of years.

Some nuclear experts believe that monitored, retrievable storage would be a much better way to handle wastes. This method involves holding wastes in underground mines or secure surface facilities where they can be watched. The storage sites would be kept open and inspected regularly. If canisters begin to leak, they could be removed for repacking. Safeguarding the wastes would be expensive and the sites might be susceptible to wars or terrorist attacks. We might need a perpetual priesthood of nuclear guardians to ensure that the wastes are never released into the environment.

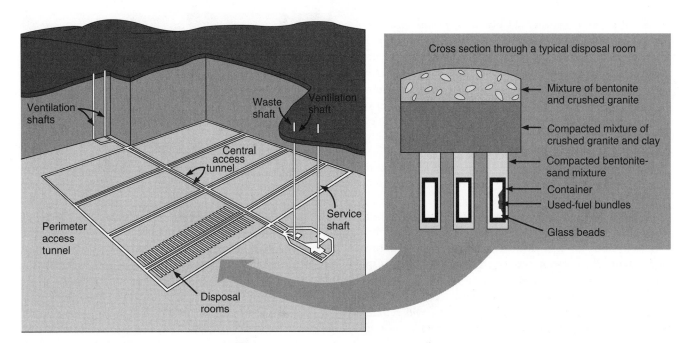

Figure 12.12 Proposal for the permanent disposal of used fuel is to bury the used fuel containers 500 to 1,000 m deep in the stable rock of the Canadian Shield. Spent fuel in the form of ceramic fuel pellets are sealed inside corrosion-resistant containers. Glass beads are compacted around the used fuel bundles and into the spaces between the fuel and the container shell. The containers are then buried in specially built vaults that are packed and sealed with special materials to further retard the migration of radioactive material.

Citizen Protests Against Nuclear Power

Opposing nuclear power has been a high priority for many conservation organizations in recent years. A federation of antinuclear alliances (e.g., Clamshell Alliance, Northern Sun Alliance) has rallied thousands of people to oppose nuclear power. Protests and civil disobedience at some sites have gone on for a decade or more, and several plants have been abandoned because there was so much opposition to them. Proponents of nuclear power argue that abandoning this energy source—especially plants that are already built—is foolish since the risks may be lower than is commonly held (fig. 12.13). Where do you stand on this issue?

Since the Chernobyl accident, public opinion in many countries has swung strongly against nuclear power. In France in 1985 at the site of the Super-Phenix breeder reactor near Lyons, more than 20,000 protesters battled police and one person was killed. In Hong Kong, more than 1 million people (20 percent of the population) signed a petition opposing the 1,200 MW reactor being built by China at Daya Bay, only 50 km from the center of this densely populated city. Hong Kong is at the

Figure 12.13 The Seabrook nuclear reactor in New Hampshire has been the site of protests and counter-protests for twenty years. Here pipefitters march in support of the plant and their jobs.

end of a narrow, rocky peninsula, so evacuation would be impossible in case of a nuclear emergency.

A few countries remain strongly committed to nuclear power. France generates 60 percent of its electricity with nuclear plants and plans to build more. The United Kingdom and Japan both depend on nuclear power for about 20 percent of their electricity. After the Chernobyl disaster, officials of the former Soviet Union declared a construction moratorium on nuclear power plants. In 1993, however, the new government of Russia announced plans to build at least thirty new nuclear stations, several of the RBMK design. Russian environmentalists denounced this plan as "unacceptable legally, ecologically, economically, and politically."

Many other countries, however, have decided to postpone building reactors, to phase out already built reactors, or simply to abandon reactors that have been built but not yet put into service. Altogether, about half the nations with nuclear power have chosen to reduce or dismantle existing plants, and few new plants have been ordered in recent years. Opponents of nuclear power claim that the nuclear fission experiment may have been a very brief episode in our history, but one whose legacy will remain with us for a long time. Proponents of nuclear power argue that it may be our only near-term alternative to fossil fuels. What do you think?

ENERGY CONSERVATION

One of the best ways to avoid energy shortages and to relieve the adverse environmental and health effects of our current energy technologies is simply to use less. Conservation offers many benefits both to society and to the environment.

Saving Energy

Many conservation techniques are relatively simple and highly cost effective. More efficient and less energy-intensive industry, transportation, and domestic practices could save large amounts of energy. Improved automobile efficiency, better mass transit, and increased railroad use for passenger and freight traffic are simple and readily available means of conserving transportation energy. In response to the 1970s oil price shocks, automobile gas-mileage averages in the United States rose from 13 mpg in 1975 to 28.8 mpg in 1988. The oil glut and falling fuel prices of the late 1980s discouraged further conservation. The 1991 average slipped to only 28.2 mpg.

Much more could be done. Prototype high-efficiency automobiles that get about 100 mpg at highway speeds and more than 65 mpg in normal city traffic already have been road tested (fig. 12.14). Raising the average fuel efficiency of the U.S. car and light truck fleet by one mile per gallon would cut oil consumption about 295,000 barrels per *day*. In one year, this would equal the total

Figure 12.14 This experimental automobile with a superefficient turbo-charged diesel engine gets over 100 mpg at a constant 40 mph and 65 mpg in city driving conditions.

amount the Interior Department hopes to extract from the Arctic National Refuge in Alaska.

Similar improvements in domestic energy efficiency have occurred in the past decade. Today's average new home uses one-half the fuel required in a house built in 1974. Energy losses can be reduced even further through better insulation, double or triple glazing of windows, thermally efficient curtains or window coverings, and by sealing cracks and loose joints. Reducing air infiltration is usually the cheapest, quickest, and most effective way of saving energy because it is the largest source of losses in a typical house. It doesn't take much skill or investment to caulk around doors, windows, foundation joints, electrical outlets, and other sources of air leakage.

For even greater savings, new houses can be built with extra thick super-insulated walls, air-to-air heat exchangers to warm incoming air, and even double-walled sections that create a "house within a house." Special double-glazed windows that have internal reflective coatings and that are filled with an inert gas (argon or xenon) have an insulation factor of R11, the same as a standard 4-inch-thick wall or ten times as efficient as a single-pane window. Superinsulated houses now being built in Sweden require 90 percent less energy for heating and cooling than the average American home.

Orienting homes so that living spaces have passive solar gain in the winter and are shaded by trees or roof overhang in the summer also helps conserve energy. Earth-sheltered homes built into the south-facing side of a slope or protected on three sides by an earth berm are exceptionally efficient energy savers because they maintain relatively constant subsurface temperatures. In addition to building more energy-efficient homes, there are many personal actions we can take to conserve energy (box 12.2).

Utility Conservation Programs

Utility companies are finding it much less expensive to finance conservation projects than to build new power plants. Pacific Gas and Electric in California and Potomac Power and Light in Washington, D.C., both have instituted large conservation programs. They have found that conservation costs about $350 per kilowatt (kw) saved. By contrast, a new nuclear power plant costs between $3,000 and $8,000 per kw of installed capacity. New coal-burning plants with the latest air pollution-control equipment cost at least $1,000 per kw. By investing $200 to $300 million in public education, home improvement loans, and other efficiency measures, a utility can avoid building a new power plant that would cost a billion dollars or more. Furthermore, conservation measures don't consume expensive fuel or produce pollutants.

Can application of this approach help alleviate energy shortages in other countries as well? Yes. Brazil, for example, could cut its electricity consumption an estimated 30 percent with an investment of 10 billion U.S. dollars. It would cost $44 billion to build new power plants to produce that much electricity. South Korea has instituted a comprehensive conservation program with energy-saving building standards, efficiency labels on new household appliances, depreciation allowances, reduced tariffs on energy-conserving equipment, and loans and tax breaks for upgrading homes and businesses. It also forbids some unnecessary uses, such as air conditioning and elevators between the first and third floors.

SOLAR ENERGY

The average amount of solar energy arriving at the top of the atmosphere is 340 watts per square meter. About half of this energy is absorbed or reflected by the atmosphere (more at high latitudes than at the equator), but the amount reaching the earth's surface is some 8,000 times all the commercial energy used each year. Although solar energy is diffuse and relatively low-temperature, there are many ways we can collect and use it.

Passive Solar Heat Collectors

Our simplest and oldest use of solar energy is **passive heat absorption,** using natural materials or absorptive structures with no moving parts to simply gather and hold heat. A glass-walled "sunspace" or greenhouse on the south side of a building, for instance, can both capture energy and provide an enjoyable living space (fig. 12.15). Incorporating massive energy-storing materials, such as brick walls, stone floors, or barrels of heat-absorbing water into buildings also collects heat to be released slowly at night.

Active Solar Heat Systems

Active solar systems generally pump a heat-absorbing, fluid medium (air, water, or an antifreeze solution) through a relatively small collector, rather than passively collecting heat in a stationary medium like masonry. Active collectors can be located adjacent to or on top of buildings rather than being built into the structure. Because they are relatively small and structurally independent, active systems can be retrofitted to existing buildings.

A flat black surface under a glass cover makes a good solar collector. A fan circulates air over the hot surface and into the house through ductwork of the type used in standard forced-air heating. Alternatively, water can be pumped through the collector to pick up heat for space heating or to provide hot water. Water heating consumes 15 percent of the United States' domestic energy budget, so savings in this area alone can be significant. A simple flat panel with about 5 sq m of surface can reach 95°C (200°F) and can provide enough hot water for an average family of four almost anywhere in the United States. In California, 650,000 homes now heat water with solar collectors. In Greece, Italy, Israel, Asia, and Africa where fuels are more expensive, up to 70 percent of domestic hot water comes from solar collectors.

Sunshine doesn't reach us all the time, of course. How can solar energy be stored for times when it is needed? There are a number of options. In a climate where sunless days are rare and seasonal variations are slight, a small, insulated water tank is a good solar energy storage system. For areas where clouds block the sun for days at a time or where energy must be stored for winter use, a large, insulated bin containing a heat-storing mass, such as stone, water, or clay, provides solar energy storage (fig. 12.15). During the summer months, a fan blows the heated air from the collector into the storage medium. In the winter, a similar fan at the opposite end of the bin blows warm air into the house. During the summer, the storage mass is cooler than the outside air, and it helps cool the house by absorbing heat. During the winter, it is warmer and acts as a heat source by radiating stored heat.

Personal Energy Efficiency: What Can You Do?

*F*or some people, home energy efficiency means designing and building a new house that incorporates proven energy conservation methods and materials. Most of us, however, have to live in houses or apartments that already exist. How can we practice home energy conservation?

Think about the major ways we use energy. The biggest home use is space heating. Heat conservation is among the simplest, cheapest, and most effective ways to save energy in the home. Easy, surprisingly effective, and relatively inexpensive measures include weatherstripping, caulking, and adding layers of plastic to windows. Insulating walls, floors, and ceilings increases the energy conservation potential. Simply lowering your thermostat, especially at night, is a proven energy and money saver.

Water heating is the second major user of home energy, followed by large electric or gas appliances, such as stoves, refrigerators, washing machines, and dryers. Careful use of these appliances can save significant amounts of energy. Consider using less hot water, lowering the thermostat of your water heater, and buying an insulating blanket designed to wrap around it. Be sure that the refrigerator door doesn't stand open, and defrost it regularly (if it is not frost-free) to reduce the ice buildup that prevents the heat exchange system from working well. Make sure your oven door has a tight seal, and plan ahead when you use it; bake several things at once rather than reheating it repeatedly. Wash your clothes in cold or cool water rather than hot. Air dry your laundry, especially in the summer when sun-dried clothes are so pleasant to wear.

Some energy-efficient appliances can save substantial amounts of energy. Air conditioners and refrigerators with better condensers and heat exchangers use one-half to one-fourth as much energy as older models. New, improved furnaces can offer 95 percent efficiency, com-

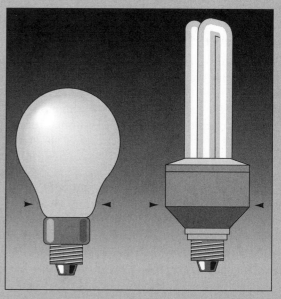

Figure 12.2.1 Comparison of an ordinary incandescent bulb with a new, compact fluorescent lamp. Both have standard screw-in bases, but the fluorescent ballast (arrows) is wider than the base and may require a socket "extender" to fit in some fixtures. A 15-watt compact fluorescent bulb puts out as much light as a 60-watt incandescent bulb and lasts about ten times as long.

pared with conventional 70 percent efficiency in older models. The pay-back period may be as little as two to three years if you trade an old wasteful appliance for a newer, more efficient one. Some light bulbs are made to save energy (box fig. 12.2.1). High-efficiency fluorescent lights emit the same amount of light but use only one-fourth as much energy as conventional incandescent bulbs. They cost ten times more than an ordinary bulb, but last ten times as long. Total lifetime savings can be $30 to $50 per lamp. You will also avoid producing about 400 kg (880 lbs) of carbon dioxide over the life of the bulb. Almost all types of appliances, large and small,

High-Temperature Solar Energy Collection

Parabolic mirrors focus light onto a single, central point (fig. 12.16). Use of parabolic reflectors to focus intense heat on a central tube containing air, water, or antifreeze produces a much higher quality (high temperature) heat than does the basic flat panel collector. Temperatures in the collection medium can reach 500°C (1,000°F).

But wouldn't it take a huge land area to build solar collectors? Not really. The entire present United States electrical output could be produced on 2,000 sq km (an area about 28 mi on a side). This is less land than would be strip-mined in a thirty-year period if all our energy came from coal or uranium. And we can put solar collectors wherever we choose (such as lands unsuited for agriculture, grazing, or habitation), whereas strip-mining

1. Drive less: make fewer trips; use telecommunications and mail instead of going places in person.
2. Use public transportation, walk, or ride a bicycle.
3. Use stairs instead of elevators.
4. Join a carpool or drive a smaller, more efficient car; reduce speeds.
5. Insulate your house or add more insulation to the existing amount.
6. Turn thermostats down in the winter and up in the summer.
7. Weatherstrip and caulk around windows and doors.
8. Add storm windows or plastic sheets over windows.
9. Create a windbreak on the north side of your house; plant deciduous trees or vines on the south side.
10. During the winter, close windows and drapes at night; during summer days, close windows and drapes if using air conditioning.
11. Turn off lights, television sets, and computers when not in use.
12. Stop faucet leaks, especially hot water.
13. Take shorter, cooler showers; install water-saving faucets and shower heads.
14. Recycle glass, metals, and paper; compost organic wastes.
15. Eat locally grown food in season.
16. Buy locally made, long-lasting materials.

are available in efficient models and brands; we simply need to shop for them.

The easiest way to save energy is to keep an eye on energy-use habits. Turning off lights, televisions, computers, and other devices when they are not in use is elementary. Another large opportunity to save energy is in transportation, which accounts for about one-third of our consumption. Car-pooling to work or school typically saves about 3,800 l (1,000 gal) of gasoline per year. A car that gets 40 mpg rather than 20 mpg will save another 1,000 l (nearly 250 gal) per year. Driving 55 mph rather than 65 mph will reduce your consumption by about 25 percent. Best of all, walk or ride a bike on short trips: it's good for you and the environment.

In some cases, the most beneficial result of living a frugal, conservative life is not so much the total amount of energy you save as the effect that this lifestyle has on you and the people around you. When you live conscientiously, you set a good example that could have a multiplying effect as it spreads through society. Making conscious ethical decisions about this one area of your life may stimulate you to make other positive decisions. You will feel better about yourself and more optimistic about the future when you make even small, symbolic gestures toward living as a good environmental citizen.

occurs wherever coal or uranium exist, regardless of other values associated with the land.

Photovoltaic Solar Energy Conversion

The photovoltaic cell offers an exciting potential for capturing solar energy in a way that will provide clean, versatile, renewable energy. This simple device has no moving parts, negligible maintenance costs,

produces no pollution, and has a lifetime equal to that of a conventional fossil fuel or nuclear power plant.

Photovoltaic cells capture solar energy and convert it directly to electrical current by separating electrons from their parent atoms and accelerating them across a one-way electrostatic barrier formed by the junction between two different types of semiconductor material (fig. 12.17). The first solar cells were made for the United States space program in 1958

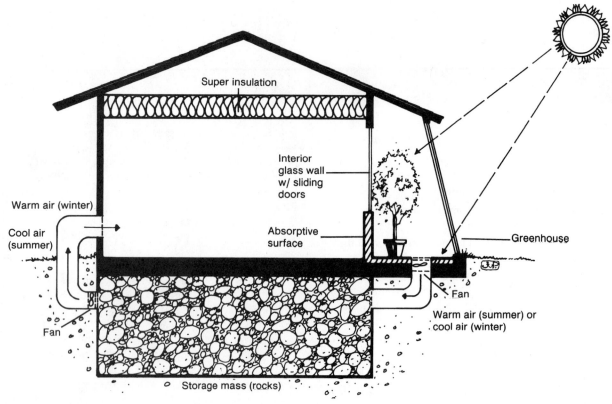

Figure 12.15 Underground massive heat storage unit. Heated air collected behind double- or triple-glazed windows is pumped down into a storage medium of rock, water, clay, or similar material, where it can be stored for a number of months.

Figure 12.16 Parabolic mirrors concentrate sunlight onto a central tube where a collecting fluid is heated to high temperatures. Steam generated by many such collectors drives an electric generator to supply municipal and industrial energy.

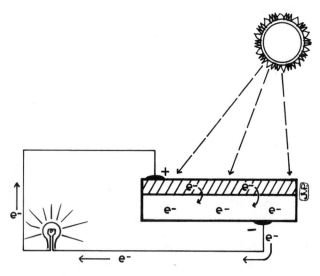

Figure 12.17 The operation of a photovoltaic cell. Boron impurities incorporated into the upper silicon crystal layers cause electrons (e-) to be released when solar radiation hits the cell. The released electrons move into the lower layer of the cell, thus creating a shortage of electrons, or a positive charge, in the upper layer and an oversupply of electrons, or negative charge, in the lower layer. The difference in charge creates an electric current in a wire connecting the two layers.

and cost $2,000 per peak watt of output. This was more than two thousand times as much as conventional energy at the time, but it was the only way to provide a renewable source of energy in outer space. Prices have fallen by a factor of ten each decade since then. By 1990 they were about $5 per watt, which makes solar energy cost-competitive with other sources in some areas.

Within a few years, photovoltaic cells could cost less than $1 per watt of generating capacity. With 15 percent efficiency and a thirty-year life, they should be able to produce electricity for around 8 cents per kilowatt hour. At that time, coal-fired steam power will probably cost about four times as much and nuclear power will likely cost ten times as much as photovoltaic cell energy.

During the past twenty-five years, the efficiency of energy capture by photovoltaic cells has increased from less than 1 percent of incident light to more than 10 percent under field conditions and more than 30 percent in the laboratory. Promising experiments are underway using exotic metal alloys, such as gallium arsenide, cadmium telluride, and silicon germanium, which are more efficient in energy conversion than silicon crystals.

You probably already use silicon photovoltaic cells. They are being built into light-powered calculators, watches, toys, photosensitive switches, and a variety of other consumer products. Japan has made this technology a priority for research and development. Already, home-roof arrays are available that provide all the electricity needed for a typical home at prices competitive with power purchased from a utility. By the end of this century, Japan plans to meet a considerable part of its energy requirements with solar power, an important goal for a country lacking fossil and nuclear fuel resources.

Think about how solar power could affect your future energy independence. Imagine the benefits of being able to build a house anywhere and having a cheap, reliable, clean, quiet source of energy with no moving parts to wear out, no fuel to purchase, and little equipment to maintain. You could have all the energy you need without commercial utility wires or monthly energy bills. Coupled with modern telecommunications and information technology, an independent energy source would make it possible to live wherever you want and yet have many of the employment and entertainment opportunities and modern conveniences available in a major city.

Storing Electrical Energy

Electrical energy is difficult and expensive to store. This is a problem for photovoltaic generation as well as other sources of electric power. Traditional lead-acid batteries are heavy and have low energy densi-

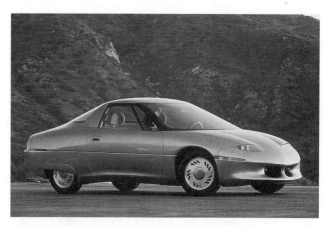

Figure 12.18 The General Motors Impact uses conventional lead-acid batteries and efficient electric motors to achieve freeway speeds. It can travel 120 miles at 55 mph before needing recharging.

ties; that is, they can store only moderate amounts of energy per unit mass or volume. A typical lead-acid battery array sufficient to store several days of electricity for an average home would cost about $5,000 and weigh 3 or 4 tons. All the components for an electric car are readily available except a cheap, lightweight, long-lasting battery. Several automobile manufacturers have experimental cars with excellent performance characteristics (fig. 12.18), but a breakthrough in battery technology is badly needed.

Other batteries exist, but all have drawbacks. Metal-gas batteries, such as the zinc-chloride cell, use inexpensive materials and have relatively high energy densities but have shorter lives than other types. Sodium-sulfur batteries have considerable potential for large-scale storage, holding twice as much energy in half as much weight as lead-acid batteries. They require an operating temperature of about 300° C (572° F), however, and are expensive to manufacture. Alkali-metal batteries have a high storage capacity but are even more expensive. Lithium batteries have very long lives and store more energy than other types, but are the most expensive of all.

Another strategy is to store energy in a form that can be turned back into electricity when needed. Surplus electricity can be used to separate water into hydrogen and oxygen gas. These gases can be liquefied (like natural gas) at very low temperatures, making them easier to store and ship than many forms of energy. They are highly explosive, however, and must be handled with great care. They can be burned in internal combustion engines, producing mechanical energy, or they can be used as fuel in fuel cells to produce more electrical energy. Hydrogen-fueled cars are already being produced. They could be very attractive in smoggy cities because they produce no carbon monoxide, smog-forming hydrocarbons, carcinogenic chemicals, or soot.

Figure 12.19 The firewood shortage in less developed countries means that women and children must spend hours each day searching for fuel. Destruction of forests and removal of ground cover result in erosion and desertification.

BIOMASS

Wood fires have been a primary energy source of heating and cooking for thousands of years. As recently as 1850, wood supplied 90 percent of the fuel used in the United States. Wood now provides less than 1 percent of the energy in most developed countries, but in many of the poorer countries of the world, wood and other biomass fuels provide up to 95 percent of all energy used (fig. 12.19). The 1,500 million cubic meters of fuelwood collected in the world each year is about half of all wood harvested and is a major cause of forest destruction.

In northern industrialized countries, wood burning has increased since 1975 in an effort to avoid rising oil, coal, and gas prices. Most of these northern areas have adequate wood supplies to meet demands at current levels, but problems associated with wood burning may limit further expansion in this use. Inefficient and incomplete burning of wood in open fireplaces and stoves produces smoke laden with fine ash and soot and hazardous amounts of carbon monoxide (CO) and hydrocarbons. In valleys where inversion layers trap air pollutants, the effluent from wood fires can be a major source of air quality degradation and health risk. Polycyclic aromatic compounds produced by burning are especially worrisome because they are carcinogenic (cancer-causing).

Highly efficient and clean-burning woodstoves are available but expensive. Brick-lined fireboxes with after-burner chambers to combust gaseous hydrocarbons do an excellent job of conserving wood while also reducing air pollution. Catalytic burners, similar to the pollution controls on cars, can be placed inside stove pipes. They burn carbon monoxide and hydrocarbons to clean emissions and to recapture heat that otherwise would escape out the chimney as unburned molecules.

Wood chips, sawdust, wood residue, and other plant materials are being used in some places in the United States and Europe as a substitute for coal and oil in industrial boilers. In Vermont, for instance, where fossil fuels are expensive and three-quarters of the land is covered by forest, 250,000 cords of unmarketable cull wood are burned annually to fuel a 50 MW power plant in Burlington. Michigan's Public Service Commission estimated that the state's surplus forest growth could provide 1,300 MW of electricity each year.

Pollution-control equipment is easier to install and maintain in a central power plant than in individual home units. Furthermore, wood burning also contributes less to acid precipitation than does coal. Because wood has little sulfur and burns at lower temperatures than coal, it produces minimal sulfuric or nitric acid. Burning wood grown as a renewable crop doesn't produce any net increase in atmospheric carbon dioxide (and, therefore, doesn't add to global warming) because all the carbon released by burning biomass was taken up from the atmosphere when the biomass was grown.

Methane from Biomass

Methane gas is a "natural" gas. It is produced by anaerobic (oxygen-free) decomposition of any moist organic material. This process occurs in swamps and rice paddies where surface waters prevent oxygen from reaching mud-dwelling bacteria. In the absence of oxygen, the bacteria produce flammable methane instead of carbon dioxide. An inexpensive, homemade **methane digester** (fig. 12.20) can produce methane very cheaply. You simply put any kind of organic wastes—manure, grass clippings, food waste, etc.—in a sealed container and provide warmth and water. Bacteria are ubiquitous enough to start the culture spontaneously.

Burning methane produced from manure provides more heat than burning the dung itself, and the sludge left over from bacterial digestion is a rich fertilizer, containing healthy bacteria as well as most of the nutrients originally in the manure. Elevated temperatures during digestion also eliminate fecal pathogens and parasites contained in human or animal wastes.

Landfills are active sites of methane production, contributing as much as 20 percent of the annual methane emissions to the atmosphere. This is a waste of a valuable resource and a threat to the environment

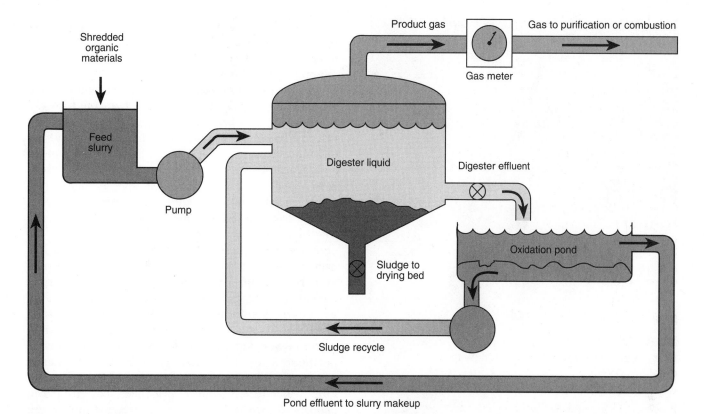

Figure 12.20 Continuous unit for converting organic material to methane by anaerobic fermentation. One kilogram of dry organic matter will produce 1–1.5 m^3 of methane, or 2,500–3,600 million calories per metric ton.
Source: *Solar Energy as a National Energy Resource,* NSF/NASA Solar Energy Panel, National Science Foundation, December 1972.

because methane absorbs infrared radiation and contributes to the greenhouse effect. Some municipalities are drilling gas wells into landfills and garbage dumps. Sewage treatment plants routinely use anaerobic digestion as a part of their treatment process, and many facilities collect the methane they produce and use it to generate heat or electricity or both for their operations. Although this technology is well developed, its utilization could be much more widespread.

Alcohol from Biomass

Ethanol (grain alcohol) and methanol (wood alcohol) are produced by anaerobic digestion of plant materials with high sugar content, mainly grain and sugar cane. Ethanol can be burned directly in automobile engines adapted to use this fuel, or it can be mixed with gasoline (up to about 10 percent) to be used in any normal automobile engine. A mixture of gasoline and ethanol is often called **gasohol.** Ethanol in gasohol raises octane ratings and is a good substitute for lead antiknock agents, the major cause of lead pollution. Furthermore, since ethanol contains more oxygen than normal gasoline, it produces fewer unburned hy-

drocarbons. In 1992, the EPA began requiring cities that fail to comply with ozone or carbon monoxide standards to supply "oxygenated fuels" during winter months. Changing over to gasohol can cause problems in some cars because it loosens gasoline residues that collect in the gas tank, gas line, and engine. Engines can be adjusted, however, to run just as well on gasohol or even straight ethanol as on other fuels.

Ethanol production could be a solution to grain surpluses and could bring a higher price for grain crops than the food market offers. It could lead to reduced dependence on gasoline, which is refined from petroleum. Brazil has instituted an ambitious national program to substitute crop-based ethanol for imported petroleum. Brazilian sugar now produces some 3 billion gallons of ethanol, and authorities plan to export much of this to the United States.

HYDROPOWER

Watermills and waterwheels have been used to grind grain, pump water, saw lumber, and do other useful chores for at least a thousand years. By 1925, hydropower dams generated 40 percent of the world's

Figure 12.21 Hydropower dams like this one in New South Wales, Australia, can prevent floods and provide useful energy, but often they destroy valuable farmland, forests, villages, and historic areas when valleys are flooded for water storage. In addition, dams can also lose large amounts of water to seepage and evaporation, or they can fail and thus pose a threat to downstream areas.

electric power. Since then, hydroelectric production capacity has grown fifteenfold, but fossil fuel use has risen so rapidly that hydropower is now only one-quarter of total electrical generation. Still, many countries produce most of their electricity from falling water. Norway, for instance, depends on hydrogeneration for 99 percent of its electricity; Brazil, 87 percent; New Zealand and Switzerland, 75 percent; and Canada and Austria, 66 percent.

Much of the hydropower development in recent years has been in enormous dams (fig. 12.21). There is a certain efficiency of scale in giant dams, and they bring pride and prestige to the countries that build them, but they can have unwanted social and environ-

mental effects as well. The Narmada Valley project in India, for instance, will drown 150,000 ha of tropical forest and displace 1.5 million people, mostly tribal minorities and low-caste hill people. China's Three Gorges project, now under construction, will displace 1.4 million people. The Akosombo Dam built on the Volta River in Ghana nearly twenty years ago displaced 78,000 people, few of whom ever found another place to settle. Those still living remain in refugee camps and temporary shelters two decades later.

Fortunately, there are alternatives to gigantic dams and destructive impoundment reservoirs. Small-scale, **low-head hydropower** technology can extract energy from small headwater dams that cause much less damage. Some modern, high-efficiency turbines can even operate on **run-of-the-river flow.** Submerged directly in the stream and small enough to not impede navigation in most cases, these turbines don't require a dam or diversion structure and can generate useful power with a current of only a few kilometers per hour. They don't interfere with fish movements, including spawning migration. **Micro-hydro generators** operate on similar principles but are small enough to provide economical power for four to six homes. Where enough rain falls to keep rivers running all year, small systems such as these enable a few families to form their own power cooperative, freeing them from dependence on large utilities and foreign energy supplies.

WIND ENERGY

The air surrounding the earth is essentially a huge storage battery for solar energy. Wind currents are the result of the uneven heating of air over land and water in equatorial zones and polar regions. The World Meteorological Organization has estimated that 20 million MW of wind power could be commercially tapped worldwide, not including contributions from windmills at sea. This is about fifty times the total present world nuclear generating capacity.

Windmills played a crucial role in the settling of the American West. The strong, steady, prairie winds provided energy to pump water that allowed agriculture to move out onto the plains. At the beginning of this century, nearly every farm or ranch west of the Mississippi River had at least one windmill. This promising technology was nipped in the bud, however, by passage of the Rural Electrification Act of 1935, which brought many benefits to rural America but effectively killed wind-power development. It is interesting to speculate what the course of history

Figure 12.22 A wind farm near San Gregorio, California. This utility-scale use of wind energy has been especially successful in California, but the Great Plains has an even larger potential for wind power than do the mountain passes of the Sierras.

might have been if we had not spent trillions of dollars on fossil fuel and nuclear energy but instead had invested that money on small-scale, renewable energy systems.

As the world's conventional fuel prices rise, interest in wind energy is resurging. Pilot projects already have demonstrated the economy of wind power. Theoretically 60 percent efficient, windmills typically produce at 35 percent efficiency under field conditions. General Electric estimates the wind could provide 14 percent of U.S. electrical demand by the year 2000 and could save 3.7 million barrels of oil per year. At least 500,000 sq km of land in the United States and Canada is suitable for wind generators (average wind speeds in excess of 25 km/hr) and is free of conflicting uses, such as cities and national parks.

Wind farms are large-scale public utility efforts to take advantage of wind power. The first major wind farm started in New Hampshire in 1981, but California quickly took the lead in wind farm development. Some 16,000 windmills now stand in rows along the crest of California's Altamont, Tehachapi, and San Gorgino passes (fig. 12.22). These windfarms produce well over 1,500 MW—the equivalent of three large nuclear reactors—or 95 percent of U.S. wind energy and 75 percent of the world's production.

Construction of wind farms is not limited to mountain ridges. The Great Plains are legendary for their strong, incessant winds. One of my colleagues claims that the prairie states and provinces could be the OPEC of wind power. Sea coasts also provide excellent wind farm sites. Offshore wind farms have been proposed, with large generators anchored a short distance out at sea where competition for real estate development is low and winds are high year-round. Collected energy would probably be stored in electrolyzed water (discussed earlier as storage for photovoltaic power) for transportation to shore.

Are there negative impacts of wind farms? They generally occupy places with wind and weather too severe for residential or other development. The wind speed at Altamont, for instance, averages 35 km/hr (22 mph) from March to September. Livestock graze undisturbed beneath the windmills. Wind farm opponents complain, however, of noise pollution (the low, rhythmic whoosh of the blades) and visual pollution in remote areas of great natural beauty. Windmills also can be a wildlife hazard when hawks, eagles, and other birds fly into the whirling blades. If California condors are ever released again into the wild, they may be threatened by windmills on their favorite mountain passes.

When a home owner or community invests independently in wind generation, the same question arises as with solar energy: What should be done about energy storage when electricity production exceeds use? Besides the storage methods mentioned earlier, many private electricity producers believe the best use for excess electricity is in cooperation with the public utility grid. When private generation is low, the public utility runs electricity through the meter and into the house or community. When the wind generator or photovoltaic systems overproduce, the electricity runs back into the grid and the meter runs backward. Ideally, the utility reimburses individuals for this electricity, for which other consumers pay the company. The Public Utilities Regulatory Policies Act requires utilities to buy power generated by small hydro, wind, cogeneration, and other privately owned technologies at a fair price. Not all utilities yet comply, but some—notably in California, Oregon, Maine, and Vermont—are purchasing significant amounts of private energy.

These alternative energy sources may not be suitable for every situation, however. The greatest security and flexibility undoubtedly will come from developing a wide range of options from which to choose. We clearly need innovative and courageous leadership from government, industry, and private citizens to bring these new energy options to the marketplace as quickly as possible.

Summary

Today, nearly 95 percent of all commercial energy is generated by fossil fuels: petroleum, coal, and natural gas (methane). Petroleum and natural gas were not used in large quantities until the beginning of this century, but supplies are already running low. Coal supplies will last several more centuries at present rates of usage, but it appears that the fossil fuel age will have been a rather short episode in the total history of humans. Nuclear power provides only about 2.5 percent of commercial energy worldwide.

The environmental damage caused by mining, shipping, processing, and using fossil fuels may necessitate cutting back on our use of these energy sources. Coal combustion is a major source of acid precipitation. It also releases heavy metals, toxic organic chemicals, and carbon dioxide.

Nuclear energy offers an alternative to many of the environmental and social costs of fossil fuels, but it introduces serious problems of its own. In the 1950s, there was great hope that these problems would be overcome and that nuclear power plants would provide energy "too cheap to meter." Recently, however, much of that optimism has been waning. Many countries are closing down existing nuclear power plants, and a growing number have pledged to remain or become "nuclear-free."

The greatest worry about nuclear power is the danger of accidents that might release the extremely hazardous radioactive materials produced in the nuclear fission reactions into the environment. The great danger represented by even small amounts of radioactivity means that many thousands of people could die from radiation exposure, cancer, and genetic defects as a result of a major nuclear power accident. Where we will put nuclear wastes and how we can ensure that they will remain safely contained for the thousands of years required for "decay" to nonhazardous levels remain serious worries. Yucca Mountain, Nevada, has been chosen for a high-level waste repository, but many experts believe that burying these toxic residues in nonretrievable storage is a mistake.

Several sustainable energy sources could reduce or eliminate our dependence on fossil fuels and nuclear energy. Exciting new technologies have been invented to use renewable energy sources. Active solar collectors can be useful in air and water heating, for instance. One of the most promising technologies is direct electricity generation by photovoltaic cells. Since solar energy is available everywhere, photovoltaic collectors could provide clean, inexpensive, nonpolluting, renewable energy, independent of central power grids or fuel-supply systems.

Biomass also may have some modern applications. In addition to direct combustion, biomass can be converted into methane or ethanol, which are clean-burning, easily storable, and transportable fuels. These alternative uses of biomass also allow nutrients to be returned to the soil and help reduce our reliance on expensive, energy-consuming artificial fertilizers.

Although conventional and alternative energy sources offer many attractive possibilities, conservation often is the least expensive and easiest solution to energy shortages. Even basic conservation efforts, such as turning off lights, can save large amounts of energy when practiced by many people. More major conservation methods, such as home insulation and energy-efficient appliances and transportation, can drastically reduce energy consumption and similarly reduce energy expenses. Our natural resources, our environment, and our pocketbooks all benefit from careful and efficient energy consumption.

Review Questions

1. What percentage of our energy is supplied by oil, gas, coal, nuclear power, and sustainable sources?
2. What are the three major categories of energy consumption in the United States? How much does each use?
3. How does our energy use compare to that of people in less-developed countries?
4. How many years will our current proven-in-place reserves of oil, gas, and coal last?
5. Describe how a nuclear power plant works as well as why they are dangerous.
6. What happened at Chernobyl?
7. How much energy could we save through conservation? Give some specific examples.
8. How does a photovoltaic collector work? How might we use them in the future?
9. Describe three methods for storing solar energy. Why is hydrogen a particularly useful fuel?
10. Summarize the potential and problems of hydropower, wind, and biomass as energy sources.

Questions for Critical or Reflective Thinking

1. Many statements in this chapter are prefaced by claims such as "many experts believe . . ." or "some people argue that. . . ." Choose a few such statements about which you are skeptical and list some reasons why you believe the experts might be wrong. What evidence would you need to be convinced?

2. Suppose that a nuclear power plant has been proposed for your neighborhood. What safeguards would you demand? Is there any way to be sure you are safe—or safe enough? What constitutes safe enough?

3. Are there any issues in this chapter in which your views have changed as a result of what you have learned? What critical information influenced you?

4. Many energy concerns involve "unknown-unknowns." (Remember our discussion of this in chapter 5?) Give some examples of unknown-unknown energy resources or problems. How can we plan for them?

5. We depend on private industry and government agencies for most of our information about energy options. How reliable do you believe these sources are? Are there reasons that they might deliberately overestimate or underestimate the resources available?

6. Picture a scenario in which you are living on half as much energy as you now use. What adjustments would you have to make? What adjustments would you want society to make? What advantages or disadvantages do you see in reduced personal energy consumption?

7. Do we have an obligation to share energy resources more equitably with people in other countries or to leave more for future generations? How shall we decide on a fair and just distribution of resources?

Key Terms

active solar systems 263
black lung disease 251
breeder reactors 257
control rods 256
fossil fuels 248
gasohol 269
low-head hydropower 270
methane digester 268
micro-hydro generators 270

parabolic mirrors 264
passive heat absorption 263
photovoltaic cells 265
pressurized water reactor (PWR) 256
run-of-the-river flow 270
strip-mining (surface-mining) 252
unconventional oil 253
wind farms 271

Suggested Readings

Chiles, J. R. "Tomorrow's Energy Today." *Audubon,* January 1990, 59. A good overview of renewable energy in the United States.

Corcoran, E. "Cleaning Up Coal." *Scientific American* 264, no. 5 (May 1991): 106. Novel market-based approaches to reducing air pollution from coal combustion are described.

Davis, G. R. *Energy for Planet Earth.* New York: Freeman, 1991. Reading on energy from the *Scientific American.* An excellent overview of our global energy resources and how we can achieve a sustainable relationship between energy use and the environment.

Drostrovsky, I. "Chemical Fuels from the Sun." *Scientific American* 265, no. 6 (December 1991): 102. A good discussion of syngas, synthetic fuels such as carbon monoxide and hydrogen, as a way to store and ship solar energy.

EPRI. "New Push for Energy Efficiency." *Electric Power Research Institute (EPRI),* April/May 1990. 4-17. Conservation from the utility perspective.

Flavin, C. "Reassessing Nuclear Power: The Fallout from Chernobyl." *Worldwatch Paper 75.* Washington, D.C.: Worldwatch Institute, 1987. A useful analysis of the consequences of the Chernobyl accident on nuclear power.

Fulkerson, W., et al. "Energy from Fossil Fuels." *Scientific American* 263, no. 3 (September 1990): 128. One of a series of articles in a special issue on energy. Also reprinted in book form.

Golay, M. W., and N. E. Todreas. "Advanced Light-Water Reactors." *Scientific American* 262, no. 4 (1989): 82. Discusses new passive safety features that can make nuclear energy safer and more attractive.

Goldsmith, E., et al. "Chernobyl: The End of Nuclear Power?" *The Economist* 16, no. 4/5 (1986): 138-209. Nuclear power from the European perspective. Useful articles on SuperPhenix, British nuclear power, ocean dumping of nuclear waste, contamination of the Irish Sea, the Windscale fire, and alternatives to nuclear energy.

Gray, C. L., Jr. and J. A. Alson. "The Case for Methanol." *Scientific American* 261, no. 5 (November 1989): 108. The authors maintain that a move to pure methanol fuel would reduce vehicular emissions of hydrocarbons and greenhouse gases and could lessen U.S. dependence on foreign energy sources.

Hamakawa, Y. "Photovoltaic Power." *Scientific American* 256, no. 4 (April 1987): 86. An excellent discussion of photovoltaics and why they work.

Häfele, W. "Energy from Nuclear Power." *Scientific American* 263, no. 3 (September 1990). Nuclear power could supply much needed energy, this author claims, if safe reactors are designed and security and waste storage problems are solved.

Hirst, E. "Boosting Energy Efficiency through Federal Action." *Environment* 33, no. 2 (March 1991): 6. Describes some actions we could take to increase our energy efficiency.

Lester, R. K. "Rethinking Nuclear Power." *Scientific American* 254, no. 3 (March 1986): 31. A good description of how a new generation of low-power, centrally fabricated nuclear reactors could be designed for inherent safety.

Louvins, A. B. "The Negawatt Revolution." *Across the Board,* September 1990, 18-23. Existing conservation technology could save three-fourths of all electricity we use today.

Moretti, P., and L. Divone. "Modern Windmills." *Scientific American* 254, no. 6 (June 1986): 110. Excellent introduction to modern wind technology and where it comes from.

Patterson, W. C. *The Plutonium Business and the Spread of the Bomb.* San Francisco: Sierra Club Books, 1984. Shows how nuclear fuel reprocessing creates a plutonium economy and provides fissionable material to countries or other groups that might want to make a bomb.

Reisner, M. "The Rise and Fall and Rise of Energy Conservation. *Amicus Journal* 9, no. 2 (Spring 1987): 22. A very readable survey of American attitudes about energy conservation.

Rosenfeld, A. H., and D. Hafernlester. "Energy-Efficient Buildings." *Scientific American* 258, no. 4 (April 1988): 78. Highly efficient homes and offices will slash energy bills and avoid building new power plants.

Schaeffer, J., et al. *Real Goods Alternate Energy Source Book.* Ukiah, Calif.: 1992. Real Goods Trading Corp. Products, graphics, instructions for setting up alternate energy and energy conservation systems.

Weinberg, C. J., and R. H. Williams. "Energy from the Sun." *Scientific American* 263, no. 3 (September 1990): 146. Describes how advances in wind, solar, thermal, and biomass technologies will soon render them cost-competitive with gasoline and coal-generated electricity.

World Resources Institute. *World Resources 1992-93.* Washington D.C.: World Resources Institute, 1992. An excellent compendium of data and case studies concerning global resources and the environment.

PART IV: TOWARD A SUSTAINABLE FUTURE

Shall we not learn from life its laws, dynamics, balances?
Learn to base our needs not on death, destruction, waste, but on
renewal? In wisdom and in gentleness learn to walk again with Eden's angels?
Learn at last to shape a civilization in harmony with the earth?

This Is the American Earth
Ansel Adams and Nancy Newhall

13
SOLID, TOXIC, AND HAZARDOUS WASTE

Wastes are only raw materials we're too stupid to use.

Arthur C. Clarke

Objectives

After studying this chapter, you should be able to:

- Identify the major components of the waste stream and describe how wastes have been—and are being—disposed of in North America and around the world.

- Explain how incinerators work, as well as the advantages and disadvantages they offer.

- Summarize the benefits, problems, and potential of recycling.

- Analyze some alternatives for reducing the waste we generate.

- Understand what hazardous and toxic wastes are and how we dispose of them.

- Evaluate the options for hazardous waste management.

- Outline some ways we can destroy or permanently store hazardous wastes.

INTRODUCTION

On August 31, 1986, the cargo ship *Khian Sea* loaded one month's production of ash (14,000 tons) from the Philadelphia municipal incinerator and set off on an odyssey that symbolizes a predicament we all share. The ship's first port of call in search of a dumping place for its noxious cargo was the poor Caribbean nation of Haiti. Four thousand tons of ash had been carried ashore by the time representatives of the environmental group Greenpeace alerted local residents to the potentially dangerous levels of arsenic, mercury, dioxins, and other toxins in the wastes. Authorities ordered the ship to take its objectionable goods elsewhere.

For twenty-four months this pariah ship wandered from port to port in the Caribbean, across to West Africa, around the Mediterranean, through the Suez Canal, past India, and over to Singapore looking for a place to dump its toxic load. Its name was changed from *Khian Sea* to *Felicia* to *Pelacano*. Its registration was transferred from Liberia to the Bahamas to Honduras in an attempt to hide its true identity, but nobody wanted it or its contents. Like Coleridge's ancient mariner, it seemed cursed to wander the oceans forever. Two years, three names, four continents, and eleven countries later, the onerous cargo was still aboard.

Then, somewhere on the Indian Ocean between Columbo, Sri Lanka, and Singapore, 14,000 tons of toxic ash disappeared. When questioned about this remarkable occurrence, the crew had no comment except that it was all gone. Everyone assumes, of course, that once the ship was out of sight of land, the ash was dumped into the ocean.

If this were just an isolated incident, perhaps it wouldn't matter too much. However, some 3 million tons of toxic and hazardous waste goes to sea every year. How much ends up in the ocean and how much is deposited in poor countries is unknown.

The problem is that we generate vast amounts of unwanted stuff every year. Places to put these wastes are becoming more and more scarce as their contents are becoming increasingly unpleasant and dangerous. No one wants it in their backyards. Vandals dump it in out-of-the-way places. Rich communities and nations send it to their impoverished neighbors. How can we stop these perilous practices? In this chapter, we will look at the kinds of waste we produce, who makes them, what problems their disposal causes, and how we might reduce our waste production and dispose of our wastes in environmentally safe ways.

SOLID WASTE

Waste is everyone's business. We all produce wastes in nearly everything we do. According to the EPA, the United States produces 11 billion tons of solid waste each year. Nearly half of that amount consists of agricultural waste, such as crop residues and animal manure, which are generally recycled into the soil on the farms where they are produced. They represent a valuable resource as ground cover to reduce erosion and fertilizer to nourish new crops, but they also constitute the single largest source of nonpoint air and water pollution in the country. About one-third of all solid wastes are mine tailings, overburden from strip mines, smelter slag, and other residues produced by mining and primary metal processing. Most of this material is stored in or near its source of production and isn't mixed with other kinds of wastes. Improper disposal practices, however, can result in serious and widespread pollution.

Industrial waste—other than mining and mineral production—amounts to some 400 million metric tons per year in the United States. Most of this material is recycled, converted to other forms, destroyed, or disposed of in private landfills or deep injection wells. About 60 million metric tons of industrial waste falls in a special category of hazardous and toxic waste, which we will discuss later in this chapter.

Municipal waste—a combination of household and commercial refuse—amounts to about 300 million metric tons per year in the United States (fig. 13.1). That's approximately 1.24 tons for each man, woman, and child every year—twice as much per capita as Europe or Japan, and five to ten times as much as most developing countries.

Figure 13.1 Bulldozers pack down trash at a municipal landfill near Niagara Falls, Canada. Rainwater percolating through unsealed landfills carries toxic pollutants into the groundwater and contaminates drinking supplies. Landfills are being closed all over the United States, but other forms of waste disposal are not being developed or are equally controversial.

The Waste Stream

Does it surprise you to learn that you generate that much garbage? Think for a moment about how much we discard every year. There are organic materials, such as yard and garden wastes, food wastes, and sewage sludge from treatment plants; junked cars; worn out furniture; and consumer products of all types. Newspapers, magazines, advertisements, and office refuse make paper one of our major wastes (fig. 13.2). In spite of recycling programs, a majority of the 200 *billion* metal, glass, and plastic food and beverage containers used every year in the United States ends up in the trash. Wood, concrete, bricks, and glass come from construction and demolition sites, as do dust and rubble from landscaping and road building. All of this varied and voluminous waste has to arrive at a final resting place somewhere.

The **waste stream** is a term that describes the steady flow of varied wastes that we all produce, from domestic garbage and yard wastes to industrial, commercial, and construction refuse. Many of the materials in our waste stream would be valuable resources if they were not mixed with other garbage. Unfortunately, our collecting and dumping processes mix and crush everything together, making separation an expensive and sometimes impossible task. In a dump or incinerator, much of the value of recyclable materials is lost.

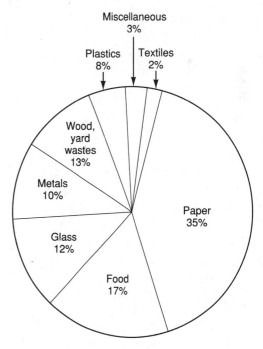

Figure 13.2 Composition of domestic waste in the United States, by weight. Plastics make up only 8 percent of the weight but up to 15 percent of the volume of our waste.

Another problem with refuse mixing is that hazardous materials in the waste stream get dispersed through thousands of tons of miscellaneous garbage. This mixing makes the disposal or burning of what might have been rather innocuous stuff a difficult, expensive, and risky business. Spray paint cans, pesticides, batteries (zinc, lead, or mercury), cleaning solvents, smoke detectors containing radioactive material, and plastics that produce dioxins and PCBs when burned are mixed willy-nilly with paper, table scraps, and other nontoxic materials. The best thing to do with household toxic and hazardous materials is to separate them for safe disposal or recycling, as we will see later in this chapter.

Waste Disposal Methods

Where are our wastes going now? In this section, we will examine some historic methods of waste disposal as well as some future options. Notice that our presentation begins with the least desirable—but most commonly used—measures and proceeds to discuss some preferable options. Keep in mind as you read this section that modern waste management reverses this order and stresses the "three R's" of reduction, reuse, and recycling before destruction and, finally, secure storage of wastes.

Open Dumps

For many people, the way to dispose of waste is to simply drop it someplace. Open, unregulated dumps are still the predominant method of waste disposal in most developing countries. The giant Third World megacities have enormous garbage problems. Mexico City, the largest city in the world, generates some 10,000 tons of trash *each day*. Until recently, most of this torrent of waste was left in giant piles, exposed to the wind and rain, as well as rats, flies, and other vermin. Manila, in the Philippines, has at least ten huge open dumps. The most notorious is called Smoky Mountain because of its constant smoldering fires (fig. 13.3). Thousands of people live and work on this 30 m (90 ft) high heap of refuse. They spend their days sorting through the garbage for edible or recyclable materials. Health conditions are abysmal, but these people have nowhere else to go. The government would like to close these dumps, but how will the residents be housed and fed? Where else will the city put its garbage?

Most developed countries forbid open dumping, at least in metropolitan areas, but illegal dumping is still a problem. You have undoubtedly seen trash accumulating along roadsides and in vacant, weedy lots in the poorer sections of town. Is this just a question of aesthetics? Consider the problem of waste oil and solvents. An estimated 200 million liters of waste motor oil are poured into the sewers or allowed to soak into

Figure 13.3 Scavengers sort through the trash at "Smoky Mountain," one of the huge metropolitan dumps in Manila, Philippines. Some 20,000 people live and work on these enormous garbage dumps. The health effects are tragic.

the ground every year in the United States. This is about five times as much as was spilled by the *Exxon Valdez* in Alaska in 1989! No one knows the volume of other solvents disposed of by similar methods.

Increasingly, these toxic chemicals are showing up in the groundwater supplies on which nearly half the people in America depend for drinking (chapter 11). An alarmingly small amount of oil or other solvents can pollute large quantities of drinking or irrigation water. One liter of gasoline, for instance, can make a million liters of water undrinkable. The problem of illegal dumping is likely to become worse as acceptable sites for waste disposal become more scarce and costs for legal dumping escalate. We clearly need better enforcement of antilittering laws as well as a change in our attitudes and behavior.

Ocean Dumping

The oceans are vast, but not so large that we can continue to treat them as carelessly as has been our habit. Every year some 25,000 metric tons (55 million lbs) of packaging, including half a million bottles, cans, and plastic containers, are dumped at sea. Beaches, even in remote regions, are littered with the nondegradable flotsam and jetsam of industrial society (fig. 13.4). About 150,000 tons (330 million lbs) of fishing gear—including more than 1,000 km (660 mi) of nets—are lost or discarded at sea each year. Environmental groups estimate that 50,000 Northern Fur seals are entangled in this refuse and drown or starve to death every year in the North Pacific alone.

Until 1988, many cities in the United States dumped municipal refuse, industrial waste, sewage, and sewage sludge in the ocean. Federal legislation now prohibits this dumping. New York City, the last to stop offshore sewage sludge disposal, finally ended this practice in 1992. Still, 60 to 80 million cubic meters of dredge spoil are disposed of at sea each year. Some people claim that the deep abyssal plain is the most remote, stable, and innocuous place to dump our wastes. Others argue that we know too little about the values of these remote places or the rare species that live there to smother them with sludge and debris.

Landfills

Over the past fifty years most American and European cities have recognized the health and environmental hazards of open dumps. Increasingly, cities have turned to **landfills**, where solid waste disposal is regulated and controlled. To decrease smells and litter and to discourage insect and rodent populations, landfill operators are required to compact the refuse and cover it every day with a layer of dirt (fig. 13.5). This method helps control pollution, but the dirt fill also takes up to 20 percent of landfill space. New methods of landfill construction are being developed to control such hazardous substances as oil, chemical compounds, toxic metals, and contaminated rainwater that seeps through piles of waste. An impermeable clay or plastic lining underlies and encloses the storage area.

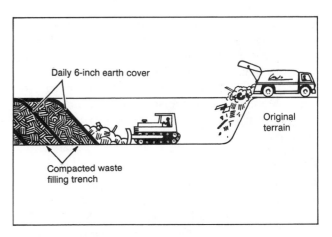

Figure 13.5 In a sanitary landfill, trash and garbage are crushed and covered each day to prevent accumulation of vermin and spread of disease. Traditionally, most landfills have not been enclosed in a waterproof lining to prevent leaching of chemicals into underground aquifers.

Drainage systems are installed in and around the liner to catch drainage and to help monitor any chemicals that may be leaking.

More careful attention is now paid to the siting of new landfills. Sites located on highly permeable or faulted rock formations are passed over in favor of sites with less leaky geologic foundations. Landfills are being built away from rivers, lakes, floodplains, and aquifer recharge zones rather than near them, as was often done in the past. More care is being given to a landfill's long-term effects so that costly cleanups and rehabilitation can be avoided.

Historically, landfills have been a convenient and relatively inexpensive waste-disposal option in most places, but this situation is changing rapidly. Rising land prices and shipping costs, as well as increasingly demanding landfill construction and maintenance requirements, are making this a more expensive disposal method. The cost of disposing a ton of solid waste in Philadelphia went from $20 in 1980 to more than $100 in 1990. Union County, New York, experienced an even steeper price rise. In 1987, it paid $70 to get rid of a ton of waste; a year later, that same ton cost $420, or about $10 for a typical garbage bag. The United States now spends about $10 billion per year to dispose of trash. By the year 2000, it may cost us $100 billion per year to dispose of our trash and garbage.

Suitable places for waste disposal are becoming scarce in many areas. Other uses compete for open space. Citizens have become more concerned and vocal about health hazards, as well as aesthetics. It is difficult to find a neighborhood or community willing to accept a new landfill. More than half of all U.S. cities will exhaust their landfills by 1995. Since 1984, when stricter financial and environmental protection

Figure 13.4 Plastic litter covers this beach in Niihau, Hawaii.

requirements for landfills took effect, more than 1,200 of the 1,500 existing landfills in the United States have closed. Many major cities already have no more dumping space. They export their trash, at enormous expense, to neighboring communities and even other states. More than half the solid waste from New Jersey goes out of state, some of it up to 800 km (500 mi) away.

Exporting Waste

As disposal costs rise and restrictions on what can be dumped become more stringent, many European and American cities and industries are sending their wastes abroad to less-developed countries, as the story of the *Khian Sea* illustrates. In 1988, the environmental organization Greenpeace identified more than fifty plans to send wastes from the United States and Europe to Africa, Latin America, and the Middle East. This is becoming a touchy political issue. Local people usually aren't told what is in the waste being dumped on their land. Provincial officials don't have the resources or the knowledge to test or regulate toxic materials. In the West African nation of Guinea, children were found playing on a huge pile of incinerator ash from Philadelphia. Nigerian officials discovered 8,000 leaking drums of radioactive waste stored in an open yard in the port city of Koko.

In 1985, two Americans were jailed in the United States for exporting 6,000 liters (1,500 gal) of toxic waste to Zimbabwe under the guise of cleaning fluid. West African countries arrested sixty-two corrupt local officials and foreigners for illegally dumping dangerous materials. In 1989, a treaty regulating international shipping of toxics was signed by 105 nations. This treaty requires that the receiving country give its permission for dumping. This doesn't stop exports completely, but it gives local authorities a chance to find out what is being dumped and to object before it happens.

It's not surprising that exports are attractive. It commonly costs $800 per barrel to dispose of toxic waste in the United States, but some African countries will take that barrel for $50. Toxic trash doesn't only move from north to south. East Germany earned $600 million in the 1970s and 1980s by accepting 5.5 million tons of household rubbish and 800,000 tons of toxic waste from its more prosperous western cousins. Other Eastern Bloc countries are waste repositories, while western countries generally are exporters.

A similar kind of garbage imperialism operates within the richer countries as well. Poor neighborhoods are much more likely to be recipients of dumps, waste incinerators, and other hazardous projects. In recent years, attention has turned to Indian reservations, which are exempt from some state and federal regulations concerning waste disposal. More

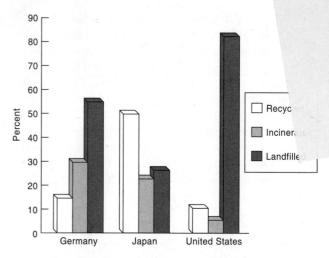

Figure 13.6 Percentages of municipal solid waste recycled, incinerated, and burned in Germany, Japan, and the United States.

than 200 proposals were made between 1990 and 1992 to store wastes on reservations. Virtually every tribe in America has been approached with these schemes.

Incineration and Resource Recovery

Landfilling is still the disposal method for the vast majority of municipal waste in the United States (fig. 13.6). Faced with growing piles of garbage and a lack of available landfills at any price, however, public officials are investigating other disposal methods. The method to which they frequently turn is burning. Another term commonly used for this technology is **energy recovery** or **waste-to-energy** because the heat derived from incinerated refuse is a useful resource. Burning garbage can produce steam used directly for heating buildings or generating electricity. Internationally, well over 1,000 waste-to-energy plants in Brazil, Japan, and Western Europe generate much needed energy while also reducing the amount that needs to be landfilled. In the United States, more than 110 waste incinerators burn 45,000 tons of garbage daily. Some of these are simple incinerators; others produce energy.

Types of Incinerators

Municipal incinerators are specially designed burning plants capable of burning thousands of tons of waste per day. In some plants, refuse is sorted as it comes in to remove unburnable or recyclable materials before combustion. This is called **refuse-derived fuel** because the enriched burnable fraction has a higher energy content than the raw trash. Another approach, called **mass burn,** is to shred everything into small pieces and then burn as much as possible (fig. 13.7). This technique avoids the expensive and unpleasant job of sorting through the garbage for nonburnable

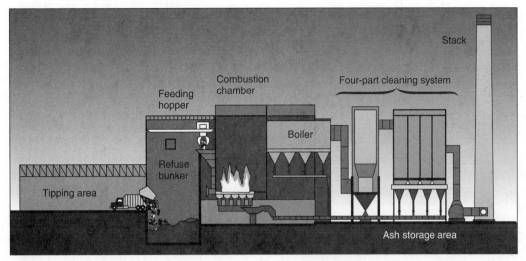

Figure 13.7 A diagram of a municipal "mass-burn" garbage incinerator. Steam produced in the boiler can be used to generate electricity or to heat nearby buildings.

materials, but it often causes greater problems with air pollution and corrosion of burner grates and chimneys.

In either case, residual ash and unburnable residues representing 10 to 20 percent of the original volume are taken to a landfill for disposal. Because the volume of burned garbage is reduced by 80 to 90 percent, disposal is a smaller task. However, the residual ash usually contains a variety of toxic components that make it an environmental hazard if not disposed of properly. Ironically, one worry about incinerators is whether enough garbage will be available to feed them. Some communities in which recycling has been really successful have had to buy garbage from neighbors to meet contractual obligations to waste-to-energy facilities. In other places, fears that this might happen have discouraged recycling efforts.

Incinerator Cost and Safety

The cost-effectiveness of garbage incinerators is the subject of heated debates. Initial construction costs are high—usually between $100 million and $300 million for a typical municipal facility. Tipping fees at an incinerator, the fee charged to haulers for each ton of garbage dumped, are often much higher than those at a landfill. As landfill space near metropolitan areas becomes more scarce and more expensive, however, landfill rates are certain to rise. It may pay in the long run to incinerate refuse so that the lifetime of existing landfills will be extended.

Environmental safety of incinerators is another point of concern. In 1988, the EPA released a report of alarmingly high levels of dioxins, furans, lead, and cadmium in incinerator ash. These toxic materials were more concentrated in the fly ash (lighter, airborne particles capable of penetrating deep into the

lungs) than in heavy bottom ash. Dioxin levels can be as high as 780 parts per billion. One part per billion of TCDD, the most toxic dioxin, is considered a health concern. All of the incinerators studied exceeded cadmium standards, and 80 percent exceeded lead standards. Proponents of incineration argue that if they are run properly and equipped with appropriate pollution-control devices, incinerators are safe to the general public. Opponents counter that neither public officials nor pollution control equipment can be trusted to keep the air clean. They argue that recycling and source reduction efforts are better ways to deal with waste problems.

The EPA, which supports incineration, acknowledges the health threat of incinerator emissions but holds that the danger is very slight. The EPA estimates that dioxin emissions from a typical municipal incinerator may cause one death per million people in seventy years of operation. Critics of incineration claim that a more accurate estimate is 250 deaths per million in seventy years.

One way to reduce these dangerous emissions is to remove batteries containing heavy metals and plastics containing chlorine before wastes are burned. Bremen, West Germany, is one of several European cities now trying to control dioxin and PCB emissions by keeping all plastics out of incinerator waste. Bremen is requiring households to separate plastics from other garbage. This is expected to eliminate nearly all dioxins and PCBs and prevent the expense of installing costly pollution-control equipment that otherwise would be necessary to keep the burners operating. Minneapolis has initiated a recycling program for the small "button" batteries used in hearing aids, watches, and calculators in an attempt to lower mercury emissions from its incinerator.

Recycling

The term "recycling" has two meanings in common usage. Sometimes we say we are *recycling* when we really are *reusing* something, such as refillable beverage containers. In terms of solid waste management, however, **recycling** is the reprocessing of discarded materials into new, useful products (fig. 13.8). Some recycling processes reuse materials for the same purposes; for instance, old aluminum cans and glass bottles are usually melted and recast into new cans and bottles. Other recycling processes turn old materials into entirely new products. Old tires, for instance, are shredded and turned into rubberized road surfacing. Newspapers become cellulose insulation, kitchen wastes become fuel pellets, and steel cans become automobiles and construction materials.

Figure 13.8 Recycling saves landfill space, energy, and raw materials. It also reduces pollution and creates jobs. Everyone can be involved.

Benefits of Recycling

Recycling is usually a better alternative to either dumping or burning wastes. It saves money, energy, raw materials, and land space, while also reducing pollution. Recycling also encourages individual awareness and responsibility for the refuse produced. Curbside pickup of recyclables costs around $35 per ton, as opposed to the $80 paid to dispose of them at an average metropolitan landfill. Some recycling programs cost nothing; they cover their own expenses with materials sales and may even bring revenue to the community.

Another benefit of recycling is that it could cut our waste volumes by 50 percent or more, drastically reducing the pressure on disposal systems. Philadelphia is investing in neighborhood collection centers that will recycle 600 tons a day, enough to eliminate the need for a previously planned, high-priced incinerator. New York City, down to one available landfill but still producing 27,000 tons of garbage a day, has set a target of 50 percent waste reduction to be accomplished by recycling office paper and household and commercial waste. New York's curbside collection service, projected to be the nation's largest, should more than pay for itself simply in avoided disposal costs.

Japan probably has the most successful recycling program in the world (fig. 13.6). Half of all household and commercial wastes in Japan is recycled while half of the rest is incinerated and half is landfilled. By comparison, the United States landfills more than 80 percent of all solid waste. Japanese families diligently separate wastes into as many as seven categories, each picked up on a different day (fig. 13.9). Would we do the same? Some authors say that Americans are too lazy to recycle. North Stonington, Connecticut, however, faced with escalating disposal costs, reduced its waste volume by two-thirds

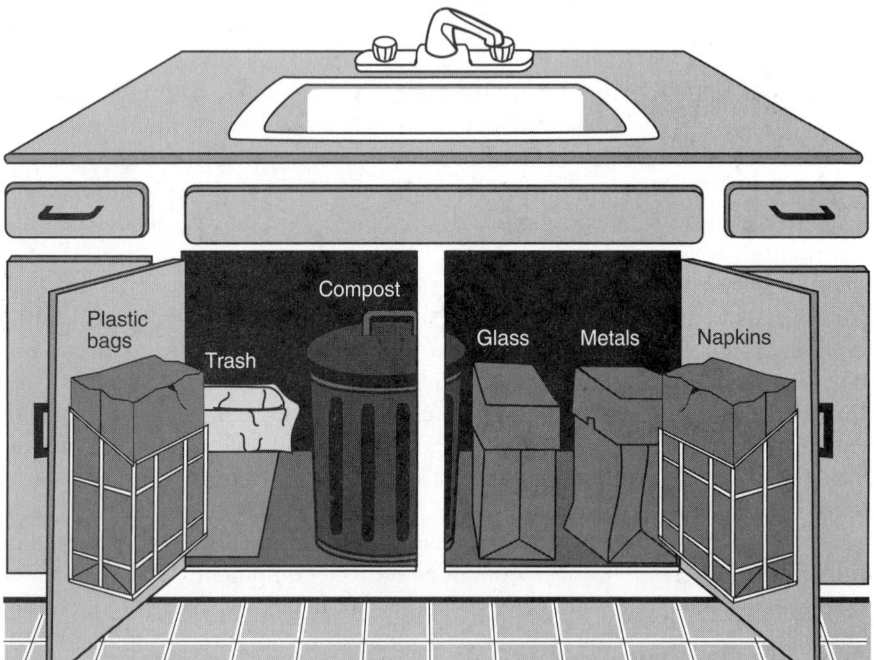

Figure 13.9 Source separation in the kitchen—the first step in a strong recycling program.

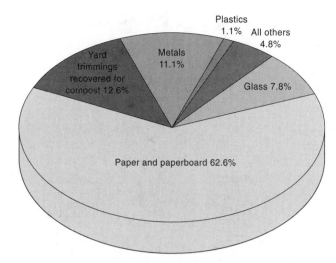

Figure 13.10 Recovery of materials from the municipal solid waste stream in the United States, 1990 (by weight).
Source: Data from U.S. Environmental Protection Agency, *EPA Journal,* 18 (3), July/August 1992.

in just two years. Overall, we already are making progress in removing material from the waste stream (fig. 13.10).

Recycling lowers our demands for raw resources. In the United States, we cut down 2 million trees every day to produce newsprint and paper products, a heavy drain on our forests. Recycling the print run of a single Sunday issue of the New York Times would spare 75,000 trees. Every piece of plastic we make reduces the reserve supply of petroleum and makes us more dependent on foreign oil. Recycling 1 ton of aluminum saves 4 tons of bauxite (aluminum ore) and 700 kg (1,540 lb) of petroleum coke and pitch, as well as keeping 35 kg (77 lb) of aluminum fluoride out of the air.

Recycling also reduces energy consumption and air pollution. Plastic bottle recycling could save 50 to 60 percent of the energy needed to make new ones. Making new steel from old scrap offers up to 75 percent energy savings. Producing aluminum from scrap instead of bauxite ore cuts energy use by 95 percent, yet we still throw away more than a million tons of aluminum every year. If aluminum recovery were doubled worldwide, more than a million tons of air pollutants would be eliminated every year.

Reducing litter is an important benefit of recycling. Ever since disposable paper, glass, metal, foam, and plastic packaging began to accompany nearly everything we buy, these discarded wrappings have collected on our roadsides and in our lakes, rivers, and oceans. Without incentives to properly dispose of beverage cans, bottles, and papers, it often seems easier to just toss them aside when we have finished using them. Litter is a costly as well as unsightly problem. We pay an estimated thirty-two cents for each

piece of litter picked up by crews along state highways, which adds up to $500 million every year. Deposits on bottles and cans have reduced littering in many states.

Creating Incentives for Recycling

In many communities, citizens have done such a good job of collecting recyclables that a glut has developed. Mountains of waste materials accumulate in warehouses (fig. 13.11) because there are no markets for them. A decade ago, you could sell old newspapers for about $20 per ton. Now, in many places, cities have to pay $20 per ton to get rid of newsprint. Much of the waste that we carefully separate for recycling ends up being mixed together and dumped in a landfill or incinerator.

Fee systems and taxes could be used to stabilize prices and create markets for recycled materials. The demand for the most easily recycled materials—paper, glass, scrap metal—are notoriously volatile. This instability is deadly for marginally capitalized programs. Guaranteed prices and policies that encourage use of recycled material would do a great deal toward making recycling successful.

Our present public policies tend to favor extraction of new raw materials. Energy, water, and raw materials are often sold to industries below their real cost to create jobs and stimulate the economy. Setting the prices of natural resources at their real cost would tend to encourage efficiency and recycling and would help create a market for used materials. Each of us can play a role in creating markets, as well. If we buy things made from recycled materials—or ask for them if they aren't available—we will help make it possible for recycling programs to succeed.

Figure 13.11 Mountains of paper, plastic, and glass are accumulated in warehouses because collection programs have been more successful than material marketing efforts. Federal and State governments need to encourage use of recycled materials. We can help by buying recycled products.

Acting Locally: Box 13.1

Building a Home Compost Pile

*H*ome composting is easy, beneficial, and educational. It takes very little effort or attention because millions of tiny microorganisms do almost all the work for you. All you have to do is supply your little helpers with a food supply, a place to live, a little water, and oxygen.

To get started, you need a supply of organic material. Grass clippings, leaves, wood chips, vegetable pealings, sawdust, seaweed, manure, squash and tomato vines, watermelon rinds, coffee grounds—even hair, paper, and old rags are all good compost material. You can simply pile everything on the ground in an out-of-the-way place, or if you prefer a tidier operation, you can build a container or bin (box fig. 13.1.1). Use boards, bricks, screen, an old barrel, or any other material that will hold the compost. Leave some spaces or holes in the side of your bin so that air can reach the lower layers—or turn the pile over frequently.

Layer material in the pile, alternating nitrogen-rich stuff such as grass, manure, and food scraps with carbon-rich substances such as cornstalks, dry leaves, straw, wood chips, and paper. A pile that is too high in carbon will take a long time to work. Too much nitrogen will produce an odor like urine or ammonia gas; it also will make the pile slimy and putrid. The ideal ratio of carbon

Figure 13.1.1 Composting is a good way to reduce yard waste, vegetable scraps, and other organic materials into useful garden mulch. Mix everything together, keep it moist and well aerated, and in a few weeks, you will have a rich, odor-free mulch. This three-bin system allows you to have batches in different stages.

to nitrogen is 30:1. Since you probably don't have the means to determine the composition of what you put in your pile, just use a variety of materials and don't overload with any one thing. If you have a very large amount of dry material—say leaves in the fall—you might want to add some extra manure or commercial fertilizer to give the compost a shot of nitrogen.

Composting

Pressed for landfill space, many cities have banned yard waste from municipal garbage. Rather than bury this valuable organic material, they are turning it into a useful product through **composting**: biological degradation or breakdown of organic matter under aerobic (oxygen-rich) conditions. The organic compost resulting from this process makes a nutrient-rich soil amendment that aids water retention, slows soil erosion, and improves crop yields. A home compost pile is an easy and inexpensive way to dispose of organic waste in an interesting and environmentally friendly way (box 13.1).

Disposable diapers contain just the right mixture of carbon and nitrogen for commercial composting. Sixteen billion soiled disposable diapers are discarded in the United States every year. Several demonstration projects for composting disposable diapers are now underway. Critics of this process argue that it would be better to use cloth diapers or to develop biodegradable ones. Determining the total life-cycle costs and environmental effects of products such as diapers

is a difficult and controversial task. See the article by Poore in the reading list for an interesting discussion of this issue.

Energy from Waste

Every year we throw away the energy equivalent of 80 million barrels of oil in organic waste in the United States. In developing countries, up to 85 percent of the waste stream is food, textiles, vegetable matter, and other biodegradable materials. Worldwide, at least one-fifth of municipal waste is organic kitchen and garden refuse. In a landfill, much of this matter is decomposed by microorganisms generating billions of cubic meters of methane (''natural gas''), which contributes to global warming if allowed to escape into the atmosphere (chapter 10). Many cities are drilling methane wells in their landfills to capture this valuable resource.

This energy-rich organic material can be burned in incinerators rather than being buried in landfills, but there are worries about air pollution from incineration. Organic wastes also can be decomposed in large,

To function well, a compost pile shouldn't be either too big or too little. If the pile is too large, oxygen doesn't get into the middle. A pile that is too small doesn't retain enough heat for the microorganisms to grow optimally. One to two meters wide and a meter (3 ft) deep is about right. Given a good nutrient supply and plenty of air, bacteria and fungi growing in the compost pile will produce a temperature of about 70°C (160°F), enough to kill most pathogens and weed seeds. Turning the pile over frequently (every week or two) will mix the components and provide enough fresh air to keep the pile working well and will prevent the sour smell of anaerobic (oxygen-starved) fermentation.

The other essential ingredient for the microorganisms is water. If you live in a rainy climate or put lots of fresh vegetables and grass clippings in your compost pile, you won't need supplemental moisture, but if you use lots of dry leaves or live in a very dry place, you may need to add some water from time to time. The compost should be moist but not saturated. Too much water blocks oxygen penetration (that's why bogs are anaerobic and organic material doesn't decompose).

How long will your compost pile take to complete its work? That depends on temperature, moisture, texture, and contents. In the summer, a few weeks should produce a dark, soft, crumbly material that smells earthy and looks like rich potting soil. Composting will even work in the winter—but slowly if you live in a cold climate. Branches, stone fruit, and large chunks of material decay very slowly. Shredding, chipping, or chopping the starting material into small pieces will speed up the process.

There are some things that you shouldn't put into your compost pile. Don't add meat, fish, cheese, butter, milk, peanut butter, oil, grease, bones, skin, or other animal products. They don't compost well, they tend to smell bad, and they attract pests and vermin. Don't add plants infected with disease or insects that could survive the compost pile's heat (examples are scale insects, rusts, smuts, etc.). Avoid plants that take too long to break down such as pine needles, eucalyptus leaves or bark, or oak leaves. The tannins and acids they contain are natural bactericides and fungicides that inhibit composting. Cat and dog feces may contain harmful pathogens that can survive composting—especially if your pile doesn't reach optimal temperature. The compost produced could be infectious.

If you avoid these materials, keep pests and vermin out, and maintain a neat, well-aerated, odor-free compost pile, you shouldn't have any complaints from your neighbors. And after a few weeks, you will have a valuable soil amendment to spread on your yard or in your garden, or to use in potting house plants.

oxygen-free digesters to produce methane under more controlled conditions than in a landfill and with less air pollution than mass garbage burning.

Anaerobic digestion also can be done on a small scale. Millions of household methane generators provide fuel for cooking and lighting for homes in China and India (chapter 12). In the United States, some farmers produce all the fuel they need to run their farms—both for heating and to run trucks and tractors—by generating methane from animal manure.

Shrinking the Waste Stream

Even better than recycling or composting is cleaning and reusing materials in their present form, thus saving the cost and energy of remaking them into something else. We do this already with some specialized items. Auto parts are regularly sold from junkyards, especially for older car models. In some areas, stained glass windows, brass fittings, fine woodwork, and bricks salvaged from old houses bring high prices. Some communities sort and reuse a variety of materials received in their dumps.

In many cities, glass and plastic bottles are routinely returned to beverage producers for washing and refilling. The reusable, refillable bottle is the most efficient beverage container we have. This is better for the environment than remelting and more profitable for local communities. A reusable glass container makes an average of fifteen round-trips between factory and customer before it becomes so scratched and chipped that it has to be recycled. Reusable containers also favor local bottling companies and help preserve regional differences.

Since the advent of cheap, lightweight, disposable food and beverage containers, many small, local breweries, canneries, and bottling companies have been forced out of business by huge, national conglomerates. These big companies can afford to ship food and beverages great distances as long as it is a one-way trip. If they had to collect their containers and reuse them, canning and bottling factories serving large regions would be uneconomical. Consequently, the national companies favor recycling rather than refilling because they prefer fewer, larger plants and don't want to be responsible for collecting and reusing containers.

In less affluent nations, reuse of all sorts of manufactured goods is an established tradition. Where most manufactured products are expensive and labor is cheap, it pays to salvage, clean, and repair products. Cairo, Manila, Mexico City, and many other cities have large populations of poor people who make a living by scavenging. Entire ethnic populations may survive on scavenging, sorting, and reprocessing scraps from city dumps.

Producing Less Waste

What is even better than reusing materials? Generating less waste in the first place. Table 13.1 describes some contributions you can make to reducing the volume of our waste stream. Industry also can play an important role in source reduction. The 3M Company has saved more than $500 million since 1975 by changing manufacturing processes, finding uses for waste products, and listening to employees' suggestions. What is waste to one division is a treasure to another.

Excess packaging of food and consumer products is one of our greatest sources of unnecessary waste. Paper, plastic, glass, and metal packaging material make up 50 percent of our domestic trash by volume. Much of that packaging is primarily for marketing and

Figure 13.12 How much more do we need? Where will we put what we already have?

Reprinted with special permission of King Features Syndicate.

has little to do with product protection (fig. 13.12). Manufacturers and retailers might be persuaded to reduce these wasteful practices if consumers ask for products without excess packaging. Communities also can have an impact. Berkeley, California; Portland, Oregon; Minneapolis and St. Paul, Minnesota; and about twenty-five other cities have passed ordinances requiring that fast-food restaurants package food in paper or other biodegradable wrappings, both to reduce litter and to protect the atmosphere. In 1990, responding to nationwide pressure, Burger King and McDonald's restaurants announced they would stop using plastic foam hamburger boxes.

Where disposable packaging is necessary, we still can reduce the volume of waste in our landfills by using biodegradable materials. Usually this means no plastics. Recently, however, plastics have become available that do break down in the environment under ideal circumstances. **Photodegradable plastics** break down when exposed to ultraviolet radiation. **Biodegradable plastics** incorporate such materials as cornstarch that can be decomposed by microorganisms. Several states have introduced legislation requiring biodegradable or photodegradable six-pack beverage yokes, fast-food packaging, and disposable diapers. However, these degradable plastics don't decompose completely but only break down to small particles. In doing so, they can release toxic chemicals into the environment. Furthermore, they make recycling less feasible and may lead people to believe that littering is okay.

HAZARDOUS AND TOXIC WASTES

The most dangerous aspect of the waste stream we have described is that it often contains highly toxic and hazardous materials that are injurious to both human health and environmental quality. We now produce and use a vast array of flammable, explosive, caustic, acidic, and highly toxic chemical substances

TABLE 13.1 *What can you do to reduce waste?*
1. Buy foods that come with less packaging; shop at farmers' markets or co-ops, using your own containers.
2. Take your own washable, refillable beverage container to meetings or convenience stores rather than use disposable ones.
3. When you have a choice at the grocery store between plastic, glass, and metal containers for the same food, buy the reusable or easier-to-recycle glass or metal.
4. When buying plastic products, pay a few cents extra for environmentally degradable varieties.
5. Separate your cans, bottles, papers, and plastics for recycling.
6. Wash and reuse bottles, aluminum foil, plastic bags, etc., for your personal use.
7. Compost yard and garden wastes, leaves, and grass clippings.
8. Write to your senators and representatives and urge them to vote for container deposits, recycling, and safe incinerators or landfills.

Source: Minnesota Pollution Control Agency.

Figure 13.13 According to the U.S. Environmental Protection Agency, industries produce about one ton of hazardous waste per year for every person in the United States. Much of this waste is dumped illegally or disposed of in environmentally dangerous ways.

for industrial, agricultural, and domestic purposes (fig. 13.13). According to the EPA, industries in the United States generate about 265 million metric tons of *officially* classified hazardous wastes each year, slightly

more than 1 ton for each person in the country. In addition, considerably more toxic and hazardous waste material is generated by industries or processes not regulated by the EPA. Shockingly, at least 40 million metric tons (22 billion lbs) of toxic and hazardous wastes are released into the air, water, and land in the United States each year (fig. 13.14).

What Is Hazardous Waste?

Legally, a **hazardous waste** is any discarded material, liquid or solid, that contains substances known to be (1) fatal to humans or laboratory animals in low doses, (2) toxic, carcinogenic, mutagenic, or teratogenic to humans or other life-forms, (3) ignitable with a flash point less than 60°C, (4) corrosive, or (5) explosive or highly reactive (undergoes violent chemical reactions either by itself or when mixed with other materials). Notice that this definition includes both toxic and hazardous materials as defined in chapter 6. Certain compounds are exempt from classification as hazardous waste if they are accumulated in less than 1 kg (2.2 lb) of commercial chemicals or 100 kg of

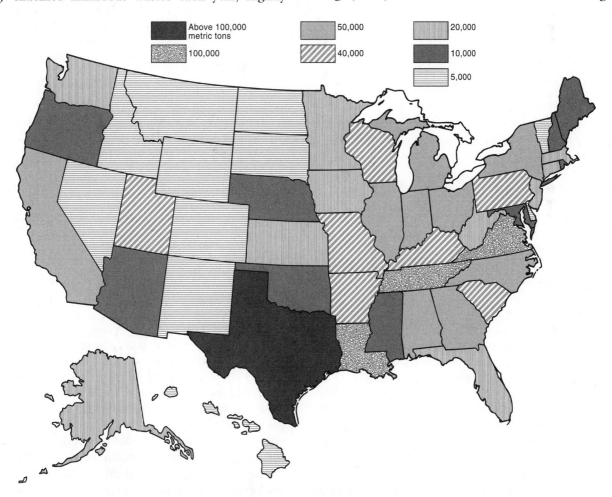

Figure 13.14 Annual toxic air pollution released in the United States. More than 300 toxic chemicals are released into the air. The 1990 Clean Air Amendments regulate these emissions for the first time.

Source: Data from Environmental Protection Agency.

Solid, Toxic, and Hazardous Waste **287**

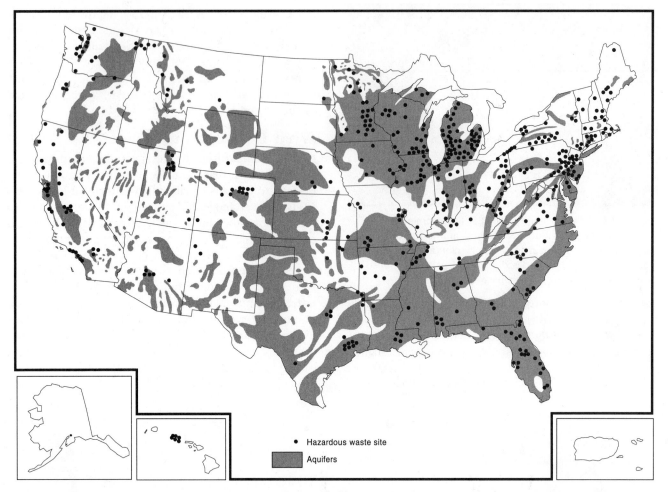

Figure 13.15 Some of the hazardous waste sites on the EPA priority cleanup list. Sites located on aquifer recharge zones represent an especially serious threat. Once groundwater is contaminated, cleanup is difficult and expensive. In some cases, it may not be possible.

Source: Data from the Environmental Protection Agency.

contaminated soil, water, or debris. Even larger amounts (up to 1,000 kg) are exempt when stored at an approved waste treatment facility for the purpose of being beneficially used, recycled, reclaimed, detoxified, or destroyed.

Hazardous Waste Disposal

Most hazardous waste is recycled, converted to non-hazardous forms, stored, or otherwise disposed of on site by the generators—chemical companies, petroleum refiners, and other large industrial facilities—so that it doesn't become a public problem. The 60 million tons per year that does enter the waste stream or the environment represent one of our most serious environmental problems. Much of this material can be a serious threat to both environmental quality and human health.

Finding ways to safely ship, handle, and dispose of this enormous volume of dangerous material is a great challenge. For years, little attention was paid to this material. Wastes stored on private property, buried, or

allowed to soak into the ground were considered of little concern to the public. An estimated 5 billion metric tons of highly poisonous chemicals were improperly disposed of in the United States between 1950 and 1975 before regulatory controls became more stringent.

Superfund Sites

The EPA estimates that there are at least 26,000 abandoned hazardous waste disposal sites in the United States. The General Accounting Office (GAO) places the number much higher, perhaps 425,000 sites when all are identified. By 1990, some 1,226 sites had been placed on the National Priority List (NPL) for cleanup with financing from the 1980 federal Superfund Act (fig. 13.15). The EPA expects that at least 2,500 sites eventually will be included on the NPL and that the total cost for cleanup will be between $8 billion and $16 billion. The GAO estimates the cost to be as high as $100 billion.

The Superfund was first established as a $1.6 billion pool, but in 1986 it was raised to $9.6 billion, far less than the $100 billion that eventually may be needed. Cleanup of individual sites has been more expensive than anticipated, and the work has progressed slowly. After a decade of work and $9 billion in expenditures, the EPA has cleaned up only 27 out of 1,226 sites on the NPL. Critics charge that 80 percent of the money has been used for administration and high-priced consultants while only a pittance has been available for actual cleanup.

What qualifies a site for placement on the NPL? These sites are considered to be especially hazardous to human health and environmental quality because they are known to be leaking or have a potential for leaking supertoxic, carcinogenic, teratogenic, or mutagenic materials (chapter 6). So far, 444 toxic pollutants have been identified at these sites. The ten most commonly found substances are lead, trichloroethylene, toluene, benzene, PCBs, chloroform, phenol, arsenic, cadmium, and chromium. These or other hazardous materials are known to have contaminated groundwater at 75 percent of the sites now on the NPL. In addition, 56 percent of these sites have contaminated surface waters, and airborne materials are found at 20 percent of the sites.

Where are these thousands of hazardous waste sites, and how did they get contaminated? Wastes come from countless sources and are disposed of in just as many ways (fig. 13.16). Chances are that one or more are near where you live. Some are large and well known; others are long forgotten—perhaps only a few barrels buried in the ground or stored in a warehouse somewhere. In this section, we will review a few of the major categories of waste sites.

Hazardous Landfills and Dumps

Many of the worst superfund sites are landfills and old dumps where industries buried metal drums full of a variety of toxic chemicals on their own or leased property. Sometimes the volumes were truly staggering. The Velsicol Chemical Company, for instance, buried at least 250,000 barrels containing about 60 million liters (16 million gal) of toxic pesticide residue in Hardeman County, Tennessee, in the 1960s. These chemicals have leaked from the barrels and contaminated groundwater aquifers used for drinking water by local residents.

Toxic landfills are frequently in or near urban areas. The revelation that homes and a school in Niagara Falls, New York, were built over a waste dump containing 20,000 metric tons of toxic chemical waste awakened many people to the dangers of improper waste disposal (box 13.2) and played a catalytic role in passage of the Superfund Act. There are thousands

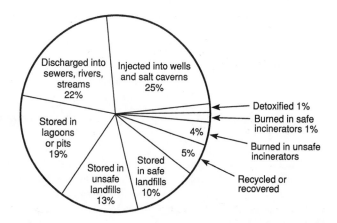

Figure 13.16 Fate of hazardous wastes in the United States. Estimated percentages of volumes of waste disposal in various legal and illegal manners.

Source: Data from the Congressional Budget Office, 1989.

of other abandoned dumps, some reported to be much larger and more dangerous than the one at Love Canal.

Waste Lagoons and Injection Wells

In many areas, liquid wastes weren't even put in barrels before they were dumped; they were simply pumped into lagoons and holding ponds or pumped into deep wells. All these techniques threaten to contaminate groundwater aquifers. The Stringfellow Acid Pit in California is a notorious example. Between 1956 and 1972, more than 120 million liters (32 million gal) of toxic chemicals were poured into shallow ponds at this site in Riverside County on the eastern edge of Los Angeles. A chemical plume is now creeping through the aquifer that supplies drinking water to 500,000 people in the Los Angeles Basin.

Warehousing and Illegal Dumping

Large quantities of hazardous wastes have simply been stacked in old warehouses and abandoned. The high concentrations of flammable materials and highly poisonous chemicals at some of these sites create a lethal combination. On April 21, 1980, a fire broke out in an old warehouse at an inactive waste treatment facility in Elizabeth, New Jersey. More than 20,000 leaking and corroded drums containing pesticides, explosives, radioactive wastes, acids, and other hazardous substances were piled in and around the building. A cloud of toxic gas drifted over heavily populated areas close to the site. Contaminated water from fighting the fire ran off into the Elizabeth River.

The site had been licensed as an incinerator facility for hazardous waste, but, in fact, the operators—some of whom have been linked with organized crime in New York and New Jersey—were simply collecting disposal fees and stacking the barrels in warehouses

The Forgotten Wastes of Love Canal

*L*ove Canal is a sixteen-acre landfill in a residential neighborhood of Niagara Falls, New York. It was intended to be the center of a budding nineteenth-century industrial empire, but the canal never was finished and finally became a dump for industrial waste. This was the beginning of a story that decades later came back to haunt the people of Niagara Falls. It cost them their homes, their health, and hundreds of millions of dollars in rescue efforts.

Early in the century, Love Canal stood empty. It was used mostly as a local swimming hole. In 1942, the city of Niagara Falls began dumping garbage there. The Hooker Chemical and Plastics Corporation, a local industry that had been in the neighborhood since the early days of the canal, also dumped chemical wastes into it. Few people lived in the area, and there was little opposition to the dumping. In April 1945, a Hooker engineer wrote in an internal memo that Love Canal was a "quagmire which will be a potential source of lawsuits." A year later, the company purchased the canal and turned it into a large-scale industrial landfill. Over the next six years, more than 20,000 metric tons of chemical wastes, including highly toxic pesticides, herbicides, and other chemicals—some of them contaminated with dioxins—were dumped into the canal.

In the 1950s, the city of Niagara Falls was growing rapidly and had surrounded the canal. Neighbors complained of foul odors and rats in the dump. It was thought that a solution to both land shortage and pollution problems would be to fill in the site and develop it for housing. In 1953, Hooker sold the site to the Board of Education and the City of Niagara for $1.00 on the condition that the company be released from any liability for injury or damage caused by the dump's contents. Homes were built on the land adjacent to the canal, and in 1954, a school and playground were built on top of the chemical dump itself (box fig. 13.2.1).

Much of the abandoned canal was a swampy, weedy gully with pools of stagnant, oily water. Children played in this wasteland, poking sticks in the black sludge that accumulated on the water and throwing rocks at the drums that floated to the surface. Parents complained that their children were burned by chemicals in the canal; dogs that roamed there developed skin diseases, and their hair fell out in clumps. Clearly, something was wrong.

In 1977, an engineering firm was hired to inspect the site and determine why basements in the area were filled with dark, smelly seepage after every rain. They discovered that the groundwater was contaminated with a variety of toxic organic chemicals. Several mothers, concerned about the health of their children, circulated a petition to close the school and adjacent playing fields. As they went from door to door, they became aware that many families had children with birth defects or chronic medical problems, such as asthma, bronchitis, continuing infections, and hyperactivity. There seemed to be an unusually high rate of miscarriages and stillbirths in the area as well. These informal surveys were dismissed by authorities as "housewife research," but on August 2, 1978, New York State ordered the emergency evacuation of all families living within two blocks of the canal.

Those people whose houses were not purchased by the state watched with mixed feelings as their neighbors departed. Suppose the house across the street from you

and on parking lots. A former employee of the company was quoted as saying: "There's millions to be made in this business. The overhead is nothing. You find a place—don't buy it, don't buy anything; just rent it and fill it up. When they catch up to you, just declare bankruptcy and get out."

The EPA estimates that as much as 90 percent of the hazardous wastes handled by private contractors is disposed of by unapproved methods. So-called midnight dumpers use a variety of illegal and highly dangerous methods to get rid of unwanted toxic wastes. Sometimes liquid wastes are emptied at night into a storm sewer or river. Some drivers pick up a tanker full of wastes and simply drive down a deserted country road with the discharge valve open. "About 60 miles is all it takes to get rid of a load," boasted one driver, "and the only way I can get caught is if the windshield wipers or the tires of the car behind me start melting."

Options for Hazardous Waste Management

What shall we do with toxic and hazardous wastes? In our homes, we can reduce our waste generation and dispose responsibly of unwanted or unusable materials (box 13.3). Collectively, our options for controlling and managing hazardous wastes are similar to those for other wastes: (1) produce less, (2) convert them to less hazardous or nonhazardous substances, or (3) put them in safe, permanent storage. Let's look more closely at some of the most promising options for hazardous waste management.

Produce Less Waste

As with other wastes, the safest and least expensive way to avoid hazardous waste problems is to avoid creating the wastes in the first place. Manufacturing processes can be modified to reduce or eliminate waste production. The 3M Company reformulated

Figure 13.2.1 Love Canal, Niagara Falls, N.Y. A housing development and a school were built on top of this unsealed hazardous waste dump. Residents' health problems forced the state to purchase and destroy homes and played a catalytic role in passage of the Superfund Act.

had been condemned but you were just outside the quarantined area and had to stay. How safe would you feel? Residents traced old streambeds that crossed the canal and showed that chemical residues came up in wet areas, sometimes blocks from the dump site. Disputes, public rallies, lawsuits, and negotiations continued for six years.

Finally, in 1988, Occidental Petroleum (the parent company of Hooker Chemical and Plastics) agreed to pay some $250 million in damages to Love Canal residents. Two years later, after twelve years of rehabilita-

tion work, the EPA concluded that four of seven areas in Love Canal are "habitable." (The other three are slated to become industrial areas or parkland.) A state-of-the-art containment system has sealed off the dump itself with thick clay walls and two clay caps. The 239 houses immediately surrounding the dump have been demolished. Some 236 previously abandoned houses in the next ring around the dumpsite were sold at bargain prices to people who didn't know or didn't care about their history. The government claims that pollution levels have been reduced enough to make the houses safe. Critics argue that the area is still dangerous and that people should not be allowed to live there. What do you think? Would you move into one of those houses? Should others be allowed to do so?

Love Canal has become a symbol of the dangers and uncertainties of toxic industrial chemicals in the environment. The tragedy is that there probably are many Love Canals, some much worse than the original one. No one knows what the total cost of our carelessness in disposing of these chemical wastes ultimately may be.

products and redesigned manufacturing processes between 1975 and 1985 to eliminate more than 140,000 metric tons of solid and hazardous wastes, 4 billion liters (1 billion gal) of wastewater, and 80,000 metric tons of air pollution each year. The company frequently found that these new processes not only spared the environment but also saved money by using less energy and fewer raw materials.

Recycling and reuse of materials also eliminates hazardous wastes and pollution. Many waste products of one process or industry are valuable commodities in another. Already, about 10 percent of the wastes that would otherwise enter the waste stream in the United States are sent to regional waste exchanges where they are sold as raw materials for use by other industries. This figure could probably be raised substantially with better waste management. In Europe, at least one-third of all industrial wastes are exchanged through clearinghouses where beneficial uses

are found. This represents a double savings: the generator doesn't have to pay for disposal, and the recipient pays little, if anything, for raw materials.

Convert to Less Hazardous Substances

Several processes are available to make hazardous materials less toxic. *Physical treatments* tie up or isolate substances. Charcoal or resin filters absorb toxins. Distillation separates hazardous components from aqueous solutions. Precipitation and immobilization in ceramics, glass, or cement isolate toxins from the environment so that they become essentially nonhazardous. One of the few ways to dispose of metals and radioactive substances is to fuse them in silica at high temperatures to make a stable, impermeable glass that is suitable for long-term storage.

Incineration is applicable to mixtures of wastes. A permanent solution to many problems, it is quick and relatively easy but not necessarily cheap—nor always

Acting Locally: Box 13.3

What To Do With Household Hazardous Wastes

Pesticides, cleaners, spot removers, disinfectants, dead batteries, paints, hobby supplies, glue, automotive products—these are just a few of the many dangerous toxic and hazardous materials in our homes. Look in storage areas around your house and you will probably find a plethora of unused and unneeded harmful or hazardous products. It's not good for you or the environment to keep all this stuff around if you don't need it anymore. But what to do with it? It is generally illegal—and certainly immoral—to dump it outside or put it into your household trash. What are our alternatives?

The first rule is to produce less waste in the first place. Buy only what you need for the job at hand rather than the large economy size, much of which will be thrown away later. Use up the last little bit or share leftovers with a friend or neighbor. If you have a half a can of paint left, it's better to put it on your wall than to dispose of it in liquid form. Did you know that the gallon of paint that you bought at the hardware store for $9.99 will cost an equal amount to dispose of as a hazardous waste?

Many common materials that you probably already have make excellent, nontoxic alternatives to expensive commercial products. Vinegar, for instance, is a good chrome or window cleaner. Lemon juice and salt will clean copper. Baking soda is an excellent cleaner and odor remover; use it to clean your oven, pots and pans, silverware, toilet, tub, and sink or to remove odors from clothes, refrigerators, or carpets. For clogged drains, use a plunger or mechanical snake rather than caustic acids

or bases. Lemon, almond, or olive oils make good furniture polish. Mix them with mineral oil if you want a thinner, penetrating polish.

Choose less toxic products. Avoid solvent-containing products labeled flammable, combustible, or explosive. Water-based or latex adhesives or paints are generally safer for you and the environment than oil or solvent-based products. Rather than buying harsh chemical paint strippers, try sandpaper, a scraper, or a heat gun. Use paint to protect wood rather than toxic chemical preservatives. Use a pump or roll-on to apply cleaning and grooming products rather than aerosol versions. Dig out dandelions by hand—or learn to appreciate their cheerful yellow flowers and tenacious personality—rather than apply chemical herbicides.

Even the most responsible shopper occasionally ends up with toxic or hazardous materials that she or he can't use. What can you do with residual stuff? Box figure 13.3.1 is a disposal guide for communities with hazardous waste collection and recycling facilities. If your community lacks these programs, try to get them established. Notice that many water-soluble materials can be safely flushed down the drain—using plenty of water—if you are hooked up to a municipal sanitary sewer system. Don't dump these materials down a storm sewer or into a septic system. Don't mix materials together—especially if they contain bleach and ammonia—a toxic gas can form. Label materials clearly. Think of the workers at the collection station who would like to know what they are handling and what they should do with it. Take materials to the collection site in sturdy, nonleaking containers, both for your sake and that of those who will be handling them.

clean—unless it is done correctly. Wastes must be heated to over 1,000°C (2,000°F) for a sufficient period of time to complete destruction. The ash resulting from thorough incineration is reduced in volume up to 90 percent and often is safer to store in a landfill or other disposal site than the original wastes.

Several sophisticated features of modern incinerators improve their effectiveness. Liquid injection nozzles atomize liquids and mix air into the wastes so they burn thoroughly. Fluidized bed burners pump air from the bottom up through burning solid waste as it travels on a metal chain grate through the furnace. The air velocity is sufficient to keep the burning waste partially suspended. Plenty of oxygen is available, and burning is quick and complete. Afterburners add to the completeness of burning by igniting gaseous hydrocarbons not consumed in the incinerator. Scrubbers and precipitators remove minerals, particulates, and other pollutants from the stack gases.

Chemical processing can transform materials so they become nontoxic. Included in this category are neutralization, removal of metals or halogens (chlorine, bromine, etc.), and oxidation. The Sunohio Corporation of Canton, Ohio, for instance, has developed a process called PCBx in which chlorine in such molecules as PCBs is replaced with other ions that render the compounds less toxic. A portable unit can be moved to the location of the hazardous wastes, eliminating the need for shipping them.

Biological waste treatment taps the great capacity of biological organisms to absorb, accumulate, and detoxify a variety of toxic compounds. Bacteria in activated sludge ponds, aquatic plants (such as water hyacinths or cattails), soil microorganisms, and other species remove toxic materials and purify effluents. Biotechnology offers exciting possibilities for finding or creating organisms to eliminate specific kinds of hazardous or toxic wastes. By using a combination of

Household HazWaste Disposal Guide

♻ = Recycle   ∪∩ = With lots of water, flush in sanitary sewer (**NOT a septic tank**)   🗑 = Dispose of dried solids in the trash   Ⓒ = Save for a household hazwaste collection

Automotive		Lawn and Garden		Home Improvement		Household Items	
Used motor oil	♻	Weed killer	Ⓒ	Latex paint	🗑	Drain and oven cleaner	Ⓒ
Auto batteries	♻	Insect killer	Ⓒ	Oil-based paint	Ⓒ	Toilet cleaner	∪∩
Transmission fluid	♻	Roach, ant poison	Ⓒ	Stain, varnish, lacquer	Ⓒ	Spot remover	Ⓒ
Brake fluid	♻	Rodent bait	Ⓒ	Paint thinner	Ⓒ	Aerosol products	Ⓒ
Antifreeze	♻ or ∪∩	Bug spray	Ⓒ	Turpentine	Ⓒ	Empty aerosols	🗑
Gasoline, fuels	Ⓒ	Fertilizer w/weed killer	Ⓒ	Furniture stripper	Ⓒ	Rubbing alcohol	∪∩
Degreasers	Ⓒ	Fertilizer (no weed killer)	🗑	Paint remover	Ⓒ	Disinfectant	∪∩
Carburetor cleaner	Ⓒ	Pool chemicals	Ⓒ	Wood preservatives	Ⓒ	Cleaner w/bleach**	∪∩
Windshield washer	∪∩	Lighter fluid	Ⓒ	Roofing tar	Ⓒ	Cleaner w/ammonia**	∪∩
				Driveway sealer	Ⓒ	Polish w/solvents*	Ⓒ
				Glue w/solvents*	Ⓒ	Glass cleaner	∪∩
				Water-based glue	🗑	Mothballs	Ⓒ
				Putty, grout, caulk	🗑	Cosmetics	🗑
				Glaze, spackle	🗑	Nail polish, remover	Ⓒ
				Concrete cleaner	Ⓒ	Empty containers	🗑

Notes: * Solvent-containing products have the words "Flammable," "Combustible," or "Contains petroleum distillates" on the labels.

** **NEVER** mix products containing bleach with those containing ammonia. A toxic gas can form!

Figure 13.3.1 Dispose of your hazardous wastes responsibly. The life you save may be your own.

Reprinted with permission from Minnesota Pollution Control Agency, 1993. Note: Disposal methods vary by state. Check with your environmental department.

classic genetic selection techniques and high-technology gene-transfer techniques, for instance, scientists have recently been able to generate bacterial strains that are highly successful at metabolizing PCBs. There are concerns about releasing such exotic organisms into the environment, however (see chapter 9). It may be better to keep these organisms contained in enclosed reaction vessels and feed contaminated material to them under controlled conditions.

Store Permanently

Inevitably, there will be some materials that we can't destroy, make into something else, or otherwise cause to vanish. We will have to store them out of harm's way. There are differing opinions about how best to do this.

Retrievable Storage. Dumping wastes in the ocean or burying them in the ground generally means that we have lost control of them. If we learn later that our dis-

posal technique was a mistake, it is difficult if not impossible to go back and recover the wastes. For many supertoxic materials, the best way to store them may be in **permanent retrievable storage.** This means placing storage containers in a secure building, salt mine, or bedrock cavern where they can be inspected periodically and retrieved, if necessary, for repacking or for transfer if a better means of disposal is developed. This technique is more expensive than burial in a landfill because the storage area must be guarded and monitored continuously to prevent leakage, vandalism, or other dispersal of toxic materials. Remedial measures are much cheaper with this technique, however, and it may be the best system in the long run.

Secure Landfills. One of the most popular solutions for hazardous wastes disposal has been landfilling. Although, as we saw earlier in this chapter, many such landfills have been environmental disasters, newer techniques make it possible to create safe, modern

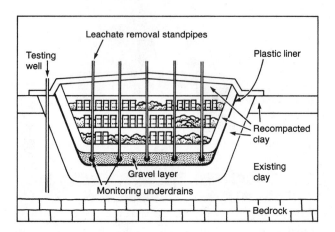

Figure 13.17 A secure landfill. Contents are enclosed by a thick plastic liner and two or more layers of impervious compacted clay. A gravel bed between the clay layers collects any lactrate, which can then be pumped out and treated. Testing wells sample for escaping contaminants.

secure landfills that are acceptable for disposing of many hazardous wastes. The first line of defense in a secure landfill is a thick bottom cushion of compacted clay that surrounds the pit like a bathtub (fig. 13.17). Moist clay is flexible and resists cracking if the ground shifts. It is impermeable to groundwater and will safely contain wastes. A layer of gravel is spread over the clay liner and perforated drain pipes are laid in a grid to collect any seepage that escapes from the stored material. A thick polyethylene liner, protected from punctures by soft padding materials, covers the gravel bed. A layer of soil or absorbent sand cushions

the inner liner and the wastes are packed in drums, which then are placed into the pit, separated into small units by thick berms of soil or packing material.

When the landfill has reached its maximum capacity, a cover much like the bottom sandwich of clay, plastic, and soil—in that order—caps the site. Vegetation stabilizes the surface and improves its appearance. Sump pumps collect any liquids that filter through the landfill, either from rainwater or leaking drums. This water is treated and purified before being released. Monitoring wells check groundwater around the site to ensure that no toxins have escaped.

Most landfills are buried below ground level to be less conspicuous; however, in areas where the groundwater table is close to the surface, it is safer to build above-ground storage. The same protective construction techniques are used as in a buried pit. An advantage to such a facility is that leakage is easier to monitor because the bottom is at ground level.

Transportation of hazardous wastes to disposal sites is of concern because of the risk of accidents. Emergency preparedness officials conclude that the greatest risk in most urban areas is not nuclear war or natural disaster but crashes involving trucks or trains carrying hazardous chemicals through densely packed urban corridors. Another worry is who will bear financial responsibility for abandoned waste sites. The material remains toxic long after the businesses that created it are gone. As is the case with nuclear wastes (chapter 12), we may need new institutions for perpetual care of these wastes (box 13.4).

Thinking Globally: Box 13.4

Toxic Waste Management in Denmark

One of the most successful systems in the world for handling hazardous wastes has been developed in Denmark. This small, heavily populated country has no "away" to which it can put wastes. They have to be in someone's backyard. The Danes have developed a joint government/industry system for collecting and disposing wastes that combines penalties for waste production with incentives for waste reduction. A nonprofit corporation has a monopoly over off-site treatment, recycling, and waste disposal facilities. There are no midnight haulers or cheap landfills. Wastes are gathered, sorted, and pretreated at a network of collection facilities. Everything is recycled, detoxified, or incinerated. Nothing is landfilled except boiler slag and some nontoxic chemical residues.

The fees charged for this service are high. It's not a competitive market. They do things right and generators pay the cost. This creates a very strong incentive for gen-

erators to reduce, recycle, or treat wastes in-house. It also makes consumer goods more expensive as increased manufacturing costs are passed on. A firm that misrepresents the contents of its wastes is charged 30 percent more than the normal fee for the first offense. The second offense brings a 60 percent increase in charges plus the costs of thorough chemical analysis of the wastes. There has never been a second offense.

The system is not perfect. There is no incentive for the waste corporation to make its operation cost-effective, since it has a monopoly and its customers have no alternative to its service. There also is little reason to develop or adopt new technologies, such as advanced plasma-gas incineration, since criteria for operation are legislatively mandated. Many people in the United States would object to the level of social control and the loss of personal freedom represented by a nationwide agency with broad powers and rigid regulations. If the alternative is to be poisoned, however, we may have to change our traditional patterns of behavior.

Summary

We produce enormous volumes of solid waste in industrialized societies, and there is an increasing problem of how to dispose of those wastes in an environmentally safe manner. In this chapter, we have looked at the character of our solid and hazardous wastes. We have surveyed the ways we dispose of our wastes and the environmental problems associated with waste disposal.

Solid wastes are domestic, commercial, industrial, agricultural, and mining wastes that are primarily nontoxic. About 80 percent of our domestic and industrial wastes are deposited in landfills; most of the rest is incinerated or recycled. Landfills are often messy and leaky, but they can be improved with impermeable clay or plastic linings, drainage, and careful siting. Incineration can destroy organic compounds, but whether incinerators can or will be operated satisfactorily is a matter of debate. Recycling is growing nationwide, encouraged by the economic and environmental benefits it brings. City leaders tend to doubt the viability of recycling programs, but successful programs have been sustained in other countries and in some American cities.

Near major urban centers, land suitable for waste disposal is becoming increasingly scarce and expensive. Costs for solid waste disposal totaled nearly $10 billion in the United States in 1990 and are expected to climb to $100 billion per year by the end of this century. A few cities now ship their refuse to other states or even other countries, but worries about toxic and hazardous material in the waste are leading to increasing resistance to shipping or storing it.

Hazardous and toxic wastes, when released into the environment, cause such health problems as birth defects, neurological disorders, reduced resistance to infection, and cancer. Environmental losses include contamination of water supplies, poisoning of the soil, and destruction of habitat. The major categories of hazardous wastes are ignitable, corrosive, reactive, explosive, and toxic. About 60 million metric tons of hazardous waste enter the waste stream each year in the United States, more than 90 percent of which is disposed of by environmentally unsound practices. Some materials that cause the most concern are heavy metals, solvents, and synthetic organic chemicals such as halogenated hydrocarbons, organophosphates, and phenoxy herbicides.

Disposal practices for solid and hazardous wastes have often been unsatisfactory. Thousands of abandoned, often unknown waste disposal sites are leaking toxic materials into the environment. Some alternative techniques for treating or disposing of hazardous wastes include not making the material in the first place, incineration, secure landfill, and physical, chemical, or biological treatment to detoxify or immobilize wastes. People are often unwilling to have transfer facilities, storage sites, disposal operations, or transportation of hazardous or toxic materials in or through their cities. Questions of safety and liability remain unanswered in solid and hazardous waste disposal.

Review Questions

1. What are solid wastes and hazardous wastes? What is the difference?
2. How much solid and hazardous waste do we produce each year in the United States? How do we dispose of the waste?
3. Why are landfill sites becoming rare around most major urban centers in the United States? What steps are being taken to solve this problem?
4. Describe some concerns about waste incineration.
5. List some benefits and drawbacks of recycling wastes. What are the major types of materials recycled from municipal waste and how are they used?
6. What is composting, and how does it fit into solid waste disposal?
7. Describe some ways that we can reduce the waste stream to avoid or reduce disposal problems.
8. List ten toxic substances in your home and how you would dispose of them.
9. What are some illegal hazardous waste disposal methods, and why are they dangerous?
10. What societal problems are associated with waste disposal? Why do people object to waste handling in their neighborhoods?

Questions for Critical or Reflective Thinking

1. A toxic waste disposal site has been proposed for the Pine Ridge Indian Reservation in South Dakota. Many tribal members oppose this plan but some favor it because of the jobs and income it will bring to an area with 70 percent unemployment. If local people choose immediate survival over long-term health, do we have a right to object or intervene?

2. There is often a tension between making environmental changes in your personal life and working for larger structural changes in society. Make a list of the arguments for and against spending time and energy sorting recyclables at home compared to working in the public arena on a bill to ban excess packaging.

3. Should industry officials be held responsible for dumping chemicals that were legal when they were dumped but are now known to be extremely dangerous? At what point can we argue that they should have known about the hazards involved?

4. Look at either the discussion of recycling or the discussion of incineration presented in this chapter. List the premises (implicit or explicit) that underlie the presentation as well as the conclusions (stated or not) that seem to be drawn from them. Do the conclusions necessarily follow from the premises?

5. Suppose that your brother or sister has decided to buy a house in Love Canal because it is selling for $20,000 below market value. What do you say to him or her?

6. Is there an overall conceptual framework or point of view in this chapter? If you were presenting a discussion of solid or hazardous waste to your class, what would be your conceptual framework?

7. Is there a fundamental difference between incinerating municipal, medical, or toxic industrial waste? Would you oppose an incinerator for one type of waste in your neighborhood but not others? Why, or why not?

8. The Netherlands incinerates much of its toxic waste out at sea by a ship-borne incinerator. Would you support this as a way to dispose of our wastes as well? What are the critical considerations for or against this approach?

Key Terms

biodegradable plastics 286
composting 284
energy recovery 280
hazardous waste 287
landfills 279
mass burn 280
permanent retrievable storage 293

photodegradable plastics 286
recycling 282
refuse-derived fuel 280
secure landfills 294
waste-to-energy 280
waste stream 277

Suggested Readings

Blumberg, L., and R. Gottlieb. *War on Waste.* Washington, D.C.: Island Press, 1989. An excellent book on all aspects of waste, from generation, to shipping, to incineration and landfilling. Includes case studies and examples.

Fairlie, S. "Long Distance, Short Life: Why Big Business Favors Recycling." *The Ecologist* 22, no. 6 (November/December 1992): 276. Industry promotes recycling rather than reuse for two main reasons: recycling is better suited to centralized, long-distance markets; and it can be used as an environmental justification for short-life products.

Grieder, W. "Hazardous Waste Exports: Changes in Sight." *EPA Journal* 16, no. 4 (1990): 46. Although progress is being made in controlling hazardous waste exports, much dangerous material still crosses boundaries. The U.S. Congress and the United Nations are considering measures to curb toxic exports.

Hall, R. H. "Poisoning the Lower Great Lakes: The Failure of U.S. Environmental Legislation." *The Ecologist* 16, no. 2/3 (1986): 118. Exposes the broader threats behind and beyond Love Canal.

Holmes, H. "Recycling Plastics." *Garbage* 3, no. 1 (January/February 1991): 32. A clear and comprehensive discussion of how recycling works, when it works.

Kunreuther, H., and R. Patrick. "Managing the Risks of Hazardous Waste." *Environment* 33, no. 3 (1991): 12. This article suggests that we need better information about the risks of hazardous waste before effective management strategies can be adopted.

Laurence, D., and B. Wynne. "Transporting Waste in the European Community: A Free Market?" *Environment* 31, no. 6 (1989): 12. As the European Community moves toward free trade, some groups fear the implications for hazardous waste transport between member nations.

Levenson, H. "Wasting Away: Policies to Reduce Trash Toxicity and Quantity." *Environment* 32, no. 2 (1990): 10. Describes policies that could reduce the quantity and toxicity of the waste we generate.

Luoma, J. "Trash Can Realities." *Audubon,* March 1990, 92. Garbage is rife with symbols of wickedness and virtue, but conventional environmental wisdom may not be so wise.

Newsday. *Rush to Burn: Solving America's Garbage Crisis.* Washington, D.C.: Island Press, 1989. Written by a team of reporters from Newsday, this book is an excellent review of our throwaway society and the effects of incineration. One of a series of excellent books on this subject published by Island Press.

Packard, V. *Waste Makers.* New York: David McKay, 1960. A classic but still highly readable account of how we have become a throwaway society.

Poore, P. "Disposable Diapers Are OK." *Garbage* 4, no. 5 (October/November 1992): 26. An interesting study both of the life-cycle costs of disposable and reusable diapers and the politics and theology of recycling.

Rathje, W. R. "The History of Garbage." *Garbage* 2, no. 5 (September/October 1990): 32. A good introduction to garbology, the science of analyzing garbage, with amazing stories about the longevity of ordinary materials in landfills.

Schneider, Paul. "Other People's Trash." *Audubon,* July/August 1991, 108. Reveals corporate schemes to dump wastes on Indian reservations.

Sekscienski, G. "Speaking of Composting." *EPA Journal* 18, no. 3 (July/August 1992): 14. A good summary of composting in an issue entirely devoted to waste disposal.

Starr, D. "Shoppers, Shoppers, Everywhere." *Audubon,* March 1990, 98. Italians use 7.5 billion plastic shopping bags a year and most of them wind up littering the landscape. This article tells how one outraged mayor took precipitous action.

Stigliani, W. M., et al. "Chemical Time Bombs: Predicting the Unpredictable." *Environment* 33, no. 4 (1991): 4. We have treated our environment as a "sink" for waste disposal. Now some areas have become poisonous wastelands. How can these chemical time bombs be detected and defused?

Thorp, L. "How to Stop Incinerators: A Community Owner's Manual." *Greenpeace* 3 (January/February 1991): 20. Lessons from the grass-roots campaign to ban the burners.

Young, J. E. "Reducing Waste, Saving Materials." *State of the World.* No. 39. Washington, D.C.: Worldwatch Institute, 1991. Describes how we can reduce waste and save resources at the same time.

14
*U*RBANIZATION AND SUSTAINABLE CITIES

The garden is the paradise of nature and the city is the paradise of culture. Or at least they could be. . . . Today, both are out of balance.

Richard Register, *Ecocity Berkeley*

O b j e c t i v e s

After studying this chapter, you should be able to:

- Distinguish between a rural village, a city, and a megacity.
- Recognize the push and pull factors that lead to urban growth.
- Appreciate the growth rate of giant metropolitan urban areas such as Mexico City, as well as the problems this growth engenders.

- Picture the living conditions for ordinary citizens in the megacities of the developing world.
- Understand the causes and consequences of noise and crowding in cities.
- Discuss options for suburban design.
- See the connection between sustainable economic development, social justice, and the solution of urban problems.

INTRODUCTION

For more than six thousand years, cities have been powerful sources of technological developments and social change, reflecting both the best and worst of human nature and human history. In many ways, the growth of cities and the emergence of civilization have been interdependent. Cities are cultural and racial melting pots in which information and technology are exchanged and resources are mobilized. Advances in art, science, education, architecture, and ethical concepts that are the human legacy for the future have come from the great cities of the world. Cities have been sources of energy, vitality, and progress; they also have been the source of pollution, crowding, disease, misery, and oppression.

Nearly half the people in the world now live in urban areas. Demographers predict that by the end of the twenty-first century 80 or 90 percent of all humans will live in cities and that some giant interconnecting metropolitan areas could have hundreds of millions of residents. In this chapter, we will look at how cities came into existence, why people live there, and what the environmental conditions of cities have been, are now, and might be in the future. Some of the most severe urban problems in the world are found in the giant megacities of the developing countries. Far more lives may be threatened by the desperate environmental conditions in these cities than by any other environmental danger we have studied. We will look at a few of those problems and possible solutions that could improve the urban environment.

URBANIZATION

Since their earliest origins, cities have been centers of education, religion, commerce, record keeping, communication, and political power (fig. 14.1). As cradles of civilization, cities have influenced culture and society far beyond their proportion of the total population. Until recently, however, only a small percentage of the world's people lived permanently in urban areas, and even the greatest cities of antiquity were

Figure 14.1 Since their earliest origins, cities have been centers of education, religion, commerce, politics, and culture. They have also been the source of pollution, crowding, disease, and misery. The Parthenon, one of the most beautiful buildings in the world, is being eaten away by air pollution, vibrations, and other urban ills of Athens, Greece.

small by modern standards. The vast majority of humanity has always lived in rural areas where farming, fishing, hunting, timber harvesting, animal herding, mining, or other natural resource-based occupations provided support.

Since the beginning of the Industrial Revolution some three hundred years ago, however, cities have grown rapidly in both size and power. In every developing country, the transition from an agrarian society to an industrial one has been accompanied by **urbanization,** an increasing concentration of the population in cities and a transformation of land use and society to a metropolitan pattern of organization. Industrialization and urbanization bring many benefits—especially to the top members of society—but they also cause many problems, as we will detail in this chapter.

What Is a City?

Just what makes up an urban area or a city? Definitions differ. The U.S. Census Bureau considers any incorporated community to be a city, regardless of size, and defines any city with more than 2,500 residents as urban. More meaningful definitions are based on *functions.* In a **rural area,** most residents depend on agriculture or other ways of harvesting natural resources for their livelihood. In an **urban area,** by contrast, a majority of the people are not directly dependent on natural resource-based occupations.

A **village** is a collection of rural households linked by culture, custom, family ties, and association with the land (fig. 14.2). A **city,** by contrast, is a differentiated community with a population and resource base large enough to allow residents to specialize in arts, crafts, services, or professions rather than natural resource-based occupations. While the rural village

often has a sense of security and connection, it also can be stifling. A city offers more freedom to experiment, to be upwardly mobile, and to break from restrictive traditions, but it can also be harsh and impersonal (fig. 14.3).

Beyond about 10 million inhabitants, an urban area is considered a supercity or **megacity.** Megacities in many parts of the world have grown to enormous size. In the United States, urban areas between Boston and Washington, D.C., have merged into a nearly continuous megacity (sometimes called Bos-Wash) containing about 35 million people. The Tokyo-Yokohama-Osaka-Kobe corridor of Japan contains nearly 50 million people. The Paris Basin (or Ille de France) encompasses nearly half the population of France, and the Antwerp-Brussels-Hamburg complex of northern Europe spreads across four countries. Architect and city planner C. A. Dioxiadis predicts that if current trends continue, the sea coasts and major river valleys of most continents will be covered with continuous strip cities, which he calls Ecumenopolises, each containing billions of people.

World Urbanization

The United States underwent a dramatic rural-to-urban shift in the nineteenth and early twentieth centuries (fig. 14.4). Now many developing countries are experiencing a similar demographic movement. In 1850, only about 2 percent of the world's population lived in cities. By 1990, 43 percent of the people in the world were urban. By the end of this century, for the first time in history, more people will live in cities than in the country. Only Africa and South Asia remain predominantly rural, but people there are swarming into cities in

Figure 14.2 A village, like this traditional hill-tribe settlement in northern Thailand, is closely tied to the land through culture, economics, and family relationships. While the timeless pattern of life here gives a great sense of identity, it can also be stifling and repressive.

Figure 14.3 A city is a differentiated community with a large enough population and resource base to allow specialization in arts, crafts, services, and professions. Although there are many disadvantages in living so closely together, there are also many advantages. Cities are growing rapidly and most of the world's population will live in urban areas if present trends continue.

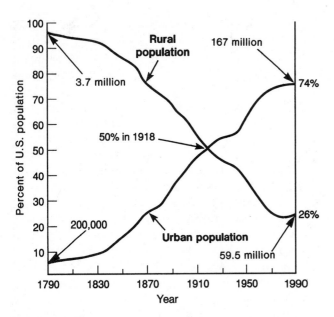

Figure 14.4 U.S. rural-to-urban shift. Percentages of rural and urban populations, 1790 to 1990.

Source: U.S. Bureau of the Census.

TABLE 14.1 Urban share of total population (percent)

Region	1950	1990	2000 (estimate)
Asia	29	31	40
Africa	15	30	42
Europe	56	75	79
Former Soviet Union	39	66	74
Latin America	41	70	79
North America	64	75	78
Oceania	61	73	73
World	29	43	48

Source: World Resources Institute.

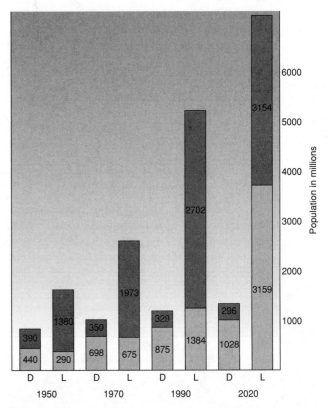

Figure 14.5 Distribution of rural and urban population in developed (D) and less-developed (L) countries. Over the next fifty years, more than 90 percent of all urban growth will take place in the less-developed countries.

Source: Data from United Nations, *Prospects of World Urbanization*, 1988, p. 28 (E.89.x111.8).

ever-increasing numbers. About three-fourths of the people in the developed countries and two-thirds of those in Latin America and East Asia already live in cities (table 14.1). Some urbanologists predict that by 2100 the whole world will be urbanized to the levels now seen in developed countries.

As figure 14.5 shows, 90 percent of the population growth over the next fifty years is expected to occur in the less-developed countries of the world. Three-quarters of that growth will be in the already over-crowded cities of the least affluent countries, such as India, China, Mexico, and Brazil. The combined population of these cities is expected to jump from its present 1 billion to nearly 4 billion by the year 2025. Meanwhile, rural populations in these countries are expected to decline somewhat as rural people migrate into the cities.

Recent urban growth has been particularly dramatic in the largest cities, especially those of the developing world. In 1900, thirteen cities had populations over 1 million; all except Tokyo were in Europe or North America (table 14.2). Presently, only one of the ten largest cities (New York) is in North America or Europe. By 1990, there were 235 metropolitan areas of more than 1 million people—an eighteenfold increase. In 1900, London was the only city with more than 5 million people; now nineteen cities have populations above 5 million. Some futurists predict that by 2025 at least four hundred cities will have populations of 1 million or more, and ninety-three super-cities each will have 5 million or more residents. Three-fourths of those cities will be in the developing nations (fig. 14.6).

The growth rate of the most rapidly expanding cities and the sizes they are predicted to reach are truly astounding (table 14.2). Many cities are growing at rates above 5 percent per year, which means they double in less than fifteen years. Mexico City, for instance, with a population of about 2.45 million in 1950, was the largest city in the world in 1990, with

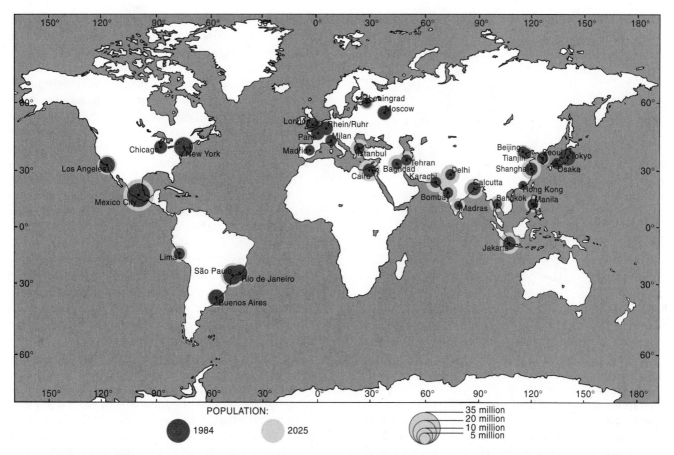

Figure 14.6 By 2025 at least 400 cities will have populations of 1 million or more and 93 supercities will have populations above 5 million. Three-fourths of the world's largest cities will be in developing countries that already have trouble housing, feeding, and employing their people.

TABLE 14.2 Top ten metropolitan regions (populations in millions)

1900		1985		2000 (Estimated)	
London	6.6	Mexico City	18.1	Mexico City	31.0
New York	4.2	Tokyo	17.2	São Paulo	25.8
Paris	3.3	São Paulo	15.9	Tokyo	24.2
Berlin	2.4	New York	15.3	New York	22.8
Chicago	1.7	Shanghai	11.8	Shanghai	22.7
Vienna	1.6	Calcutta	11.0	Beijing	19.9
Tokyo	1.5	Buenos Aires	10.9	Los Angeles	17.1
Saint Petersburg	1.4	Rio de Janeiro	10.4	Bombay	16.8
Philadelphia	1.4	Seoul	10.4	Calcutta	16.7
Manchester	1.3	Bombay	10.1	Jakarta	16.6

Source: United Nations.

19.4 million residents. Now growing at a rate of about two thousand people per day (fig. 14.7), it is expected to exceed 26 million by the year 2000, an elevenfold increase in fifty years. If that growth trend continues, Mexico City could have 100 million inhabitants by the middle of the next century.

Can a city function with that many people? Can it supply food, water, sanitation, transportation, jobs, housing, police protection, fire control, electricity, education, and other public services necessary to sustain a civilized way of life? Adding 750,000 new people annually to Mexico City amounts to building a

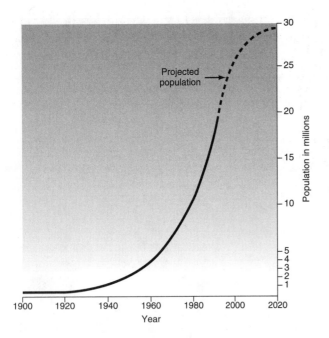

Figure 14.7 Growth of Mexico City metropolitan area, 1900–2020.

Source: U.N. Population Council.

new city the size of Baltimore or San Francisco every year. This growth is occurring, as is most urban growth in the world, in a country with a sagging economy, an unstable government, and a high foreign debt load. The environmental costs (air and water pollution, soil erosion, solid and hazardous waste accumulation, noise and congestion) as well as the human costs (lack of housing, jobs, transportation, health care, social services, and living space) are tragic.

CAUSES OF URBAN GROWTH

Urban populations grow in two ways: by natural increase (more births than deaths) and by immigration. Natural increase is fueled by improved food supplies, better sanitation, and advances in medical care that reduce death rates and cause populations to grow both within cities and in the rural areas around them (chapter 4). In Latin America and East Asia, natural increase is responsible for two-thirds of urban population growth. In Africa and West Asia, immigration is the largest source of urban growth. Immigration to cities can be caused both by **push factors** that force people out of the country and by **pull factors** that draw them into the city.

Immigration Push Factors

People migrate to cities for many reasons. In some areas, the countryside is overpopulated and simply can't support more people. The "surplus" population is forced to migrate to cities in search of jobs, food, and housing. Not all rural-to-urban shifts are caused by overcrowding in the country, however. In some

places, economic forces or political, racial, or religious conflicts drive people out of their homes. The countryside may actually be depopulated by such demographic shifts. The United Nations estimated that in 1992 at least 10 million people fled their native country and that another 30 or 40 million were internal refugees within their own country, displaced by political, economic, or social instability. Many of these refugees end up in the already overcrowded megacities of the developing world.

Land tenure patterns and changes in agriculture also play a role in pushing people into cities. The same pattern of agricultural mechanization that made farm labor obsolete in the United States early in this century is spreading now to developing countries. Furthermore, where land ownership is concentrated in the hands of a wealthy elite, subsistence farmers are often forced off the land so it can be converted to grazing lands or monoculture cash crops (chapter 8). Speculators and absentee landlords let good farm land sit idle that otherwise might house and feed rural families.

Immigration Pull Factors

Even in the largest and most hectic cities, many people are there by choice, attracted by the excitement, vitality, and chance to meet others like themselves. Cities offer jobs, housing, entertainment, and freedom from the constraints of village traditions. Possibilities exist in the city for upward social mobility, prestige, and power not ordinarily available in the country. Cities support specialization in arts, crafts, and professions for which markets don't exist elsewhere.

Modern communications also draw people to cities by broadcasting images of luxury and opportunity. An estimated 90 percent of the people in Egypt, for instance, have access to a television set. The immediacy of television makes city life seem more familiar and attainable than ever before. We see pictures of beggars and homeless people on the streets of teeming Third World cities and generally assume they have no other choice, but many of these people want to be in the city. In spite of what appears to be dismal conditions, living in the city may be preferable to what the country had to offer.

Government Policies

Government policies often favor urban over rural areas in ways that both push and pull people into the cities. Developing countries commonly spend most of their budgets on improving urban areas (especially around the capital city where leaders live), even though only a small percentage of the population lives there or benefits directly from the investment. This gives the major cities a virtual monopoly on new jobs, housing, education, and opportunities, all of which bring in rural people searching for a better life.

In Peru, for example, Lima accounts for 20 percent of the country's population but has 50 percent of the national wealth, 60 percent of the manufacturing, 65 percent of the retail trade, 73 percent of the industrial wages, and 90 percent of all banking in the country. Similar statistics pertain to São Paulo, Mexico City, Manila, Cairo, Lagos, Bogotá, and a host of other cities.

Governments often manipulate exchange rates and food prices for the benefit of more politically powerful urban populations but at the expense of rural people. Importing lower-priced food pleases city residents, but local farmers then find it uneconomical to grow crops. As a result, more people leave rural areas to become part of a large urban work force, keeping wages down and industrial production high. Zambia, for instance, sets maize prices below the cost of local production to discourage farming and to maintain a large pool of workers for the mines. Keeping the currency exchange rate high stimulates export trade but makes it difficult for small farmers to buy the fuels, machinery, fertilizers, and seeds that they need. This depresses rural employment and rural income while stimulating the urban economy. The effect is to transfer wealth from the country to the city.

CURRENT URBAN PROBLEMS

Large cities in both developed and developing countries face similar challenges in accommodating the needs and by-products of dense populations. The problems are most intense, however, in rapidly growing cities of developing nations.

The Developing World

Ninety percent of the human population growth in the next century is expected to occur in the developing world, mainly in Africa, Asia, and South America (see fig. 14.5). Almost all of that growth will occur in cities—especially the largest cities—which already have trouble supplying food, water, housing, jobs, and basic services for their residents. The unplanned and uncontrollable growth of those cities causes serious urban environmental problems. Let's examine some of them. Keep in mind that these problems are not limited to Third World cities, but that their intensity is magnified when compared to wealthy cities of the developed countries.

Traffic and Congestion

A first-time visitor to a supercity—particularly in a less-developed country—is often overwhelmed by the immense crush of pedestrians and vehicles of all sorts that clog the streets. The noise, congestion, and confusion of traffic make it seem suicidal to venture onto the street (fig. 14.8). Cairo, for instance, is one of the most densely populated cities in the world. Traffic is chaotic almost all the time. People commonly spend

Figure 14.8 Motorized rickshaws, motor scooters, bicycles, street vendors, and pedestrians all vie for space on the crowded streets of Delhi. The heat, noise, smells, and sights are overpowering. In spite of the difficulties, however, many people here lead reasonably happy lives.

three or four hours each way commuting to work from outlying areas. Calcutta also has monumental traffic problems. The Howrah Bridge over the Hooghly River carries 30,000 motor vehicles and 500,000 pedestrians daily. Gigantic traffic jams occur day and night and there is almost always a wait to cross over.

Air Pollution

The dense traffic (commonly old, poorly maintained vehicles), smoky factories, and use of wood or coal fires for cooking and heating often create a thick pall of air pollution in the world's supercities. Lenient pollution laws, corrupt officials, inadequate testing equipment, ignorance about the sources and effects of pollution, and lack of funds to correct dangerous situations usually exacerbate the problem. What is its human toll? An estimated 60 percent of Calcutta's residents are thought to suffer from respiratory diseases linked to air pollution. Lung cancer mortality in Shanghai is reported to be four to seven times higher than rates in the countryside. Mexico City, which sits in a high mountain bowl with abundant sunshine, little rain, high traffic levels, and frequent air stagnation, has one of the highest levels of photochemical smog (chapter 10) in the world.

Sewer Systems and Water Pollution

Few cities in developing countries can afford to build modern waste treatment systems for their rapidly growing cities. The World Bank estimates that only 35 percent of urban residents in developing countries have satisfactory sanitation services. The situation is especially desperate in Latin America, where only 2 percent of urban sewage receives any treatment. In Egypt, Cairo's sewer system was built about fifty years

Crowding, Stress, and Crime in Cities

*L*arge cities in America clearly experience more crimes per capita than do smaller towns and rural areas. In 1985, according to the Federal Bureau of Investigation, cities with populations above 250,000 had an average murder rate more than twice the national average of 8.0 per 100,000. Box table 14.1 shows that larger cities had approximately three times the murder rate of either rural areas or cities under 100,000 residents, and they had six times as many murders per capita as towns under 10,000 people.

Some people believe that the sheer size of big cities makes them violent and inhumane, that constantly being surrounded by too many people (box fig. 14.1.1) causes nervous overload and leads to mental illness, deviant behavior, and cultural breakdown. It is argued that the anonymity and stress of the urban environment underlie much of the hostility, paranoia, criminality, drug abuse, and other abnormal or antisocial behaviors that plague modern urban societies.

One line of evidence linking crowding to abnormal behavior comes from studies of laboratory animals living in very high population densities. When rats or mice are given unlimited food and water and allowed to reproduce to very

Figure 14.1.1 Crowds of people jam the streets in Calcutta. Although prices are high, living conditions are crowded, and air quality is bad, many people still want to come here because of the job opportunities, entertainment, education, and vitality of the city.

high numbers in a confined space, they frequently exhibit a condition called "stress-related disease."

Whether this says anything about human biology and human behavior, however, remains to be seen. Humans are

BOX TABLE 14.1 Annual crime rate per 100,000 residents in United States cities					
City population	Murder	Rape	Robbery	Assault	Property crimes
Above 250,000	18.6	74	670	526	8,596
50,000–100,000	5.9	38	185	320	5,898
Under 10,000	2.9	17	39	192	3,946
Rural areas	5.2	18	15	160	1,900

ago to serve a population of 2 million people. It is now being overwhelmed by more than 11 million people. Less than one-tenth of India's three thousand towns and cities have even partial sewage systems and water treatment facilities. Some 150 million of India's urban residents lack access to sanitary sewer systems. In Colombia, the Bogotá River, 200 km (125 mi) downstream from Bogotá's 5 million residents, still has an average fecal bacterial count of 7.3 million cells per liter, more than 7,000 times the safe drinking level and 3,500 times higher than the limit for swimming.

Some 400 million people or about one-third of the population in developing world cities do not have safe drinking water, according to the World Bank.

Where people have to buy water from merchants, it often costs one hundred times as much as piped city water and may not be safe to drink after all. Many rivers and streams in Third World countries are little more than open sewers, and yet they are all that poor people have for washing clothes, bathing, cooking, and—in the worst cases—for drinking (chapter 11). Diarrhea, dysentery, typhoid, and cholera are widespread diseases in these countries, and infant mortality is tragically high (chapter 6).

Housing

The United Nations estimates that at least 1 billion people—20 percent of the world's population—live in crowded, unsanitary slums of the central cities and in much more complex than rodents, and are able to adapt to or change their environment in ways that rats cannot. People can live successfully in very high densities if that is considered normal in their culture, if they have a choice of where and how they live, and if they have a sense of hope for the future. Boats carrying refugees from German concentration camps to Israel in 1945 were so crowded that disease and conflict would seem to be inevitable. However, these people were going from much worse to much better conditions, and they survived very well.

In some societies, people always are in direct physical contact with others. They find privacy simply by withdrawing mentally and emotionally from what is going on around them. Cultures that are adjusted to crowded conditions develop behavioral traits, such as soft voices, polite speech, and confrontation avoidance, that minimize conflict. Hong Kong, for instance, is one of the most densely populated cities in the world. More than 5 million people live in approximately 65 sq km (220 sq mi), yet they have one of the lowest crime rates in the world. By contrast, metropolitan Los Angeles has one-one hundredth the density of Hong Kong but ten times the murder rate and one hundred times the violent crime rate.

It is often suggested that western, urban, technological societies are more violent and crime-prone than others, but statistics do not bear this out. Many rural, sparsely populated countries are just as violent as industrial, densely crowded countries. The three highest murder rates in the world (number per 100,000 people per year) are in Lesotho (140), the Bahamas (23), and Guyana (22). None of these countries is highly urbanized, but all have a history of foreign domination, high poverty levels, and cultural and racial conflict.

By contrast, some of the lowest murder rates in the world are in the more densely populated and highly urbanized countries. Indonesia, South Korea, and Japan have less than two murders per 100,000 people; Denmark, England, Malaysia, Hong Kong, the Philippines, and Singapore each have between two and three murders per 100,000. Clearly, the residents of the slums and shantytowns of Manila, Jakarta, Hong Kong, and Kuala Lumpur are subjected to crowding, poverty, pollution, noise, squalor, and other stressful aspects of urban living that are at least as bad as the worst parts of American cities, and yet they don't seem to have nearly as much violence as western countries have. There must be other factors to consider.

Some sociologists argue that neither the absolute size nor density of cities cause most crime and violence but rather the concentration of poverty and social problems. They point out that it was mostly people at the bottom of the socio-economic spectrum who migrated from other countries or moved into the city during the rural-to-urban shift at the beginning of this century. Furthermore, the least successful and, therefore, least mobile people remained in the city when the upper and middle classes fled to the suburbs after the Second World War. The result has been an accumulation of people who are trapped in a "culture of poverty" from a continuing cycle of family violence, drug abuse, low self-esteem, hopelessness, undereducation, unemployment, and neglect.

What can we do about crime and violence in the city? Clearly, we all would benefit if the stresses in urban environments caused by pollution, noise, litter, squalor, and congestion were reduced. We also should try to make decent housing, jobs, and education available to all. Perhaps we also need to find ways to divide the city into smaller neighborhoods, encourage diversity in population and housing, reduce the total population size and density, and provide opportunities for advancement and hope for the future so that people won't feel so alienated and angry.

the vast shantytowns and squatter settlements that ring the outskirts of most Third World cities. Around 100 million people have no home at all. In Bombay, for example, it is thought that half a million people sleep on the streets, sidewalks, and traffic circles because they can find no other place to live (fig. 14.9). In Brazil, perhaps 3 million "street kids" who have run away from home or been abandoned by their parents live however and wherever they can. This is surely a symptom of a tragic failure of social systems.

Slums are generally legal but inadequate multi-family tenements or rooming houses, either custom built to rent to poor people or converted from some other use. The chals of Bombay, for example, are high-rise tenements built in the 1950s to house immi-grant workers. Never very safe or sturdy, these dingy, airless buildings are already crumbling and often collapse without warning. Eighty-four percent of the families in these tenements live in a single room; half of those families consist of six or more people. Typically, they have less than 2 square meters of floor space per person and only one or two beds for the whole family. They may share kitchen and bathroom facilities down the hall with fifty to seventy-five other people. Even more crowded are the rooming houses for mill workers where up to twenty-five men sleep in a single room only 7 meters square. Because of this crowding, household accidents are a common cause of injuries and deaths in Third World cities, especially to children. Charcoal braziers or kerosene stoves used

Figure 14.9 In Bombay, as many as half a million people sleep on the streets because they have no other place to live. Ten times as many live in crowded, dangerous slums and shantytowns throughout the city.

Figure 14.10 A shantytown or spontaneous settlement on the outskirts of Mexico City. Millions of people live with no water, power, or sanitation in *colonias* such as this.

in crowded homes are a routine source of fires and injuries. With no place to store dangerous objects beyond the reach of children, accidental poisonings and other mishaps are a constant hazard.

Shantytowns are settlements created when people move onto undeveloped lands and build their own houses. Shacks are built of corrugated metal, discarded packing crates, brush, plastic sheets, or whatever building materials people can scavenge. Some shantytowns are simply illegal subdivisions where the landowner rents land without city approval. Others are spontaneous or popular settlements or **squatter towns** where people occupy land without the owner's permission. Sometimes this occupation involves thousands of people who move onto unused land in a highly organized, overnight land invasion, building huts and laying out streets, markets, and schools before authorities can root them out (fig. 14.10). In other cases, shantytowns just gradually "happen."

Called barriads, barrios, favelas, or turgios in Latin America, bidonvillas in Africa, or bustees in India, shantytowns surround every megacity in the developing world. They are not an exclusive feature of poor countries, however. Some 200,000 immigrants live in the colonias along the southern Rio Grande in Texas. Only 2 percent have access to adequate sanitation. Most live in conditions as awful as you would see in any Third World city. Smaller enclaves of the poor and dispossessed can be found in most American cities.

The problem is magnified in less-developed countries. Nouakchott, Mauritania, the fastest growing city in the world, is almost entirely squatter settlements and shantytowns. It has been called "the world's largest refugee camp." About three-quarters of the residents of Addis Ababa, Ethiopia, or Luanda, Angola,

live in squalid refugee camps. Two-thirds of the population of Calcutta live in unplanned squatter settlements, and nearly half of the 19.4 million people in Mexico City live in uncontrolled, unauthorized shantytowns also called colonias. Many governments try to clean out illegal settlements by bulldozing the huts and sending riot police to drive out the settlers, but the people either move back in or relocate in another shantytown elsewhere.

These popular but unauthorized settlements usually lack sewers, clean water supplies, electricity, and roads. Often the land on which they are built was not previously used because it is unsafe or unsuitable for habitation. In Bhopal, India, and Mexico City, for example, squatter settlements were built next to deadly industrial sites. In Rio de Janeiro, La Paz (Bolivia), Guatemala City, and Caracas (Venezuela), they are perched on landslide-prone hills. In Bangkok, thousands of people live in shacks built over a fetid tidal swamp. In Lima (Peru), Khartoum (Sudan), and Nouakchott, shantytowns have spread onto sandy deserts. In Manila, 20,000 people live in huts built on towering mounds of garbage and burning industrial waste in city dumps.

As desperate and inhumane as conditions are in these slums and shantytowns, many people do more than merely survive there. They keep themselves clean, raise families, educate their children, find jobs, and save a little money to send home to their parents. They learn to live in a dangerous, confusing, and rapidly changing world and have hope for the future. The people have parties; they sing and laugh and cry. They are amazingly adaptable and resilient. In many ways, their lives are no worse than those in the early industrial cities of Europe and America a century ago. Perhaps continuing development will bring better conditions to cities of the Third World as well.

The Developed World

For the most part, the rapid growth of central cities that accompanied industrialization in nineteenth- and early twentieth-century Europe and North America has now slowed or even reversed. London, for instance, once the most populous city in the world, has lost nearly 2 million people, dropping from its high of 8.6 million in 1939 to about 6.7 million now. While the greater metropolitan area surrounding London has been expanding to about 10 million inhabitants, it is now only the twelfth largest city in the world.

Many of the worst urban environmental problems of industrialized countries have been substantially reduced in recent years. Air and water quality have improved greatly. Working conditions and housing are better for most people. Improved sanitation and medical care have reduced or totally eliminated many of the worst communicable diseases that once were the scourge of cities. Dispersal of the people and businesses to the suburbs has significantly reduced crowding and congestion. Still, city life is generally regarded as being more stressful than life in the country or suburbs (see box 14.1). Part of the stress is related to the high noise levels in busy cities (box 14.2).

The major problems now facing cities of the United States tend to be associated with decay and blight. Businesses and jobs have fled to the suburbs along with the middle class. Tax revenues are down. The city infrastructure—streets, sewers, schools, and housing—is getting old and dilapidated. Drug dealers, pawn shops, bars, sleazy rooming houses, graffiti, broken windows, trash, and abandoned cars have become symbols of the worst parts of inner cities. Freeways make it much easier for suburbanites to get into and out of the city, but they also destroy housing, cause noise, increase air pollution, and create vast concrete landscapes that divide neighborhoods and make travel difficult for people without cars.

Many of the social problems of central cities are a result of concentrated minority populations—the poor, elderly, handicapped, unemployed, homeless, or otherwise socially depressed members of society—who were left behind when the middle and upper classes moved to the suburbs. Unemployment in some inner-city areas is routinely 50 percent or higher, and jobs that are available generally are menial, minimum-wage work with little prospect for advancement. Homelessness is an increasing problem that is full of depressing statistics. Welfare agencies estimate that 3.5 million people are homeless in the United States and ten to twenty times that number live in substandard housing. Most of the homeless are in large cities, and about one-third are now families. Of the single females living on the street, perhaps three-fourths are mentally disturbed and would formerly have been

Figure 14.11 A "bag lady" carrying all her possessions on her back looks for shelter, food, and safety in an affluent American city. Many people fall through the safety net of social services designed to help them.

cared for in hospitals or institutions (fig. 14.11). Of the single, homeless men, two-thirds are elderly poor, disabled, drug addicts, or alcoholics.

The increasing concentration of poor and disadvantaged people in cities began at a time when federal assistance was being cut by 80 percent and city tax revenues were declining. The rising statistics of violence, drug abuse, child neglect, and unwanted pregnancies are symptoms of poverty and decay in the cities.

How do we break the cycle of poverty, ignorance, and lack of opportunity in which inner-city residents often seem to be trapped? How do we deal with the homeless people, especially those who are unable or barely able to take care of themselves? Are the negative aspects of large cities doomed to increase, or can they be dealt with constructively? Consider Detroit and Watts; they not only survived their crises, but have been improved.

TRANSPORTATION AND CITY GROWTH

Transportation has always played an essential role in city development. Most cities are located at crossroads, river fords, seaports, junctions of major rivers, agricultural centers, or other sites where travelers were brought together and merchandise could be traded, bought, or sold. Transportation is needed to bring building materials, energy sources, and food into the city and to remove wastes. Most older cities have been remodeled and reshaped repeatedly by the changing transportation systems that provide their life-giving circulation.

Before the Industrial Revolution, most cities were compact and densely populated (fig. 14.12a). Except for a few broad ceremonial boulevards, streets were narrow, twisting, and often barely passable for

Noise

*E*very year since 1973, the U.S. Department of Housing and Urban Development has conducted a survey to find out what city residents dislike about their environment. And every year the same factor has been named most objectionable. It is not crime, pollution, or congestion; it is noise—something that reaches every part of the city every day.

We have known for a long time that prolonged exposure to noises, such as loud music or the roar of machinery, can result in hearing loss. Evidence now suggests that noise-related stress also causes a wide range of psychological and physiological problems ranging from irritability to heart disease. An increasing number of people are affected by noise in their environment. By age forty, nearly everyone in America has suffered hearing deterioration in the higher frequencies. An estimated 10 percent of Americans (24 million people) suffer serious hearing loss, and the lives of another 80 million people are significantly disrupted by noise.

What is noise? There are many definitions, some technical and some philosophical. What is music to your ears might be noise to someone else. Simply defined, noise pollution is any unwanted sound or any sound that interferes with hearing, causes stress, or disrupts our lives. Sound is measured either in dynes, watts, or decibels (box table 14.2). Note that decibels (db) are logarithmic; that is, a 10 db increase represents a tenfold increase in sound energy.

City noises come from many sources. Traffic is generally the most omnipresent noise. Cars, trucks, and buses create a roar that permeates nearly everywhere in the city. Near airports, jets thunder overhead, stopping conversation, rattling dishes, sometimes even cracking walls. Jackhammers rattle in the streets; sirens pierce the air; motorcycles, lawnmowers, snowblowers, and chain saws create an infernal din; and music from radios, TVs, and loudspeakers fills the air everywhere.

The sensitivity and discrimination of our hearing is remarkable. Normally, humans can hear sounds from 16 hertz to 20,000 hertz (cycles per second). A young child whose hearing has not yet been damaged by excess noise can hear the whine of a mosquito's wings at the window when less than one quadrillionth (1×10^{-15}) of a watt per sq cm is reaching the eardrum.

Prolonged exposure to sounds above about 90 decibels can permanently damage the sensitive mechanism of the inner ear. By age thirty, most Americans have lost 5 db of sensitivity and can't hear anything above 16,000 hertz (Hz); by age sixty-five, the sensitivity reduction is 40 db for most people, and all sounds above 8,000 Hz are lost. By contrast, in the Sudan, where the environment is very quiet, even seventy-year-olds have no significant hearing loss.

Extremely loud sounds—above 130 db, the level of a loud rock band or music heard through earphones at a high setting—actually can destroy sensory nerve endings, causing aberrant nerve signals that the brain interprets as a high-pitched whine or whistle. You may have experienced ringing ears after exposure to very loud noises. Coffee, aspirin, certain antibiotics, and fever also can cause ringing sensations, but they usually are temporary.

A persistent ringing is called tinnitus. It has been estimated that 94 percent of the people in the United States suffer some degree of tinnitus. For most people, the ringing is noticeable only in a very quiet environment, and we rarely are in a place that is quiet enough to hear it. About thirty-five out of one thousand people have tinnitus severely enough to interfere with their lives. Sometimes the ringing becomes so loud that it is unendurable, like shrieking brakes on a subway train. Unfortunately, there is not yet a treatment for this distressing disorder.

---

wheeled traffic. Transportation was by sailboat, horse-drawn vehicles, or on foot. The wealthiest citizens usually lived in the middle of the city close to the centers of power and prestige, while the poorer people lived on the outskirts. Housing was mainly multistoried apartment complexes over street-level shops.

Automobiles brought many changes to cities. The middle class moved to single-family housing developments that filled in open spaces between transit lines (fig. 14.12c). The wealthy moved out of town to rural estates and satellite cities. Business offices remained downtown, but shopping centers sprang up at the nodes where major streets intersected transit lines. In the country, small towns and crossroads stores that had been built at about five-mile intervals to accommodate horse-and-buggy travelers dwindled when people were able to travel to the county seat or the nearest big city to do their shopping. The development of chain stores and brand-name merchandise was largely due to this new mobility of shoppers.

The decision to build freeways has been called the most important land-use decision ever made in the United States. Freeways have profoundly reshaped where we live, work, and shop, and how we get from place to place. Shopping malls located at freeway interchanges have largely replaced downtown department stores and have become the urban centers of expanding rings of suburbs around major cities (fig. 14.12d). Freeways allow us to travel with

BOX TABLE 14.2 Sources and effects of noise

Source	Sound pressure (dynes/cm^2)	Decibels (db)	Power at ear (watts/cm^2)	Effects
Shotgun blast (1 m)		150	10^{-1}	Instantaneous damage
Stereo headphones (full volume)	2,000	140		Hearing damage in 30 sec
50 hp siren (at 100 m)		130	10^{-3}	Pain threshold
Jet takeoff (at 200 m)	200	120		Hearing damage in 7.5 min
Heavy metal rock band		110	10^{-5}	Hearing damage in 30 min
Power mower, motorcycle	20	100		Damage in 2 hr
Heavy city traffic		90	10^{-7}	Damage in 8 hr
Loud classical music	2	80		OSHA 8-hour standard
Vacuum cleaner		70	10^{-9}	Concentration disrupted
Normal conversation	0.2	60		Speech disrupted
Background music		50	10^{-11}	
Bedroom	0.02	40		Quiet
Library		30	10^{-13}	
Soft whisper	0.002	20		Very quiet
Leaves rustling in the wind		10	10^{-15}	Barely audible
Mosquito wings at 4 m	0.0002	0		Hearing threshold youth 1,000–4,000 Hz

One of the first charges to the EPA when it was founded in 1970 was to study noise pollution and to recommend ways to reduce the noise in our environment. Standards have since been promulgated for noise reduction in automobiles, trucks, buses, motorcycles, mopeds, refrigeration units, power lawnmowers, construction equipment, and airplanes. The EPA is considering ordering that warnings be placed on power tools, radios, chain saws, and other household equipment. The Occupational Safety and Health Agency also has set standards for noise in the workplace that have considerably reduced noise-related hearing losses.

Noise is still all around us. In many cases, the most dangerous noise is that to which we voluntarily subject ourselves. Perhaps if people understood the dangers of noise and the permanence of hearing loss, we would have a quieter environment.

greater privacy, freedom, convenience, and speed (usually) than a mass transit system.

But freeways also bring many problems. They have torn through neighborhoods, choked cities with traffic, enormously increased energy consumption, and caused pollution, noise, and urban sprawl (fig. 14.13). Some of the most contentious U.S. environmental battles of the past two decades concerned freeways planned to cut across residential neighborhoods, parks, scenic and historic areas, or farmlands. Many people were initiated into environmental activism through their opposition to a particular stretch of freeway that threatened an area they cared about.

Los Angeles epitomizes the modern postindustrial city. Its multiple suburban centers are linked by a 600-mile-long network of freeways. It has been estimated that two-thirds of downtown Los Angeles and one-third of the total metropolitan area is devoted to roads, parking lots, service stations, and other automobile-related uses. Five million vehicles crowd onto the roads and highways each day, causing about 85 percent of both the air pollution and urban noise in the metropolitan area. Most Los Angeles freeways no longer have morning rush in one direction and evening rush in the other; traffic now comes to a standstill morning and evening in both directions on most freeways, while busy intersections have slowdowns throughout the day and into the night. It can take three hours to drive 20 miles at peak traffic periods. Between commuting to work and driving to

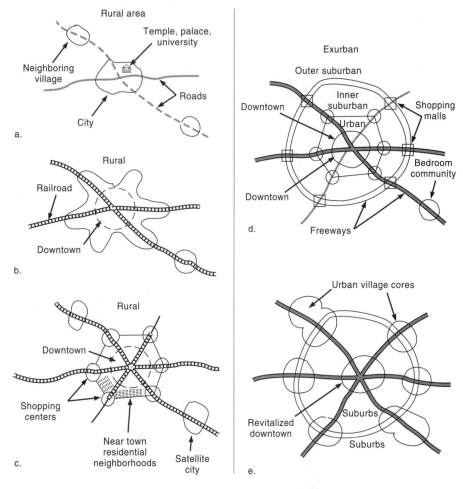

Figure 14.12 Patterns of city development from preindustrial (*a*) to polycentric postindustrial (*e*). Note how housing areas follow rail lines in early industrial stage (*b*). Introduction of the automobile allows people to move away from the rail lines (*c*). Freeways divert business to shopping malls and encourage growth of suburban rings beyond the central city (*d*). Finally, these urban village cores develop a full range of services and amenities that rival the central core (*e*). Removal of heavy manufacturing from downtown allows gentrification and revitalization.

Figure 14.13 Freeways make it easy for suburban residents to get into and out of the city, but consume land, bring pollution into the city, create noise and visual blight, and destroy neighborhoods.

beaches, the mountains, or for shopping, many people spend a substantial portion of their lives on the freeways.

What's next now that we seem to have saturated our urban areas with automobiles? Some cities are using modern mass transit systems to redesign where people live and work. Shopping centers are developing into urban villages with a complete set of services and facilities (fig. 14.12*e*). The old downtown, freed of the congestion and pollution of heavy industry, is undergoing revitalization as a cultural center. In the next section we will look at some ideas for how cities might be reconceived and rebuilt as better places to live.

CITY PLANNING

City planning has a long history. Many of the earliest cities in Mesopotamia, India, and Egypt were built in an orderly and civilized fashion. The Greek city-states of the first millennium B.C. reached a level of architectural beauty and rational city planning that has rarely been surpassed. Miletus and Pergamum on the Ionian coast of Turkey, for instance, and Alexandria on the Nile delta in Egypt, were splendid examples of Greek city planning. They were beautifully situated, with an orderly, unified design, magnificent buildings, and a pleasant mix of public and private spaces.

Residents of these cities regarded them as oases of comfort, convenience, safety, and delight in an otherwise cruel and dangerous world. Cities were seen as elysian refuges protected by a sacred wall from the chaos of evil spirits, wild animals, and dangerous people in the profane world outside. City temples were regarded as homes for the gods; their gardens were sanctuaries of beauty and sacred order. As centers of education, religion, political power, commerce, communication, and technology, cities offered many advantages over life in the country.

The Greeks considered optimum city size to be 30,000 to 50,000 people. This size was large enough to support a theater, university, library, temple, lively

Figure 14.14 Pierre L'Enfant's plan for Washington D.C. is considered one of the best urban designs in North America. Spacious boulevards, gracious traffic circles, and grand ceremonial malls create open space and interesting patterns.

Source: Produced by the U.S. Geological Survey for the Library of Congress.

marketplace, and an active political and intellectual life, yet small enough to draw food from the surrounding region and compact enough so that a person could walk out into the country within a few minutes from anywhere in the city. These planned cities resembled efficient ecosystems in which inhabitants participated in exchanges of matter and energy with their environment.

In the sixteenth and seventeenth centuries, following the rediscovery of Greek and Roman principles of city planning, Rome, London, Paris, Berlin, Vienna, and Brussels all were rebuilt with spacious boulevards, parks, and imposing public buildings. The design of these European capitals is reflected in Pierre L'Enfant's plan for Washington, D.C., which is one of the most notable city plans in the United States (fig. 14.14). Washington's rectangular street grid is cut by diagonal avenues converging on the Capitol and Executive Mansion, forming numerous squares, circles, and triangles that make interesting spaces for pocket parks. The broad, tree-shaded streets and impressive vistas created by this plan make Washington a grand setting for the national capital.

Garden Cities and New Towns

The twentieth century has seen numerous experiments in building **new towns** for society at large that try to combine the best features of the rural village and the modern city. One of the most influential of all urban planners was Ebenezer Howard (1850–1929),

who not only wrote about ideal urban environments but also built real cities to test his theories. In his *Garden Cities of Tomorrow,* written in 1898, Howard proposed that the congestion of London could be relieved by moving whole neighborhoods to **garden cities** separated from the central city by a greenbelt of forests and fields. Howard was the first modern urban planner to advocate comprehensive land-use planning and to reintroduce the Greek concept of organic growth and human scale to the city.

In the early 1900s, Howard worked with architect Raymond Unwin to build two cities called Letchworth and Welwyn Garden just outside of London. Interurban rail transportation provided access to these cities. Houses were clustered in "superblocks" surrounded by parks, gardens, and sports grounds. Streets were curved. Safe and convenient walking paths and overpasses protected pedestrians from traffic. Businesses and industries were screened from housing areas by vegetation. Each city was limited to about 30,000 people to facilitate social interaction. Housing and jobs were designed to create a mix of different kinds of people and to integrate work, social activities, and civic life. Trees and natural amenities were carefully preserved, and the towns were laid out to maximize social interactions and healthful living. Care was taken to meet residents' psychological needs for security, identity, and stimulation.

Letchworth and Welwyn Garden each have seventy to one hundred people per acre. This is a true urban density, about the same as New York City in the early 1800s and five times as many people as most suburbs today. By planning the ultimate size in advance and choosing the optimum locations for housing, shopping centers, industry, transportation, and recreation, Howard believed he could create a hospitable and satisfying urban setting while protecting open space and the natural environment. He intended to create parklike surroundings that would preserve small-town values and encourage community spirit in neighborhoods.

Letchworth and Welwyn Garden were the first of thirty-two new towns established in Great Britain, which now house about 1 million people. The Scandinavian countries have been especially successful in building garden cities (box 14.3). The former Soviet Union also built about two thousand new towns. Some are satellites of existing cities, and others are entirely new communities far removed from existing urban areas.

Cities of the Future

Is the expansion of cities—especially onto agricultural land—a waste of land and resources? Some people feel that growing cities decrease local agricultural resources and corrupt once pristine countryside and quaint rural towns. Others see this as a

Thinking Globally: Box 14.3

Tapiola and Farsta

Tapiola, Finland, and Farsta, Sweden, are outstanding examples of successful new towns built according to Ebenezer Howard's ideals for garden cities. Tapiola, built in 1951 as a suburb of Helsinki, occupies a beautiful setting along the shores of the Gulf of Finland. It is divided into three distinct neighborhoods separated by greenbelts. About 20 percent of the housing is high-rise apartments, and the rest is mostly single-family terrace and row houses nestled among rocky hills and lush evergreen forests (box fig. 14.3.1). Building designs harmonize. Housing is socially integrated and cooperatively managed. Socioeconomic status of the occupants isn't apparent from the outside.

Parks and playgrounds radiate from a central plaza where a striking town center is located. Walkways lead to carefully landscaped residential areas. Pedestrians need never cross streets at grade level. Industrial centers are unobtrusively situated and screened by vegetation so that people can work close to where they live and yet not suffer impacts of noise, pollution, and heavy traffic.

Farsta, Sweden, was built in 1954, 9 mi south of Stockholm and beyond the greenbelt enclosing the metropolitan area. Farsta is centered around a pedestrian shopping mall with an underground parking garage and a rapid railroad station with express service to the central station in downtown Stockholm. About half of all adult residents work in Farsta. Most of the rest commute by train to Stockholm. Housing is primarily in high-rise apartment buildings close to the central plaza, but single-family houses are also available. Parks and recreation areas surround the village center. No one walks far to school, work, shopping areas, or playgrounds. About 36

percent of all trips are by foot, 27 percent are by public transit, and 7 percent are by bicycle, while only 30 percent are by car—even though most families own at least one automobile. The town has a boat harbor and is located in a nature reserve. Within a few minutes of leaving home, residents can be in a pristine forest of spruce, fir, and birch.

Both Tapiola and Farsta have international reputations for beauty and a high quality of life. Because housing uses much less space in these cities than in a typical American suburb, more land is available for open space, parks, and recreation areas. Their higher density also supports neighborhood shops, mass transit, and other urban community services not commonly available in American suburbs.

Planned communities following the theories of Ebenezer Howard also have been built in the United States, but most plans have been based on personal automobiles rather than public transit. In the 1920s, Lewis Mumford, Clarence Stein, and Henry Wright drew up plans that led to the establishment of Radburn, New Jersey, and Chatham Village (near Pittsburgh). Reston, Virginia, and Columbia, Maryland, both were founded in the early 1960s and are widely regarded as the most successful attempts to build new towns of their era (box fig. 14.3.2). Another movement to build new towns according to Howard's principles has sprung up in the 1990s. Towns such as Seaside in northern Florida, Kentlands in the Maryland suburbs of Washington, and Laguna West near Sacramento cluster houses to save open space and create a sense of community. Commercial centers are located within a few minutes walk of most houses, and streets are designed to encourage pedestrians and to provide places to gather and visit.

Figure 14.3.1 Tapiola, Finland, occupies a beautiful setting on the Baltic Ocean near Helsinki. High-rise apartments and cluster housing sit in parklike open spaces. Mass transit provides easy access to the city.

Building a New Town: Finland's New Garden City by Heikli Von Hertzen and Paul Sprdiregen, MIT Press © 1971, 1973.

Figure 14.3.2 By clustering buildings together, Reston, Virginia has retained a high percentage of open space as public commons. Pedestrian walkways are separated from roads to provide a safe, pleasant way to get anywhere in the village.

Figure 14.15 The central core of most large cities in the industrialized world is dominated by the gleaming glass box skyscraper. Is this high-technology "built environment" a liberation from natural constraints or a dehumanizing machine for living?

healthy opportunity to decongest central cities and build smaller, more pleasant, more livable communities that have the benefits of both technological cities and rural villages. Perhaps the demographic shifts of the past forty years have merely been a transition between nineteenth-century industrial cities and garden cities of tomorrow. The polycentric network of urban villages in surrounding metropolitan areas may be a step toward more human-scale communities.

An alternative to spreading the population across a wide area of the countryside is to build upward. This model, which depends strongly on technology, has been called the **technopolis, vertical city,** or **city of the future.** The central hub of most big American cities now is dominated by skyscrapers and a highly technological environment (fig. 14.15). The emerging supercities of the Third World are also moving toward this style, in part because of its association with wealth, power, and progress.

There have been many proposals for gigantic, vertical supercities based on skyscrapers and completely artificial environments. An example was Buckminster Fuller's proposed mile-high geodesic dome to cover most of Manhattan. He also designed a hollow, tetrahedronal building two hundred sto-

ries high that would float on the ocean and could house 1 million people. It could be built in stages, he suggested, like a beehive, with trailerlike modules inserted in successive layers into a mountainous, open-truss framework. Some architects have talked seriously about the eventual possibility of erecting 500-story, mile-high buildings! The psychological and physiological effects of living 5,000 feet above street level can only be imagined.

Japan is now building eighteen new high-technology cities intended to be centers for economic and scientific growth in the next century. With names like Teletopia, Agripolis, and New Media City, these regional research, education, and marketing centers will have innovative housing, enclosed shopping malls, and high-technology communication and transportation systems. They will concentrate on leading-edge research and industries, such as fifth-generation supercomputers, biotechnology, lasers, ceramics, and bioelectronics. Some high-tech cities may be giant, floating structures similar to Buckminster Fuller's visionary proposals. Others may consist of a maze of tunnels and chambers entirely underground, not unlike a giant ant nest, opening onto huge twenty-story-deep air wells. This plan would conserve energy and preserve scarce surface air; however, many technical and psychological problems must be overcome.

Suburban Redesign

Most people in the United States now live in suburbs and will probably continue to do so in the foreseeable future. What can be done to make them more humane and environmentally sustainable?

Although suburbs have many advantages, they also have disadvantages. Most suburbs are too spread out for people to meet and interact, except with their immediate neighbors. Residential streets are empty during much of the day. Suburbs often have no sense of community and can be places of alienation and loneliness, just as cities often are. Although many urban problems have been eliminated, so have many of the finer aspects of city living. Suburbs, for instance, lack the activities, energy, and diversity that make cities exciting and dynamic. They have very limited artistic, cultural, and educational opportunities, compared to cities.

The low population density of the suburb makes public transport prohibitively expensive and private automobiles essential. Carpools for work, school, and extracurricular activities help fill group transportation needs. The uniform, single-family, detached houses of the suburbs offer few options to those who don't fit into the traditional, middle-class, nuclear family structure. As children grow up and leave home, parents often find the house bigger than they need. Single-parent families, households of single adults, the elderly, and the poor tend to have difficulty finding affordable suburban housing that fits their needs.

What can or should be done to create ideal suburban environments? Obviously not everyone in America will want to leave established suburban communities for new garden communities or wilderness utopias. Nor should they. Abandoning existing buildings and civic infrastructures would be a terrible waste. We need, instead, to find ways to remodel and revitalize existing suburban cities, reduce their problems, and adapt to the changing needs of their residents.

How can suburbs be redesigned to make them more diverse, flexible, and energy efficient? Ten proposals listed below might give them some of the better aspects of both the rural village and the big city.

1. Limit city size or organize them in modules of 30,000 to 50,000 people, large enough to be a complete city but small enough to be a community. Provide a greenbelt of agricultural and recreational land around the city to limit growth while promoting efficient land use. By careful planning and cooperation with neighboring regions, a city of 50,000 people can have real urban amenities such as museums, performing arts centers, schools, hospitals, etc.

2. Determine in advance where development will take place. This protects property values and prevents chaotic development in which the least desirable uses drive out the most desirable ones. It also recognizes historical and cultural values, agricultural resources, and such ecological factors as impact on wetlands, soil types, groundwater replenishment and protection, and preservation of esthetically and ecologically valuable sites.

3. Turn shopping malls into real city centers that invite people to stroll, meet friends, or listen to a debate or a street musician (fig. 14.16). If there aren't one hundred places for an impromptu celebration, a place isn't a real city. Another test of a city is a vital night life. Design city spaces with sidewalk cafes, pocket parks, courtyards, balconies, and porticoes that shelter pedestrians, bring people together, and add life and security to the street. Restaurants, theaters, shopping areas, and public entertainment that draw people to the streets generate a sense of spontaneity, excitement, energy, and fun.

4. Locate everyday shopping and services so people can meet daily needs with greater convenience, less stress, less automobile dependency, and less use of time and energy. This might be accomplished by encouraging small-scale commercial development in or close to residential areas. Perhaps we should

Figure 14.16 Many cities have redesigned core shopping areas to be more "user friendly." Pedestrian shopping malls, such as this closed street in Boulder, Colorado, create space for entertainment, dining, chance encounters, and building urban community in the best sense.

once again have "mom and pop" stores on street corners or in homes.

5. Increase jobs in the community by locating offices, light industry, and commercial centers in or near suburbs, or by enabling work at home via computer terminals. These alternatives save commuting time and energy and provide local jobs. There are also concerns, however, about work-at-home employees being exploited in low-paying "sweat-shop" conditions by unscrupulous employers. Some safeguards may be needed.

6. Encourage walking or the use of small, low-speed, energy-efficient vehicles (microcars, motorized tricycles, bicycles, etc.) for many local trips now performed by full-size automobiles. Creating special traffic lanes, reducing the number or size of parking spaces, or closing shopping streets to big cars might encourage such alternatives.

7. Promote more diverse, flexible housing as alternatives to conventional, detached, single-family houses. "In-fill" building between existing houses saves energy, reduces land costs, and might help provide a variety of

living arrangements. Allowing owners to turn unused rooms into rental units provides space for those who can't afford a house and brings income to retired people who don't need a whole house themselves. Allowing single-parent families or groups of unrelated adults to share housing and to use facilities cooperatively also provides alternatives to those not living in a traditional nuclear family. One of the great "discoveries" of urban planning is that mixing various types of housing—individual homes, townhouses, and high-rise apartments—can be attractive if buildings are esthetically arranged in relation to one another.

8. Create housing "superblocks" that use space more efficiently and foster a sense of security and community. Widen peripheral arterial streets and provide pedestrian overpasses so traffic flows smoothly around residential areas; then reduce interior streets within blocks to narrow access lanes with speed bumps and barriers to through traffic so children can play more safely. The land released from streets can be used for gardens, linear parks, playgrounds, and other public areas that will foster community spirit and encourage people to get out and walk. Cars can be parked in remote lots or parking ramps, especially where people have access to public transit and can walk to work or shopping.

9. Make cities more self-sustainable by growing food locally, recycling wastes and water, using renewable energy sources, reducing noise and pollution, and creating a cleaner, safer environment. A greenbelt of agricultural land and forestland around the city provides food and open space as well as such valuable ecological services as purifying air, supplying clean water, and protecting wildlife habitat and recreation land.

10. Invite public participation in decision making. Emphasize local history, culture, and environment to create a sense of community and identity. Create local networks in which residents take responsibility for crime prevention, fire protection, and home care of children, the elderly, the sick, and the disabled. Coordinate regional planning through metropolitan boards that cooperate with but do not supplant local governments.

SUSTAINABLE DEVELOPMENT IN THE THIRD WORLD

What can be done to improve conditions in Third World cities? While the advantages of garden cities and advanced transportation systems would greatly improve conditions in the developing world, the overcrowded cities of Third World countries have more basic problems to solve first. Among the immediate needs are housing, clean water, sanitation, food, education, health care, and basic transportation for their residents.

Some countries, recognizing the need to use vacant urban land, are redistributing unproductive land or closing their eyes to illegal land invasions. Indonesia, Peru, Tanzania, Zambia, Mexico, and Pakistan have learned that squatter settlements make a valuable contribution to meeting national housing needs. Squatters' rights are being upheld in some cases, and such services as water, sewers, schools, and electricity are being provided to the settlements (fig. 14.17). Some countries intervene directly in land distribution and land prices. Tunisia, for instance, has a "rolling land bank" to buy and sell land. This strong and effective program controls urban land prices and reduces speculation and unproductive land ownership.

Many planners argue that social justice and sustainable economic development are answers to the urban problems we have discussed in this chapter. If people have the opportunity and money to buy better housing, adequate food, clean water, sanitation, and other things they need for a decent life, they will do so. Democracy, security, and improved economic conditions help in slowing population growth and reducing rural-to-city movement. An even more important measure of progress may be institution of a social welfare "safety net" guaranteeing that old or sick people will not be abandoned and alone.

Some countries have accomplished these goals even without industrialization and high incomes. Sri Lanka, for instance, has lessened the disparity between the city cores and the peripheral areas of the

Figure 14.17 In this *colonia* on the outskirts of Mexico City, residents work with the government to bring in electric power and install water and sewer lines. Like many developing countries, Mexico has come to recognize that helping people help themselves is the best way to improve urban living.

country. Giving all people equal access to food, shelter, education, and health care eliminates many incentives for interregional migration. Both population growth and city growth have been stabilized, even though the per capita income is only $800 per year. China has done something similar on a per capita income around $300 per year.

Whether sustained, environmentally sound economic development is possible for a majority of the world's population remains one of the most important and most difficult questions in environmental science. The unequal relationship between the richer "Northern" countries and their impoverished "Southern" neighbors is a major part of this dilemma. Some people argue that the best hope for developing countries may be to "delink" themselves from the established international economic systems and develop direct south-south trade based on local self-sufficiency, regional cooperation, barter, and other forms of nontraditional exchange that are not biased in favor of the richer countries.

Summary

A rural area is one in which a majority of residents are supported by methods of harvesting natural resources. An urban area is one in which a majority of residents are supported by manufacturing, commerce, or services. A village is a rural community. A city is an urban community with sufficient size and complexity to support economic specialization and to require a higher level of organization and opportunity than is found in a village.

Urbanization in the United States over the past two hundred years has caused a dramatic demographic change. A similar shift is occurring in most parts of the world. Only Africa and South Asia remain predominantly rural, and cities are growing rapidly there as well. By the end of this century, we expect that more than half the world's people will live in urban areas. Most of that urban growth will be in the supercities of the Third World. A century ago only thirteen cities had populations above 1 million; now there are 235 such cities. In the next century, that number will probably double again, and three-fourths of those cities will be in the Third World.

Cities grow by natural increase (births) and migration. People move into the city because they are "pushed" out of rural areas or because they are "pulled" in by the advantages and opportunities of the city. Huge, rapidly growing cities in the developing world often have appalling environmental conditions. Among the worst problems faced in these cities are traffic congestion, air pollution, inadequate or nonexistent sewers and waste disposal systems, water pollution, and housing shortages. Millions of people live in slums and shantytowns where conditions would crush any but the strongest spirit, yet these people raise families, educate their children, learn new jobs and new ways of living, and have hope for the future.

The problems of developed world cities tend to be associated with decay and blight. Over the past fifty years, many in the urban middle and upper class moved to the suburbs, leaving the old, very poor, handicapped, and economically marginal people in the inner city. Increasing joblessness and poverty in the inner city create a cycle of poverty from which it is difficult to escape. Still, there are ways that we can improve cities in both the developed and the developing world to make them healthier, safer, more environmentally sound, socially just, and culturally fulfilling than they are now.

Review Questions

1. What is the difference between a city and a village and between rural and urban?
2. How many people now live in cities, and how many live in rural areas worldwide?
3. What changes in urbanization are predicted to occur in the next fifty years, and where will that change occur?
4. Identify the ten largest cities in the world. Has the list changed in the past fifty years? Why?
5. When did the United States pass the point at which more people live in the city than the country? When will the rest of the world reach this point?
6. Describe the current conditions in Mexico City. What forces contribute to its growth?
7. Describe the difference between slums and shantytowns.
8. Why are urban areas in U.S. cities decaying?
9. How has transportation affected the development of cities? What have been the benefits and disadvantages of freeways?
10. Describe some ways that American cities and suburbs could be redesigned to be more ecologically sound, socially just, and culturally amenable.

Questions for Critical or Reflective Thinking

1. Picture yourself living in a rural village or a Third World city. What aspects of life there would you enjoy? What would be the most difficult for you to accept?
2. Are there fundamental differences between the lives of homeless people in First World and Third World cities? Where would you rather be?
3. A city could be considered an ecosystem. Using what you learned in chapters 2 and 3, describe the structure and function of a city in ecological terms.
4. Extrapolating from laboratory animals to humans is always a difficult task. How would you interpret the results of laboratory experiments on stress and crowding in terms of human behavior?
5. Weigh the costs and benefits of automobiles on modern American life. Would we have been better off if the internal combustion engine had never been invented?
6. Boulder, Colorado, has been a leader in controlling urban growth. One consequence is that housing costs have skyrocketed and poor people have been driven out. If you lived in Boulder, would you vote for additional population limits? What do you think is an optimum city size?
7. Ten proposals are presented in this chapter for suburban redesign. Which of them would be appropriate or useful for your community? Try drawing up a plan for the ideal design of your neighborhood.
8. How much do you think the richer countries are responsible for conditions in the developing countries? How much have people there brought on themselves? What role should or could we play in remedying their problems?

Key Terms

city 299
city of the future 313
garden cities 311
megacity 299
new towns 311
pull factors 302
push factors 302
rural area 299

shantytowns 306
slums 305
squatter towns 306
technopolis 313
urban area 299
urbanization 299
vertical city 313
village 299

Suggested Readings

Bookchin, M. *The Limits of the City*. Montreal: Black Rose, 1986. What makes a livable city?

Doxiadis, C. A. *Anthropolis: A City for Human Development* and the proceedings of a symposium with René Dubos, Erik Erikson, Margaret Mead, et al. New York: Norton, 1974. An idealistic plan for reorganizing cities into urban villages.

Dunkle, T. "The Sound of Silence." *Science* 3, no. 3 (April 1982): 30. A highly readable account of tinnitus, ringing in the ears from noise damage.

Herbes, J. *The New Heartlands: America's Flight Beyond the Suburbs*. New York: Time Books, 1986. Good analysis of the growth of regional cities.

Howard, E. *Garden Cities of Tomorrow*. London: Farber and Farber, 1902. A classic in city planning that sparked the new town movement. Many of the features of modern suburbs feature designs of Howard and his chief architect, Raymond Unwin. Unfortunately, they do not usually include the total community design envisioned by these pioneers.

Huth, M. J. *The Urban Habitat: Past, Present and Future*. Chicago: Nelson-Hall, 1970. A review of the urbanization phenomenon from antiquity to present, urban planning in Europe and America, and a formula for a more humane urban America.

Kozol, Jonathan. *Rachel and Her Children*. New York: Crown Publishers, 1988. A vivid, personal account of the plight of homeless people in New York.

Kropotkin, P. *Fields, Factories and Workshops or Industry Combined with Agriculture and Brainwork with Manual Work*. First published in Boston, 1899. Revised edition London: Putnam, 1913. Sociological and economic intelligence of the first order founded on Kropotkin's specialized competence as a geographer and his passion as a communist anarchist. Especially emphasizes planning for undeveloped areas.

Le Corbusier. *Urbanisme*. Paris, 1924. Translation: *The City of Tomorrow and Its Planning*. New York:

Dover Press, 1930. Suggestions for a mechanical metropolis with widely spaced skyscrapers and multiple-decked traffic ways. One of the most influential books of its generation.

Livermash, R. "Human Settlements." *World Resources 1990–1991.* Washington, D.C.: World Resources Institute, 1990. A good overview of urban problems and solutions in developing countries.

Lowe, M. D. "Rethinking Urban Transport." In *State of the World.* Washington, D.C.: Worldwatch Institute, 1991. Describes approaches to mass transit and environmentally friendly transport systems.

McHarg, I. *Design with Nature.* Philadelphia: Natural History Press, 1969. A classic of landscape architecture that pioneered a holistic approach and consideration of natural features in urban design.

Mumford, Lewis. *The City in History: Its Origins, Its Transformations, and Its Prospects.* New York: Harcourt, Brace & World, 1961. A comprehensive, classic review of city planning through history.

Park, R. *Human Communities: The City and Human Ecology.* Glenco: Free Press, 1952. The origin of the idea of human ecology.

Register, R. 1992. *Ecocity Berkeley.* Berkeley, Calif.: Urban Ecology Institute. Suggestions on building healthy and humane cities.

Sargent, F. O., et al. *New Rural Environmental Planning for Sustainable Communities.* Washington, D.C.: Island Press, 1991. A new edition of a classic guide to development and planning in harmony with nature.

Teaford, J. C. *The Twentieth-Century American City.* Baltimore: Johns Hopkins Press, 1984. Puts urban problems into historical context.

Todd, N. J., and J. Todd. *Bioshelters, Ocean Arks, City Farming.* San Francisco: Sierra Club Books, 1985. An excellent discussion of appropriate technology and ecological design from New Alchemy Institute in Massachusetts.

Tolstoy, L. N. 1881. *What Then Shall We Do?* An account of the census of 1880 and of conditions in Moscow during the age of industrialism. One of the seminal books in social philosophy.

Van der Ryn, S., and P. Calthorpe. *Sustainable Communities.* San Francisco: Sierra Club Books, 1986. An excellent description of principles for building (and rebuilding) ecologically sound cities.

Vininy, D. R., Jr. "The Growth of the Core Regions of the Third World." *Scientific American* 252, no. 4 (April 1985): 42. A good discussion of the causes and effects of megacity growth.

Wachs, M., and M. Crawford, eds. *The Car and the City: The Automobile, the Built Environment, and Daily Urban Life.* Ann Arbor, Mich.: University of Michigan Press, 1990. An interdisciplinary perspective on urban development and change detailing the impact of the automobile on the form and functioning of the city.

World Bank. *The Urban Edge.* Vol. 14. Washington, D.C.: World Bank, 1990. Annual review of urban development in the Third World.

15
TOWARD A SUSTAINABLE FUTURE

"And the people asked Him, " 'What then shall we do?' "

Luke 3:10

Objectives

After studying this chapter you should be able to:

■ Understand the ethical and philosophical bases of environmental protection along with some alternate beliefs about our place in nature.

■ Recognize opportunities for making a difference through the goods and services we choose as well as the limits of green consumerism.

■ Discuss the spectrum of environmental groups and the varied tactics they employ to bring about social change.

■ Appreciate the need for sustainable development and explain how nongovernmental groups work toward this goal.

■ Describe how green politics and government function nationally and internationally to help protect the earth.

■ Formulate your own philosophy and action plan for what you can and should do to create a better world and a sustainable environment.

INTRODUCTION

What will our future be? Are we headed toward crisis and disaster, or will we enjoy a happier, more fulfilling, more creative life than is possible now? In many ways, the choices we make in managing our resources and our relationships with others determine what our lives—and those of our children—will be in the future (fig. 15.1). We have examined many pressing environmental problems in this book, as well as some ways that these problems could be overcome.

Figure 15.1 What will our environmental future be? In many ways, the choices we make new in managing resources and working with others to protect and restore our environment will determine what our lives and those of our children will be in years to come.

Are you pessimistic, optimistic, or a bit of both about our prospects for the future? Considering the rate at which we are depleting resources and abusing environmental systems, there is reason to worry that we may trigger a catastrophic downward spiral of environmental, economic, and social decay that will make the future bleak and dismal. On the other hand, we have made some encouraging progress toward eliminating pollution, poverty, and resource waste. Perhaps there is reason to hope that human ingenuity, resourcefulness, and enterprise will continue to stabilize population, conserve resources, preserve biodiversity, and foster cooperation to bring about a healthier, happier, more just, and more meaningful life for everyone.

Whether you tend to be optimistic, pessimistic, or somewhere between these extremes, most of us agree that current trends are only indications of what *may* be, depending on the courses we choose to follow, not predictions of what *must* be. In this chapter, we will examine some ethical and philosophical reasons for protecting nature as well as ways people are working to create a better society and a sustainable environment.

ENVIRONMENTAL ETHICS AND PHILOSOPHIES

Maybe the most important enterprise of civilization—after basic survival—is asking what constitutes the "good life" and how we, as moral beings, ought to behave. Environmental ethics is a special branch of philosophy that widens our circle of inquiry to include our relationships with and obligations to species other than our own. Some environmentalists expand our moral consideration to include abiotic components of ecosystems or even whole landscapes, ecosystems, and nature itself.

Rights, Duties, and Obligations

Many philosophers hold that only humans can be moral agents, capable of behaving morally or immorally and subject to duties, rights, and responsibilities. Of course, not all humans at all times are able to form moral judgments, exercise willpower and resolve to carry out decisions, or be accountable for their actions. Children, the mentally retarded, mentally ill, and others who lack full use of reason are considered moral subjects who have moral interests and rights of their own even though their moral obligations are diminished.

The anthropocentric (human-centered) view holds that only humans have intrinsic value (value in and of themselves). In this perspective, other species or nonliving objects are outside the range of moral consideration and have only instrumental value—that is, they are valuable only as tools to carry out human intentions. Many environmentalists prefer a more biocentric (life-centered) view or ecocentric view that considers both living and nonliving components of the environment to have interests, values, and goods of their own, independent of their usefulness to humans. This view of moral consideration greatly expands the possible range of our duties and obligations to other beings (fig. 15.2). It raises many questions about our right to take resources, kill other organisms, and modify the environment to suit our purposes without regard to the needs of other species or biological communities.

Stewardship and Relationships

Rather than focus on rights and duties, some philosophers argue that we should learn to be stewards or caretakers of the environment for our own benefit and that of other species and future generations. Holders of this view claim that our intelligence, foresight, and technology give us both power and responsibility to manage resources and create an agreeable environment. To do otherwise is both arrogant and foolish. Ecofeminists, on the other hand, claim that assumption of privileges, rights, and duties by male-centered societies has led to oppression of women, children, minorities, and nature. They contend that we should focus more on relationships—even kinship—both with other people and nature if we are to create a kinder, gentler, more sustainable world.

An interesting application of scientific stewardship is the growing field of **environmental restoration:**

Figure 15.2 Do other organisms have inherent values and rights? These chickens are treated as if they are merely egg-laying machines. Many people argue that we should treat them more humanely.

repairing environmental damage caused by human neglect (box 15.1). Although many restoration projects have successfully restored damaged habitats and endangered species, we will never be able to re-create completely natural conditions. There are worries that successful environmental restoration could be used to legitimate further damage or as an excuse for business-as-usual. As author Seth Zuckerman points out, "we can't switch wetlands around like living room furniture" to replace those destroyed by development projects. We must avoid the arrogance of assuming that we can rearrange or remake nature in any way, place, or time we wish.

Modernism and Technology

Strong undercurrents of antimodernism and fear of science and technology run through much of the environmental movement. Many of us yearn for a golden Arcadian past when, we suppose, life was simpler, easier, and more satisfying. We decry the fast pace and anonymity of modern life and the destructive forces of technology. Some people advocate a return to an agrarian or pastoral way of life. A few even suggest going back to hunting and gathering as a way to save both the environment and ourselves.

Few of us, though, would willingly give up the conveniences and comforts of modern life. Think of what life was actually like for your ancestors one hundred or one thousand or ten thousand years ago. Would you really want to trade places? Ironically, many environmentalists use the latest cutting-edge technology such as desktop publishing, electronic bulletin boards, fax machines, modems, photocopiers, and jet travel to criticize modernity. Is it valid to

Restoration of the Bermuda Cahow

*T*he cahow is a seabird endemic (restricted) to Bermuda and adjacent islands off the East Coast of North America (box fig. 15.1.1). A member of the petrel family, related to albatrosses, shearwaters, and other wide-ranging seabirds, cahows once formed dense, noisy colonies that fed on the rich fisheries around the island. When European sailors first landed on Bermuda four hundred years ago, cahows were abundant. Like many endemic island species, the ground-nesting cahow had never experienced predation and had no defenses against the pigs, goats, and rats introduced by the first settlers. Overhunting and habitat destruction further decimated the species. By the late 1600s—about the same time that the last Dodo was killed on the Island of Mauritius in the Indian Ocean—cahows had disappeared from Bermuda.

For three centuries the cahow was assumed to be extinct. In 1951, though, scientists were delighted to find a few living cahows on some tiny islands in the Bermuda harbor. A protection and recovery program was begun immediately. The most successful and significant aspect of this project was the establishment of a sanctuary on the 6 ha (15 acre) Nonsuch Island, which has become a symbol of the potential for environmental restoration.

Nonsuch was a near desert after centuries of abuse, neglect, and habitat destruction. All the native flora and fauna were gone along with most of its soil. This was a case of re-creating nature rather than merely protecting what was left. Sanctuary superintendent David Wingate, who has devoted his entire professional life to this project, has brought about a miraculous transformation of this barren little plot of land. Reestablishing a viable population of cahows provided the incentive and funds for rebuilding an entire biological community.

The first step in restoration was to reintroduce native vegetation and re-create habitat. More than 5,000 native tree and shrub seedlings were planted. Initial progress was slow as trees struggled to get a foothold; once the forest knit itself into a dense thicket that deflected the salt spray and powerful ocean winds, however, the natural community quickly began to reestablish itself. The benefits of indigenous species became apparent in 1987 when Hurricane Emily roared across Bermuda. Up to 70 percent of non-native trees were uprooted or snapped off by gale-force winds, littering streets and bringing down

Figure 15.1.1 The Bermuda cahow or hook-billed petrel was thought to be extinct for nearly three centuries. A breeding population has been reestablished, however, and the restored sanctuary created to protect them is helping many other species as well.

powerlines. The dense, low-profile, native trees on Nonsuch were barely touched by the winds. Demands soared for hurricane-adapted species to replace those lost along streets and in gardens.

Just providing habitat for the cahows was not enough, however, to restore the population. Each pair lays only one egg per year and only about half survive under ideal conditions. It takes eight to ten years for fledglings to mature, giving the species a low reproductive potential. They also compete poorly against the more common long-tailed tropic birds that steal nesting sites and destroy cahow eggs and fledglings. Special underground burrows were built with baffled entrances designed to admit only cahows. Young birds were hand-raised by humans to ensure a proper diet and protection.

By 1990, the cahow population had rebounded to nearly fifty nesting pairs. It is too early to know if this is enough to be stable over the long term, but the progress to date is encouraging. Perhaps more important than rebuilding this single species is that the island has become a living museum of precolonial Bermuda that benefits many species besides the most famous resident. It is a heartening example of what can be done with vision, patience, and some hard work.

use the tools and resources of industrial society to object to those same tools and resources? Perhaps our challenge is to find a balance in eliminating the worst aspects of modern life while retaining the best (fig. 15.3).

Deep, Shallow, Social, or Progressive?

Norwegian philosopher Arnae Naess criticizes the superficial commitment of environmentalists who claim to be green but are quick to compromise and who do little to bring about fundamental change. He characterizes this as **shallow ecology** and contrasts it with what he calls **deep ecology,** which calls for a profound shift in our attitudes and behavior. An equally radical but more humanist philosophy called **social ecology** is proposed by Murray Bookchin, who draws on the communatarian anarchism of the Russian geographer Peter Kropotkin.

Among the tenants of both these "dark green" philosophies are voluntary simplicity; rejection of anthropocentric attitudes; intimate contact with nature; decentralization of power; support for cultural and biological diversity; a belief in the sacredness of nature; and direct personal action to protect nature, improve the environment, and bring about fundamental societal change. They differ mainly on the necessity for personal freedom and communal interaction.

Both Naess and Bookchin criticize mainstream environmental groups who favor pragmatic work within the broadly agreed-upon social agenda and the established political system to bring about incremental,

progressive reform rather than radical revolution. Most progressives naturally object to being called shallow, light, or superficial; they prefer instead to characterize their position as moderate rather than radical, or reformist rather than revolutionary. They tend to view the deep ecologists and anarchists as unrealistic zealots who would rather be righteous than effective.

Aside from name-calling, there are some important philosophical questions in this debate. One of the questions is whether we face an environmental crisis that requires radical action or only "problems" that can best be overcome by working within established channels. Another important question is whether it is better to overturn society to explore new ways of living, or to work for progressive change within existing political, economic, and social systems. Finally, is it more important to work for personal perfection, or collective improvement? There may be no single answer to any of these questions; it's good to have people working in many different ways to find solutions. Later in this chapter we will look more closely at some mainstream and radical environmental groups and the tactics they espouse.

GREEN CONSUMERISM

A prime reason for our destructive impacts on the earth is our consumption of resources. Technology has made consumer goods and services cheap and readily available in the richer countries of the world. As you already have seen throughout this book, we in the industrialized world use resources at a rate out of proportion to our percentage of the population. If everyone in the world were to attempt to live at our level of consumption, given current methods of production, the results would surely be disastrous. In this section we will look at some options for consuming less and reducing our environmental impacts.

How Much Is Enough?

The first question we should ask ourselves is whether we really need to consume so much. How much ought we leave for other people and future generations? Although advertising urges us to buy and discard more, are our lives really better or simply more complicated and stressful with more stuff? Shopping in many industrialized countries has become a major pastime. The average American spends more than two hours every week shopping for goods other than food. We own twice as many cars, drive two-and-a-half times as far, use twenty times as much plastic, and travel twenty-five times as far by air as did our parents in 1950.

To avoid waste production, we can practice "precycling," making environmentally sound decisions at the store and reducing waste before we buy. We are already making good progress in this area, but there is

Figure 15.3 What aspects of modernity and technology do we want to save, and what can be discarded?

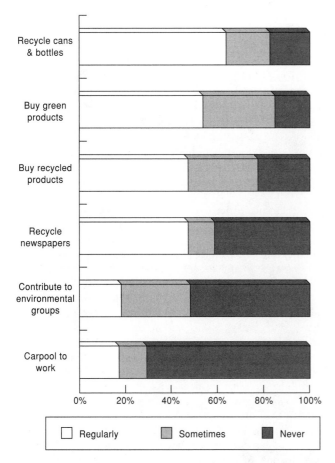

Figure 15.4 A random sample of adults was asked the following question: here are some things that people do for the environment. Is each something that you do regularly, sometimes, or never?

Source: Random sample from *Time*. 24 December 1990, p. 50

much more we could do (fig. 15.4). Table 15.1 offers some suggestions for reducing our consumption levels by buying only what we need. Box 15.2 shows what can be accomplished even in the Third World with a little planning, education, and collective action.

Shopping for "Green" Products

Obviously, we can never reduce our consumption levels to zero. We can, however, make sound, informed decisions about the products we do buy to select those that do the least environmental damage in production, use, and disposal. And as consumers demand environmentally friendly products, manufacturers, food producers, and merchants are moving more and more to humane, safer, and more sustainable consumer items. Although each of our individual choices makes a small impact, collectively they can be important.

With surveys showing nine out of ten of us willing to pay more for products and packaging that don't harm the environment, many merchandisers are moving quickly to cash in on green consumerism (box 15.3).

TABLE 15.1 Reducing consumption

Purchase less
- Ask yourself whether you really need more stuff.
- Avoid buying things you don't need or won't use.
- Use items as long as possible (and don't replace them just because some new gizmo becomes available).
- Use the library instead of purchasing every book you read.
- Make gifts from materials already on hand, or give nonmaterial gifts.

Avoid excess packaging
- Carry reusable bags when shopping and refuse bags for small purchases.
- Buy items in bulk or with minimal packaging; avoid single-serving foods.
- Choose packaging that can be recycled or reused.

Don't use disposable items
- Use cloth diapers, napkins, handkerchiefs, and towels.
- Bring a washable cup to meetings; use washable plates and utensils rather than single-use items.
- Buy pens, razors, flashlights, and cameras with replaceable parts.
- Choose items built to last and have them repaired; you will save materials and energy while providing jobs in your community.

Conserve energy
- Walk, bicycle, or use public transportation.
- Turn off (or avoid turning on) lights, water, heat, and air conditioning when possible.
- Put up clotheslines or racks in the backyard or basement to avoid using a clothes dryer.
- Carpool and combine trips to reduce car mileage.

Save water
- Water lawns and gardens only when necessary.
- Use water-saving devices and fewer flushes with toilets.
- Don't leave water running when washing hands, food, dishes, and teeth.

Used by permission. Based on material by Karen Oberhauser, Bell Museum Imprint, University of Minnesota, 1992.

Green sales and services are projected to nearly quintuple, from $1.8 billion in 1990 to $8.8 billion in 1995. Some of the claims made by green marketers are of questionable validity, however. Consumers must look closely to avoid "green scams." Many terms used in advertising are vague and have little meaning. For example:

■ "Nontoxic" suggests that a product has no harmful effects on humans. Since there is no legal definition of the term, however, it can have many

Curitiba, An Environmental Showcase

Curitiba, a Brazilian city of about two million people located on the Atlantic coast about 650 km (400 mi) southwest of Rio de Janeiro, has acquired a worldwide reputation for its innovative urban planning and environmental protection policies. The architect of this remarkable program is mayor Jaime Lerner, who began in 1962 as a student leader protesting the proposed destruction of Curitiba's historic downtown center. Elected mayor nine years later, he has worked for more than two decades to preserve the city and make it a more beautiful and habitable place. His success has thrust the city into the international spotlight as a shining example of what conservation and good citizenship can do to improve urban environments, even in Third World cities.

The heart of Curitiba's environmental plan is education for both children and adults. Signs posted along roadways proclaim "50 kg of paper equals one tree" and "recycle; it pays." School children study ecology along with Portuguese and math. With the assistance of children who encourage their parents, the city has successfully instituted a complex recycling plan that requires careful separation of different kinds of materials. The mayor calculates that 1,200 trees per day are saved by paper recycled in this program. "Imagine if the whole of Brazil did this with an urban population 60 times greater than Curitiba," the mayor exclaims, "we could save 26 million trees per year!"

Another area in which Curitiba is setting an example is transportation. Faced with a population that tripled in two decades, bringing increasing levels of traffic congestion and air pollution, Curitiba had a transportation dilemma. The choices were either to bulldoze freeways through the historic heart of the city or to institute mass transit. The city chose mass transit. Now more than three quarters of the city's population leave their automobiles at home every day and take special express buses to work. The system is so successful that ridership has increased from 25,000 passengers a day 20 years ago to more than 1.25 million per day now. The result is not only less congestion and pollution, but major energy savings.

Other measures include a limit on building height and construction of an industrial park outside city boundaries to clear the air and reduce congestion. Maximum use is made of all materials and buildings. Worn-out buses become city training centers, an old military fort is a cultural center, and a gunpowder depot is now a theater. Water and energy conservation are practiced widely. Even litter—a ubiquitous component of most Brazilian cities—is absent in Curitiba. People are so imbued with city pride that they keep their surroundings spotless.

Although many residents initially were skeptical of this environmental plan, now a remarkable 99 percent of the city's inhabitants would not want to live anywhere else. The World Bank uses Curitiba as an example of what can be done through civic leadership and public participation to clean up the urban environment. Some people claim that Curitiba, with its cool climate and high percentage of European immigrants, may be a special case among Brazilian cities. Mayor Lerner claims that Curitiba has no special features except concern, creativity, and communal efforts to care for its environment. Could you start a similar program in your hometown?

---

meanings. How nontoxic is the product? And to whom? Substances not poisonous to humans still can harm other organisms.

- "Biodegradable," "recyclable," "reusable," or "compostable" may be technically correct but not signify much. Almost everything will biodegrade *eventually,* although it may take thousands of years. Similarly, almost anything is potentially recyclable or reusable; the real question is whether there are programs to do so in your community. If the only recycling or composting program for a particular material is half a continent away, this claim has little practical meaning.

- "Natural" is another vague and often abused term. Many natural ingredients—lead or arsenic, for instance—are highly toxic. Synthetic materials are not necessarily more dangerous or environmentally damaging than those created by nature.

- "Organic" can connote different things in different places. Some states have standards for organic food, but others do not. On products such as shampoos and skin-care products, "organic" may have no significance at all. Most detergents and oils are organic chemicals whether they are synthesized in a laboratory or found in nature. Many such products are unlikely to have pesticide residues anyway.

- "Environmentally friendly," "environmentally safe," and "won't harm the ozone layer" are often empty claims. Since there are no standards to define these terms, anyone can use them. How much energy and nonrenewable materials are used in manufacture, shipping, or use of the product? How much waste is generated, and how will it be disposed of when no longer functional? One product may well be more environmentally benign than another, but be careful who makes this claim.

Acting Locally: Box 15.3

Green Business and Environmental Jobs

*C*an environmental protection and resource conservation be a strategic advantage in business? Many companies think so. An increasing number are jumping on the environmental bandwagon, and most large corporations now have an environmental department. A few are beginning to explore integrated programs to design products and manufacturing processes to minimize environmental impacts. Called "design for the environment," this approach is intended to avoid problems at the beginning rather than require companies to deal with them later on a case-by-case basis. In the long run, executives believe, this will save money and make their business more competitive in future markets. The alternative is to face increasing pollution-control and waste-disposal costs—now estimated to be more than $100 billion per year for all American businesses—as well as be tied up in expensive litigation and administrative proceedings.

The market for pollution-control technology and know-how is also expected to be huge. Cleaning up the former East Germany alone is expected to cost some $200 billion. Many companies are positioning themselves to cash in on this enormous market. Right now Germany and Japan appear to be the leaders in the pollution-control field because they have had more stringent laws than America for many years, giving them more experience in reducing effluents.

The rush to green up business is good news for those looking for jobs in environmentally related fields, which are predicted to be among the fastest growing areas of employment for the next few years. The federal government alone projects a need to hire some ten thousand people per year in a variety of environmental disciplines (box fig. 15.3.1). How can you prepare yourself to enter this market? The best bet is to get some technical training: environmental engineering, analytical chemistry, microbiology, ecology, limnology, groundwater hydrology, and computer science all have great potential. At a recent conference on jobs in the environment, an EPA representative told my colleagues and me that a chemical engineer with a graduate degree and some experience in an environmental field could practically name her or his salary. Some other very good possibilities are environmental law and business administration, both rapidly expanding fields.

For those who aren't inclined toward technical fields, there are still opportunities for environmental careers. A good liberal arts education will help you develop skills such as communication, critical thinking abilities, balance, vision, flexibility, and caring that should serve you well. Large companies need a wide variety of people; small companies need a few people who can do many things well.

Figure 15.3.1 Many interesting, well-paid jobs are opening up in environmental fields. Here an environmental technician uses a video probe to inspect a residential sewer line. Can you picture yourself in a job such as this?

If you are interviewing with a company that bills itself as a good environmental citizen, ask whether it accepts and complies with the Valdez Principles, a set of standards for corporate responsibility drawn up after the wreck of the *Exxon Valdez* in 1989. These principles are:

- *Biosphere protection:* Companies will minimize the release of any pollutant that may damage the air, water, or earth, including those that contribute to the greenhouse effect, depletion of the ozone layer, acid rain, and smog.
- *Sustainable natural resource use:* Companies will make sustainable use of renewable natural resources, such as water, soils, and forests, including protection of wildlife habitat, open spaces, and wilderness, and preservation of biodiversity.
- *Reduction and disposal of waste:* Companies will minimize waste, especially hazardous waste, and recycle whenever possible. All waste will be disposed of safely and responsibly.

- *Wise energy use:* Companies will make every effort to use environmentally safe and sustainable energy sources and invest in energy efficiency and conservation.
- *Risk reduction:* Companies will minimize environmental and health and safety risks to employees and local communities by employing safe technologies and preparing for emergencies.
- *Marketing of safe products and services:* Companies will sell products or services that minimize adverse environmental impacts and that are safe for consumer use.
- *Damage compensation:* Companies will take responsibility for any harm caused to the environment through cleanup and compensation.

- *Disclosure:* Companies will disclose to employees and community any incidents that cause environmental harm or pose health or safety risks.
- *Environmental directors and managers:* At least one member of the board of directors will be qualified to represent environmental interests, and the company will fund a senior executive position for environmental affairs.
- *Assessment and annual audit:* Companies will conduct annual self-evaluation of progress in implementing these principles and make results of independent environmental audits available to the public.

Blue Angels and Green Seals

Products that claim to be environmentally friendly are being introduced at twenty times the normal rate for consumer goods. To help consumers make informed choices in their shopping, several national programs have been set up to carry out independent, scientific analysis of life-cycle environmental impacts of major products. Germany's Blue Angel, begun in 1978, is the oldest of these programs. Endorsement is highly sought after by producers since environmentally conscious shoppers have shown that they are willing to pay more for products they know have minimum environmental impacts. To date, more than two thousand products display the Blue Angel symbol, ranging from recycled paper products, energy-efficient appliances, and phosphate-free detergents, to refillable dispensers.

Similar programs are being proposed in every Western European country as well as in Japan and North America. Some are autonomous, nongovernmental efforts, like the United States' new Green Seal program (managed by the Alliance for Social Responsibility in New York) or the Good Earthkeeping Seal bestowed by Good Housekeeping. Others are quasi-governmental organizations such as the Canadian Environmental Choice programs (fig. 15.5). The best of these programs attempt "cradle-to-grave" life-cycle analysis (fig. 15.6) that evaluates material and energy inputs and outputs at each stage of manufacture, use, and disposal of the product. While you need to consider your own situation in making choices, the information supplied by these independent agencies is generally more reliable than self-made claims from merchandisers.

Limits of Green Consumerism

To quote Kermit the Frog, "it's not easy being green." Even with the help of endorsement programs, doing the right thing from an environmental perspec-

(a)

(b)

Figure 15.5 American (*a*) and Canadian (*b*) symbols that will be used to indicate products that are "environmentally superior."

ᴍ—Offical mark of Environment Canada.

tive may not be obvious. Often we are faced with complicated choices. Do the social benefits of buying rainforest nuts justify the energy expended in transporting them here, or would it be better to eat only locally grown products? In switching from freon propellants to hydrocarbons, we spare the stratospheric ozone but increase hydrocarbon-caused smog. By choosing reusable diapers over disposable ones, we decrease the amount of material going to the landfill, but we also increase water pollution, energy consumption, and pesticide use (cotton is one of the most pesticide-intensive crops grown in the United States).

When the grocery store clerk asks you, "paper or plastic?" you probably choose paper and feel environmentally virtuous. Right? Everyone knows that plastic

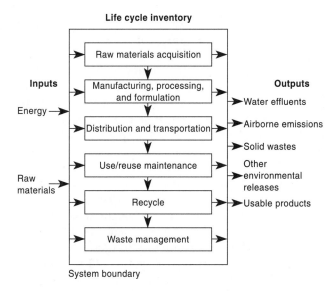

Life cycle inventory

Inputs

Energy →

Raw materials →

Raw materials acquisition

Manufacturing, processing, and formulation

Distribution and transportation

Use/reuse maintenance

Recycle

Waste management

System boundary

Outputs

→ Water effluents

→ Airborne emissions

→ Solid wastes

→ Other environmental releases

→ Usable products

Figure 15.6 At each stage in its life cycle, a product receives inputs of materials and energy and produces outputs of materials or energy that move to subsequent phases and outputs of wastes that are released into the environment. The relationship between inputs and outputs and the six stages of a life cycle inventory is shown.

is made by noxious chemical processes from nonrenewable petroleum or natural gas. Paper from naturally growing trees is a better environmental choice, isn't it? Well, not necessarily. In the first place, papermaking consumes water and causes much more water pollution than does plastic manufacturing. Paper mills also release air pollutants including foul-smelling sulfides and captans as well as highly toxic dioxins.

Furthermore, the brown paper bags used in most supermarkets are made mostly from virgin paper. Recycled fibers aren't strong enough for the weight they must carry. This paper comes mainly from intensively managed, heavily fertilized, clear-cut agroforestry plantations. It takes a great deal of energy to pulp wood and dry newly made paper. Paper is also heavier and bulkier to ship than plastic. Although the polyethylene used to make a plastic bag contains many calories, in the end, paper bags are generally more energy-intensive to produce and market than plastic ones.

If both paper and plastic go to a landfill in your community, the plastic bag takes up less space. It doesn't decompose in the landfill, but then, neither does the paper in an air-tight, water-tight landfill. If paper is recycled but plastic is not, then the paper bag may be the better choice. If you are lucky enough to have both paper and plastic recycling, the plastic bag is probably a better choice since it recycles more easily and produces less pollution in the process. The best choice of all is to bring your own reusable cloth bag.

Complicated, isn't it? We often must make decisions without complete information, but it's important to make the best choices we can. Don't automatically assume that your neighbors are wrong if they reach conclusions different than yours. They may have valid considerations of which you are unaware. The truth is that simple black-and-white answers often don't exist.

Paying Attention to What's Important

There can be many benefits to taking personal responsibility for our environmental impact. Recycling, buying green products, and other environmental good works not only set good examples for your friends and neighbors; they also strengthen your sense of involvement and commitment in valuable ways. There are limits, however, to how much we can do individually through our buying habits and personal actions to bring about the fundamental changes needed to save the earth. Green consumerism generally can do little about larger issues of global equity, chronic poverty, oppression, and the suffering of millions of people in the Third World. There is a great danger that exclusive focus on problems such as paper or plastic bags will divert our attention from the real need to change basic institutions.

During the 1970s, many people joined the back-to-the-land movement. Abandoning mass society, they moved to places like the mountains of North Carolina or remote valleys in Vermont where they experimented with self-sufficient lifestyles, living in converted school buses, teepees, or geodesic domes; growing their own food; cutting firewood; carrying water; and experimenting with alternative, utopian living. Such a lifestyle can be satisfying and morally uplifting but may contribute little in the long-run to helping other people live better lives.

Remaining a part of mass society, living an ordinary life, and devoting time and energy to bringing about important structural changes may ultimately be more valuable than withdrawing to "do your own thing." Rather than sort recyclables for which there is no market, for instance, your time might be better spent working for legislation to prevent excess packaging and to create markets for recycled materials. As in many other cases, there is no simple choice between these alternatives. Some of us will focus primarily on living sustainably, others will work to bring about change in the system, while still others will find some balance between these two options.

COLLECTIVE ACTIONS

While a few exceptional individuals can be effective working alone to bring about change, most of us find it more productive and more satisfying to work with others (fig. 15.7). Collective action multiplies your power. You get encouragement and useful information

Figure 15.7 Students cleaning up the beach in Santa Cruz, California, after a tanker collision. Ordinary people can make a contribution to improving the environment. Doing so can be personally rewarding and can foster public spirit.

from meeting regularly with others who share your interests. It's easy to get discouraged by the slow pace of change; having a support group helps maintain your interest and enthusiasm. You should realize, however, that there is a broad spectrum of environmental and social action groups. Some will suit your particular interests, preferences, or beliefs more than others. In this section, we will look at some environmental organizations as well as options for getting involved.

Student Environmental Groups

A number of organizations have been established to teach ecology and environmental ethics to elementary and secondary school students, as well as to get them involved in active projects in cleaning up their local community. Groups such as Kids Saving the Earth or Eco-Kids Corps are an important way to reach this vital audience. Environmental education in the classroom is also an important topic that we don't have room to cover here in any detail.

Most of you reading this book may be more interested in organizations that you can join and ways that you can become involved yourself. The largest student environmental group in North America is the Student Environmental Action Coalition (SEAC). Formed in 1988 by students at the University of North Carolina at Chapel Hill, SEAC has grown rapidly to more than 30,000 members in some 500 campus environmental groups. SEAC is both an umbrella organiza-

tion and a grass-roots network that functions as an information clearinghouse as well as a training center for student leaders. Member groups undertake a diverse spectrum of activities ranging from politically neutral recycling promotion to confrontational protests of government or industrial projects (fig. 15.8). National conferences bring together thousands of activists who share tactics and inspiration while also having fun. If there isn't a group on your campus, why not look into organizing one?

Another important student organizing group is the network of Public Interest Research Groups active on most campuses in the United States. While not focused exclusively on the environment, the PIRGs usually include environmental issues in their priorities for research. By becoming active, you could probably introduce environmental concerns to your local group if they are not already working on problems of importance to you.

One of the most important skills that you are likely to learn in either SEAC or other groups committed to social change is how to organize. Organizing is a dynamic process that must constantly adapt to changing conditions. Some basic principles apply in most situations, however (table 15.2). Remember that you are not alone. Others share your concerns and want to work with you to bring about change; you just have to find them (see box 15.4). There is power in numbers. As Margaret Mead once said, "Never doubt that a small, highly committed group of individuals can change the world; indeed, it is the only thing that ever has."

Using the communications media to get your message out is an important part of the modern environmental movement. Table 15.3 suggests some important considerations in planning a media campaign.

Figure 15.8 Imaginative costumes and street theater can get your message across in a humorous, non-threatening way that is attractive to the media if carefully executed. Here, a talking trash can tells students the benefits of recycling.

TABLE 15.2 Organizing an environmental campaign

1. What do you want to change? Are your goals realistic within the time and resources you have available?

2. What and who will be needed to get the job done? What resources do you have now, and how can you get more?

3. Who are the stakeholders in this issue? Who are your allies and constituents? How can you make contact with them?

4. How will your group make decisions and set priorities? Will you operate by consensus, majority vote, or informal agreement?

5. Have others already worked on this issue? What successes or failures did they have? Can you learn from their experience?

6. Who has the power to give you what you want or to solve the problem? Which individuals, organizations, corporations, or elected officials should be targeted by your campaign?

7. What tactics will be effective? Using the wrong tactics in a given situation can alienate people and be worse than taking no action at all.

8. Are there social, cultural, or economic factors that should be recognized in this situation? Will the way you dress, talk, or behave offend or alienate your intended audience? Is it important to change your appearance or tactics to gain support?

9. How will you know when you have succeeded? How will you evaluate the possible outcomes?

10. What will you do when the battle is over? Is yours a single issue organization, or will you want to maintain the interest, momentum, and network you have established?

Source: Based on material from Claire Greensfelder and Mike Roselle, "Grassroots Organizing for Everyone," in Call to Action, *edited by Brad Erickson, Sierra Club Books, 1990.*

Acting Locally Box 15.4

Cleaning Up the Nashua River

The Nashua River meanders for about 90 km (55 mi) through a heavily industrialized region in central Massachusetts and southern New Hampshire before joining the Merrimack River near the town of Nashua, New Hampshire. For years the river was so badly polluted by paper mill effluents, printing inks, municipal wastes, and agricultural runoff that it was virtually an open sewer. It ran a different color every day, depending on what was being dumped into it. Great globs of toxic yellow-orange sludge often covered the surface. Foul smells drifted through nearby communities and dead fish floated gently down the stream.

In 1962, Marion Stoddard moved to Groton, Massachusetts, not far from the Nashua River (box fig. 15.4.1). Disgusted by the water's condition, she decided to organize her neighbors to begin cleaning it up. The first step was to identify who cared about the river and how they might pool their efforts. A Nashua River Clean-up Committee was formed (and later reorganized into the Nashua Watershed Association to include land-use issues). Next, local, state, regional, and federal agencies were contacted to find out about plans for the river and relevant statutes and regulations.

An important weapon in this campaign was provided by the Massachusetts Clean Water Act, which provided for public hearings at which citizens could comment on water-quality standards. With a little community organizing

Figure 15.4.1 Marion Stoddard on the Nashua, a river no one wanted.

and publicity, hundreds of citizens were mobilized to attend hearings and voice their demands for clean water. A reclassification of the river resulted in new stringent standards for pollution control and wastewater treatment. Local industries complained, but most of the costs were paid by federal grants.

Another key to success was the ability to get widely different people to work together. The general manager of one paper mill was persuaded to serve on the association board of directors together with zealous environmentalists and conservative farmers. A broad-based coalition of private citizens, labor unions, business leaders, and politicians were persuaded that having clean water made good economic sense. In an unheard-of partnership, the U.S. Army, local communities, and two state governments worked together to sponsor clean-up days in which tons of trash and garbage were dragged from the river.

The end result was spectacularly successful. Six new wastewater treatment plants were built. A 2,400 ha (6,000 acre) greenway lines the river bank to protect the watershed and provide for public recreation. The river now runs clean and clear; people once again use it for swimming, fishing, and boating. Property values have risen and new companies have been attracted by high environmental quality and community spirit. The river that had been given up for dead is once more alive and well. The EPA recognized Marion Stoddard and the Watershed Management Association for their environmental leadership. It didn't take great technical knowledge or wealth to do what they did; just a concern for nature, perseverance, savvy use of the media, some organizing skills, and a willingness to work together for the common good. You could do the same.

TABLE 15.3 Using the media to influence public opinion

Shaping opinion, reaching consensus, electing public officials, and mobilizing action are accomplished primarily through the use of the communications media. To have an impact in public affairs, it is essential to know how to use these resources. Here are some suggestions:

1. *Assemble a press list.* Learn to write a good press release by studying books from your public library on press relations techniques. Get to know reporters from your local newspapers and TV stations.

2. *Appear on local radio and TV talk shows.* Get experts from local universities and organizations to appear.

3. *Write letters to the editor, feature stories, and news releases.* You may include black-and-white photographs. Submit them to local newspapers and magazines. Don't overlook weekly community shoppers and other "freebie" newspapers, which usually are looking for newsworthy material.

4. *Try to get editorial support from local newspapers, radio, and TV stations.* Ask them to take a stand supporting your viewpoint. If you are successful, send a copy to your legislator and to other media managers.

5. *Put together a public service announcement and ask local radio and TV stations to run it (preferably not at 2 A.M.).* Your library or community college may well have audiovisual equipment that you can use. Cable TV stations usually have a public service channel and will help with production.

6. *If there are public figures in your area who have useful expertise, ask them to give a speech or make a statement.* A press conference, especially in a dramatic setting, often is a very effective way of attracting attention.

7. *Find music stars or media personalities to support your position.* Ask them to give a concert or performance, both to raise money for your organization and to attract attention to the issue. They might like to be associated with your cause (fig. 15.9).

8. *Hold a media event that is photogenic and newsworthy.* Clean up your local river and invite photographers to accompany you. Picket the corporate offices of a polluter, wearing eye-catching costumes and carrying humorous signs. Don't be violent, abusive, or obnoxious; it will backfire on you. Good humor usually will go farther than threats.

9. *If you hear negative remarks about your issue on TV or radio, ask for free time under the Fairness Doctrine to respond.* Stations need to do a certain amount of public service to justify relicensing and may be happy to accommodate you.

10. *Ask your local TV or newspaper to do a documentary or feature story about your issue or about your organization and what it is trying to do.* You will not only get valuable free publicity, but you may inspire others to follow your example.

Figure 15.9 Actor Martin Sheen joins local activists in a protest in East Liverpool, Ohio, site of the largest hazardous waste incinerator in the United States. About 1000 people marched to the plant to pray, sing, and express their opposition. Involving celebrities draws attention to your cause. A peaceful, well-planned rally builds support and acceptance in the broader community.

Mainline Environmental Organizations

Among the oldest, largest, and most influential environmental groups in the United States are The National Wildlife Federation, World Wildlife Fund, The Audubon Society, Sierra Club, The Izaak Walton League, Friends of the Earth, Greenpeace, Ducks Unlimited, and The Wilderness Society. Sometimes known as the "group of 10," these organizations are criticized by radical environmentalists for their tendency to compromise and cooperate with the establishment. Although many of these groups were militant—even extremist—in their formative stages, they now tend to be more staid and conservative. Members are mostly passive and know little about the inner workings of the organization, joining as much for publications or social aspects as for their stands on environmental issues.

Still, these groups are powerful and important forces in environmental protection. Their mass membership, large professional staffs, and long history give them a degree of respectability and influence not found in newer, smaller groups. The Sierra Club, for instance, with nearly half a million members and chapters in nearly every state, has a national staff of about four hundred, an annual budget approaching $20 million, and twenty full-time professional lobbyists in Washington. These national groups have become a potent force in Congress, especially when they band together to pass specific legislation such as the Alaska National Interest Lands Act or the Clean Air Act.

In a survey that asked congressional staff and officials of government agencies to rate the effectiveness of groups that attempt to influence federal policy on pollution control, the top five were national environmental organizations. In spite of their large budgets and important connections, the American Petroleum Institute, the Chemical Manufacturers Association, and the Edison Electric Institute ranked far behind these environmental groups in terms of influence.

Some environmental groups, such as the Environmental Defense Fund (EDF), The Nature Conservancy (TNC), National Resources Defense Council (NRDC), and The Wilderness Society (WS), have limited contact with ordinary members except through their publications. They depend on a professional staff to carry out the goals of the organization through litigation (EDF and NRDC), land acquisition (TNC), or lobbying (WS). Although not often in the public eye, these groups can be very effective because of their unique focus. The Nature Conservancy buys land of high ecological value that is threatened by development. With more than $50 million in cash purchases and $44 million in donated lands, it now owns some nine hundred parcels of land, the largest privately owned system of nature sanctuaries in the world. Altogether, it has saved more than 1 million ha (2.47 million acres) of land in its forty-year history.

Broadening the Environmental Agenda

The environmental movement in the United States tends to be overwhelmingly white, middle-class, and suburban. Few blue-collar workers and even fewer minorities are involved, especially in leadership or professional positions. This is unfortunate in several ways. The movement will probably never be successful until it can put together a broad-based coalition of support. Furthermore, poor people are often most seriously affected by toxic waste dumps, noise, urban blight, and air and water pollution (fig. 15.10). They should be included in discussions of how to remedy these problems. Unfortunately, most environmental groups have acquired a reputation for caring about plants, animals, and wilderness areas more than about people. Until we can convince everyone that they have a stake in environmental protection, we are unlikely to fully achieve our goals.

Radical Environmental Groups

A striking contrast to the mainline conservation organizations are the "direct action" groups, such as Earth First!, Sea Shepherd, and a few other groups that form either the "cutting edge" or the "radical fringe" of the environmental movement, depending on your outlook. Often associated with the deep ecology philosophy and bioregional ecological perspective (table 15.4), the main tactics of these groups are civil

Figure 15.10 The great Louisiana Toxics March mobilized local residents from the notorious stretch of the Mississippi River between New Orleans and Baton Rogue known as "cancer alley" because of its high concentration of petro-chemical industries and environmental illnesses. Environmental justice combines elements of the civil rights movement with environmental protection.

disobedience and attention-grabbing actions, such as guerrilla street theater, picketing, protest marches, road blockades, and other demonstrations. Many of these techniques are borrowed from the civil rights movement and Mahatma Gandhi's nonviolent civil disobedience. While often more innovative than the mainstream organizations, pioneering new issues and new approaches, the tactics of these groups can be controversial.

Members of Earth First! perched in Douglas firs marked for felling in Oregon, chained themselves to a giant tree-smashing bulldozer in Texas to prevent forest clearing, and blockaded roads being built into wilderness areas (fig. 15.11). In some cases, the protests can be humorous and lighthearted, as when the Earth First! members dressed in bear costumes to protest grizzly bear management policies in Yellowstone National Park.

A more problematic tactic is **monkey wrenching** or environmental sabotage, a concept made popular by author Edward Abbey's book *The Monkey Wrench*

TABLE 15.4 Test your bioregional knowledge

1. Trace the water you drink from precipitation to tap.
2. How many days until the moon is full (plus or minus a couple of days)?
3. Describe the soil type around your home.
4. What were the primary subsistence techniques of the culture(s) that lived in the area before you?
5. Name five native edible plants in your bioregion and their season(s) of availability.
6. From what direction do winter storms generally come in your region?
7. Where does your garbage go?
8. How long is the growing season where you live?
9. On what day of the year are the shadows the shortest where you live?
10. Name five trees in your area. Are any of them native? If you can't name names, describe them.
11. Name five resident and any migratory birds in your area.
12. What is the land-use history by humans in your bioregion during the past century?
13. What primary geological event/process influenced the landform where you live?
14. What species have become extinct in your area?

15. What are the major plant associations in your region?
16. From where you are reading this, point north.
17. What spring wildflower is consistently among the first to bloom where you live?
18. What kinds of rocks and minerals are found in your bioregion?
19. Were the stars out last night?
20. Name some beings (nonhuman) that share your place.
21. Do you celebrate the turning of the summer and winter solstice? If so, how do you celebrate?
22. How many people live next door to you? What are their names?
23. How much gasoline do you use a week, on the average?
24. What energy costs you the most money? What kind of energy is it?
25. What developed and potential energy resources are in your area?
26. What plans are there for massive development of energy or mineral resources in your bioregion?
27. What is the largest wilderness area in your bioregion?

From Deep Ecology: Living as if Nature Mattered *by Bill Devall and George Sessions, copyright 1985, published by Gibbs Smith, Publisher. Reprinted by permission.*

Figure 15.11 Earth First! cofounder Mike Roselle, foreground, and others block a bulldozer building a road into de-facto wilderness in the Siskiyon National Forest in Oregon. Environmentalists have borrowed civil-disobedience tactics from the civil rights movement.

Gang. Among the actions advocated by some Earth First!ers are driving large spikes in trees to protect them from loggers, vandalizing construction equipment, pulling up survey stakes for unwanted developments, and destroying billboards.

This raises some difficult ethical questions. While each of us has a duty to resist immoral authority and destruction of nature, at what point should we take action? Suppose that in spiking trees you endanger loggers or mill workers; what moral responsibility do you bear? On a pragmatic level, does violent action engender or justify counter violence? Do the ends justify the means?

Sea Shepherd, a radical offshoot from Greenpeace, is determined to stop the killing of marine mammals. Convinced that peaceful protests were not working, its members have taken direct action by sinking whaling vessels and destroying machinery at whale processing stations. Greenpeace criticized this destructive approach, perhaps because of memories of the sinking of its ship *Rainbow Warrior* by the French secret service in New Zealand in 1985. Greenpeace and other critics of ecotage see these acts as environmental terrorism that gives all environmentalists a bad name. Sea Shepherd's Paul Watson says, "Our aggressive nonviolence is just doing what must be done, according to the dictates of our own conscience." Earth First!'s Dave Foreman (sometimes described as a kamikaze environmentalist) says, "Extremism in defense of Mother Earth is no vice." What do you think? Do ends justify means?

Anti-environmental Backlash

Many people whose interests are threatened by proposals for restricted resource use, pollution control, or wilderness preservation are organizing to resist these changes. This anti-environmental movement rallies supporters under the banner of "wise use" or "multiple use" of public lands. Ranchers, loggers, miners, industrialists, and land developers who believe their access to resources is threatened or who face expensive changes in the way they do business form the core group. Off-road vehicle enthusiasts, hunters, and others who resent restrictions placed on their use of public lands add their support. Not a few people join simply because they believe that the social changes are occurring too rapidly or going too far. Some conservatives, for instance, question the morality of liberal environmentalists, equating biocentric ecology with heathenism and social change with communism.

The largest numbers of adherents to the wise-use movement are generally found in the Western states, where public land ownership is high, rugged individualism and independence are valued, and the federal government is regarded with suspicion and hostility. Their central policy statement is contained in *The Wise Use Agenda* written by conservative activists Ron Arnold and Alan Gottlieb. It lists twenty-five goals, including opening all public lands—including wilderness areas and national parks—to mineral, energy, and timber production, eliminating protection of "nonadaptive species" such as spotted owls and California condors, and allowing motorized recreation anywhere on all public lands.

Wise-use groups are often generously funded by oil, mineral, and timber corporations. Workers join in because they believe that their jobs are jeopardized by environmental restrictions. A few violent types threaten to form vigilante posses, turning the tactics of intimidation, monkey wrenching, and "direct action" used by radical environmentalists back on their adversaries. So far, this anti-green movement is a fringe group without much national power, but environmentalists should not underestimate the anger, alienation, and resistance that underlie this reactionary force.

GLOBAL ISSUES

At the first United Nations Conference on the Environment in 1972 in Stockholm, Prime Minister Indira Gandhi of India declared that "poverty is the greatest danger to the environment." Twenty years of research and subsequent international conferences have shown the truth of that statement. In 1992, the United Nations Conference on Environment and Development (*see* box 1.3) focused on the plight of the poor and the need for sustainable development to alleviate poverty and provide ways to avoid environmental destruction.

The needs are tremendous. The United Nations estimates that about 1 billion people worldwide live in acute poverty and lack access to secure food supplies, safe drinking water, education, health services, infrastructure (roads, markets, etc.), land, credit, and jobs. Even more—around 1.5 billion—are exposed to dangerous air pollution, especially in urban areas and in homes where smoky fires provide the only means of cooking. Nearly one-third of all humans—about 1.8 billion—have inadequate sanitation, leading to spread of infectious diseases that debilitate the poor and prevent them from working effectively to improve their conditions.

The connection between poverty and the environment is a vicious circle. The poor are both victims and agents of environmental damage. Having to meet urgent short-term needs for survival, they are forced to "mine" their natural capital through excessive tree cutting, poor farming and grazing practices that lead to erosion and nutrient depletion, and overharvesting of fisheries and wildlife. The consequent degradation of the land results in further impoverishment of the people who depend on that land for survival (fig. 15.12). As the World Bank *World Development Report* says, "Without adequate environmental protec-tion, human development is undermined; without human development, environmental protection will fail."

Sustainable Development

What is **development?** In human terms, it is an improvement in the well-being of people. Raising living standards and improving health, education, and quality of life are goals of development programs. Economic growth can contribute to these programs but is not sufficient in itself. Increasing equality of opportunity, ensuring political and civil rights, and providing opportunities for people to reach their full human potential are also part of the development package; environmental protection is essential as well.

What is **sustainable development?** It is development that lasts. We realize that we are reaching limits in the extent to which we can extract resources and dispose of wastes in destructive, nonrenewable ways. There are no longer new places to go if we use up resources or foul the environment with our garbage. In economic terms, we have to learn to live on the interest from our environmental capital rather than use up the principle. In other words, we have to live on renewable resources in ways that will last into future generations.

Our use of nonrenewable resources and destructive technologies to meet our present needs raise questions of **intergenerational justice** or fairness to future generations. The environmental damage we do now limits our children's ability to meet their own needs (*see* chapter 1). They inherit the degraded environment and reduced prospects that we leave to them. On the other hand, the resources that we invest now in the form of education, learning skills, inventing technology, improving soil fertility, reforestation, etc. can improve the lives of future generations. Intergenerational justice means not only leaving resources for the future but also leaving a better situation for those who follow us.

Achieving Our Goals

How can we accomplish these goals? The developing countries of the world need access to more efficient, less polluting technologies and to learn from the successes and failures of the developed countries (table 15.5). Technology transfer is essential. Environmental and resource protection benefit both the richer countries of the world and the poorer ones. The costs of this protection should, therefore, be shared in some equitable way. Maurice Strong, chair of the Earth Summit, estimates that development aid from the richer countries should be some $150 billion per year, while internal investments in environmental protection by developing countries will need to be about twice that amount.

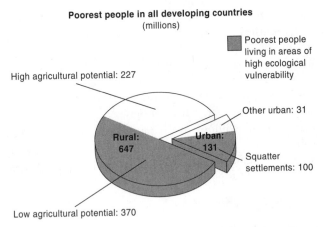

Poorest people in all developing countries
(millions)

Poorest people living in areas of high ecological vulnerability

High agricultural potential: 227

Other urban: 31

Rural: 647

Urban: 131

Squatter settlements: 100

Low agricultural potential: 370

Figure 15.12 A total of some 470 million people, or 60 percent of the developing world's 780 million *poorest* people, live in rural or urban areas of high ecological vulnerability—areas where ecological destruction or severe environmental hazards threaten their well-being.

Source: Data from H. Jeffrey Leonard in *Environment and the Poor,* ed. by Richard E. Feinberg and Valeriana Kallab. Copyright © 1989 Transaction Publishers.

TABLE 15.5 An agenda for sustainable development

1. Remove subsidies that encourage excessive use of fossil fuels, water, pesticides, minerals, and logging.

2. Clarify rights to own and manage lands, forests, and fisheries to give people access to the means of production and security to invest in long-range improvements.

3. Improve sanitation, clean water, education (especially for girls), family planning, agricultural extension, credit, and research that helps the poor.

4. Empower, educate, and involve local people in their own long-term interests.

5. Provide rural jobs, access to markets, housing, and other reforms that encourage people to remain in villages rather than to crowd into already overstressed urban areas.

6. Protect natural habitat and biodiversity along with critical watersheds, wetlands, forests, and other fragile ecological areas.

International Nongovernmental Organizations

The rise in international environmental organizations in recent years has been phenomenal. At the Stockholm Conference in 1972, only a handful of environmental groups attended, almost all from First World countries. Twenty years later, at the Rio Earth Summit, more than 30,000 individuals representing several thousand environmental groups, many from the Third World, held a global Ecoforum to debate issues and form alliances for a better world (*see* box 1.3). We call these groups working for social change **nongovernmental organizations (NGOs).** They have become a powerful aspect of environmental protection.

Some NGOs are located primarily in the more highly developed countries of the North and work mainly on local issues. Others are headquartered in the North but focus their attention on the problems of developing countries in the South. Still others are truly global with members in active groups in many different countries. A few are highly professional, combining private individuals with representatives of government agencies on quasi-governmental boards or standing committees with considerable power. A few are at the fringes of society, sometimes literally voices crying in the wilderness. Many work for political change, more specialize in gathering and disseminating information, and a few undertake direct action to protect a specific resource.

Public education and consciousness-raising using protest marches, demonstrations, civil disobedience, and other participatory public actions and media events are generally important tactics for these groups. Greenpeace, for instance, carries out well-publicized confrontations with whalers, seal hunters, toxic-waste dumpers, and others who threaten very specific and visible resources (fig. 15.13). Greenpeace may well be the largest environmental organization in the world, with some 2.5 million contributing members.

Figure 15.13 Greenpeace activists try to stop the killing of whales by placing themselves between the whaling ship and its quarry. In 1985, they were narrowly missed by a harpoon fired directly over their heads by a Russian whaler.

In contrast to these highly visible groups, others choose to work behind the scenes, but their impact may be equally important. Conservation International has been a leader in debt-for-nature swaps to protect areas particularly rich in biodiversity. It also has some interesting initiatives in economic development seeking products made by local people that will provide income along with environmental protection.

The tagua nut project is a good example of this approach. Tagua palms grow in tropical South America. The nut has a hard, smooth, white interior that looks and feels like ivory. Some four thousand indigenous people have been organized to harvest and carve this "vegetable ivory" on a sustainable basis. More than a quarter million buttons have been sold to clothing manufacturers in the United States and Europe, offering an alternative to destructive resource harvesting that had been the only means of support for local people.

GREEN GOVERNMENT AND POLITICS

While winning the hearts and minds of the public and influencing policymakers are important and gratifying work, it is sometimes necessary to dive into the nitty-gritty of politics or administration to bring about meaningful change. In this section, we will explore how these systems function, as well as some ways in which you can get involved.

Green Politics

In many countries with a parliamentary form of government, "green" parties based on environmental interests have become a political force in recent years. The largest and most powerful Green party is *Die Grünen* in Germany, which controls about 10 percent of the seats in the *Bundestag* or parliament (fig. 15.14). It is a grass-roots, egalitarian, council-style movement, committed to participatory democracy, environmental protection, and a fundamental transformation of society. An essential premise is that ceaseless industrial growth destroys both people and the environment and must, therefore, be stopped. The Greens have formed coalitions with antinuclear weapons protesters, feminists, human rights advocates, and other public interest groups, but have struggled continuously over whether to work with or oppose the ruling centrist party. Political realists argue that they could be more effective within a larger coalition; idealists vow never to compromise their principles.

The majority-rule political system in the United States makes it extremely difficult to start a new political party at the national level. Green candidates have been successful at the local or state level, however. In 1990, Alaska became the first state to give a Green party official standing. Jim Sykes, the Green gubernatorial candidate, received 3.2 percent of the total vote, enough to guarantee the party a spot on future ballots. National groups such as the League of Conservation Voters and "green committees of correspondents" work to introduce environmental issues

Figure 15.14 Delegates to the Green party convention in Germany debate the political, social, economic, and environmental platform. Although the Greens lost almost all their seats in the National Parliament in 1990, they remain a force in local politics.

into party politics and support candidates who share their environmental concerns. Campus Greens are strong at many colleges and universities. The four key values they espouse are (1) ecological wisdom, (2) peace and social justice, (3) grass-roots democracy, and (4) freedom from violence.

National Legislation

There are many opportunities for individuals to express their opinions and make their wishes known in conventional politics. Figure 15.15 shows the pathway that a piece of legislation follows in the United States, from inception to being signed into law by the president. The idea for a law can originate from an ordinary citizen, as well as from an elected official. After introduction, the bill is sent to the appropriate committee for hearings. Sometimes the hearing process is very extensive and may include field hearings in which ordinary citizens have a chance to express their opinions (fig. 15.16).

Contact your legislator to find out if there will be field hearings in your city. If they are scheduled, go even if you don't intend to speak. It is an educational experience to see how the process works. If you intend to testify, try to coordinate your presentation with others who share your position. Well-organized, factual testimony can make a good impression and have a positive impact both on elected officials and the public.

Almost all of the advance work in getting a bill ready for introduction and in guiding it through the hearing process is done by the legislative staff. These people are key in the success or failure of bills in which you are interested. It is well worth getting to know them and working with them. You could become

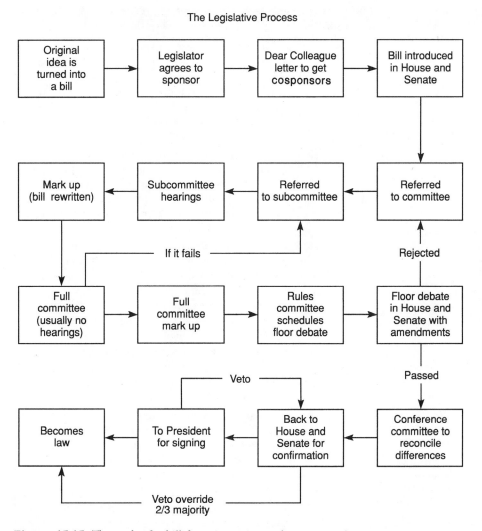

The Legislative Process

Original idea is turned into a bill → Legislator agrees to sponsor → Dear Colleague letter to get cosponsors → Bill introduced in House and Senate

Mark up (bill rewritten) ← Subcommittee hearings ← Referred to subcommittee ← Referred to committee ← Bill introduced in House and Senate

Rejected → back to Referred to committee

Full committee (usually no hearings) → Full committee mark up → Rules committee schedules floor debate → Floor debate in House and Senate with amendments

If it fails → Referred to subcommittee

Passed

Veto → Becomes law ← To President for signing ← Back to House and Senate for confirmation ← Conference committee to reconcile differences

Veto override 2/3 majority → Becomes law

Figure 15.15 The path of a bill from inception to becoming a law.

Figure 15.16 Citizens line up to speak at a legislative hearing. Speaking about public issues is a right that few people in the world enjoy. By getting involved in the legislative process, you can be informed and have an impact on governmental policy.

a legislative aide. Being in the "corridors of power" is an exciting and educational experience.

Few of us actually get directly involved in the legislative process, but you can make your wishes and opinions known by communicating with your elected representatives. Table 15.6 gives some suggestions for how to write effectively. Most representatives get little mail, so a single, well-written letter from a constituent can make a difference. Even on controversial issues, a representative might get only fifty to one hundred letters. Since each congressional district includes about half a million people, the legislator tends to assume that each letter represents the views of five thousand to ten thousand constituents. Your voice has great impact! If you don't have time to write, you can call the local or Washington office. You probably will not talk directly to your representative or senator, but your opinion will be registered by an aide. Another option is to send a telegram. Western Union has a night rate of only a few dollars for short political opinion messages.

The Courts

In the 1960s and 1970s, the judicial branch of government was highly responsive to environmental and social concerns. Much of this activist philosophy has been lost through recent court appointments, but citizens still can seek relief in the courts from unjust laws or actions in two ways. One way is to bring a civil suit, asking for payment of damages that were caused by a private individual, corporation, or governmental agency and that injured you or your property. The other way is to ask for a judgment by the court about the constitutionality of laws passed by Congress or the adequacy and legality of regulations established by an administrative agency. If the court finds a law or regulation to be improper, it can issue an injunction to stop implementation or application of that law or regulation. It also can order an agency to rewrite

TABLE 15.6 How to write to your elected officials

1. Address your letter properly:
   a. *Your representative:*
      The Honorable_____,
      House Office Building
      Washington, D.C. 20515
      Dear Representative _____,
   b. *Your senators:*
      The Honorable_____,
      Senate Office Building
      Washington, D.C. 20510
      Dear Senator _____,
   c. *The president:*
      The President
      The White House
      1600 Pennsylvania Avenue, N.W.
      Washington, D.C. 20500
      Dear Mr. President,

2. Tell who you are and why you are interested in this subject. Be sure to give your return address.

3. Always be courteous and reasonable. You can disagree with a particular position, but be respectful in doing so. You will gain little by being shrill, hostile, or abusive.

4. Be brief. Keep letters to one page or less. Cover only one subject, and come to the point quickly. Trying to cover several issues confuses the subject and dilutes your impact.

5. Write in your own words. It is more important to be authentic than polished. Don't use form letters or stock phrases provided by others. Speak or write from your own personal experiences and interests. Try to show how the issue affects the legislator's own district and constituents.

6. If you are writing about a specific bill, identify it by number (for instance, H.R. 321 or S.123). You can get a free copy of any bill or committee report by writing to the House Document Room, U.S. House of Representatives, Washington, D.C. 20515 or the Senate Document Room, U.S. Senate, Washington, D.C. 20510.

7. Ask your legislator to vote a specific way, support a specific amendment, or take a specific action. Otherwise you will get a form response that says: "Thank you for your concern. Of course I support clean air, pure water, apple pie, and motherhood."

8. If you have expert knowledge or specifically relevant experience, share it. But don't try to intimidate, threaten, or dazzle your representative. Don't pretend to have vast political influence or power. Legislators quickly see through artifice and posturing; they are professionals in this field.

9. If possible, include some reference to the legislator's past action on this or related issues. Show that you are aware of his or her past record and are following the issue closely.

10. Follow up with a short note of thanks after a vote on an issue that you support. Show your appreciation by making campaign contributions or working for candidates who support issues important to you.

11. Try to meet your senators and representatives when they come home to campaign, or visit their office in Washington if you are able. If they know who you are personally, you will have more influence when you call or write.

12. Join with others to exert your combined influence. An organization is usually more effective than isolated individuals.

---

its regulations and may even direct what the new regulations should be.

There are several problems that make lawsuits difficult for ordinary citizens who want to protect the environment or some other public resource:

1. You have to show that the action or law you oppose is illegal; that may be hard to do.
2. You have to establish that you have standing in court (a right to be heard). To do so, you must prove that you are directly affected, sometimes a difficult requirement.
3. Suits are expensive; a major suit might cost hundreds of thousands of dollars in legal fees, court fees, witnesses, etc.

4. It may take years before a suit is finally settled, and by that time it may be too late to save the resource.
5. You have to prove that the defendant (an agency, corporation, or individual) is responsible for the harm that you allege. A corporation can admit that they produce a toxic chemical, but you have to show, beyond reasonable doubt, that it was *their specific* chemical that caused the problem.

In spite of these difficulties, the courts often have been the most successful place for environmental organizations to change how we manage our environment. More than one hundred public-interest law firms in the United States specialize in social and environmental issues. Several environmental organizations, such as the Environmental Defense Fund, the National Resources Defense Council, and the Sierra Club Legal Foundation, act primarily or exclusively through litigation (bringing suits in court).

The Executive Branch

Of the roughly 14 million federal employees in the United States, the vast majority work in the executive branch. Many of these civil servants are in the administrative agencies that carry out and enforce the laws passed by the legislature. They monitor, manage, control, and protect our resources and our environment. Figure 15.17 shows the major agencies and branches of government that have environmental responsibilities. The Environmental Protection Agency (EPA) is often regarded as the main guardian of environmental quality. It has responsibility for regulating air and water pollution, solid and hazardous wastes, toxic substances, noise, radiation, and certain pesticides.

The Department of the Interior manages the national parks, wildlife refuges, wild rivers, historic sites, BLM lands, the Bureau of Reclamation, and the Bureau of Mining. It is by far the largest land manager in the country. The Department of Agriculture administers the national forests, the Agricultural Research Service, the Soil Conservation Service, and the Food Safety Inspection Service. The Department of Health administers the Public Health Service and the Food and Drug Administration. The Department of Labor is responsible for occupational safety and health. The Department of Energy is responsible for nuclear energy and fossil fuels.

These administrative agencies represent a tremendous bureaucracy, which often is ponderous and unresponsive, but then, we have a big country, and running it is a big job. For the most part, our civil service is made up of dedicated, professional people. Our system has been criticized for responding incrementally to problems. It may make improvements more slowly than we would like in some situations, but it also makes mistakes more slowly. If it had the power to make sudden leaps, it might make them in the wrong direction.

Environmental Impact Statements

One of the most useful tools for those concerned about environmental protection is a provision in the National Environmental Policy Act of 1970 requiring all federal agencies to prepare an **environmental impact statement (EIS)** analyzing the effects of any major program that it plans to undertake. As we discussed earlier in this chapter, it often is difficult for citizens to prove they are personally injured by a pollutant or an action of the government. It usu-

ally is not difficult, however, to show that an agency has prepared an inadequate EIS or has failed to consider important environmental consequences of their actions.

Intervening in administrative proceedings has become much easier since the Freedom of Information Act (FOIA), passed in 1974, requires federal agencies to make public all minutes of meetings, correspondence, and other official documents. This means that you can find out what an agency considered when preparing an EIS, who officials talked to, and why they decided as they did. It often is fairly easy to show that undue consideration was given to commercial interests or that environmental values were overlooked or deliberately ignored. In some cases, simply asking for an EIS will prompt an agency to reconsider a harmful project. In other cases, you may have to ask the courts to order one. Often the delay caused by litigation over an EIS gives time to organize other opposition. Sometimes just generating a discussion of the adverse effects of a project is enough to kill it. Letters to your legislators are considered private correspondence and cannot be made public under the FOIA.

International Environmental Treaties and Conventions

No international body has authority to pass binding environmental laws, nor are there international agencies with power to regulate resources on a global scale. An international court sits at the Hague in the Netherlands, but it has little power to enforce its decisions. Powerful nations can simply ignore the court, as the United States did when it was found in violation of international law after it mined the harbors of Nicaragua in 1983. Still, there is movement toward environmental protection on a worldwide scale through special covenants, treaties, and multilateral agreements.

In spite of these difficulties, several successful international conventions and treaties have been passed. The 1979 Convention on Long-Range Transboundary Air Pollution was the first multilateral agreement on air pollution and the first environmental accord involving all the nations of Eastern and Western Europe and North America. The 1989 Accord on Chlorofluorocarbon Emissions (chapter 10) was achieved remarkably quickly to address the threat to the stratospheric ozone shield. It was the first treaty between the industrialized North and the developing South. Recognizing the need for economic expansion by the less-developed countries, this treaty requires greater reductions from the bigger polluters than the smaller ones and a sharing of new technology to ease the effects of change on poorer nations.

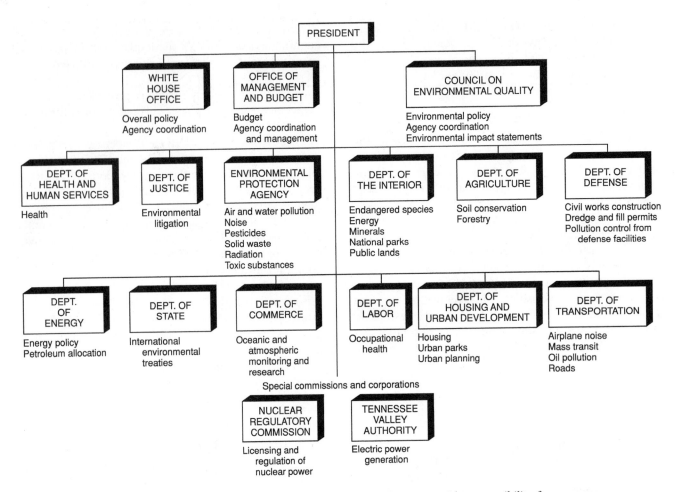

Figure 15.17 Major agencies of the executive branch of the federal government with responsibility for resource management and environmental protection.

Source: Data from U.S. General Accounting Office.

The Antarctic Treaty of 1961 reserves the Antarctic continent for peaceful scientific research and bans all military activities in the region. Proposed exploitation of Antarctic mineral and energy resources raises important questions about regulating the global commons. In 1992, the United States proposed dividing up Antarctic resources on a "first come, first served" basis, arguing that those who invest in developing and harvesting the resource deserve to enjoy the profits. The less-developed nations resisted, claiming that these resources are a common heritage of all humankind that should benefit everyone, not just those wealthy enough to afford the technology to harvest them. They see this as the first application of a new economic order, in which we share more equally in resource benefits. A compromise was finally reached in which it was agreed that no commercial activities will be allowed in Antarctica for fifty years. Many environmentalists believe that this last pristine continent should be set aside as an international wilderness area.

The U.N. Conference on Environment and Development in 1992 considered a wide range of international environmental issues. Among the most important of these are the convention on biodiversity, which calls for the conservation and sustainable use of biological diversity together with a fair sharing of the wealth produced from these resources. Only the United States, among all the world's nations, refused to sign because the administration claimed that benefit-sharing provisions were incompatible with existing intellectual property rights, such as ownership of patents or copyrightable information.

A convention on global climate change was also expected to be a centerpiece of the Earth Summit. Due to objections of the United States over timetables and specific commitments to limit emissions, only a vague and general document passed. Still, many important issues were discussed and the contacts made in this process, both government-to-government and person-to-person networking, may turn out to have been its most important contribution (fig. 15.18). The "Earth Charter" or "Rio Declaration" adopted by the

TABLE 15.7 Key principles of the Earth Charter

1. Every state has a sovereign right to exploit its own resources in accordance with its own policies, providing it doesn't harm the environment elsewhere.

2. All peoples have a right to sustainable development.

3. Environmental protection must be an integral part of development.

4. Unsustainable patterns of production and consumption must be reduced and appropriate demographic policies must be promoted.

5. People have a right to information and the opportunity to participate in political processes.

6. Polluters should pay for their environmental impacts through internalization of costs and the use of appropriate taxes, charges, and incentives for environmental protection.

Source: Adopted by the United Nations Conference on Environment and Development, Rio de Janeiro, Brazil, 1992.

Figure 15.18 A little girl helps former Brazilian President Fernando Collor de Mello plant a tree on the grounds of the Rio Center as part of the Earth Summit in 1992. Although the treaties passed at this meeting were less powerful than many of us would prefer, like this sapling and the children who nurture it, they have great potential.

delegates contains some important principles on both environmental protection and sustainable development (table 15.7).

All history consists of successive excursions from a single starting point, to which man returns again and again to organize yet another search for a durable scale of values.

Aldo Leopold, *The Sand County Almanac*

Summary

Current trends are not predictions of what must be, but rather warnings of what may be, depending on the course we follow. There are many definitions of what constitutes the "good life" and how we ought to behave. An anthropocentric view holds that only humans can be moral agents or moral subjects, all other beings having only instrumental value. A biocentric or ecocentric view considers that other creatures and even nonliving objects have goods and interests of their own. Another perspective holds that our intelligence and foresight create obligations for stewardship or caretaking of the environment on behalf of other creatures and future generations. Ecofeminists argue that we should focus on relationships and kinship rather than duties, rights, or obligations.

Deep ecology is a philosophy that calls for voluntary simplicity, rejection of anthropocentrism, decentralization, and direct personal action. Progressive environmentalists, on the other hand, advocate working within existing systems to bring about gradual reform. There is much we can do through our individual choices of goods and services, but we must be careful to avoid green scams and fraudulent marketing. Independent, scientific, lifetime analysis can help us make informed decisions.

There are many opportunities for collective action to bring about social change. Student environmental groups offer chances to network with others, learn useful organizing techniques, do good work, and have fun. The large national "mainline" environmental organizations have a degree of respectability, power, and influence unmatched by smaller, independent groups. Monkey wrenching or direct action is a highly controversial tactic espoused by some radical environmentalists. An anti-environmental backlash from the conservative far right threatens to use these same tactics in retaliation.

Sustainable development promises to alleviate acute poverty while also protecting the environment. Many nongovernmental organizations (NGOs) as well as intergovernmental treaties and conventions work toward these twin goals. In many countries, green politics espouses four key values of ecological wisdom, peace and social justice, grass-roots democracy, and freedom from violence. While it is difficult to introduce new national parties in the United States, there are opportunities to work within the legislative, judicial, and administrative agencies to bring about change.

Review Questions

1. What is the difference between anthropocentric and biocentric philosophies?
2. What are deep and shallow ecology? Why do progressives object to being called shallow ecologists?
3. List some things you can do to reduce resource consumption.
4. Discuss some of the limits to green consumerism and how you can spot fraudulent or meaningless green marketing claims.
5. Review the trade-offs between paper and plastic bags. Which is better in your view?
6. List ten key issues in organizing an environmental campaign or using media to influence public opinion.
7. What are the ten largest, oldest, mainline environmental organizations in the United States.
8. Describe and evaluate the tactics used by radical environmental groups. Do you subscribe to monkey wrenching?
9. Define sustainable development and explain why it is needed.
10. List the four key values of the Green party.
11. Explain how the legislative, judicial, and administrative branches of government work to protect the environment.
12. Compare the intent and effectiveness of three international environmental conventions or treaties mentioned in this chapter.

Questions for Critical or Reflective Thinking

1. An article in a major environmental journal criticized a group of kayakers who paddled down a river in the Arctic National Wildlife Refuge to protest proposed oil drilling there. As they stood on the beach in their polypropylene-neoprene-Gortex paddling outfits, next to their plastic kayaks and nylon tents, waiting for a charter plane to pick them up on the first leg of a trip of several thousand miles back home, they deplored our excessive use of petroleum. Was this hypocritical? When do the ends justify the means?
2. Are we merely part of nature, or do we occupy some special position in the scheme of things? Is it arrogant to assume that we have a right or obligation to manage nature? Summarize the arguments for and against anthropocentrism, biocentrism, stewardship, and ecofeminism.
3. What could be wrong with environmental restoration? Why would some people object to repairing or improving the environment? Suppose that to restore a viable population of Bermuda cahows, it becomes necessary to kill a large number of long-tailed tropic birds to reduce competition. Is this valid?
4. Do you support the conservative, mainstream approach to conservation followed by the mainline environmental groups, or do you prefer the more challenging, innovative approaches of more radical groups? What are the advantages and disadvantages of each?
5. Surely monkey wrenching just for the thrill of it is wrong, but at some point all of us must believe that resistance to protect nature is justified. Where would that point be for you and how far would you go in resistance? Would you condone violence to protect life?
6. Is sustainable development possible? When will we come up against the limits to further growth in population, resource consumption, and conversion of natural ecosystems to suit human needs?
7. Do you agree that sustainable development has the potential to simultaneously reduce poverty *and* protect the environment? What responsibility do those of us in rich nations bear toward the rest of the world?
8. Choose one of the ambiguous dilemmas presented in this chapter such as whether we have a right to modify nature or kill other species at will. What additional information would you need to choose between alternatives? How should competing claims or considerations be weighed in an ambiguous issue such as this?
9. There are good reasons to be pessimistic about what we are doing to our environment, but there also are signs of progress and reasons for optimism. Where are you on this spectrum?
10. The Talmud says, "If not now, when?" How might this apply to environmental science?

deep ecology 322
development 334
environmental impact statement (EIS) 339
environmental restoration 320
intergenerational justice 334

monkey wrenching 332
nongovernmental organizations (NGOs) 335
shallow ecology 322
social ecology 322
sustainable development 334

Suggested Readings

Anderson, B. N. *Ecologue: The Environmental Catalogue and Consumer's Guide for a Safe Earth.* New York: Prentice Hall, 1990. An illustrated guide to green products with interesting tips and recommendations.

Benjamin, M., and A. Freedman. *Bridging the Global Gap: A Handbook to Linking Citizens of the First and Third Worlds.* Cabin John, Md.: Seven Locks Press, 1989. Discusses personal actions to bring about social change and lists many sources for socially conscious travel, consumerism, and action.

Bookchin, M. "A Philosophical Naturalism." *Society and Nature* 1, no. 2 (1992): 60. A discussion of humanity's place in nature by the founder of social ecology in a special issue on ecological philosophy.

Caldwell, L. K. *International Environmental Policy.* 2d ed. Durham, N.C.: Duke University Press, 1991. Excellent analysis of international diplomacy, treaties, and conventions for environmental management.

Durning, A. T. "Long on Things, Short on Time." *Sierra* 78, no. 1 (January/February 1993): 60. We can—and must—make do with less stuff while learning to measure wealth not in dollars but in hours.

Eckersley, R. *Environmentalism and Political Theory: Toward an Ecocentric Approach.* Albany: State University of New York Press, 1992. A thoughtful discussion of environmental political theory.

Elkington, J., et al. *The Green Consumer.* New York: Penguin Books, 1990. An extensive guide to green products with detailed information and many sources.

Erickson, B., ed. *Call to Action: Handbook for Ecology, Peace, and Justice.* San Francisco: Sierra Club Books, 1990. Short but useful articles by a wide variety of activists on organizing for social change.

Frankel, C. "Trouble With Green Products." *Buzzworm* 3, no. 6 (November/December 1991): 37. People say they will pay more for environmentally friendly products, but small businesses have a hard time breaking into the market.

Goldsmith, E., et al. "Whose Common Future?" *The Ecologist* 22, no. 4 (July/August 1992): 122. A special issue devoted to a critique of development and management policies for the global commons.

Gottlieb, A. M. *The Wise Use Agenda.* Bellevue, Wash.: Free Enterprise Press, 1989. Lists the priorities of the wise-use movement and of member organizations.

Haas, P. M., et al. "Appraising the Earth Summit." *Environment* 34, no. 8 (October 1992): 6. A balanced discussion of the successes and failures of UNCED and the Global Forum with a summary of major documents signed in Rio.

Hansen, P. "Canada's Green Plan." *Canada Today/d'aujourd'hui* 22, no. 1 (1991): 3. A special issue explaining the comprehensive, nationwide plan to protect and improve Canada's environment following the guidelines of the Brundtland Commission.

Hawken, P. "The Ecology of Commerce." *INC.,* April 1992, 93–100. An interesting discussion of green business by the founder of Smith & Hawken garden products and Erewhon natural foods.

Homer-Dixon, T. F., et al. "Environmental Change and Violent Conflict." *Scientific American* 268, no. 2 (February 1993): 38. Suggests that competition for resources is causing increasing conflict between nations and ethnic groups.

Leonard, H. J., et al. *Environment and the Poor: Development Strategies for a Common Agenda.* New Brunswick: Transaction Books, 1989. An excellent overview of connections between environment and poverty with detailed case studies from the Third World.

Livernash, R. "The Growing Influence of NGOs in the Developing World." *Environment* 34, no. 5 (June 1992): 12. A good overview of Third World NGOs.

MacNeill, J., et al. *Beyond Interdependence: Meshing the World's Economy and the Earth's Ecology.* Oxford: Oxford University Press, 1991. A Trilateral Commission Report on the need for continued growth and sustainable development.

McCormick, J. *Reclaiming Paradise: The Global Environmental Movement*. Bloomington: Indiana University Press, 1989. A history of the environmental movement that is truly worldwide in scope.

Merchant, C. *Radical Ecology: The Search for a Livable World*. London: Routledge, 1992. An excellent survey of radical environmental politics and ethics, including deep ecology, social ecology, green politics, ecofeminism, and spiritual ecology.

Naess, A. "Deep Ecology and Ultimate Premises." *Society and Nature* 1, no. 2 (1992): 108. A reprint of Naess's seminal platform on deep ecology together with responses to critics.

Nilsen, R. *Helping Nature Heal: An Introduction to Environmental Restoration*. Berkeley, Calif.: Whole Earth Catalog/Ten Speed Press, 1991. Many case studies and practical examples of restoration from all over the world.

Norton, B. G. *Toward Unity Among Environmentalists*. Oxford: Oxford University Press, 1991. A history of environmentalism and environmental policies in America.

O'Callaghan, K. "Whose Agenda for America?" *Audubon* 94, no. 5 (September/October 1992): 80. An environmentalist looks at the wise-use movement and what it wants.

Plant, C., and J. Plant, eds. *Green Business: Hope or Hoax?* Santa Cruz, Calif.: New Society Publishers, 1991. Articles by many authors on green consumerism, green scams, and alternatives for a better world.

Poole, W. "Neither Wise Nor Well." *Sierra* 77, no. 6 (November/December 1992): 58. An environmentalist goes undercover at the wise-use convention. Who are those guys, anyway?

Post, J. E. "Managing As If the Earth Mattered." *Business Horizons* 34, no. 4 (July/August 1991): 32. An overview of business challenges in an environmental age.

Robin, V. "How Much Is Enough?" *In Context* 26 (Summer 1990): 62. Part of a special issue on green consumerism and fulfilling lifestyles for a small planet.

Sachs, W. "Environment and Development: The Story of a Dangerous Liaison." *The Ecologist* 21, no. 6 (November/December 1991): 252. An interesting critique of both development theory and environmentalism.

Smith, A. A. *Campus Ecology: A Guide to Assessing Environmental Quality and Creating Strategies for Change*. Los Angeles: Living Planet Press, 1993. A guide from the Student Environmental Action Coalition with many examples and case studies of things you can do for a better environment on your campus.

Street, P. *Global Environmental Change: Human and Policy Dimensions*. Oxford: Butterworth-Heinemann Ltd., 1991. Papers on environmental policy published in cooperation with the United Nations University.

Taylor, D. "The Environmental Justice Movement." *EPA Journal* 18, no. 1 (March/April 1992): 23. Part of a special issue on environmental racism.

Webster, D. "Sweet Home Arkansas." *Outside*, January 1992. Describes how a group of determined activists are resisting toxic dumping in their backyards. Reprinted in the *Utne Reader*, July/August 1992, together with an article on "Nimbymania."

World Bank. *World Development Report 1992* Oxford: Oxford University Press, 1992. An authoritative overview of world environment, poverty, and sustainable development.

World Resources Institute. *World Resources 1992-93*. Washington, D.C.: World Resources Institute, 1992. An excellent source of data and analysis of world resources. This year's edition has a special section on international nongovernmental organizations.

CANDIDE

▼

or Optimism

by Voltaire

I had hoped, said Pangloss, to reason a while with you concerning effects and causes, the best of all possible worlds, the origin of evil, the nature of the soul, and pre-established harmony.

At these words, the door was slammed in their faces. Pangloss, Candide, and Martin, as they returned to their little farm, passed a good old man who was enjoying the cool of the day at his doorstep under a grove of orange trees.

You must possess, Candid said to the man, an enormous and splendid property.

I have only twenty acres, he replied; I cultivate them with my children, and the work keeps us from three great evils, boredom, vice, and poverty.

You are perfectly right, said Pangloss; for when man was put into the Garden of Eden, he was put there *ut operatur eum* so that he should work it; this proves that man was not born to take his ease.

Let's work without speculating, said Martin; it's the only way of rendering life bearable.

The whole little group entered into this laudable scheme; each one began to exercise his talents. The little plot yielded fine crops. Cunegonde was, to tell the truth, remarkably ugly; but she became an excellent pastry cook. Paquette took up embroidery; the old woman did laundry. Everyone, down even to Brother Giroflee did something useful; he became a very adequate carpenter, and even an honest man; and Pangloss sometimes used to say to Candide:—All events are linked together in the best of possible worlds; for, after all, if you had not been driven from a fine castle by being kicked in the backside for love of Miss Cunegonde, if you hadn't been sent before the Inquisition, if you hadn't traveled across America on foot, if you hadn't lost all your sheep from the good land of Eldorado, you couldn't be sitting here eating candied citron and pistachios.

That is very well put, said Candide, but we must cultivate our garden.

Reprinted from Voltaire, Candide *or Optimism, A Norton Critical Edition, Translated and edited by Robert M. Adams, by permission of W. W. Norton & Company, Inc. Copyright © 1966 by W. W. Norton & Company, Inc.*

APPENDIX A: WORLD MAP

▼

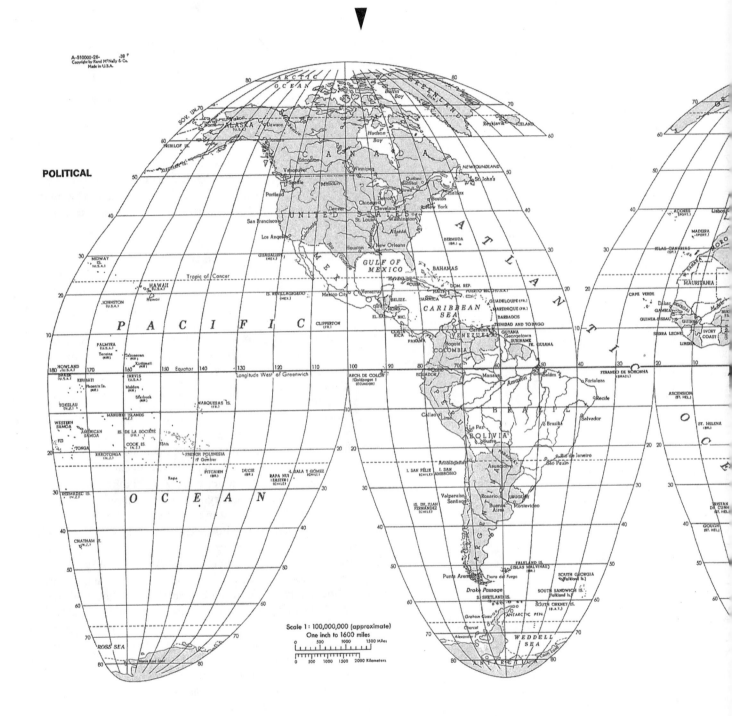

Appendix A. Goode's Hololosine Equal Area Projection. It is impossible to portray a perfect representation of the spherical surface of the Earth on a flat piece of paper. Most projections exaggerate the size and importance of the Northern Hemisphere with respect to the South. This projection gives a more realistic portrayal of the relative size of the continents.

From *Goode's World Atlas*. Map © 1993 by Rand McNally R. L. 93–S–82

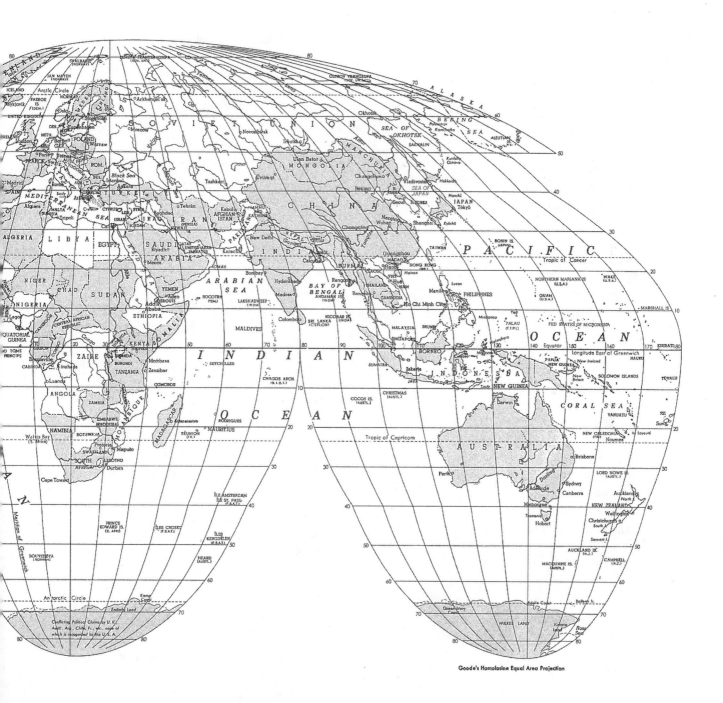

Goode's Homolosine Equal Area Projection

APPENDIX B: UNITS OF MEASUREMENT METRIC/ENGLISH CONVERSIONS

▼

Length

1 meter = 39.4 inches = 3.28 feet = 1.09 yard

1 foot = 0.305 meters = 12 inches = 0.33 yard

1 inch = 2.54 centimeters

1 centimeter = 10 millimeter = 0.394 inch

1 millimeter = 0.001 meter = 0.01 centimeter = 0.039 inch

1 fathom = 6 feet = 1.83 meters

1 rod = 16.5 feet = 5 meters

1 chain = 4 rods = 66 feet = 20 meters

1 furlong = 10 chains = 40 rods = 660 feet = 200 meters

1 kilometer = 1,000 meters = 0.621 miles = 0.54 nautical miles

1 mile = 5,280 feet = 8 furlongs = 1.61 kilometers

1 nautical mile = 1.15 mile

Area

1 square centimeter = 0.155 square inch

1 square foot = 144 square inches = 929 square centimeters

1 square yard = 9 square feet = 0.836 square meters

1 square meter = 10.76 square feet = 1.196 square yards = 1 million square millimeters

1 hectare = 10,000 square meters = 0.01 square kilometers = 2.47 acres

1 acre = 43,560 square feet = 0.405 hectares

1 square kilometer = 100 hectares = 1 million square meters = 0.386 square miles = 247 acres

1 square mile = 640 acres = 2.59 square kilometers

Volume

1 cubic centimeter = 1 milliliter = 0.001 liter

1 cubic meter = 1 million cubic centimeters = 1,000 liters

1 cubic meter = 35.3 cubic feet = 1.307 cubic yards = 264 U.S. gallons

1 cubic yard = 27 cubic feet = 0.765 cubic meters = 202 U.S. gallons

1 cubic kilometer = 1 million cubic meters = 0.24 cubic mile = 264 billion gallons

1 cubic mile = 4.166 cubic kilometers

1 liter = 1,000 milliliters = 1.06 quarts = 0.265 U.S. gallons = 0.035 cubic feet

1 U.S. gallon = 4 quarts = 3.79 liters = 231 cubic inches = 0.83 imperial (British) gallons

1 quart = 2 pints = 4 cups = 0.94 liters

1 acre foot = 325,851 U.S. gallons = 1,234,975 liters = 1,234 cubic meters

1 barrel (of oil) = 42 U.S. gallons = 159 liters

Mass

1 microgram = 0.001 milligram = 0.000001 gram

1 gram = 1,000 milligrams = 0.035 ounce

1 kilogram = 1,000 grams = 2.205 pounds

1 pound = 16 ounces = 454 grams

1 short ton = 2,000 pounds = 909 kilograms

1 metric ton = 1,000 kilograms = 2,200 pounds

Temperature

Celsius to Fahrenheit °F = (°C × 1.8) + 32

Fahrenheit to Celsius °C = (°F – 32) ÷ 1.8

Energy and Power

1 erg = 1 dyne per square centimeter

1 joule = 10 million ergs

1 calorie= 4.184 joules

1 kilojoule = 1,000 joules = 0.949 British Thermal Units (BTU)

1 megajoule = MJ = 1,000,000 joules

1 kilocalorie = 1,000 calories = 3.97 BTU = 0.00116 kilowatt-hour

1 BTU= 0.293 watt-hour

1 kilowatt-hour = 1,000 watt-hour = 860 kilocalories = 3,400 BTU

1 horsepower = 640 kilocalories

1 quad = 1 quadrillion kilojoules = 2.93 trillion kilowatt-hours

APPENDIX C: ENVIRONMENTAL OPPORTUNITIES FOR STUDENTS

Here is a list of some organizations that you can contact to become more involved in social or environmental issues:

Access
50 Beacon Street, Boston, MA 02108
(617) 720-JOBS
Gathers information on public interest and nonprofit internships and career opportunities.

American Friends Service Committee (AFSC)
1501 Cherry Street, Philadelphia, PA 19102
(215) 421-7000
AFSC is a Quaker organization that provides material assistance to grass-roots groups in Asia, Africa, Latin America, the Middle East, and poor communities in the United States. AFSC also coordinates cross-cultural education programs, foreign policy programs, and human rights campaigns.

Amigos de las Americas
5618 Star Lane, Houston, TX 77057
(800) 231-7796 or (713) 782-5290
Volunteers lead public health projects in Latin America and the Caribbean. Each volunteer receives training and raises the funds needed for his or her project during the school year and then works with a partner in a village for six to eight weeks during the summer. Projects include teaching dental hygiene, distributing eyeglasses, helping to build latrines, and providing immunization against disease.

Center for Global Education
c/o Augsburg College
731 21st Avenue S., Minneapolis, MN 55454
(612) 330-1159
Coordinates travel seminars designed to introduce participants to the realities of life in Mexico, Central America, the Caribbean, and the Philippines. Costs range between $900 and $2,000. Some scholarships are available.

Cool It!
National Wildlife Federation
1400 16th Street, NW, Washington, DC 20036
(202) 797-5435
Regional Cool It! coordinators work with students over the phone and conduct on-site workshops to help establish models of environmentally sound practices. Other services include a newsletter, job bank, and speakers bureau.

Earth Day Resources
116 New Montgomery Street, Suite 530, San Francisco, CA 94105
(800) 727-8619 or (415) 495-5987
Maintains a network of grass-roots organizations committed to environmental solutions at the local level. Offers materials such as lesson plans, fact sheets, and guidance for recycling and rideshare programs.

Earthwatch
P.O. Box 403, Watertown, MA 02272
(617) 926-8200
Earthwatch links volunteers with scientists and scholars on academic research expeditions in the United States and abroad. You might help to study rock art in Italy or help to make a documentary movie on Brazilian festivals. Costs range between $600 and $2,500.

Eco-Net
18 DeBoom Street, Dept. GM, San Francisco, CA 94107
(415) 442-0220
An on-line computer network of more than one hundred electronic bulletin boards that deal with environmental issues and job opportunities.

The Environmental Careers Organization, Inc.
286 Congress Street, Dept. GM, Boston, MA 02210
(617) 426-4375
Places college students and recent graduates in short-term paid internships with environmental groups.

Environmental Project on Central America (EPOCA)
Earth Island Institute
300 Broadway, Suite 28, San Francisco, CA
94133-3312
(415) 788-3666
Sponsors reforestation brigades in Nicaragua.
Brigadistas plant tens of thousands of trees to
provide wind blocks and to prevent erosion and
flooding on Nicaragua's farmland.

Gemquest
Global Exchange Motivators, Inc.
Montgomery County Intermediate Unit Building,
Montgomery Ave. and Paper Mill Road,
Erdenheim, PA 19118
(215) 233-9558
Organizes study and travel programs that promote
cultural understanding through intensive contact
between travelers and their hosts. Includes
homestays and the possibility for university study
abroad.

Global Action Plan
84 Yerry Hill Road, Woodstock, NY 12498
(914) 679-4830; fax (914) 679-4834
Provides workbook and support for the formation
of small groups or "eco teams" that meet to learn
about simple and effective ways to reduce their
impact on the environment. Each group quantifies
their results, such as water saved each month,
and adds the data to the results of other eco
teams around the world.

Global Exchange
2940 16th Street, Room 307, San Francisco, CA
94103
(415) 255-7296
Creates tours designed to help people become
more involved in Third World development
efforts. Tours go to Central and South America,
the Caribbean, and southern Africa.

Green Corps
1724 Gilpin Street, Denver, CO 80218
(303) 355-1881
An intensive training program in grass-roots
organizing, fund-raising, media, and campaign
skills, followed by a year using your skills in a real
life situation.

Habitat for Humanity
419 West Church Street, Americus, GA 31709
(912) 924-6935
Volunteers donate labor, money, and materials in
this nonprofit Christian housing ministry to build
and renovate homes in the United States and
abroad.

International Bicycle Fund (IBF)
4247 135th Place SE, Bellevue, WA 98006-1319
Sponsors bicycle tours to Africa to encourage
people-to-people contact and to offer education
about the cultures, histories, and economies of
the peoples visited. Costs are about $1,000 plus
airfare.

International Development Exchange (IDEX)
777 Valencia Street, San Francisco, CA 94110
(415) 621-1494
IDEX brings together school and church groups in
the United States with grass-roots development
projects in the Third World.

MADRE
121 W. 27th Street, Room 301, New York, NY
10001
(212) 627-0444
MADRE works to build friendships between
women in Central America, the Caribbean, and
the United States.

Mennonite Central Committee
21 S. 12th Street, Akron, PA 17501
(717) 859-1151
Volunteers work on health, education, social
services, and community development projects in
more than fifty countries.

Overseas Development Network (ODN)
P.O. Box 1430, Cambridge, MA 02238
(617) 868-3002
A network of college activists, ODN organizes
educational and fund-raising events dealing with
Third World development issues.

Partners for Global Justice
4920 Piney Branch Road, NW, Washington, DC
20011
(202) 723-8273
Partners for Global Justice empowers U.S. citizens
to influence public policy effectively. Volunteers
spend one year in a Third World country and one
year in the United States working with
cooperatives and community organizations.

Public Interest Research Groups
The Fund for Public Interest Research
29 Temple Place, Boston, MA 02111
(617) 292-4800
The PIRGs organize student groups, campaign on
a variety of environmental and social justice
issues, and provide numerous canvassing and
fund-raising jobs to recent college graduates.

Student Conservation Association
1800 N. Kent Street, Suite 1260, Arlington, VA 22209
(703) 524-2441
Volunteers work with stewardship and conservation programs on public land and in national parks. For example, you might help with a botany research project in Yellowstone National Park.

Student Environmental Action Coalition (SEAC)
P.O. Box 1168, Chapel Hill, NC 27514
(919) 967-4600
SEAC publishes a monthly newsletter for its members and sponsors more than seventy gatherings across the country for students interested in environmental justice. Through the Campus Ecology Project, students can assess environmental quality and create strategies for change on their own campuses.

SERVAS International
11 John Street, Suite 706, New York, NY 10038
(212) 267-0252
SERVAS is an international network of people interested in peace issues. SERVAS does not organize or lead trips. Instead, SERVAS compiles national directories of people in more than seventy countries who are interested in meeting travelers for a short homestay or just for coffee.

APPENDIX D: ENVIRONMENTAL ORGANIZATIONS

▼

This is a sampling of nongovernmental national and international environmental organizations. For the more complete, annually updated Conservation Directory, send $15 to the National Wildlife Federation, 1400 Sixteenth Street NW, Washington, D.C., 20036. The directory includes many professional associations and state or local organizations not listed here. Your local library may have a copy. Most of the organizations publish a magazine or newsletter.

African Wildlife Foundation, 1717 Massachusetts Avenue NW, Washington, DC 20036 (202-265-8393). Finances and operates wildlife conservation projects in Africa.

American Committee for International Conservation, c/o Mike McCloskey, Sierra Club, 330 Pennsylvania Avenue SE, Washington, DC 20003 (202-547-1144). An association of nongovernmental organizations concerned with international conservation.

American Farmland Trust, 1920 N Street NW, Suite 400, Washington, DC 20036 (202-659-5170). Works for preservation of family farms and for soil conservation.

American Museum of Natural History, Central Park West at 79th Street, New York, NY 10024 (212-769-5000). Conducts natural history research and publishes educational material.

Appalachian Mountain Club, 5 Joy Street, Boston, MA 02108 (617-523-0636). Sponsors trail maintenance, outdoor education, and recreational hikes and climbs. Operates a mountain hut system on the Appalachian trail.

Bioregional Institute, 233 Miramar Street, Santa Cruz, CA 95060 (408-425-0264). Works toward the restoration of an ecological harmony with the land. Many bioregional organizations have excellent newsletters.

Center for Environmental Education, 1725 DeSales Street NW, Suite 500, Washington, DC 20036 (202-429-5609). A nongovernmental, nonprofit organization that sponsors programs for protection of endangered species.

Center for Science in the Public Interest, 1501 16th Street, NW, Washington, DC 20036 (202-332-9110). National consumer advocacy organization that focuses on health and nutrition.

Clean Water Action Project, 317 Pennsylvania Avenue SE, Washington, DC 20003 (202-547-1196). Works for clean water and for protection of natural resources. Conducts voter education and public awareness projects.

Conservation International, 1015 18th Street NW, Suite 1000, Washington, DC 20036 (202-429-5660). Buys land or trades foreign debt for land set aside for nature preserves in developing countries.

Cousteau Society, Inc., 930 W. 21st Street, Norfolk, VA 23517 (804-627-1144). Produces television films, lectures, books, and research on ocean quality and other resource issues.

Ducks Unlimited, Inc., One Waterfowl Way, Long Grove, IL 60047 (312-438-4300). Perpetuates waterfowl by purchasing and protecting wetland habitat.

Earth Island Institute, 300 Broadway, Suite 28, San Francisco, CA 94133. A clearinghouse for international information on environmental and resource issues. Founded by David Brower to bring together sources of conservation action and news.

Environmental Defense Fund, Inc., 257 Park Avenue South, New York, NY 10010 (212-505-2100). Protects environmental quality and public health through litigation and administrative appeals.

Environmental Law Institute, 1616 P Street NW, Suite 200, Washington, DC 20036 (202-328-5150). Sponsors research and education on environmental law and policy.

Food and Agriculture Organization of the United Nations, Via delle Terme di Caracalla, Rome 00100 Italy (Tele: 57971). A special agency of the United Nations established to improve food production, nutrition, and the health of all people.

Friends of the Earth, 530 Seventh Street SE, Washington, DC 20003 (202-543-4312). Affiliated with groups in thirty-two countries around the world. Works to form public opinion and to influence government policies to protect nature.

Fund for Animals, Inc., 200 West 57th Street, New York, NY 10019 (212-246-2096). Advocacy group for humane treatment of all animals.

Greater Yellowstone Coalition, P.O. Box 1874, 420 West Mendenhall, Bozeman, MT 59715 (406-586-1593). A small but influential organization dedicated to the preservation and protection of the Greater Yellowstone ecosystem.

Greenpeace USA, 1436 U Street NW, Washington, DC 20009 (202-462-1177). Worldwide organization that works to halt nuclear weapons testing, to protect marine animals, and to stop pollution and environmental degradation.

Humane Society of the United States, 2100 L Street NW, Washington, DC 20037 (202-452-1100). Dedicated to the protection of both domestic and wild animals.

Institute for Food and Development Policy, 1885 Mission Street, San Francisco, CA 94103. An international research and education center working for social justice and environmental protection.

Institute for Social Ecology, P.O. Box 89, Plainsfield, VT 85667. Offers courses for credit through Goddard College and workshops on social ecology, ecofeminism, urban design, and ecological planning.

International Alliance for Sustainable Agriculture, 1701 University Avenue SE, Minneapolis, MN 55414. An alliance of organic farmers, researchers, consumers, and international organizations dedicated to sustainable agriculture.

International Union for Conservation of Nature and Natural Resources (IUCN), Avenue du Nont-Blanc CH-1196 Gland, Switzerland (022-64-71-81). Promotes scientifically based action for the conservation of wild plants, animals, and resources.

Izaak Walton League of America, Inc., 1401 Wilson Boulevard, Level B, Arlington, VA 22209 (703-528-1818). Educates the public on land, water, air, wildlife, and other conservation issues.

Land Institute, Route 3, Salina, KS 67401 (913-823-5376). Carries out research on perennial species, prairie polycultures, and sustainable agriculture. Has training programs and conferences.

League of Conservation Voters, 2000 L Street NW, Suite 804, Washington, DC 20036 (202-785-8683). A nonpartisan national political campaign committee that strives to elect environmentally responsible public officials. Publishes an annual evaluation of voting records of Congress.

National Audubon Society, 950 Third Avenue, New York, NY 10022 (212-832-3200). One of the oldest and largest conservation organizations, Audubon has many educational and recreational programs as well as an active lobbying and litigation staff.

National Parks and Conservation Association, 1015 31st Street NW, Washington, DC 20007 (202-944-8530). A private nonprofit organization dedicated to preservation, promotion, and improvement of our national parks.

National Wildlife Federation, 1400 Sixteenth Street NW, Washington, DC 20036 (202-797-6800). Specializes in wildlife conservation but recognizes the importance of habitat and other resources to all living things. More than 5 million members.

Natural Resources Defense Council, Inc., 122 East 42nd Street, New York, NY 10168 (212-949-0049). An environmental organization that monitors government agencies and brings legal action to protect the environment.

(The) Nature Conservancy, 1815 North Lynn Street, Arlington, VA 22209 (703-841-5300). Works with state and federal agencies to identify ecologically significant natural areas. Manages a system of over 1000 nature sanctuaries nationwide.

New Alchemy Institute, Box 432, Woods Hole, MA 02543 (617-563-2655). Conducts research and provides education on sustainable agriculture, aquaculture, and bioshelters.

Planet/Drum Foundation, P.O. Box 31251, San Francisco, CA 94131. Promotes wise use of resources and new attitudes toward nature.

Population Reference Bureau, 777 14th Street NW, Suite 800, Washington, DC (202-639-8040). Gathers, interprets, and publishes information on social, economic, and environmental implications of world population dynamics. Excellent data source.

Project Lighthawk, P.O. Box 8136, Santa Fe, NM 87504 (505-982-9656). Uses light aircraft for aerial surveys of forest condition, range management, and other conservation issues. Offers a unique vantage point to environmentalists.

Rainforest Action Network, 300 Broadway, Suite 28, San Francisco, CA 94133 (415-398-4404). Shares offices with Earth Island Institute and focuses on actions designed to save rainforests around the world.

Resources for the Future, 1616 P Street NW, Washington, DC 20036 (202-328-5000). Conducts research and provides education about natural resource conservation issues.

Rodale Institute, 222 Main Street, Emmaus, PA 18049 (215-967-5171). A leading research institute for organic farming and alternative crops. Publishes magazines, books, and reports on regenerative farming.

Save-the-Redwoods League, 114 Sansome Street, Room 605, San Francisco, CA 94104 (415-362-2352). Buys land, plants trees, and works with state and federal agencies to save redwood trees.

Scientists' Institute for Public Information, 355 Lexington Avenue, New York, NY 10017 (212-661-9110). Enlists scientists and other experts in public information programs and public policy forums on a variety of environmental issues.

Sea Shepherd Conservation Society, P.O. Box 7000-S, Redondo Beach, CA 90277 (213-373-6979). An international marine conservation action program. Carries out field campaigns to call attention to and stop wildlife destruction and resource misuse.

Sierra Club, 730 Polk Street, San Francisco, CA 94109 (415-776-2211). Founded in 1892 by John Muir and others to explore, enjoy, and protect the wild places of the earth. Conducts outings, educational programs, volunteer work projects, litigation, political action, and administrative appeals. Has one of the most comprehensive programs of any conservation organization.

Student Conservation Association, Inc., Box 550, Charlestown, NH 03603 (603-826-5206). Coordinates environmental internships and volunteer jobs with state and federal agencies and private organizations for students and adults.

United Nations Environment Programme, P.O. Box 30552, Nairobi Kenya; and United Nations, Rm. DC2-0803, New York, NY 10017 (212-963-8138). Coordinates global environmental efforts with United Nations agencies, national governments, and nongovernmental organizations.

Waldebridge Ecological Centre, Worthyvale Manor Farm, Camelford, Cornwall P132 9TT England (Tele: 0840 212711). Conducts research and education on a variety of environmental issues. Publishes *The Ecologist,* an excellent global environmental journal.

Wilderness Society, 1400 I Street NW, 10th Floor, Washington, DC 20005 (202-842-3400). Dedicated to preserving wilderness and wildlife in America.

World Resources Institute, 1735 New York Avenue NW, Washington, DC 20006 (202-638-6300). A policy research center that publishes excellent annual reports on world resources.

Worldwatch Institute, 1776 Massachusetts Avenue NW, Washington, DC 20036 (202-452-1999). A nonprofit research organization concerned with global trends and problems. Publishes excellent periodic reports and annual summaries.

Zero Population Growth, Inc., 1601 Connecticut Avenue NW, Washington, DC 20009 (202-332-2200). Advocates of worldwide population stabilization.

GLOSSARY

A

Abiotic Environmental factors that are nonliving components of ecosystems.

Acid precipitation The deposition of wet acidic solutions or dry acidic particles from the air; commonly known as acid rain but also includes acid fog, snow, etc.

Active solar system A heating system that pumps a heat-absorbing, fluid medium (air, water, or an antifreeze solution) through a relatively small collector.

Acute effects A single exposure to a toxin that results in an immediate health crisis.

Aerobic respiration The intracellular breakdown of sugar or other organic compounds in the presence of oxygen that releases energy and produces carbon dioxide and water.

Aerosols Small particles or droplets suspended in a gas.

Aesthetic degradation Undesirable changes in chemical or physical characteristics of the atmosphere, such as noise, odor, and light pollution, that are not life-threatening but that reduce the quality of life.

Age-specific death rates Mortality rates for specific age classes.

Agricultural revolution The shift from hunting and gathering for food to raising domesticated animals and cultivating crops.

Albedo A description of a surface's reflective properties.

Allergens Substances that activate the immune system and cause an allergic response; may not be directly antigenic themselves but may make other materials antigenic.

Alpine The high, treeless biogeographic zone of mountains that consists of slopes above the timberline.

Ambient air The air that surrounds us.

Amino acid An organic compound containing an amino group and a carboxyl group; amino acids are the units or building blocks that make peptide and protein molecules.

Anaerobic respiration The incomplete intracellular breakdown of sugar or other organic compounds in the absence of oxygen that releases some energy and produces organic acids and/or alcohol.

Anemia Low levels of hemoglobin due to iron deficiency or lack of red blood cells.

Annual A plant that lives for a single growing season.

Anthracite Hard, dense, rocklike coal that has a glassy texture, does not make coke, and burns with a nonluminous flame.

Anthropocentrism The belief that humans hold a special place in nature; being centered primarily on humans and human affairs.

Antigens Substances that stimulate the production of and react with specific antibodies.

Aquifers Porous, water-bearing layers of sand, gravel, and rock below Earth's surface; reservoirs for groundwater.

Arithmetic growth A pattern of growth that increases at a constant amount per unit time, such as 1, 2, 3, 4 or 1, 3, 5, 7.

Artesian well The result of a pressurized aquifer intersecting the surface or being penetrated by a pipe or conduit, from which water gushes without being pumped; also called a spring.

Asphyxiants Chemicals that exclude oxygen or actively interfere with oxygen uptake and distribution; includes inert chemicals, such as nitrogen gas or halothane, that can displace oxygen and fill enclosed spaces.

Asthma A distressing disease characterized by shortness of breath, wheezing, and bronchial muscle spasms.

Atom The smallest unit of matter that has the characteristics of an element; consists of three main types of subatomic particles: protons, neutrons, and electrons.

Autotroph An organism that synthesizes food molecules from inorganic molecules by using an external energy source, such as light energy.

B

Barrier islands Low, narrow, sandy islands that form offshore from a coastline.

BAT The Best Available Technology; the best pollution control technology available; the Clean Water Act effectively negates this category by stipulating that equipment must be economically feasible.

Bedrock Unbroken, solid rock overlain by sand, gravel, or soil.

Bill A piece of legislation introduced in Congress and intended to become law.

Bioaccumulation The selective absorption and concentration of molecules by cells.

Biocentric preservation A philosophy of preserving nature for its own sake.

Biocentrism The belief that all creatures have rights and values; being centered on nature rather than humans.

Bioregionalism Organization of human activities according to natural geographic or ecological boundaries and associations. This philosophy emphasizes a sense of place and living within the resources of one's local ecosystem.

Biodegradable plastics Plastics that can be decomposed by microorganisms.

Biogeochemical cycles Movement of matter within or between ecosystems; caused by living organisms, geological forces, or chemical reactions. The cycling of nitrogen, carbon, sulfur, oxygen, phosphorus, and water are examples.

Biogeographical area An entire ecosystem and its associated land, water, air, and wildlife resources.

Biological or biotic factors Organisms and products of organisms that are part of the environment and potentially affect the life of other organisms.

Biological community The populations of plants, animals, and microorganisms living and interacting in a certain area at a given time.

Biological oxygen demand (BOD) A standard test for measuring the amount of dissolved oxygen utilized by aquatic microorganisms over a five-day period.

Biological resources The earth's organisms.

Biomagnification Increase in concentration of certain stable chemicals (e.g., heavy metals or fat-soluble pesticides) in successively higher trophic levels of a food chain or web.

Biomass The total mass or weight of all the living organisms in a given population or area.

Biomass fuel Organic material produced by plants, animals, or microorganisms that can be burned directly as a heat source or converted into a gaseous or liquid fuel.

Biomass pyramid A metaphor or diagram that explains the relationship between the amounts of biomass at different trophic levels.

Biome A broad, regional type of ecosystem characterized by distinctive climate and soil conditions and a distinctive kind of biological community adapted to those conditions.

Biosphere The zone of air, land, and water at the surface of the earth that is occupied by organisms.

Biota All organisms in a given area.

Biotic Pertaining to life; environmental factors created by living organisms.

Biotic potential The maximum reproductive rate of an organism, given unlimited resources and ideal environmental conditions. Compare with environmental resistance.

Birth control Any method used to reduce births, including celibacy, delayed marriage, contraception, methods that prevent implantation of fertilized zygotes, and induced abortions.

Bituminous coal Soft coal that has high contents of volatile gases and ash.

Black lung disease Inflammation and fibrosis caused by accumulation of coal dust in the lungs or airways. *See* respiratory fibrotic agents.

Bog An area of waterlogged soil that tends to be peaty; fed mainly by precipitation; low productivity; some bogs are acidic.

Boreal forest A broad band of mixed coniferous and deciduous trees that stretches across northern North America (and also Europe and Asia); its northernmost edge, the **taiga,** intergrades with the arctic **tundra.**

BPT The Best Practical Control Technology; the best technology for pollution control available at reasonable cost and operable under normal conditions.

Breeder reactors High-density, high-pressure fission reactors that produce more fuel than they consume.

Bronchitis An inflammation of bronchial linings that causes persistent cough, copious production of sputum, and involuntary muscle spasms that constrict airways.

C

Cancer Invasive, out-of-control cell growth that results in malignant tumors.

Captive breeding Raising plants or animals in zoos or other controlled conditions to produce stock for subsequent release into the wild.

Carbohydrate An organic compound consisting of a ring or chain of carbon atoms with hydrogen and oxygen attached; examples are sugars, starches, cellulose, and glycogen.

Carbon cycle The circulation and reutilization of carbon atoms, especially via the processes of photosynthesis and respiration.

Carbon monoxide (CO) Colorless, odorless, nonirritating but highly toxic gas produced by incomplete combustion of fuel, incineration of biomass or solid waste, or partially anaerobic decomposition of organic material.

Carbon sink Places of carbon accumulation, such as in large forests (organic compounds) or ocean sediments (calcium carbonate); carbon is thus removed from the carbon cycle for moderately long to very long periods of time.

Carcinogens Substances that cause cancer.

Carnivores Organisms that mainly prey upon animals.

Carrying capacity The maximum number of individuals of any species that can be supported by a particular ecosystem on a long-term basis.

Cash crops Those crops produced for sale, especially luxury crops that bring high prices on foreign markets.

Cellular organelle An intracellular structure specialized for a particular function; examples are nucleus, mitochondria, chloroplasts, Golgi apparatus or dictyosomes, lysosomes, endoplasmic reticulum, and others.

Cellular respiration The process in which a cell breaks down sugar or other organic compounds to release energy used for cellular work; may be **anaerobic** or **aerobic** depending on the availability of oxygen.

Chain reaction A self-sustaining reaction in which the fission of nuclei produces subatomic particles that cause the fission of other nuclei.

Chemical bond The force that holds molecules together.

Chloroplasts Chlorophyll-containing organelles in eukaryotic organisms; sites of photosynthesis.

Chronic effects Long-lasting results of exposure to a toxin; can be a permanent change caused by a single, acute exposure or a continuous, low-level exposure.

Chronic food shortages Long-term undernutrition and malnutrition; usually caused by people's lack of money to buy food or lack of opportunity to grow it themselves.

City A differentiated community with a sufficient population and resource base to allow residents to specialize in arts, crafts, services, and professional occupations.

Clearcut Cutting every tree in a given area, regardless of species or size; an appropriate harvest method for some species; can be destructive if not carefully controlled.

Climate A description of the long-term pattern of weather in a particular area.

Climax community A relatively stable, long-lasting community reached in a successional series; usually determined by climate and soil type.

Closed canopy A forest where tree crowns spread over 20 percent of the ground; has the potential for commercial timber harvests.

Coal gasification The heating and partial combustion of coal to release volatile gases, such as methane and carbon monoxide; after pollutants are washed out, these gases become efficient, clean-burning fuel.

Coal washing Coal technology that involves crushing coal and washing out soluble sulfur compounds with water or other solvents.

Coastal Zone Management Act Legislation of 1972 that gave federal money to thirty seacoast and Great Lakes states for development and restoration projects.

Co-composting Microbial decomposition of organic materials in solid waste into useful soil additives and fertilizer; often, extra organic material in the form of sewer sludge, animal manure, leaves, and grass clippings are added to solid waste to speed the process and make the product more useful.

Co-evolution Simultaneous evolution of two species in a mutual relationship in which each species influences the characteristics of the other.

Cofactor Nonprotein components needed by enzymes in order to function; often minerals or vitamins.

Congeneration The simultaneous production of electricity and steam or hot water in the same plant.

Cold front A moving boundary of cooler air displacing warmer air.

Coliform bacteria Bacteria that live in the intestines (including the colon) of humans and other animals; used as a measure of the presence of feces in water or soil.

Commensalism A symbiotic relationship in which one member is benefited and the second is neither harmed nor benefited.

Common law A body of unwritten principles based on custom and the precedents of previous legal decisions.

Community ecology The study of interactions of all populations living in the ecosystem of a given area.

Competition A struggle for scarce resources such as food, living space, or sunlight between members of a single species or between different species.

Competitive exclusion A principle recognizing that when two or more populations are competing for the same resource, one population will eventually dominate and cause the other population to dwindle, become extinct, or adapt to an ecological niche that reduces or eliminates competition.

Complexity (ecological) Ecological complexity is the number of species at each trophic level and the number of trophic levels in a community.

Compound A molecule made up of two or more kinds of atoms held together by chemical bonds.

Condensation The aggregation of water molecules from vapor to liquid or solid when the saturation concentration is exceeded.

Condensation nuclei Tiny particles that float in the air and facilitate the condensation process.

Conifers Needle-bearing trees that produce seeds in cones.

Conifer forests Forests composed of cone-bearing, generally evergreen trees that have needles or scales rather than broad leaves.

Conservation of matter In any chemical reaction, matter changes form; it is neither created nor destroyed.

Consumer An organism that obtains energy and nutrients by feeding on other organisms or their remains. *See also* heterotroph.

Consumption The fraction of withdrawn water that is lost in transmission or that is evaporated, absorbed, chemically transformed, or otherwise made unavailable for other purposes as a result of human use.

Contour plowing Plowing along hill contours; reduces erosion.

Control rods Neutron-absorbing material inserted into spaces between fuel assemblies in nuclear reactors to regulate fission reaction.

Convection currents Rising or sinking air currents that stir the atmosphere and transport heat from one area to another. Convection currents also occur in water; *see* spring overturn.

Conventional pollutants The seven substances (sulfur dioxide, carbon monoxide, particulates, hydrocarbons, nitrogen oxides, photochemical oxidants, and lead) that make up the largest volume of air quality degradation; identified by the Clean Air Act as the most serious threat of all pollutants to human health and welfare; also called criteria pollutants.

Coral reefs Prominent oceanic features composed of hard, limy skeletons produced by coral animals; usually formed along edges of shallow, submerged ocean banks or along shelves in warm, shallow, tropical seas.

Core region The primary industrial region of a country; usually located around the capital or largest port; has both the greatest population density and the greatest economic activity of the country.

Coriolis effect The influence of friction and drag on air layers near Earth; deflects air currents to the direction of Earth's rotation.

Corridors Small reserves of natural habitat that link larger reserves so that species can move from one area to another.

Cover crops Plants, such as rye, alfalfa, or clover, that can be planted immediately after harvest to hold and protect the soil.

Criteria pollutants *See* conventional air pollutants.

Critical factor The single environmental factor closest to a tolerance limit for a given species at a given time. *See* limiting factors.

Critical thinking The ability to distinguish between fact and opinion and to weigh an argument in order to decide what to believe and do.

Croplands Lands used to grow crops.

Crude birth rate The number of births in a year divided by the midyear population.

Crude death rates The number of deaths per thousand persons in a given year; also called crude mortality rates.

Cultural eutrophication An increase in biological productivity and ecosystem succession caused by human activities.

Cyclone collectors A gravitational settling device that removes heavy particles from an effluent stream.

D

Deciduous plants Trees and shrubs that shed their leaves at the end of the growing season.

Decline spiral A catastrophic deterioration of a species, community, or whole ecosystem; accelerates as functions are disrupted or lost in a downward cascade.

Decomposers Fungi and bacteria that break complex organic material into smaller molecules.

Deep ecology An integrated combination of beliefs, life-style, and activism based on voluntary simplicity, decentralization, personal freedom, the sacredness of nature, and direct action in environmental protection.

Degradation (of water resource) Deterioration in water quality due to contamination or pollution; makes water unsuitable for other desirable purposes.

Delta Fan-shaped sediment deposit found at the mouth of a river.

Demographic transition A pattern of falling death and birth rates in response to improved living conditions; could be reversed in deteriorating conditions.

Demography Vital statistics about people: births, marriages, deaths, etc.; the statistical study of human populations relating to growth rate, age structure, geographic distribution, etc., and their effects on social, economic, and environmental conditions.

Denitrifying bacteria Free-living soil bacteria that convert nitrates to gaseous nitrogen and nitrous oxide.

Density-dependent factors Factors for which mortality and reproductive failure are linked to population size.

Density-independent factors Factors that influence mortality or reproductive failure regardless of population size.

Dependency ratio The number of nonworking members compared to working members for a given population.

Desertification The process of denuding and degrading a once-fertile land that initiates a desert-producing cycle.

Deserts Areas with too little available moisture for any but the hardiest and most highly drought-adapted plants to grow.

Developing countries Countries that now are undergoing social and economic development.

Development An improvement in the well-being of people.

Dew point The temperature at which condensation occurs for a given concentration of water vapor in the air.

Dieback A sudden population decline; also called a population crash.

Discharge The amount of water that passes a fixed point in a given amount of time; usually expressed as liters or cubic feet of water per second.

Disease A deleterious change in the body's condition in response to destabilizing factors, such as nutrition, chemicals, or biological agents.

Dissolved oxygen (DO) content Amount of oxygen dissolved in a given volume of water at a given temperature and atmospheric pressure; usually expressed in parts per million (ppm).

Diversity (species diversity, biological diversity) The number of species present in a community (species *richness*), as well as the relative *abundance* of each species.

DNA Deoxyribonucleic acid; the long, double helix molecule in the nucleus of cells that contains the genetic code and directs the development and functioning of all cells.

Dominant plants Those plant species in a community that provide a food base for most of the community; usually take up the most space and have the largest biomass.

Drip irrigation Uses pipe or tubing perforated with very small holes to deliver water one drop at a time directly to the soil around each plant. This conserves water and reduces soil waterlogging and salinization.

E

Ecofeminism A philosophy that applies feminist critical analysis and perspective to ecological issues. In this view, oppression of both women and nature by patriarchial society are similar and have related destructive effects.

Ecojustice Justice in the social order and integrity in the natural order.

Ecological niche The functional role and position of a species (population) within a community or ecosystem, including what resources it uses, how and when it uses the resources, and how it interacts with other populations.

Ecological succession A progression of communities replacing each other at a given site over a period of time;

eventually a long-lasting **climax community** may result. *See also* primary succession and secondary succession.

Ecology The study of environmental factors and how organisms interact with them.

Economic development A rise in real income *per person;* usually associated with new technology that increases productivity or resources.

Economic growth An increase in the total wealth of a nation; if population grows faster than the economy, there may be real economic growth, but the share per person may decline.

Ecosystem A specific biological community and its physical environment interacting in an exchange of matter and energy.

Ecosystem restoration Reinstate an entire community of organisms to as near its natural condition as possible.

Ecotage Direct action (guerrilla warfare) or sabotage in defense of nature; also called monkeywrenching.

Ecotourism A combination of adventure travel, nature appreciation, and cultural exploration in wild settings.

Edge effect Occurs when the transitional zone between two communities provides habitat for a different set of organisms or provides habitat that can be used by members of both adjacent communities.

Electrostatic precipitators The most common particulate controls in power plants; fly ash particles pick up an electrostatic surface charge as they pass between large electrodes in the effluent stream, causing particles to migrate to the collecting plate.

Element A molecule composed of one kind of atom; cannot be broken into simpler units by chemical reactions.

Emigration The movement of members from a population.

Emission standards Regulations for restricting the amounts of air pollutants that can be released from specific point sources.

Emphysema An irreversible, obstructive lung disease in which airways become permanently constricted and alveoli are damaged or destroyed.

Endangered species A species considered to be in imminent danger of extinction.

Energy The capacity to do work (i.e., to change the physical state or motion of an object).

Energy efficiency A measure of energy produced compared to energy consumed.

Energy pyramid A representation of the loss of useful energy at each step in a food chain.

Energy recovery Incineration of solid waste to produce useful energy.

Environment A combination of external biological and nonbiological factors that influence the life of a cell or organism.

Environmental ethics A search for moral values and ethical principles in human relations with the natural world.

Environmental Impact Statement An analysis, required by provisions in the National Environmental Policy Act of 1970, of the effects of any major program a federal agency plans to undertake; also called an EIS.

Environmentalism A social movement concerned with the health of the entire environment, both natural and built by humans.

Environmental resistance All the limiting factors that tend to reduce population growth rates and set the maximum allowable population size or carrying capacity of an ecosystem.

Environmental resources Anything an organism needs that can be taken from the environment.

Environmental restoration Repairing environmental damage caused by human neglect.

Environmental science The study of the complex interactions of human populations with matter and energy resources; it incorporates aspects of the natural and social sciences, business, law, technology, and other fields.

Enzymes Molecules, usually proteins or nucleic acids, that act as catalysts in biochemical reactions.

Epidemiology The study of the distribution and causes of disease and injuries in human populations.

Epiphyte A plant that grows on a substrate other than the soil, such as the surface of another organism.

Erosion Soil removal by the abrasive forces of wind, water, or ice.

Estuary Enclosed or semi-enclosed body of water that forms where a river enters the ocean and creates an area of mixed fresh water and ocean water.

Eukaryotic cell A cell containing a membrane-bounded nucleus and membrane-bounded organelles.

Eutrophic Rivers and lakes rich in organisms and organic material (eu = good; trophic = nutrition).

Evaporation The process in which a liquid is changed to vapor (gas phase).

Evergreens Coniferous trees and broad-leaved plants that retain their leaves year-round.

Evolution The process of gradual change in genetic characteristics, physical form, and behavior of a species as the individuals in a population that are best suited for the existing environmental conditions survive and reproduce more successfully than others.

Exhaustible resources Generally considered the Earth's geologic endowment; minerals, nonmineral resources, fossil fuels, and other materials present in fixed amounts in the environment.

Existence value The importance we place on just knowing that a particular species or a specific organism exists.

Exponential growth Growth at a constant rate of increase per unit of time; can be expressed as a constant fraction or exponent. *See* geometric growth.

External costs Expenses, monetary or otherwise, borne by someone other than the individuals or groups who use a resource.

Extinction The irrevocable elimination of species; can be a normal process of the natural world as species out-compete or kill off others or as environmental conditions change.

Extirpate To destroy totally; extinction caused by direct human action, such as hunting, trapping, etc.

F

Family planning Controlling reproduction; planning the timing of birth and having as many babies as are wanted and can be supported.

Famines Acute food shortages characterized by large-scale loss of life, social disruption, and economic chaos.

Fauna All of the animals present in a given region.

Fecundity The physical ability to reproduce.

Feral A domestic animal that has taken up a wild existence.

Fermentation (alcoholic) A type of anaerobic respiration that yields carbon dioxide and alcohol; used in commercial fermentation processes, including production of raised bakery dough products and alcoholic beverages.

Fertility Measurement of actual number of offspring produced through sexual reproduction; usually described in terms of number of offspring of females since paternity can be difficult to determine.

Fibrosis The general name for accumulation of scar tissue in the lung.

Filters A porous mesh of cotton cloth, spun glass fibers, or asbestos-cellulose that allows air or liquid to pass through but holds back solid particles.

Fire-climax community An equilibrium community maintained by periodic fires; examples include grasslands, chapparal shrubland, and some pine forests.

First law of thermodynamics States that energy is conserved: it is neither created nor destroyed under normal conditions.

First World The industrialized capitalist or market economy countries of Western Europe, North America, Japan, Australia, and New Zealand.

Flora All of the plants present in a given region.

Flue-gas scrubbing Treating combustion exhaust gases with chemical agents to remove pollutants. Spraying crushed limestone and water into the exhaust gas stream to remove sulfur is a common scrubbing technique.

Fluidized bed combustion High pressure air is forced through a mixture of crushed coal and limestone particles lifting the burning fuel and causing it to move like a boiling fluid. Fresh coal and limestone are added continuously to the top of the combustion bed while ash and slag are drawn off below.

Food aid Financial assistance intended to boost less developed countries' standards of living.

Food chain A linked feeding series; in an ecosystem, the sequence of organisms through which energy and materials are transferred, in the form of food, from one trophic level to another.

Food surpluses Excess food supplies.

Food web A complex, interlocking series of individual food chains in an ecosystem.

Forest management Scientific planning and administration of forest resources for sustainable harvest, multiple use, regeneration, and maintenance of a healthy biological community.

Fossil fuels Fuels derived from the residues of plants and animals changed by heat and pressure deep below the earth's surface.

Founder bottleneck Serious long-term genetic problems of a species caused when too few individuals remain to provide genetic diversity for future populations.

Fourth world The world's poorest nations, both countries without market economies or central planning, and indigenous communities within wealthy nations.

Fresh water Water other than seawater; covers only about 2 percent of Earth's surface, including streams, rivers, lakes, ponds, and water associated with several kinds of wetlands.

Frontier An unexploited natural area at the leading edge of human settlement.

Frontier mentality The idea that the world has an unlimited supply of resources for human use regardless of the consequences to natural ecosystems and the biosphere.

Fuel assembly A bundle of hollow metal rods containing uranium oxide pellets; used to fuel a nuclear reactor.

Fuel-switching A change from one fuel to another.

Fuelwood Branches, twigs, logs, wood chips, and other wood products harvested for use as fuel.

Fugitive emissions Substances that enter the air without going through a smokestack, such as dust from soil erosion, strip mining, rock crushing, construction, and building demolition.

Fungi One of the five kingdom classifications; consists of nonphotosynthetic, eukaryotic organisms with cell walls, filamentous bodies, and absorptive nutrition.

G

Gaia hypothesis A theory that the living organisms of the biosphere form a single complex interacting system that creates and maintains a habitable Earth; named after Gaia, the Greek Earth mother goddess.

Gamma rays Very short wavelength forms of the electromagnetic spectrum.

Garden city A new town with special emphasis on landscaping and rural ambience.

Gasohol A mixture of gasoline and ethanol.

Gene A unit of heredity; a segment of DNA nucleus of the cell that contains information for the synthesis of a specific protein, such as an enzyme.

Gene banks Storage for seed varieties for future breeding experiments.

General fertility rate Representation of population age structure and fecundity; crude birth rate multiplied by the percentage of fecund women (between approximately fifteen and forty-four years of age) by 1,000.

Genetic assimilation The disappearance of a species as its genes are diluted through crossbreeding with a closely related species.

Geometric growth Growth that follows a geometric pattern of increase, such as 2, 4, 8, 16, etc. *See* exponential growth.

Germ plasm Genetic material that may be preserved for future agricultural, commercial, and ecological values (plant seeds or parts or animal eggs, sperm, and embryos).

Ghandian sufficiency A view that the world has "enough for everyone's need, but not enough for anyone's greed."

Grasslands Lands on which ecological conditions prevent forest growth and maintain prairie or herbaceous communities.

Green revolution Dramatically increased agricultural production brought about by "miracle" strains of grain; usually requires high inputs of water, plant nutrients, and pesticides.

Groundwater Water held in gravel deposits or porous rock below Earth's surface; does not include water or crystallization held by chemical bonds in rocks or moisture in upper soil layers.

Gully erosion Removal of layers of soil, creating channels or ravines too large to be removed by normal tillage operations.

H

Habitat Describes the place or set of environmental conditions in which an organism lives.

Half-life (of radioisotopes) The length of time required for half the nuclei in a sample to change into another isotope.

Hazardous Dangerous due to high flammability, explositivity, or chemical reactivity.

Hazardous chemicals Dangerous chemicals, including flammables, explosives, irritants, sensitizers, acids, and caustics; may be relatively harmless in diluted concentrations.

Hazardous waste Any discarded material containing substances known to be toxic, mutagenic, carcinogenic, or teratogenic to humans or other life-forms; ignitable, corrosive, explosive, or highly reactive alone or with other materials.

Health A state of complete physical, mental, and social well-being, not merely the absence of disease or infirmity.

Heat A form of energy transferred from one body to another because of a difference in temperatures.

Heat capacity The amount of heat energy that must be added or subtracted to change the temperature of a body; water has a high heat capacity.

Heat of vaporization The amount of heat energy required to convert water from a liquid to a gas.

Herbivore An organism that eats only plants.

Heterotroph An organism that is incapable of synthesizing its own food and, therefore, must feed upon organic compounds produced by other organisms.

High quality energy Energy useful in carrying out work because it is intense, concentrated, and high temperature.

Homeostasis Maintaining a dynamic, steady state in a living system through opposing, compensating adjustments.

Homestead Act Legislation passed in 1862 allowing any citizen or applicant for citizenship over twenty-one years old and head of a family to acquire 160 acres of public land by living on it and cultivating it for five years.

Host organism An organism that provides lodging for a parasite.

Human ecology The study of the interactions of humans with the environment.

Human resources Human wisdom, experience, skill, and enterprise.

Humus Sticky, brown, insoluble residue from the bodies of dead plants and animals; gives soil its structure, coating mineral particles and holding them together; serves as a major source of plant nutrients.

Hydrologic cycle Describes the circulation of water as it evaporates from the earth, condenses, falls back to the earth's surface and moves into rivers, lakes, and seas.

Hypothyroidism Listlessness and other metabolic symptoms caused by low thyroid hormone levels.

I

Igneous rocks Crystalline minerals solidified from molten magma from deep in Earth's interior; basalt, rhyolite, andesite, lava, and granite are examples.

Inbreeding depression In a small population, an accumulation of harmful genetic traits (through random mutations and natural selection) that lowers viability and reproductive success of enough individuals to affect the whole population.

Industrial revolution The development of new technologies in the early 1800's, like steam powered engines that have given humans tremendous power to change our environment and the way that we live in it.

Industrial timber Trees used for lumber, plywood, veneer, particleboard, chipboard, and paper; also called roundwood.

Infiltration The process of water percolation into the soil and pores and hollows of permeable rocks.

Inflammatory response A complex series of interactions between fragments of damaged cells, surrounding tissues, circulating blood cells, and specific antibodies; typical of infections.

Informal economy Small-scale family businesses in temporary locations outside the control of normal regulatory agencies.

Insolation Incoming solar radiation.

Instrumental value Value or worth of objects that satisfy the needs and wants of moral agents. Objects that can be used as a means to some desirable end.

Intangible resources Abstract commodities, such as open space, beauty, serenity, genius, information, diversity, and satisfaction.

Integrated pest management IPM is an ecologically-based pest control strategy that relies on natural mortality factors, such as natural enemies, weather, cultural control methods, and carefully applied doses of pesticides.

Internal costs The expertises (monetary or otherwise) borne by those who use a resource.

Internalizing costs Planning so that those who reap the benefits of resource use also bear all the external costs.

Intergenerational justice The concept that we are obligated to leave resources and a better situation for those who follow us.

Interplanting The system of planting two or more crops, either mixed together or in alternating rows, in the same field; protects the soil and makes more efficient use of the land.

Interspecific competition In a community, competition for resources between members of *different* species.

Intraspecific competition In a community, competition for resources among members of the *same* species.

Ionizing radiation High-energy electromagnetic radiation or energetic subatomic particles released by nuclear decay.

Ions Atoms that have picked up or lost extra electrons, leaving them with a net negative or positive charge.

Irritants Corrosives (strong acids), caustics (alkaline reagents), and other substances that damage biological tissues on contact.

Irruptive growth *See* Malthusian growth.

Island effects Reductions in species diversity caused by reduction in ecosystem area.

Isotopes Different forms of the same element.

J

J curve A growth curve that depicts exponential growth; called a **J** curve because of its shape.

Jet streams Powerful winds or currents of air that circulate in shifting flows; similar to oceanic currents in extent and effect on climate.

K

Kinetic energy The energy contained in moving objects.

Known resources Minerals or other useful environmental materials or services that are identified and partially mapped; may not be environmentally or socially acceptable or economically feasible to exploit.

Kwashiorkor A widespread human protein deficiency disease resulting from a starchy diet low in protein and essential amino acids.

L

Landfills Land disposal sites for solid waste; operators compact refuse and cover it with a layer of dirt to minimize rodent and insect infestation, wind-blown debris, and leaching by rain.

Land rehabilitation A utilitarian program to repair damage and make land useful to humans.

Landscape ecology Ecological study that deals with understanding the effects of topography and soils on the development of biological communities.

Law of diminishing returns At some stages of economic development adding more workers or more labor will not increase productivity; instead, average output and standard of living decrease with each additional worker.

LD50 A chemical dose lethal to 50 percent of a test population.

Less developed countries (LDC) Nonindustrialized nations characterized by low per capita income, high birth and death rates, high population growth rates, and low levels of technological development.

Life expectancy The average number of years lived by a group of individuals after reaching a given age; the probable number of years of survival for an individual of a given age.

Life span The longest period of life reached by a type of organism.

Lignite Soft, brown to black coal of low BTU value with original plant components still discernible.

Limiting factors Chemical or physical factors that limit the existence, growth, abundance, or distribution of an organism. The **principle of limiting factors** states that for each physical factor in the environment, there are both minimum and maximum limits (**tolerance limits**) beyond which a particular species cannot survive. The single factor closest to a tolerance limit for a given species at a given time is the **critical factor.**

Lipid A nonpolar organic compound that is insoluble in water but soluble in solvents, such as alcohol and ether; includes fats, oils, steroids, phospholipids, and carotenoids.

Liquid metal fast breeder A nuclear power plant that converts Uranium 238 to Plutonium 239; thus, it creates more nuclear fuel than it consumes. Because of the extreme heat and density of its core, the breeder uses liquid sodium as a coolant.

Logistic growth Growth rates regulated by internal and external factors that establish an equilibrium with environmental resources; species may grow exponentially when resources are unlimited but slowly as the carrying capacity is reached. *See* **S** curve.

Longevity The length or duration of life; compare to survivorship.

Low-head hydropower Small-scale hydro technology that can extract energy from small headwater dams; causes much less ecological damage.

Low quality energy Energy that is difficult to gather and to use because it is diffuse, dispersed, and at a low temperature.

M

Majority world The poorest 80% of humanity which consumes less than half of the world's resources.

Malignant tumor A mass of cancerous cells that have left their site of origin, migrated through the body, invaded normal tissues, and are growing out of control.

Malnourishment A nutritional imbalance caused by lack of specific dietary components or inability to absorb or utilize essential nutrients.

Malthusian growth A population explosion followed by a population crash; also called irruptive growth.

Man and biosphere program (MAB) A program that encourages indigenous peoples to manage and live in peripheral zones around protected areas.

Marasmus A widespread human protein deficiency disease caused by a diet low in calories and protein or imbalanced in essential amino acids.

Marine Living in or pertaining to the sea.

Market equilibrium The dynamic balance between supply and demand under a given set of conditions in a "free" market (one with no monopolies or government interventions).

Marsh Wetland without trees; in North America, this type of land is characterized by cattails and rushes.

Mass burn Incineration of unsorted solid waste.

Matter Something that occupies space and has mass.

Megacity *See* megalopolis.

Megalopolis Also known as a megacity or supercity; megalopolis indicates an urban area with more than ten million inhabitants.

Megawatt (MW) Unit of electrical power equal to one thousand kilowatts or one million watts.

Metabolism All the energy and matter exchanges that occur within a living cell or organism; collectively, the life processes.

Methane digester A sealed container used to generate methane by microbial degradation of organic waste such as manure, grass clippings, and garbage.

Micro-hydro generators Small power generators that can be used in low-level rivers to provide economical power for four to six homes, freeing them from dependence on large utilities and foreign energy supplies.

Milankovitch cycles Periodic variations in tilt, eccentricity, and wobble in Earth's orbit; Milutin Milankovitch suggested that it is responsible for cyclic weather changes.

Milpa agriculture An ancient farming system in which small patches of tropical forests are cleared and perennial polyculture agriculture practiced and is then followed by many years of fallow to restore the soil; also called swidden agriculture.

Mineral A naturally occurring, inorganic, crystalline solid with definite chemical composition and characteristic physical properties.

Mixed perennial polyculture Growing a mixture of different perennial crop species (where the same plant persists for more than one year) together in the same plot; imitates the diversity of a natural system and is often more stable and more suitable for sustainable agriculture than monoculture of annual plants.

Molecule A combination of two or more atoms.

Monkey wrenching A controversial tactic of environmental preservation through vandalism and sabotage of construction equipment.

Monoculture agroforestry Intensive planting of a single species; an efficient wood production approach, but one that encourages pests and disease infestations and conflicts with wildlife habitat or recreation uses.

Monsoon A seasonal reversal of wind patterns caused by the different heating and cooling rates of the oceans and continents.

Montane coniferous forests Coniferous forests of the mountains consisting of belts of different forest communities along an altitudinal gradient.

Moral agents Beings capable of making distinctions between right and wrong and acting accordingly. Those whom we hold responsible for their actions.

Moral subjects Beings that are not capable of distinguishing between right or wrong or that are not able to act on moral principles and yet are susceptible of being wronged by others. This category assumes some rights or inherent values in moral subjects that gives us duties or obligations towards them.

Morbidity Illness or disease.

More developed countries (MDC) Industrialized nations characterized by high per capita incomes, low birth and death rates, low population growth rates, and high levels of industrialization and urbanization.

Mortality Death rate in a population; the probability of dying.

Mulch Protective ground cover, including manure, wood chips, straw, seaweed, leaves, and other natural products, or synthetic materials, such as heavy paper or plastic, that protect the soil, save water, and prevent weed growth.

Multiple use Many uses that occur simultaneously; used in forest management; limited to mutually compatible uses.

Mutagens Agents, such as chemicals or radiation, that damage or alter genetic material (DNA) in cells.

Mutation A change, either spontaneous or by external factors, in the genetic material of a cell; mutations in the gametes (sex cells) can be inherited by future generations of organisms.

Mutualism A symbiotic relationship between individuals of two different species in which both species benefit from the association.

N

NAAQS National Ambient Air Quality Standard; federal standards specifying the maximum allowable levels (averaged over specific time periods) for regulated pollutants in ambient (outdoor) air.

Natality The production of new individuals by birth, hatching, germination, or cloning.

Natural history The study of where and how organisms carry out their life cycles.

Natural increase Crude death rate subtracted from crude birth rate.

Natural resources Goods and services supplied by the environment.

Natural selection The mechanism for evolutionary change in which environmental pressures cause certain genetic combinations in a population to become more abundant; genetic combinations best adapted for present environmental conditions tend to become predominant.

Neo-Malthusian Modern advocates of the Malthusian point of view that human populations will grow to the maximum carrying capacity of the environment or until conditions become intolerable.

Net energy yield Total useful energy produced during the lifetime of an entire energy system minus the energy used, lost, or wasted in making useful energy available.

Neurotoxins Toxic substances, such as lead or mercury, that specifically poison nerve cells.

Neutron A subatomic particle, found in the nucleus of the atom, that has no electromagnetic charge.

Never to be developed countries (NDC) Some of the poorest countries are so environmentally devastated that they will probably never reach modern standards of living and may cease to function as nations.

Newly industrialized countries (NICs) Countries like Indonesia, Thailand, Malaysia, and Brazil that are now industrializing.

New towns Experimental urban environments that seek to combine the best features of the rural village and the modern city.

Nihilists Those who believe the world has no meaning or purpose other than a dark, cruel, unceasing struggle for power and existence.

Nitrate-forming bacteria Bacteria that convert nitrites into compounds that can be used by green plants to build proteins.

Nitrite-forming bacteria Bacteria that combine ammonia with oxygen to form nitrites.

Nitrogen cycle The circulation and reutilization of nitrogen in both inorganic and organic phases.

Nitrogen-fixing bacteria Bacteria that convert nitrogen from the atmosphere or soil solution into ammonia that can then be converted to plant nutrients by nitrite- and nitrate-forming bacteria.

Nitrogen oxides Highly reactive gases formed when nitrogen in fuel or combustion air is heated to over 650° C (1,200° F) in the presence of oxygen or when bacteria in soil or water oxidize nitrogen-containing compounds.

Noncriteria pollutants See unconventional air pollutants.

Nongovernmental organizations A term referring collectively to pressure and research groups, advisory agencies, political parties, professional societies, and other groups concerned about environmental quality, resource use, and many other issues.

Nonpoint sources Scattered, diffuse sources of pollutants, such as runoff from farm fields, golf courses, construction sites, etc.

Nonrenewable resources Materials or services from the environment that are not replaced or replenished by natural processes at a rate comparable to our use of the resource; a resource depleted or exhausted by use.

Northern coniferous forest See boreal forest.

North/south division A way of seeing the world economy recognizing that most countries in the northern hemisphere are wealthier and have greater control of world resources than countries in the southern hemisphere.

Nuclear fission The radioactive decay process in which isotopes split apart to create two smaller atoms.

Nuclear fusion A process in which two smaller atomic nuclei fuse into one larger nucleus and release energy; the source of power in a hydrogen bomb.

Nucleic acids Large organic molecules made of nucleotides that function in the transmission of hereditary traits, in protein synthesis, and in control of cellular activities.

Nucleus The center of the atom; occupied by protons and neutrons. In cells, the organelle that contain the chromosomes (DNA).

Numbers pyramid A diagram showing the relative population sizes at each trophic level in an ecosystem; usually corresponds to the biomass pyramid.

O

Oceanic islands Islands in the ocean; formed by breaking away from a continental landmass, volcanic action, coral formation, or a combination of sources; support distinctive communities.

Ocean shorelines Rocky coasts and sandy beaches along the oceans; support rich, stratified communities.

Offset allowances A controversial component of air quality regulations that allows a polluter to avoid installation of control equipment on one source with an "offset" pollution reduction at another source.

Old soils Soils from which rainwater has carried away most of the soluble aluminum and iron and has left behind clay and rust-colored oxides.

Oligotrophic Condition of rivers and lakes that have clear water and low biological productivity (oligo = little; trophic = nutrition); are usually clear, cold, infertile headwater lakes and streams.

Omnivore An organism that eats both plants and animals.

Open canopy A forest where tree crowns cover less than 20 percent of the ground; also called woodland.

Open range Unfenced, natural grazing lands; includes woodland as well as grassland.

Open system A system that exchanges energy and matter with its environment.

Optimum The most favorable condition in regard to an environmental factor.

Orbital The space or path in which an electron orbits the nucleus of an atom.

Organic compounds Complex molecules organized around skeletons of carbon atoms arranged in rings or chains; includes biomolecules, molecules synthesized by living organisms.

Overburden Overlying layers of noncommercial sediments that must be removed to reach a mineral or coal deposit.

Overnutrition Receiving too many calories.

Overshoot The extent to which a population exceeds the carrying capacity of its environment.

Oxygen cycle The circulation and reutilization of oxygen in the biosphere.

Oxygen sag Oxygen decline downstream from a pollution source that introduces materials with high biological oxygen demands.

Ozone A highly reactive molecule containing three oxygen atoms; a dangerous pollutant in ambient air. In the stratosphere, however, ozone forms an ultraviolet absorbing shield that protects us from mutagenic radiation.

P

Pacific Coast coniferous forests The several kinds of cool, moist forest ecosystems from Alaska to northern California. *See* temperate rain forest.

Parabolic mirrors Curved mirrors that focus light from a large area onto a single, central point, thereby concentrating solar energy and producing high temperatures.

Parasite An organism that lives in or on another organism, deriving nourishment at the expense of its host, usually without killing it.

Parent material Undecomposed mineral particles and unweathered rock fragments beneath the subsoil; weathering of this layer produces new soil particles for the layers above.

Particulate material Atmospheric aerosols, such as dust, ash, soot, lint, smoke, pollen, spores, algal cells, and other suspended materials; originally applied only to solid particles but now extended to droplets of liquid.

Parts per billion (ppb) Number of parts of a chemical found in one billion parts of a particular gas, liquid, or solid mixture.

Parts per million (ppm) Number of parts of a chemical found in one million parts of a particular gas, liquid, or solid mixture.

Parts per trillion (ppt) Number of parts of a chemical found in one trillion (10^{12}) parts of a particular gas, liquid, or solid mixture.

Passive heat absorption The use of natural materials or absorptive structures without moving parts to gather and hold heat; the simplest and oldest use of solar energy.

Pasture Enclosed domestic meadows or managed grazing lands.

Patchiness Within a larger ecosystem, the presence of smaller areas that differ in some physical conditions and thus support somewhat different communities; a diversity-promoting phenomenon.

Pathogen An organism that produces disease in a host organism, an alteration of one or more metabolic functions in response to the presence of the organism.

Peat Deposits of moist, acidic, semidecayed organic matter.

Pellagra Lassitude, torpor, dermatitis, diarrhea, dementia, and death brought about by a diet deficient in tryptophan and niacin.

Peptides Two or more amino acids linked by a peptide bond.

Perennial A plant that survives and reproduces each year.

Perennial species Plants that grow for more than two years.

Permafrost A permanently frozen layer of soil that underlies the arctic tundra.

Permanent retrievable storage Long-term storage of supertoxic materials in secure buildings, salt mines, or bedrock caverns where they can be inspected and retrieved for repacking or for transfer if a better storage process is developed.

Pest Any organism that reduces the availability, quality, or value of a useful resource.

Pesticide Any chemical that kills, controls, drives away, or modifies the behavior of a pest.

pH A value that indicates the acidity or alkalinity of a solution on a scale of 0 to 14, based on the proportion of H^+ ions present.

Photochemical oxidants Products of secondary atmospheric reactions. *See* smog.

Photodegradable plastics Plastics that break down when exposed to sunlight or to a specific wavelength of light.

Photosynthesis The biochemical process by which green plants and some bacteria capture light energy and use it to produce chemical bonds. Carbon dioxide and water are consumed while oxygen and simple sugars are produced.

Photosynthetic efficiency The efficiency with which plants collect sunlight and produce organic compounds.

Photovoltaic cell An energy-conversion device that captures solar energy and directly converts it to electrical current.

Physical or abiotic factors Nonliving factors, such as temperature, light, water, minerals, and climate, that influence an organism.

Phytoplankton Microscopic, free-floating, autotrophic organisms that function as producers in aquatic ecosystems.

Pioneer species In primary succession on a terrestrial site, the plants, lichens, and microbes that first colonize the site.

Plankton Primarily microscopic organisms that occupy the upper water layers in both freshwater and marine ecosystems; photosynthetic protists comprise the **phytoplankton;** nonphotosynthetic protists and small invertebrate animals comprise the **zooplankton.**

Poachers Those who hunt wildlife illegally.

Point sources Specific locations of highly concentrated pollution discharge, such as factories, power plants, sewage treatment plants, underground coal mines, and oil wells.

Pollution To make foul, unclean, dirty; any physical, chemical, or biological change that adversely affects the health, survival, or activities of living organisms or that alters the environment in undesirable ways.

Polycentric complex Cities with several urban cores surrounding a once dominant central core.

Poorer industrialized nations (PINs) Countries, such as those in Eastern Europe, that are economically impoverished but have a history of industry and a trained work force.

Population A group of individuals of the same species occupying a given area.

Population crash A sudden population decline caused by predation, waste accumulation, or resource depletion; also called a dieback.

Population explosion Growth of a population at exponential rates to a size that exceeds environmental carrying capacity; usually followed by a population crash.

Population hurdle A need for investment in infrastructure and social services in a rapidly growing population that prevents the capital investment necessary for real economic development.

Population momentum A potential for increased population growth as young members reach reproductive age.

Potential energy Stored energy that is latent but available for use.

Power The rate of energy delivery; measured in horsepower or watts.

Predation The act of feeding by a predator.

Predator An organism that feeds directly on other organisms in order to survive; live-feeders, such as herbivores and carnivores.

Pressurized water reactor (PWR) A nuclear reactor in which high pressure water circulates between the radioactive core and a secondary cooling cycle that generates steam to power an electric turbine.

Prevention of significant deterioration A clause of the Clean Air Act that prevents degradation of existing clean air; opposed by industry as an unnecessary barrier to development.

Price elasticity A situation in which supply and demand of a commodity respond to price.

Primary pollutants Chemicals released directly into the air in a harmful form.

Primary sewage treatment A process that removes solids from sewage before it is discharged or treated further.

Primary standards Regulations of the 1970 Clean Air Act; intended to protect human health.

Primary succession An ecological succession that begins in an area where no biotic community previously existed.

Principle of competitive exclusion A result of natural selection whereby two similar species in a community occupy different ecological niches, thereby reducing competition for food.

Producer An organism that synthesizes food molecules from inorganic compounds by using an external energy source; most producers are photosynthetic.

Production frontier The maximum output of two competing commodities at different levels of production.

Productivity The amount of biological matter or biomass produced in a given area during a given unit of time.

Prokaryotic Cells that do not have a membrane-bounded nucleus or membrane-bounded organelles.

Promoters Agents that are not carcinogenic but that assist in the progression and spread of tumors; sometimes called cocarcinogens.

Pronatalist pressures Influences that encourage people to have children.

Proteins Chains of amino acids linked by peptide bonds.

Public trust A doctrine obligating the government to maintain public lands in a natural state as guardians of the public interest.

Pull factors (in urbanization) Conditions that draw people from the country into the city.

Push factors (in urbanization) Conditions that force people out of the country and into the city.

R

Radioactive An unstable isotope that decays spontaneously and releases subatomic particles or units of energy.

Radioactive decay A change in the nuclei of radioactive isotopes that spontaneously emit high-energy electromagnetic radiation and/or subatomic particles while gradually changing into another isotope or different element.

Radionucleides Isotopes that exhibit radioactive decay.

Rain forest A forest with high humidity, constant temperature, and abundant rainfall (generally over 380 cm [150 in.] per year); can be tropical or temperate.

Rain shadow Dry area on the downwind side of a mountain.

Rangeland Grasslands and open woodlands suitable for livestock grazing.

Recently industrialized countries (RICs) Countries like the "Asian Tigers" of South Korea, Taiwan, and Hong Kong that have recently become industrialized.

Recharge zone Area where water infiltrates into an aquifer.

Recycling Reprocessing of discarded materials into new, useful products; not the same as reuse of materials for their original purpose, but the terms are often used interchangeably.

Red tide A population explosion or bloom of minute, single-celled marine organisms called dinoflagellates.

Billions of these cells can accumulate in protected bays where the toxins they contain can poison other marine life.

Reduced tillage systems Systems, such as minimum till, conserv-till, and no-till, that preserve soil, save energy and water, and increase crop yields.

Refuse-derived fuel Processing of solid waste to remove metal, glass, and other unburnable materials; organic residue is shredded, formed into pellets, and dried to make fuel for power plants.

Regenerative farming A less intensive style of farming that uses little or no inorganic fertilizer, pesticides, water, machinery, and fossil fuel energy; yields may be lower, but so are the input costs of supplies and material.

Regulations Rules established by administrative agencies; regulations can be more important than statutory law in the day-to-day management of resources.

Relative humidity At any given temperature, a comparison of the actual water content of the air with the amount of water that could be held at saturation.

Relativists Those who believe moral principles are always dependent on the particular situation.

Renewable resources Resources normally replaced or replenished by natural processes; resources not depleted by moderate use; examples include solar energy, biological resources such as forests and fisheries, biological organisms, and some biogeochemical cycles.

Residence time The length of time a component, such as an individual water molecule, spends in a particular compartment or location before it moves on through a particular process or cycle.

Resilience The ability of a community or ecosystem to recover from disturbances.

Resistance (inertia) The ability of a community to resist being changed by potentially disruptive events.

Resource partitioning In a biological community, various populations sharing environmental resources through specialization, thereby reducing direct competition. *See also* **ecological niche**.

Respiratory fibrotic agents Special class of irritants, including chemical reagents and particulate materials, that damages the lungs, causing scar tissue formation that lowers respiratory capacity.

Rill erosion The removing of thin layers of soil as little rivulets of running water gather and cut small channels in the soil.

Risk Probability that something undesirable will happen as a consequence of exposure to a hazard.

Risk assessment Evaluation of the short-term and long-term risks associated with a particular activity or hazard; usually compared to benefits in a cost-benefit analysis.

RNA (ribonucleic acid) A nucleic acid used for transcription and translation of the genetic code found on DNA molecules.

Ruminant animals Cud-chewing animals, such as cattle, sheep, goats, and buffalo, with multichambered stomachs in which cellulose is digested with the aid of bacteria.

Runoff The excess of precipitation over evaporation; the main source of surface water and, in broad terms, the water available for human use.

Run-of-the-river flow Ordinary river flow not accelerated by dams, flumes, etc. Some small, modern, high-efficiency turbines can generate useful power with run-of-the-river flow or with a current of only a few kilometers per hour.

Rural area An area in which most residents depend on agriculture or the harvesting of natural resources for their livelihood.

S

Sagebrush Rebellion A coalition of cattlemen, miners, loggers, developers, farmers, politicians, and others who wanted to see more local control over land management and natural resources.

Salinity Amount of dissolved salts (especially sodium chloride) in a given volume of water.

Salinization A process in which mineral salts accumulate in the soil, killing plants; occurs when soils in dry climates are irrigated profusely.

Saltwater intrusion Movement of saltwater into freshwater aquifers in coastal areas where groundwater is withdrawn faster than it is replenished.

Scavenger An organism that feeds on the dead bodies of other organisms.

S curve A curve that depicts logistic growth; called an **S** curve because of its shape.

Secondary pollutants Chemicals modified to a hazardous form after entering the air or are formed by chemical reactions as components of the air mix and interact.

Secondary recovery technique Pumping pressurized gas, steam, or chemical-containing water into a well to squeeze more oil from a reservoir.

Secondary sewage treatment Bacterial decomposition of suspended particulates and dissolved organic compounds that remain after primary sewage treatment.

Secondary standards Regulations of the 1970 Clean Air Act intended to protect materials, crops, visibility, climate, and personal comfort.

Secondary succession Succession on a site where an existing community has been disrupted.

Second law of thermodynamics With each successive energy transfer or transformation in a system less energy is available to do work.

Second World The industrialized, socialist, centrally planned economy nations of Eastern Europe and the former Soviet Union and its allies.

Secure landfill A solid waste disposal site lined and capped with an impermeable barrier to prevent leakage or leaching. Drain tiles, sampling wells, and vent systems provide monitoring and pollution control.

Sedimentary rock Deposited material that remains in place long enough or is covered with enough material to compact into stone; examples include shale, sandstone, breccia, and conglomerates.

Sedimentation The deposition of organic materials or minerals by chemical, physical, or biological processes. Sediments can be transported from their source to their place of deposition by gravity, wind, water, or ice. If subjected to sufficient heat, pressure, or chemical reactions, sediments can solidify into sedimentary rock.

Selective cutting Harvesting only mature trees of certain species and size; usually more expensive than clearcutting, but it is less disruptive for wildlife and often better for forest regeneration.

Seriously undernourished Those who receive less than 80 percent of their minimum daily caloric requirements; are likely to suffer permanently stunted growth, mental retardation, and other social and developmental disorders.

Shallow ecology A term used by more radical groups to describe mainstream environmentalists who are more willing to compromise with government and industry.

Shantytowns Settlements created when people move onto undeveloped lands and build their own shelter with cheap or discarded materials; some are simply illegal subdivisions where a landowner rents land without city approval; others are land invasions.

Sheet erosion Peeling off thin layers of soil from the land surface; accomplished primarily by wind and water.

Sinkholes A large surface crater caused by the collapse of an underground channel or cavern; often triggered by groundwater withdrawal.

Sludge Semisolid mixture of organic and inorganic materials that settles out of wastewater at a sewage treatment plant.

Slums Legal but inadequate multifamily tenements or rooming houses; some are custom built for rent to poor people, others are converted from some other use.

Smog The term used to describe the combination of smoke and fog in the stagnant air of London; now often applied to photochemical pollution products or urban air pollution of any kind.

Social ecology A humanist environmental philosophy based on social justice and environmental protection.

Social justice Equitable access to resources and the benefits derived from them; a system that recognizes inalienable rights and adheres to what is fair, honest, and moral.

Soil A complex mixture of weathered mineral materials from rocks, partially decomposed organic molecules, and a host of living organisms.

Species A population of morphologically similar organisms that can reproduce sexually among themselves but that cannot produce fertile offspring when mated with other organisms.

Species diversity The number and relative abundance of species present in a community.

Species recovery plan A plan for restoration of an endangered species through protection, habitat management, captive breeding, disease control, or other

techniques that increase populations and encourage survival.

Squatter towns Shantytowns that occupy land without owner's permission; some are highly organized movements in defiance of authorities; others grow gradually.

Stability In ecological terms, a dynamic equilibrium among the physical and biological factors in an ecosystem or a community; relative homeostasis.

Stable runoff The fraction of water available year round; usually more important than total runoff when determining human uses.

Statutory law Rules passed by a state or national legislature.

Steady-state economy Characterized by low birth and death rates, use of renewable energy sources, recycling of materials, and emphasis on durability, efficiency, and stability.

Stewardship A philosophy that holds that humans have a unique responsibility to manage, care for, and improve nature.

Strategic minerals Materials a country cannot produce itself but which it uses for essential materials or processes.

Stratosphere The zone in the atmosphere extending from the tropopause to about 50 km (30 mi) above Earth's surface; temperatures are stable or rise slightly with altitude; has very little water vapor but is rich in ozone.

Streambank erosion Washing away of soil from banks of established streams, creeks, or rivers, often as a result of the removal of trees and brush along streambanks or cattle damage to the banks.

Stress Physical, chemical, or emotional factors that place a strain on an animal. Plants also experience physiological stress under adverse environmental conditions.

Stress-related disease *See* stress-shock.

Stress-shock A loose set of physical, psychological, and/or behavioral changes thought to result from the stress of excess competition and extreme closeness to other members of the same species.

Strip-farming Planting different kinds of crops in alternating strips along land contours; when one crop is harvested, the other crop remains to protect the soil and prevent water from running straight down a hill.

Strip-mining Removing surface layers over coal seams using giant, earth-moving equipment; creates a huge open pit from which coal is scooped by enormous surface-operated machines and transported by trucks; an alternative to deep mines.

Structure (ecological) Ecological structure refers to distribution patterns of individuals and populations within a community as well as relations between one community and its neighbors.

Subbituminous coal Banded, black, and fairly soft low-energy coal that has woody layers commonly visible.

Subsidence A settling of the ground surface caused by the collapse of porous formations that result from withdrawal of large amounts of groundwater, oil, or other underground materials.

Subsoil A layer of soil beneath the topsoil that has lower organic content and higher concentrations of fine mineral particles; often contains soluble compounds and clay particles carried down by percolating water.

Succession A gradual progression of species making up the biological community on a particular site as the organisms modify their own environment and make it suitable for some species and not for others.

Sulfur dioxide A colorless, corrosive gas directly damaging to both plants and animals.

Supercity *See* megacity or megalopolis.

Supply/demand curve The relationship between the available supply of a commodity or service and its price.

Surface mining Some minerals are also mined from surface pits. *See* strip-mining.

Surface tension A condition in which the water surface meets the air and acts like an elastic skin.

Survivorship The percentage of a population reaching a given age or the proportion of the maximum life span of the species reached by any individual.

Sustainable agriculture An ecologically sound, economically viable, socially just, and humane agricultural system. Stewardship, soil conservation, and integrated pest management are essential for sustainability.

Sustainable development Using renewable resources in harmony with ecological systems to produce a rise in real income per person and an improved standard of living for everyone.

Sustained yield Utilization of a renewable resource at a rate that does not impair or damage its ability to be fully renewed on a long-term basis.

Swamp Wetland with trees, such as the extensive swamp forests of the southern United States.

Swidden agriculture *See* milpa agriculture.

Symbiosis The intimate living together of members of two different species; includes **mutualism, commensalism,** and, in some classifications, **parasitism.**

Synergism The combined effects of two or more factors on a system.

Synergistic effects When an injury caused by exposure to two environmental factors together is greater than the sum of exposure to each factor individually.

Systemic A condition or process that affects the whole body; many metabolic poisons are systemic.

T

Taiga The northernmost edge of the boreal forest, including species-poor woodland and peat deposits; intergrading with the arctic tundra.

Tailings Mining waste left after mechanical or chemical separation of minerals from crushed ore.

Technological optimism A view of the world as a place of abundance and opportunity in which the beneficial aspects of technology will outweigh its problems.

Tectonic plates Huge blocks of Earth's crust that slide around slowly, pulling apart to open new ocean basins or crashing ponderously into each other to create new larger landmasses.

Temperate deciduous forests Forests of trees which produce lush growth during warm, humid summers and then shed their leaves during cold winters.

Temperate rain forest The cool, dense, rainy forest of the northern Pacific coast; enshrouded in fog much of the time; dominated by large conifers.

Temperature A measure of the speed of a typical atom or molecule in a substance.

Teratogens Chemicals or other factors that specifically cause abnormalities during embryonic growth and development.

Terracing Shaping the land to create level shelves of earth to hold water and soil; requires extensive hand labor or expensive machinery, but it enables farmers to farm very steep hillsides.

Territoriality An area surrounding an organism's home site or nesting site that the organism actively defends, especially against members of its own species; the territory may be occupied and defended by an individual, a mated pair, or a social group, depending on species.

Tertiary sewage treatment The removal of inorganic minerals and plant nutrients after primary and secondary treatment of sewage.

Thermal plume A plume of hot water discharged into a stream or lake by a heat source, such as a power plant.

Thermodynamics A branch of physics that deals with transfers and conversions of energy.

Thermodynamics, first law Energy can be transformed and transferred, but cannot be destroyed or created.

Thermodynamics, second law With each successive energy transfer or transformation, less energy is available to do work.

Thermoplastics Soft plastics composed of single-chain, unlinked polymers, such as polyethylene, polypropylene, polyvinylchloride, polystyrene, and polyester, that can be remelted and reformed to make useful products.

Thermoset polymers Hard plastics composed of cross-linked molecular networks, such as acrylic, phenolic, or epoxy resins, that cannot be remelted or recycled.

Third World Less developed countries that are not capitalistic and industrialized (First World) or centrally-planned socialist economies (Second World); not intended to be derogatory.

Threatened species While still abundant in parts of its territorial range, this species has declined significantly in total numbers and may be on the verge of extinction in certain regions or localities.

Timberline In mountains, the highest-altitude edge of forest that marks the beginning of the treeless alpine tundra.

Tolerance limits *See* limiting factors.

Tool making revolution The first use of technology in wood or stone tools for defense, building fires, and obtaining food.

Topsoil The first true layer of soil; layer in which organic material is mixed with mineral particles; thickness ranges from a meter or more under virgin prairie to zero in some deserts.

Tornado A violent storm characterized by strong swirling winds and updrafts; tornadoes form when a strong cold front pushes under a warm, moist air mass over the land.

Total fertility rate The number of children born to an average woman in a population during her entire reproductive life.

Total growth rate A measure of population growth which includes immigration and emigration as well as births and deaths.

Toxins Poisonous chemicals that react with specific cellular components to kill cells or to alter growth or development in undesirable ways; often harmful, even in dilute concentrations.

Tragedy of the commons Depletion or degradation of a commons, or publicly-held resource, to which everyone has access but no one has ownership or a sense of responsibility.

Transitional zone A zone in which populations from two or more adjacent communities meet and overlap.

Transpiration The evaporation of water from plant surfaces, especially through stomates.

Trauma Injury caused by accident or violence.

Trombe wall An interior, heat-absorbing wall; may be water-filled glass tubes that absorb heat rays and let light into interior rooms.

Trophic level A step in the movement of energy through an ecosystem; an organism's feeding status in an ecosystem.

Tropical forests Can be wet, lush rain forests like those in the Amazon basin, or seasonally dry deciduous forests of the Central American Pacific coast.

Tropical rain forest A species-rich biome type; consists of stratified communities of broad-leaved shrubs, trees, epiphytes, lianas, numerous insects, birds, and other animals; abundant rainfall, year-round warm to hot temperatures, and old and nutrient-poor soils.

Tropical seasonal forest Semi-evergreen or partly deciduous forests tending toward open woodlands and grassy savannas dotted with scattered, drought-resistant tree species; distinct wet and dry seasons, hot year-round.

Tsunami Giant seismic sea swells that move rapidly from the center of an earthquake; they can be 10 to 20 meters high when they reach shorelines hundreds or even thousands of kilometers from their source.

Tundra Treeless arctic or alpine biome characterized by cold, harsh winters, a short growing season, and potential for frost any month of the year; vegetation includes low-growing perennial plants, mosses, and lichens.

U

Unconventional air pollutants Toxic or hazardous substances, such as asbestos, benzene, beryllium, mercury, polychlorinated biphenyls, and vinyl chloride, not listed in the original Clean Air Act because they were not released in large quantities; also called noncriteria pollutants.

Unconventional oil Oil that can be extracted from unconventional deposits like tar sands.

Undernourished Those who receive less than 90 percent of the minimum dietary intake over a long-term time period; they lack energy for an active, productive life and are more susceptible to infectious diseases.

Undiscovered resources Potential supplies of a mineral or other useful material believed to exist based on history, scientific theory, or general knowledge of geology, biology, or geography of an unexplored area.

Universalists Those who believe that some fundamental ethical principles are universal and unchanging. In this vision, these principles are valid regardless of the context or situation.

Utilitarianism A philosophy that regards an action as right if it produces the greatest good for the greatest number of people.

Upwelling Convection currents within a body of water that carry nutrients from bottom sediments toward the surface.

Urban area An area in which a majority of the people are not directly dependent on natural resource-based occupations.

Urbanization An increasing concentration of the population in cities and a transformation of land use to an urban pattern of organization.

Utilitarian conservation Saving resources because of their value to humans; to provide the greatest good for the greatest number for the longest time; a philosophy argued by Theodore Roosevelt and Gifford Pinchot.

V

Vertical stratification The vertical distribution of specific subcommunities within a community.

Village A collection of rural households linked by culture, custom, and association with the land.

Visible light A portion of the electromagnetic spectrum that includes the wavelengths used for photosynthesis.

Vitamins Organic molecules essential for life that we cannot make for ourselves; we must get them from our diet; they act as enzyme cofactors.

Volatile organic compounds Organic chemicals that evaporate readily and exist as gases in the air.

Vulnerable species Species at risk of extinction because of human activities or because they are naturally rare.

W

Waste stream The steady flow of varied wastes, from domestic garbage and yard wastes to industrial, commercial, and construction refuse.

Water cycle The recycling and reutilization of water on Earth, including atmospheric, surface, and underground phases and biological and nonbiological components.

Waterlogging Water saturation of soil that fills all air spaces and causes plant roots to die from lack of oxygen; a result of overirrigation.

Water pollution Any physical, biological, or chemical change in water quality that adversely affects living organisms or makes water unsuitable for desired uses.

Water table The top layer of the zone of saturation; undulates according to the surface topography and subsurface structure.

Wealthy industrialized nations (WINs) Industrialized nations with strong economies, like the Western European countries.

Weather A description of the physical conditions of the atmosphere (moisture, temperature, pressure, and wind).

Weathering Changes in rocks brought about by exposure to air, water, changing temperatures, and reactive chemical agents.

Wetlands Ecosystems of several types in which rooted vegetation is surrounded by standing water during part of the year. *See also* swamps, marshes, bogs.

Wilderness An area of undeveloped land affected primarily by the forces of nature; an area where humans are visitors who do not remain.

Wilderness Act Legislation of 1964 recognizing that leaving land in its natural state may be the highest and best use of some areas.

Wildlife Plants, animals, and microbes that live independently of humans; plants, animals, and microbes that are not domesticated.

Wildlife refuges Areas set aside to shelter, feed, and protect wildlife; due to political and economic pressures, refuges often allow hunting, trapping, mineral exploitation, and other activities that threaten wildlife.

Windbreak Rows of trees or shrubs planted to block wind flow, reduce soil erosion, and protect sensitive crops from high winds.

Wind farms Large numbers of windmills concentrated in a single area; usually owned by a utility or large-scale energy producer.

Withdrawal A description of the total amount of water taken from a lake, river, or aquifer.

Woodland A forest where tree crowns cover less than 20 percent of the ground; also called open canopy.

Work The application of force through a distance; requires energy input.

World conservation strategy A system of maintaining essential ecological processes, preserving genetic diversity, and ensuring that utilization of species and ecosystems is sustainable.

World Ocean The interconnected world seas and oceans.

X

X ray Very short wavelength in the electromagnetic spectrum; can penetrate soft tissue; although it is useful in medical diagnosis, it also damages tissue and causes mutations.

Y

Yellowcake The concentrate of 70 to 90 percent uranium oxide extracted from crushed ore.

Young soils Soils that haven't weathered much and, therefore, are rich in soluble minerals from parent rocks, such as silicon, iron, and aluminum.

Z

Zero population growth (ZPG) The number of births at which people are just replacing themselves; also called the replacement level of fertility.

Zone of aeration Upper soil layers that hold both air and water.

Zone of leaching The layer of soil just beneath the topsoil where water percolates, removing soluble nutrients that accumulate in the subsoil; may be very different in appearance and composition from the layers above and below it.

Zone of saturation Lower soil layers where all spaces are filled with water.

CREDITS

LINE ART AND TEXT

Box Fig 1.2.1

From Ricki Lewis, *Beginnings of Life.* Copyright © 1992 Wm. C. Brown Communications, Inc., Dubuque, Iowa. All Rights Reserved. Reprinted by permission.

Fig 3.4

From: *ECOLOGY* 3/e by Robert E. Ricklefs. Copyright © 1990 by W.H. Freeman and Company. Reprinted with permission.

Fig 3.16

Reprinted with the permission of Macmillan Publishing Company from *Communities and Ecosystems,* Second Edition by Robert H. Whittaker. Copyright © 1975 by Robert H. Whittaker.

Fig 4.5

Figure modified with permission from *Population Bulletin,* vol. 18, no. 1, 1985, Population Reference Bureau, Inc.

Fig 4.8

From W.L. Langer, "The Next Assignment," *American Historical Review, vol. 63, no. 1, (1958) pp. 283-305. Copyright © 1958 American Historical Association. Used by permission.*

Figs 5.14 & 5.15

Reprinted from *Beyond the Limits* copyright 1992 by Meadows, Meadows and Randers. With permission from Chelsea Green Publishing Co., Post Mills, VT, and the Canadian Publishers, McClelland & Stewart, Toronto.

Fig 5.16

From *Environment,* "Ecology Meets Economics," by R.A. Carpenter and J.A. Dixon, June 1985. Reprinted with permission of the Helen Reid Educational Foundation. Published by Heldref Publications, 1319 Eighteenth St., N.W., Washington, D.C. 20036-1802. Copyright © 1985.

Fig 6.6

Reprinted with permission from *The New England Journal of Medicine,* May 8, 1986, p. 1226

Fig 6.14

From Paul Slovic, "Perception of Risk" in *Science,* Vol. 236, April 17, 1987, p. 280. Copyright 1987 by the AAAS. Used by permission.

Box 9.6

From Ricki Lewis, *Life.* Copyright © 1992 Wm. C. Brown Communications, Inc., Dubuque, Iowa. All Rights Reserved. Reprinted by permission.

Fig 9.8

From Fisheries and Agriculture Science 47, *Canadian Offshore Fishery Atlas, ed. by D.J. Scarrat, 1982 (also in State of Canada's Environment, fig. 8.5 on p. 8-13, Minister of Environment Canada. Used by permission.*

Fig 9.15

From Norman Myers, "Threatened Biotas: 'Hotspots' in Tropical Forests," *Environmentalist,* Vol. 8, No. 3:1-20, 1988; and Vol. 10, No. 4:243-256, 1990. By permission of the Environmentalist.

Fig 10.12

Courtesy of General Motors Corporation.

Fig 12.10

Courtesy of Northern States Power Company, Minneapolis.

ILLUSTRATOR CREDITS

Bowring Cartographics: 7.7, 8.12, 9.16

Diphrent Strokes, Inc.: 2.3, 2.7, 2.10, 2.11, 2.14, 2.18, Box 4.3, 4.14, 4.16, 6.6, 6.14, 8.7, 11.16, Box 12.2.1, 12.6, 13.2, 13.7, 14.5, 14.7

Don Luce: Box 3.1.1, 3.13, 7.11

Illustrious, Inc.: Box 1.2.1

Don Luce/Marjorie C. Leggitt: 2.12

Margorie C. Leggitt: 2.8, 2.9, 3.5, 3.6, 3.8, 3.10, Box 7.4.1, 7.10, 9.2, 9.3

Norman Frisch: 12.5

Laurie O. Keefe: 9.6

Rolin Graphics: 1.2, 1.3, 2.2, 2.6, 2.15, 2.16, 3.2, 3.3, 3.4, 3.7, 3.15, 3.16, Box 4.1.1, 4.2, 4.3, 4.4, 4.5, 4.6, 4.7, 4.8, 4.9, 4.10, 4.11, 4.12, 4.19, 5.2, 5.3, 5.7, 5.9, 5.10, 5.12, 5.13, 5.14, 5.15, 5.16, 5.17, 6.3, 6.13, 7.3, 7.6, 7.13, 7.16, 7.17, 8.3, Box 8.3.1, Box 9.2(A), 9.5, 9.7, 9.8, 9.12, 9.14, 9.15, 9.19, Box 10.1.1, Box 10.1.2, Box 10.2.2, Box 11.1.1, Box 11.2.1, 11.3, Box 11.3.1, 11.4, 11.14, 12.1, 12.8, Box 13.3.1, 13.6, 13.10, 13.14, 14.6, 14.14, 15.4, 15.6, 15.12

PHOTO CREDITS

Part Openers

1 and 4: Photograph by Ansel Adams, Copyright © 1993 by the Trustees of the Ansel Adams Publishing Rights Trust. All Rights Reserved.

2 and 3: Ansel Adams/National Archives Still Picture Branch.

Chapter 1

1.1: Science VU/Visuals Unlimited

1.4: N.R. Farbman, Life Magazine © Time Warner Inc.

1.5: © IFA/Peter Arnold, Inc.

1.6: The Bettmann Archive

1.7: © Robert Lindsey/Photo Researchers, Inc.

1.9: Courtesy of the Bancroft Library, University of California, Berkeley

1.10: AP/Wide World Photos

BOX 1.3.1: © Mark Ludak/Impact Visuals

1.11: © Ilene Perlman/Stock Boston

1.13: The Bettmann Archive

Chapter 2

2.1: Photo by David Swanlund, courtesy of the Save-the-Redwoods League

2.4: © E. H. Newcomb & W. P. Wergin, University of Wisconsin/ Biological Photo Service

2.5: © Carl Purcell/Photo Researchers, Inc.

BOX 2.2.2: © Laurence Pringle/Photo Researchers, Inc.

2.17: © J. Brinton/Visuals Unlimited

Chapter 3

3.1: © John D. Cunningham/Visuals Unlimited

3.9: © Karl Maslowski/Photo Researchers, Inc.

3.11: © Lynwood M. Chace/Photo Researchers, Inc.

BOX 3.3.1: © V. Ahmadjian and J. B. Jacobs/Visuals Unlimited

3.12: © D. Wilder/Tom Stack and Associates

3.14: Dr. Carl W. Bollwinkel

3.17: © John D. Cunningham/Visuals Unlimited

3.18: © Ron Spomer/Visuals Unlimited

3.19: © Joe McDonald/Visuals Unlimited

3.20: © Doug Sokell/Visuals Unlimited

Chapter 4

4.1: © R. Frerck/Odyssey Productions

BOX 4.21: © Steve Maines/Stock Boston

4.18A: © James Shaffer

4.18B, C, F: © Bob Coyle

4.18D: © Hank Morgan/Photo Researchers, Inc.

4.18E: © Ray Ellis/Photo Researchers, Inc.

Chapter 5

5.1: © Jerry Irwin

5.4: © Lowell Georgia/Photo Researchers, Inc.

5.5: © Audrey Lang/Valan Photos

5.6: © T. Kitchen/Valan Photos

5.8: © John D. Cunningham/Visuals Unlimited

BOX 5.1.1: © Frank J. Miller/Photo Researchers, Inc.

BOX 5.2.1: © Michael Dwyer/Stock Boston

Chapter 6

6.1: © Walter Frerck/Odyssey Production

6.2: Courtesy of Dr. Stanley Erlandsen

6.4: WHO/World Bank/R. Witlin

6.5: AP/Wide World

Chapter 7

7.1: © Allan Tannenbaum/Sygma

7.2: © Chester Higgins, Jr./Photo Researchers, Inc.

BOX 7.1.1: © Audrey Topping/Photo Researchers, Inc.

7.4: © Omikron/Photo Researchers, Inc.

7.5: U.S.D.A.

7.8: © Victor Englebert/Photo Researchers, Inc.

7.9: © Pam Hickman/Valan Photos

7.12: Soil Conservation Serv., U.S. Dept. of Agriculture

7.14: Library of Congress

7.15: © Michael Dwyer/Stock Boston

BOX 7.21: Terry Gips with "International Alliance for Sustainable Agriculture"

7.18: © Wolfgang Kaehler

7.19: © Joe Monroe/Photo Researchers, Inc.

Chapter 8

8.1: © Chip & Jill Isenhart/Tom Stack and Associates

8.4: Chippewa Valley Museum

8.5: © Doug Vargas/The Image Works

8.6: Wm. P. Cunningham

BOX 8.1.1: © Randy Hyman

BOX 8.2.1: Courtesy of Dan Janzen

8.8: © Greg Vaughn/Tom Stack and Associates

8.9: © Milton Rand/Tom Stack & Associates

8.10: © William and Marcy Levy/Photo Researchers, Inc.

8.11: © Eastcott/Momatiuk/The Image Works

8.13: © John D. Cunningham/Visuals Unlimited

8.14: © Lee Battaglia/Photo Researchers, Inc.

8.15: Wm. P. Cunningham

8.16: © Gregory Dimijian/Photo Researchers, Inc.

8.17: © Jay Pasachoff/Visuals Unlimited

8.18: © Alain Compost/WWF International

Chapter 9

9.1: © Gunter Ziesler/Peter Arnold, Inc.

BOX 9.1.1: Wm. P. Cunningham

9.4: © Robert C. Simpson/Valan Photos

9.9: © Tom McHugh/Photo Researchers, Inc.

9.10: © Patrick Forestier/Gamma Liaison

9.11: © Lynn Funkhouser/Peter Arnold, Inc.

9.18: Courtesy of Robert O. Bieeregaard/Smithsonian Institution

Chapter 10

10.1: © Robert and Linda Mitchell

10.5: © 1988 Ted Spiegel/Black Star

10.7: M. A. Cunningham

10.9: © Simon Fraser/Photo Researchers, Inc.

10.10: © John D. Cunningham/Visuals Unlimited

10.14: © Klaus Reisinger/Black Star

Chapter 11

11.1: © Peter B. Kaplan/Photo Researchers, Inc.

11.5: © Joe Monroe/Photo Researchers, Inc.

11.6: © Robert and Linda Mitchell

11.8: U.S.G.S.

BOX 11.2B: © David Turnley/Black Star

11.10: Tim McCabe Soil Conservation Service

11.11: © Roger A. Clark, Jr./Photo Researchers, Inc.

11.12: © Frans Lantiny/Photo Researchers, Inc.

BOX 11.4.1: © 1975 Aileen and W. Eugene Smith/Black Star

11.15: Wm. P. Cunningham

Chapter 12

12.2: © W. Ormerod/Visuals Unlimited

12.13: © Arthur Grace/Sygma

12.14: Volvo

12.16: © Peter Menzel/Stock Boston

12.18: Courtesy of General Motors

12.19: © Gary J. James/Biological Photo Service

12.21: © R. Frerck/Odyssey Productions

12.22: © Peter Menzel/Stock Boston

Chapter 13

13.1: © Thomas Kitchen/Tom Stack & Associates

13.3: © Fred McConnaughey/Photo Researchers, Inc.

13.4: © Ken Sakamoto/Black Star

13.8: © Steve Skloot/Photo Researchers, Inc.

13.11: © Mike Brisson

13.13: © Greenlar/The Image Works

BOX 13.2.1: Andy Levin

Chapter 14

14.1: © Robert Frerck/Odyssey Productions

14.2: Wm. P. Cunningham

14.3: © R. Frerck/Odyssey Productions

14.8: © Robert and Linda Mitchell

14.9: © Robert Frerck/Odyssey Productions

14.10: Wm. P. Cunningham

BOX 14.1.1: © Carl Purcell/Words & Pictures

14.11: © Sally Myers

14.13: © Tom Myers

BOX 14.3.1: Building of New Town, p. 44

BOX 14.3.2: Wm. P. Cunningham

14.15: © Wolfgang Kaehler

14.16: © Tom E. Myers/Tom Stack and Associates

14.17: Wm. P. Cunningham

Chapter 15

15.1: © Sam Kittner/Greenpeace

15.2: © Grant Heilman/Grant Heilman Photography, Inc.

BOX 15.3.1: © Frank Pedrick/The Image Works

15.7: © Chuck Abbott/Photo Researchers, Inc.

15.8: Wm. P. Cunningham

BOX 15.4.1: © Melanie S. Freeman/The Christian Science Monitor July 21, 1987

15.9: © Piet van Lier/Impact Visuals

15.10: © Sam Kittner/Greenpeace

15.11: © David Cross/Impact Visuals

15.13: © 1989 Morgan/Greenpeace

15.14: © DPA/Photoreporters, Inc.

15.16: © Bob Daemmrich/The Image Works

15.18: Reuters/Bettmann Archives

NAME AND SUBJECT INDEX

▼

A

Abbey, Edward, 332
Accidents as health hazard, 120
Acid precipitation, **217, 218,** 219
Acids as water pollutant, 240-41
Acquired characteristics, theory of, 50
Active solar systems, **263,** *266*
Acute effects of toxic chemicals, **126**
Adams, Ansel, 1, 90, 133, 275
Adaptation, 48-49
    symbiotic relationships and co-, 53,
      55
Aerosol, **212**
Aesthetic and cultural benefits of
    biological resources, 186-88
Africa, population growth and food
    production in sub-Sahara, *141*
Agenda 21, 13 (box)
Agricultural revolution, 6-7
Agriculture. *See also* Food
    cash crops, 142
    energy resources for, 148,
      149 (table)
    fertilizer for, 147-48
    greenhouse effect and, 207
    increasing outputs of, 148-50
    major food crops, 139 (table)
    output, 134
    photosynthetic efficiency and, 43
    soil resources and, 134, 142-47
    stewardship values in, *5*
    sustainable, 150-56
    urbanization and, 311, 313
    waste generated by, 276-77
    water conservation by, 238
    water resources used by, 147, 230,
      231, 233 (box)
AIDS (acquired immune deficiency
    syndrome), 116
Air pollution, 205, 208-23
    control of, 219-22
    conventional "criteria" pollutants
      and, 209-13
    effects of, 215-19
    emission permits, 101, 102 (box)
    future trends in, 222-23
    indicator species for, 54
    indoor, 213-15
    primary, secondary, and fugitive
      emissions causing, 209
    radon as, 214-15 (box)

    toxic, in United States, *287*
    unconventional "noncriteria"
      pollutants and, 213
    as urban problem, 303
Air resources. *See also* Air pollution
    atmosphere and climate, 26, 205,
      206
    greenhouse effect and, 206,
      207-8 (box)
    ozone-shield depletion, 211-12 (box)
Alaska, 176
Alien species, 57, 194, 195-96 (box)
Allergens, **117**
Amazon Basin, deforestation and
    murder in, 165-66 (box)
Ambient air, **209**
Ancient forests, 168-69
Animals. *See also* Livestock
    production; Species; Wildlife
    competition and, 53
    overharvesting products from, 192
    as pets and in scientific trade, 193
    territoriality of, **53**
    testing toxic chemicals on, 124-25
Antarctica, international treaty on, 340
Antigens, **117**
Aquaculture, *140*
Aquatic ecosystems, 60. *See also*
    Marine ecosystems
    effect of air pollutants on, 217-18
    energy pyramid in, *31*
    food chains in, 30, *52*
    succession in, 55, *56*
Aquifers, *228,* 232, 233 (box), 245
Aral Sea, destruction of, 235-36 (box)
Arnold, Ron, 333
Asphyxiants, **117**
Asthma, **216**
Atmosphere, 26, 205-6
    carbon dioxide in, 35, *36,* 207
    greenhouse effect in, **206,**
      207-8 (box)
    solar energy and, 205, *206*
Atoms, **22,** *23*
Automobile
    air pollution caused by, 222
    effect of, on cities, 308-9, *310*
    electrical, 267
    emission-control system, *221*
    energy conservation in, 262

B

Bacteria, nitrogen-fixing, 36-37, *38*
Bangladesh, Grammeen Banks, 105-6
    (box)
Bennett, Hal Zina, 112
Bennett, Hugh, 146
Bermuda, environmental restoration in,
    321 (box)
Best available, economically achievable
    technology (BAT), **245**
Best practical control technology
    (BPT), **245**
Beyond the Limits (Meadows), 99-100
Bioaccumulation, **123**
Biodegradable plastics, **286**
Biodiversity. *See* Species diversity
Biogeographical area, **199,** *200*
Biological community, **27,** 42-57. *See*
    also Biomes
    climax, 28, 56-57
    competition in, 52-53
    complexity of, 43-45
    ecological niche in, 49-51
    ecological succession in, 55-57
    introduced species and change in,
      57
    limiting factors and tolerance limits
      in, 48
    natural selection and adaptation in,
      48-49, 50 (box)
    populations and species in, 42, *43*
    predation in, 51-52
    productivity of, 42-43, *44*
    species diversity in, 43
    structure of, 45-46, *47*
    symbiosis in, 53, 54 (box), 55
    in tropical rainforests, 46-47 (box)
Biological oxygen demand (BOD), **240**
Biological resources, 183-204
    alien species as threat to, 195-96
      (box)
    benefits of, 185-88
    destruction of, 188-90
    ecotourism and, 187-88 (box)
    extent of, 184-85
    human activities and destruction of,
      190-97
    management of, 197-202
Biological study, 21, *22*
Biological waste treatment, 292-93
Biomagnification, **123**

Biomass, **29,** 268-69
  alcohol from, 269
  methane produced from, 268-69
  production rate of, in biological
    communities, 42-43
Biomass energy, 249
Biomass pyramid, 31, *32*
Biomes, **57-60**
  aquatic, 60
  deserts, 57-58, *59*
  forests, 59, *60*
  grasslands, 58, *59*
  map of world, *58*
  types of, by temperature and
    moisture gradient, *58*
Bioregion, testing personal knowledge
  of, 332 (table)
Biosphere preserves, 176, *201, 202.*
    See also Parks and preserves
Biotic potential, **65**
Birds
  ecological niche of, 50, *51*
  habitat restoration for Bermuda
    cahow, 321 (box)
  introduced species of, 194
  species diversity among, 43
  trade in rare, 193
Birth control, **82**
  current methods of, 82, *83, 84*
  new developments in, 83-86
  personal plan for, 84-85 (box)
Birth defects, 117-18
Birth rates, 71-72
  birth dearth and, 79-80
  factors affecting, 77-79
Bison, 172-73 (box), 197
Black lung disease, **251**
Blood flukes, 115
Blue Angel symbol, 326
Botswana, 105, 171
Boulding, Kenneth, 93, 95, 96
Brazil
  biological reserves in, 201
  deforestation of, 164, 165-66 (box)
  environmental cleanup in Cubatao,
    223
  land tenure in, 178-79
  urban planning and conservation in
    Curitiba, 324 (box)
  water resources of, 229
Breast-feeding and child survival, 114
    (box)
Breeder reactors, **257-60**
Bronchitis, **215**
Brundtland, Gro Harlem, 12, 13
Bubonic plague in 14th-century
    Europe, 17, 68, 116
Buffalo Commons, proposed U.S.
    project of, 172-73 (box)
Buildings and monuments, acid
    precipitation effects on, 219

Bureau of Land Management (BLM),
    U.S., 173, 176
Bush, George, administration of, 13-14
    (box)

C

Canada
  acid precipitation over, *218*
  Environmental Choice program in,
    326
  fossil fuel resources in, *254*
  laws protecting biological resources
    in, 197-98
  nuclear reactors in, 256, 257
  old-growth forests in, 168-69
  unconventional oil in, 253, *254*
Cancer, **118**
  dietary fat and breast, *121*
  ozone depletion and skin, 212 (box)
  risk assessment of, 128-29
Cannon, Walter, 28
Carbohydrates, 23
Carbon, 23
Carbon cycle, **35,** *36*
Carbon dioxide in atmosphere, 35, *36,*
    207, 210
Carbon monoxide, *209,* 210
Carbon sinks, **35–36**
Carcinogens, **118,** 125
Carnivores, **30**
Carrying capacity, **65**
  increasing, 98, *99*
Carson, Rachel, 11, *12*
Cash crops, **142**
Cells, 23-24
Cellular respiration, **27**
Central Arizona Project, 235
Chemical processing of hazardous
    substances, 292
Chemicals
  as health hazard, 116-18 (*see also*
    Hazardous waste; Toxic
    chemicals)
  movement and fate of, in
    environment, *122*-23
  organic, 23, 241-42
  as water pollutants, 240-42
Chernobyl nuclear disaster, 258-59
    (box)
Children
  disease and programs for survival
    of, 113-14 (box)
  environmental hazards causing birth
    defects in, 117-18
  infant mortality, 77-78, 81-82, 114
  mortality of, 81-82
  problems of undernourishment for,
    137, *138*
China
  air pollution in, 222

food supplies and famine in, 136
    (box)
  population stabilization programs
    in, 76-77 (box)
  soil erosion in, 146, 147 (table)
  water transfer projects in, 235
Chlorofluorocarbons (CFCs), 210,
    211-12 (box), 339
Chloroplasts, *23,* 26
Christianity and attitudes toward
    nature, 9-10
Chronic effects of toxic chemicals, **126**
Chronic food shortages, **136-**37
Cities, 298-318
  causes of urban growth, 302-3
  current problems of, 303-9
  defining, **299**
  planning (*see* City planning)
  sustainable development of Third
    World, 315-16
  transportation and growth of,
    307-10
  urbanization, 298-302
  world's largest, 301 (table)
Citizen protests against nuclear power,
    261-62. *See also* Collective action
City of the future, **313**
City planning, 310-15
  of ancient cities, 310-11
  in Brazil, 324 (box)
  future, 311-13
  garden cities and new towns, 311
  suburbs and, 313-15
  in Sweden and Finland, 312 (box)
Clark, Collin, 64
Clarke, Arthur C., 276
Clean Air Act of 1990, 102, 209, 213
Clean Water Act of 1972, 242, 245
Clear-cut forests, **169,** *170*
Clements, F. E., 28
Climate
  greenhouse effect and, 206,
    207-8 (box), 340
  solar energy and, 205, *206*
Climax community, 28, **56-57**
Closed canopy forest, **161**
Closed community, 46, *47*
  on Isle Royale, 67-68 (box)
Coal, 35, 248
  mining of, 251-52
  resources and reserves of, 251, *254*
Coevolution, **52**
Collective action, 327-33. *See also*
    Politics, environmental
  anti-environmental backlash and, 333
  broadening environmental agenda
    with, 331, *332*
  example of personal action, 329-30
    (box)
  mainline environmental
    organizations, 331

against nuclear power, 261-62
organizing environmental campaign, 329 (table)
radical environmental organizations, 331-33
student environmental groups, 328
using media to influence public opinion, 330 (table)
writing elected officials, 338 (table)
Colonialism, population and, 81, *82*
Colorado River, 235, 236
Commensalism, **53**
Committee on the Status of Endangered Wildlife in Canada (COSEWIC), 197
Commoner, Barry, 11
Community. *See* Biological community
Competition, **52-53**
ecological niche and, 50, *51*, 53
Competitive exclusion, **53**
Complexity, **43-45**
Composting, **284-85** (box)
Compounds, **22**
organic, **23**
Computer models of resource use, 99, *100*
Condoms, *83, 85*
Cone of depression, 232
Conifer forests, **59-**60
Conservation
economic development and, 177
of energy, 262-63
of matter, **24-25**
movement for resource, 10, 11
strategies for world, 177 (table)
of water resources, 237-38
Conservation International, 335
Consumerism, 10, 11 (table), **322-27**
Blue Angels and Green Seals for, 326
jobs and green, 325-26 (box)
limits of green, 326-27
personal responsibility for, 327
questioning, 322-23
reducing, 323 (table)
shopping for green products, 323-24
Consumers in ecosystems, **30**
Consumption of water resources, **230**
Contour plowing, **152**
Control rods, **256**
Conventional "criteria" air pollutants, 209-13
Convention of International Trade in Endangered Species (CITES), 192, 198-202
Convergent evolution, 49
Coral reefs, 43
Corridors, habitat, *200*
Costa Rica
dry tropical forest restoration in, 167-68 (box)

ecotourism in, *187*
medicinal products from forests of, 186
Cost/benefit analysis, **103-4**
Costs, internal and external, 101
Courts, United States, 337-39
Crime as urban problem, 304-5 (box)
Criteria air pollutants, **209-**13
Critical thinking, **4-5** (box)
Crops
cash, 142
energy requirements for producing food, 148, 149 (table)
land available for growing, 149, 150 (table)
major food, 139 (table)
monoculture, 146
trap, 156
world production of, *140*
Crowding as urban problem, 304-5 (box)
Crude birth rate, **71**
in China, 76
Crude death rate, **72-73**

D

Daly, Herman E., 93
Dams and reservoirs, 236-37, 269-70
Dark reactions, photosynthetic, 26, *27*
Darwin, Charles, 49
DDT, 54, 242
bioaccumulation and biomagnification of, *123*
as threat to biological resources, 196
Death rates, 72-73
Debt-for-nature swaps, 168, 335
Decline spiral, **189**
Decomposers in ecosystems, *30,* **31**
Deep ecology, **12, 322**
Degradation of water quality, **230**
Demand, 96-97
Demographic transition, **80-82**
ecojustice and, 81
infant and child mortality and, 81-82
optimistic view of, 80-81
pessimistic view of, 81
phases of, *80*
social justice view of, 81
Demography, human, **71-77**
Denmark, toxic waste management in, 294 (box)
Density-dependent factors (population biology), **66**
Density-independent factors (population biology), **66**
Deoxyribonucleic acid (DNA), 22, 117, 124
Dependency ratio, **76**
Desalination, 234
Desertification, 171

Deserts, **57-58**, *59*
Developing countries, **16**. *See also* Third World
Development, **334**. *See also* Economic development; Sustainable development
Diaphragm (birth control), *84*
Dieback, population, **65**, 86
Diet, health and, 119, 120-21
Dioxin, 281
Discount rates, **102-3**
Disease, **112**. *See also* Health
air pollution and, 215-16
bubonic plague in 14th-century Europe, 17, 68, 116
child survival rates and, 113-14 (box)
coal mining and, 251-52
infectious organisms and, 113-16
malnutrition/undernutrition and, 137, 138, 139
Minamata, 241 (box)
as predator, 66, 194-96
water-borne, 237, 239, 243-44
Domestic water use, 231
Domination of nature, 9-10
Dose/response curves of toxic chemicals, *125, 126*
Drift nets, 191
Drinking water, 232-34
desalination to create, 234
groundwater supplies of, 232, 233 (box)
pollution of, 148, 239
shortages of, 232
as urban problem, 303-4
Dubos, Renee, 12, 13
Duvall, Bill, 12

E

Earth
diversity of life on, 2
Goode's Homolosine Equal Area Projection map of, *346-47*
as superorganism, 28 (box)
Earth Charter, key principles of, 341 (table)
Earth First!, 331, *333*
Earth Summit of 1992, 12, 13-14 (box), 340, 341 (table)
Eckholm, Erick, 134
Ecological cycles, 31-39
Ecological equivalents, *49*
Ecological niche, **49-51**
competition and, 50, *51,* 53
Ecological pyramids, 31, *32*
Ecological succession, **55-57**
Ecology, 21, **27-31**
shallow, deep, and social views of, 12, 322

Economic development, 16, **92**
  conservation and, 177
  demographic transition and, 80-82
  effect of growing human population
    on, 97-98
  limits to, 99-101
  natural resource use and level of,
    95-96
  supply and demand curves at three
    levels of, *97*
  sustainable, 12, 107, 108 (table),
    334-35
  World Bank and international, 104-6
Economic growth, **92**
Economics, 91-111
  as context, 91-93
  economic development and
    resource use, 95-96
  limits to economic and population
    growth, 99-101
  population, technology, and
    resource scarcity, 96-99
  resource economics, 101-7
  resources, reserves, and, 93-95
  sustainable development, 107-8
Ecosystem, **27-29**
  disruption of, 188-89
  food chains and food webs in, *29,
    30*
  management of, 199
  open and closed, 46-47
  productivity of, 29, 42-43, *44*
  trophic levels of, 30-31
Ecotone, **46,** *47*
Ecotourism, **187**-88 (box)
Ehrlich, Paul, 99, 134
Eisenhower, Dwight, 255
Electromagnetic fields (EMF), health
    hazards of, 119-20 (box)
Electromagnetic spectrum, *26*
Electrons, 22, *23*
Electrostatic precipitators, **220**
Elements, 22
Elephant, *177*, 192
Emigration, population demographics,
    and, 76-77
Emphysema, **215,** *216*
Endangered species, 177, *188*, 192,
    197-98
  in Hawaii, *199*
  recovery programs for, 198
Endangered Species Act of 1973, 169,
    197-98
Energy, **24**
  capture of, by photosynthesis, 26,
    27
  cellular respiration and release of, 27
  nutritional requirements for human,
    137-38
  solar, 25-26
  thermodynamics and transfers of, 25

types and quality of, 24
  units of, 249 (table)
Energy pyramid, 31, *32*
Energy recovery, **280**
Energy resources, 248-74
  agricultural uses of, 148, 149 (table)
  biomass as, 268-69
  coal as, 251-52
  conservation of, 262-63,
    264-65 (box)
  from fuelwood, 163
  hydropower as, 269-70
  major sources of, 248-49
  major uses of, 250
  natural gas as, 253-55
  nuclear power as, 255-62
  oil as, 252-53
  per capita consumption of, 249
  personal plan for saving, 265 (box)
  solar, 263-67
  from waste, 284-85
  wind as, 270-71
Environment, **3**
  natural, technological, social, and
    cultural components of, *3*
  poverty and degradation of, 15,
    333-34
Environmental campaign, organizing,
    329 (table)
Environmental Defense Fund, 331
Environmental ethics, 319-22, 333
Environmental future, 16-18, 319-44
  city planning and, 311-13
  collective actions, 327-33
  communal cooperation and social
    justice in, 17
  environmental restoration in, 321
    (box)
  ethics and philosophies for, 319-22
  global issues in, 333-36
  government and politics in, 336-41
  green consumerism and, 322-27
  human population in, 86
  introduction to, 319
  optimism on, 16-17
  pessimism on, 16
Environmental impact statements (EIS),
    339
Environmental indicators, 48
Environmentalism, 11-12. *See also*
    Collective action
  anti-environmental backlash, 333
  deep, shallow, and social, 322
  green consumerism, 322-27
  personal perspective on, 18 (box)
  student opportunities in, 349-51
Environmental organizations, 11, 12,
    322, 352-54. *See also* Collective
    action
  mainline, 331
  nongovernmental, 335-36

radical, 331-33
  student, 328
Environmental problems, 4-5
Environmental protection
  jobs in, 106-7
  market-based incentives for, 101-2
    (box)
  public support for, 4, *5*
Environmental resistance, population,
    and, 66
Environmental restoration, **320,**
    321 (box)
Environmental science, introduction to,
    2-20
  attitudes toward nature, 9-10, 12
  central questions of, 4
  conservation and environmentalism,
    10-12
  critical thinking and, 4-5 (box)
  Earth Summit of 1992, 12, 13-14
    (box)
  environment, components of, *3*
  environmental futures, 16-18
    environmental problems and, 4-5
  personal environmental perspective,
    18 (box)
  poor vs. rich and, 12-16
  resource use, technology, and
    development, 6-9
  scientific method and, 7-8 (box)
Environmental standards, setting,
    128-30
Enzymes, **24**
Equilibrium communities, 57
Erosion, **144**-47
  in select river basins, 146 (table)
  in United States, 145-46
  worldwide, 146-47
Ethics, environmental, 319-22, 333
Ethiopia, 105
  soil erosion in, 146
Eurasian water milfoil, 195-96
Europe
  acid precipitation effects on forests
    of, 218, *219*
  air pollution in eastern, 222-23
  famine and plague of 14th century
    in, 17, 68
Evolution, **42,** 48-49, 50 (box)
  coevolution, 52
Executive branch, United States
  government, 339, *340*
Exhaustible resources, **93**
Existence value, **188**
Exotic species, 57, 194, 195-96 (box)
Exponential growth of population, 65
Exports of waste to Third World, 276,
    280
External costs, **101**
Extinction of species, **183,** 188-90
  current rate of, 189-90

human-caused, 189, 190-97
mass, 189
natural causes of, 189
species endangered or threatened
by, *177, 188,* 192, *197-98*
Stone Age hunters and, 6
Extractive forest reserve, 166

F

Family planning, **82-86**
in China, 76-77 (box)
current methods of, 82, *83, 84*
new developments in, 83, *85*
traditional, 82
Famine, **135,** 136 (box)
Fecundity, **71**
Fertility, **71**
birth rates and, 71-72
control of, 82-86
environmental factors affecting
species', 49
factors affecting, 77, *78*
Fertilizers, agricultural, 147-48
Fibrosis, **216**
Filariasis, 115
Filters, **220**
Financial institutions and development
microlending at Grammeen Bank,
105-6 (box)
World Bank and large-scale projects,
104-6, 166
Finland, city planning in, 312 (box)
Fire-climax communities, 57
First law of thermodynamics, **25**
First World, **15,** 16
fossil fuel consumption in, 249
north/south division between poor
and, 12, 14
quality of life indicators for, 15
(table)
rural and urban population
distribution in, *300*
ten richest countries in, 14
urban problems of, 307
Fish
deforestation and threats to, 170
as food source, 140, 185
laws regulating harvesting of, 197
overharvesting of, 191-92
Flue gas desulfurization, 221
Fluidized bed combustion, 220-21
Food, 134-42
biological resources and, 185
chronic shortages of, 136-37
daily caloric input, worldwide, 137
(table)
energy requirements for producing,
148, 149 (table)
famine and lack of, 135, 136 (box)

greenhouse effect and production
of, 207
Green Revolution and increased
production of, 150
international trade in, 141-42
nutritional requirements, 137-39
overharvesting sources of, 190-91
personal production of, 156 (box)
production of, 134-35, 140-41 (*see
also* Agriculture)
wild animals as source of, 171-72
world resources of, 139-41
Food, Agriculture, and Conservation
Act of 1990, U.S., 168
Food aid, **142**
Food and Drug Act, U.S., 128
Delaney Clause, 126
Food chain, **29,** *30*
Food guide pyramid, *137*
Food web, **29,** *30*
Foreign aid
for conservation, 168
food aid, 142
Foreman, Dave, 333
Forest(s), 161-70
as carbon sinks, 35-36
conifer, 59, *60*
distribution of, 161, *162*
effects of acid precipitation on, 217,
219
management of, 163-64
northern, 168-70
old-growth ancient, *160,* 168-69
products from, 162-63, 166, 185-86
temperate deciduous, 59
tropical, 60, 164-68 (*see also*
Tropical forests)
vegetation zones of world's, *162*
Forest management, **163-64**
Fossil fuels, **248-49**
coal, 251-52
natural gas, 253-55
oil, 252-53
U.S. and Canadian deposits of, *254*
Founder bottleneck, **201**
Fourth World, 15
Freeways, 308-9, *310*
Fresh water, 232-34. *See also* Drinking
water
depletion of groundwater supplies
of, 232, 233 (box)
increasing supplies of, 234-37
replenishing supplies of, 228-29
shortages of, 232
subsidence and saltwater infiltration
of, 234
Frontier economy and level of resource
use, 95
Fuelwood, **163,** 249, *268*
Fugitive emissions, **209**
Fungicides, 155

Fur industry, *192*
Future. *See* Environmental future

G

Gaia hypothesis, 28
Gandhian sufficiency, **17**
Gap analysis, **199**
Garden cities, **311,** 312 (box)
Gardening, 156 (box)
Gasohol, **269**
Gastrointestinal infections, 113
Genetic assimilation, **196-97**
Georgescu-Roegen, Nicholas, 93
Germany, Blue Angel symbols in, 326
Giardia protozoan, *113*
Gila National Forest, New Mexico, 175
Gleason, H. A., 28
Global citizenship, 12
Global Environmental Monitoring
System (GEMS), 223
Glossary, 355-71
GOBI child survival program, 113-14
(box)
Goode's Homolosine Equal Area
Projection map, *346-47*
Gottlieb, 333
Government policy. *See* Politics,
environmental; Public policy
Gradient of diversity, 43
Grammeen Bank, microlending at,
105-6 (box)
Grasslands, **58,** *59*
Great Britain, nuclear power in, 256
Great Lakes, threat of introduced
species into, 195-96 (box)
Green consumerism, 322-27
jobs and, 325-26 (box)
limits of, 326-27
personal responsibility and, 327
products of, 323-24
questioning consumerism, 322-23
seals for, 326
Greenhouse effect, **206**
countries contributing most to, *208*
gases responsible for, *207*
global climate change due to, 207-8
(box)
species extinction due to, 194
Greenhouse gases, 35, *36*
Greenpeace, 276, 280, 333, *335*
Green political party, 336
Green Revolution, **150**
Green Seal program, *326*
Gross domestic product (GDP), 96
Gross national product (GNP), 96
percent of world population and,
100
Ground cover, 152-53
Groundwater, 35, **228**
depleting supplies of, 232, 233 (box)

pollution of, *244-45*, 278, 290
radon pollution in, 215
Growth monitoring of children, 113 (box)
Guanacaste National Park, Costa Rica, 167-68 (box)
Guest workers, 76

H

Habitat, **49**
corridors, *200*
extinction due to loss of, 189-90, 194
protection of, 199-201
restoration of, for Bermuda cahow, 321 (box)
in tropical rainforests, 46-47 (box)
Haiti, 146-47, 178
Halogens as air pollutants, 210, 211-12 (box)
Hardin, Garret, 81
Hawaii, endangered species in, *199*
Hazardous, defined, **116**
Hazardous waste, 286-94
air pollution and, *287*
disposal of, 288-89
disposal sites in United States, *288*
fate of, in United States, *289*
household, 292-93 (box)
legally defining, **287-88**
Love Canal, New York, 290-91 (box)
management options for, 290-94
warehousing and illegal dumping of, 289-90
water pollution and, 242, 244
Health, **112-32**. *See also* Disease
assessment and acceptance of risks to, 127-28
chemicals as hazard to, 116-18
children, survival of, and, 113-14 (box)
diet and, 119, 120, 121 (table)
effects of air pollution on, 215-16
electromagnetic fields as hazard to, 119-20
infectious organisms as hazard to, 112-16
major environmentally-related problems of, 113 (table)
noise as hazard to, 308 (box)
nuclear radiation and, 259 (box)
pesticides and problems of, 155
physical agents, trauma, and stress as hazard to, 119-20
public policy and environmental standards for, 128-30
toxicity, measuring, 124-27
toxins, minimixing effects of, 123-24

toxins, movement, distribution, and fate of, 122-23
toxins, natural and synthetic, as hazards to, 119
ultraviolet radiation from ozone depletion and, 211, 212
water resources and, 239, 243-44
Heat, **24**
Heller, Walter, 91
Herbicides, 155
Herbivores, **30**
High-quality energy, **24**
High-Temperature, Gas-Cooled Reactor (HTGCR), 256-57
Hightower, Jim, 227
Historic Roots of Our Ecological Crisis, The (White), 9
Homelessness, 307
Homeostasis, **28**
Hooker Chemical and Plastics Corporation, 290-91 (box)
Household hazardous wastes, managing, 292-93 (box)
Housing
energy conservation in, 262-63, 264-65 (box)
as First World problem, 307
radon pollution in, 214, *215*
as Third World urban problem, 304-6
toxic chemicals in, 117
Howard, Ebenezer, 311
Human activities, 6-9
agricultural revolution, 6-7
biological losses due to, 190-97
industrial revolution, 7-9
species extinction caused by, 189
tool-making revolution, 6
Human health. *See* Health
Human population, 64-89
in cities (*see* Cities)
economic development and demographic transition in, 80-82
effects of, on economic development and technology advances, 97-98
family planning and fertility control of, 76-77 (box), 82-86
food resources required by, 148-50
future of, 86
gross national product and percent of, *100*
growth of, 69-71, 77-80 (*see also* Population growth, human)
history of, 66-69
human demography and, 71-77
introduction to, 64-65
population biology and, 65-66, 67-68 (box)
Human resources, **93**

Hunger, 135-39
Hunting and gathering, 6
laws regulating, 197
overhunting, 190-91
Hurricanes, environmental restoration, and, 321 (box)
Hydrocarbons, 212-13
controlling emissions of, 221-22
Hydrologic cycle, **32**, *34, 35*, 228
Hydropower, 249, 269-70

I

Immigration push and pull factors, 302
Immune system, 117
Immunization programs for children, 114 (box)
Incineration of wastes, 280-81, 291-92
India
land tenure in, 178
Narmada River dams, 105, 270
population growth and family size in, 73-74 (box)
urban problems in, *303,* 304, 305, *306*
water resources in, *232,* 243
Indicator species, 48
lichens as, 54
Indoor air pollution, 213-15
Industrial and commercial products from biological resources, 162-63, 166, 185-86
Industrial economy and level of resource use, 95-96
Industrial Revolution, 7-9
Industrial timber, **162**
Industry
energy use by, *250*
hazardous waste produced by, 287, 290-91 (box)
solid waste produced by, 277
water conservation by, 238
water pollution caused by, 240, 241 (box)
water use by, 231
Infanticide, 77 (box)
Infant mortality, 81-82
population growth and, 77-78, 114
Influenza, 115, 116
Injection wells, 289
Insecticides, 155
Insects
biotic potential of, 65
disease carried by, 114-15
Insolation, **206**
Instituto Nacional de Biodiversidad (INBIO), 186
Intangible assets, distributing, 104
Intangible resources, **94**
Integrated pest management (IPM), **153,** 156

Inter-American Development Bank (IDB), 166
Intergenerational justice, 101-3, **334**
Internal costs, **101**
Internalizing costs, **101**
International trade, 106
    in animal and plant resources, 192-93
    in foods, 141-42
    in ivory, 192, 199
International Union for the Conservation of Nature and Natural Resources (IUCN), 176, 177 (table)
Interspecific competition, 50, 52-53
Intraspecific competition, 52-53
Introduced species, 57, 194, 195-96 (box)
Iodine deficiency, 138-39
Irrigation, 147
    center-pivot sprinklers, *231, 233*
Irritants, **116**
Isle Royale National Park, wolves and moose populations on, 67-68 (box)
Ivory trade, 192, 199

J

Japan
    city planning in, 313
    Minamata mercury poisoning in, 241 (box)
    recycling in, 282
Jobs in environmental protection, 106-7, 325-26 (box)
Justice. *See* Social justice

K

Kennedy, John F., 92
Kenya, ecotourism in, 187
Keystone species, 52
Khana study on population growth, 73-74 (box)
Kinetic energy, **24**
Krill, *184*
Kropotkin, Peter, 322
Kuwait, 252
Kwashiorkor, **138**

L

Landfills, **279-80**
    of hazardous waste, 289, 290-91 (box)
    secured, for hazardous waste, 293, *294*
Land reform, 178-79
Land resources, 160-82
    characteristics of world, 161
    decisions about, 160

forests, 161-70 (*see also* Forest(s))
    national parks, 174-75
    ownership of, and land reform, 178-79
    rangelands, 170-74
    wilderness areas, 175-76
    world parks and preserves, 176-77
Land use
    categories of, *161*
    conversion of farm land to non-agricultural purposes, 144
    decisions about, 160, 308-9
    increasing, for agricultural purposes, 149-50
Larmarck, Jean Baptiste de, 50
Law of diminishing returns, **98**
LD50 dosage, **125**
Lead poisoning, 196, 240
Legislation, United States environmental, 336-37
    Clean Air Act of 1990, 102, 209, 213
    Clean Water Act of 1972, 242, 245
    Endangered Species Act of 1973, 169, 197-98
    Food, Agriculture, and Conservation Act of 1990, U.S., 168
    Food and Drug Act, U.S., 126, 128
    path of bill becoming law, *337*
    Superfund Act of 1980, 288
    Wilderness Act of 1964, 175
L'Enfant, Pierre, 311
Leopold, Aldo, 17, 160, 175
Less-developed countries (LDCS), 16. *See also* Third World
Lichens, 54 (box)
Life expectancy, **73**
    demographic implications of increased, 74-76
    in select world regions, *74*
Life span, **73**
Lighting, energy conservation in, *264*
Light reactions, photosynthetic, 26, *27*
Limiting factors, **48**
Limits to Growth (Meadows), 99-100
Lipids, 23
Livestock production, 139-40, 170-74
    overgrazing and desertification, 170-71
    in tropical forests, *164,* 165-66 (box)
Living organism(s), 21-41. *See also* Biological resources
    biological studies of, 21, *22*
    cells as fundamental units of, 23-24
    Earth as, 28
    energy and, 24, 25-27
    material cycles and life processes of, 31-39
    matter and, 22-23, 24-25
    species, communities, and ecosystems of, 27-31

temperature range of, 25
Los Angeles, California, as urban area, 309-10
Love Canal, New York, hazardous waste disaster of, 290-91 (box)
Lovelock, James, 28
Low-head hydropower, **270**
Low-quality energy, **24**

M

MacArthur, Robert, 44
Malaria, 114-15
Malaysia, water resources in, 243
Malignant tumors, **118**
Malnourishment, 113, 136-37, **138**
Malthus, Thomas, on population growth, 16, 69-70, 81, 98
Man and Biosphere (MAB) program, **201,** 202
Man and Nature (Marsh), 11
Mao Zedong, 76, 136
Marasmus, **138**
Marine ecosystems
    krill, *184*
    predation in, *52*
    saltwater wetlands, 60
Market equilibrium, **97**
Markets
    efficiencies of, and technological development, 97
    supply and demand in, 96-97
Marsh, George Perkins, 11
Marshall, Bob, 175
Marx, Karl, on population growth, 69, *70*
Mass burn, **280,** *281*
Mass transit, 310
Materials cycles, 31-39
    carbon cycle, 35-36
    nitrogen, 36-38
    phosphorus, 38-39
    water, 32-35
Matter
    basic forms of, 22-23
    conservation of, 24-25
Meadows, Donella, 99
Measurement units, metric and English conversions, 348
Meat as food source, 139-40
Medicinal drugs from biological resources, 185 (table), *186*
Megacity, **299,** *301*
Mendes, Francisco ''Chico,'' murder of, 165-66 (box)
Metabolism, **24**
    of toxic chemicals, 124
Metacognition, 4-5 (box)
Metals as air pollutant, 210
Methane, 207, 253-55, 268-69, 284
Methane digester, **268,** *269*

Metric/English conversions, 348
Mexico
    population demographics in, 74, *75*
    Sian Ka'an Reserve in, 201, 202
Mexico City
    air pollution in, 222
    population of, *64*, 300–301, *302*
    urban problems of, *306*
Meyers, Norman, 185
Micro-hydro generators, **270**
Microlending, 105–6 (box)
Migration
    famine and, 135
    human population demographics
        and, 76–77
    push and pull factors of, 302
    of species, 46
Milk as food source, 139–40
Mill, John Stuart, 93, 107
Minamata disease, 241 (box)
Minerals, nutritional requirements for,
    138–39
Mining
    coal, 251–52
    waste produced by, 277
    water pollution caused by, 240
Modernism, 320–22
Molecules, **22–23**
Monkey wrenching, **332–33**
Monoculture, soil erosion and, 146
Monoculture agroforesty, **164**
Moose, wolves and, on Isle Royale,
    67–68 (box)
Morbidity, **113**
More-developed countries (MDCs), 16.
    See also First World
Mortality, 72–73
    infant and child, 81–82
Mt. Pinatubo, 207
Muir, John, *11*
Municipal waste, 277, *280. See also*
    Solid waste
Murie, Adolph, 67
Mutagens, **117**, 125

N

Nashua River, New England, cleanup
    of, 329–30 (box)
National forests, United States, 11, 169,
    175
National parks, United States, 11
    current problems of, 174–75
    Isle Royale, 67–68 (box)
    origin of, 174
    Yellowstone, 199, *200*
    Yosemite, *174*, 175
National Priority List (NPL) of
    hazardous waste sites, 288–89
National Resources Defense Council,
    331

National Wilderness System, United
    States, 175–76
Natural gas, 220, 249, 253–55
    reserves of, 253, *255*
    unconventional sources of, 253, 255
    U.S. and Canadian deposits of *254*
Natural increase of population, **73**
Natural resources, **93**
    categories of, *94*, 95
    computer models of use of, 99, *100*
    conservation of, 100–101
    conservation of, movement for, 10,
        11
    consumption of, 10, 11 (table)
    economic context for use of, 91–93
    economic development and use of,
        95–96
    economics of, 101–7
    exhaustible, renewable, and
        intangible, 93–94
    population, technology, and scarcity
        of, 96–99
    soil as, 142–43
Natural selection, 42, 48–**49**, 50 (box)
Nature, attitudes toward, 9–10
Nature Conservancy, 331
Nature preserves, 199–202. *See also*
    Parks and preserves
Nematodes, 115
Neo-Malthusian thought, **16, 70,** 81
Neurotoxins, **117**
Neutrons, 22, *23*
Never-to-be-developed countries
    (NDCs), 16
Newhall, Nancy, 1, 90, 133, 275
Newly industrializing countries (NICs),
    16
New towns, **311**
Nitrogen cycle, **36**, *37*, 38
Nitrogen oxides, *209*, **210**
    control of, 221
Noise
    as health hazard, 119, 308–9 (box)
    sources and effects of, 309 (table)
Noncriteria air pollutants, 213
Nongovernmental organizations
    (NGOs), **335–36**
Nonpoint sources of water pollution,
    239, 242
Norplant birth control, 83, *85*
Northern spotted owl, 198
North vs. South division of wealth and
    power, 12–15
Nuclear power, 220, 255–60
    alternative reactor designs, 256–57
    breeder reactors, 257–60
    Chernobyl disaster, 258–59 (box)
    citizen protests against, 261–62
    managing radioactive wastes from,
        260, *261*

reactor fuel, 255–56
    types of reactors in use, 256
Nuclear radiation, 25, 54
Nucleic acids, 23
Numbers pyramid, 31, *32*
Nutritional requirements, human,
    137–39
    energy needs as, 137–38
    food guide pyramid as, *137*
    minerals, 138–39
    per capita daily caloric input,
        worldwide, 137 (table)
    proteins, 138
    vitamins, 139

O

Oceans
    as carbon sinks, 35–36
    dumping waste in, 276, 278–79
    fish (*see* Fish)
    greenhouse effect and rising levels
        of, 207
    oil spills in, 239, 253
    pollution in, 239
Ogallala Aquifer, 233 (box)
Oil, 35, 248, 252–53
    ocean spills of, 239, 253
    resources and reserves of, *252*–53
    unconventional, 253
    waste, 278
Old-growth forests, *160,* 168–69
Omnivores, **30**
Onchocerciasis (river blindness), 115,
    116
Only One Earth (Dubos, Ward), 13
Open canopy forest, **161**
Open community, 46, *47*
Open dumps, 278
    of hazardous waste, 289, 290–91
        (box)
Open range, **170**
Oral rehydration therapy (ORT),
    113–14 (box)
Organic compounds, **23**
Organisms. *See* Living organism(s)
Organization for Economic
    Cooperation and Development
    (OECD), 15
Our Common Future, 12, 13
Overgrazing, 170–71, 173–74
Overharvesting, 190–91
Overhunting, 190
Overnutrition, 138
Overshoot carrying capacity, **65**
Oxygen-demanding wastes in water,
    239, *240*
Ozone, **213**
Ozone shield, 26
    depletion of, 211–12 (box)

P

Paper products vs. plastic products, 327
Parabolic mirrors, **264-65**, *266*
Parasites, **51**, 53, 115
Parks and preserves, 176-77, 199-202
  national parks in Costa Rica, 167-68 (box)
  national parks in United States, 11, 67-68 (box), 174-75
  wildlife habitat protection in, 199-201
Particulate material, *209*, **212**
  removal techniques for, 220
Passive heat absorption, **263**
Pasture, **170**
Patchiness, **45**
Pathogens, **51**
  human health and, 113-16
  as predators, 66, 194-96
Pellagra, **139**
Permanent retrievable storage of hazardous waste, **293**
Persistence of toxic chemicals, 123
Pesticides, **154-55** (box). *See also* DDT
  in diet, 119
  pests controlled by, 115, 155
  pollution from, 154, 155, 196
  resistance to, 155
Pests
  biological control of, 186
  extinctions due to control of, 193
  integrated pest management for control of, **153,** 156
Pets, 193
pH, **33**
Philosophies, environmental, 319-22
Phosphorus cycle, 38, *39*
Photochemical oxidants, **213**
Photodegradable plastics, **286**
Photosynthesis, *23,* **25-26**
  energy capture by, 26, *27*
Photosynthetic efficiency, **43**
Photovoltaic cells, **265,** *266,* 267
Physical agents as health hazards, 119-20
Physical treatments of hazardous substances, 291
Pinchot, Gilbert, 11
Pioneer species, **55**
Plankton, **52**
Planned communities, 312 (box)
Plants. *See also* Species
  areas important to species diversity of, *200*
  competition and, *53*
  diseases of, 194, 196
  effects of air pollution on, 216-17
  as food source, 185
  introduced species of, 194
  native, as food source, 150

overharvesting wild, 192
photosynthesis in, *23*
Plastics, 286
  paper vs., 327
Plutonium, 258, 260
Point sources of water pollution, **239**
Political economies of world nations, 15-16
Politics, environmental, 336-41. *See also* Collective action; Environmentalism; Environmental organizations
  courts and, 337-39
  environmental impact statements and, 339
  executive branch and, 339, *340*
  green political parties, 336
  influencing national legislation, 336-37
  international treaties and conventions, 339-41
Pollution. *See also* Air pollution; Hazardous waste; Water pollution
  biological resource losses caused by, 196
  charges for units of, 101-2 (box)
Poorer industrialized nations (PINs), 16. *See also* Second World
Poor nations, 14-15. *See also* Third World
Population, **27.** *See also* Human population
Population biology, 65-66
  biotic potential, 65
  density-dependent and density-independent factors, 66
  growth to stable population, 65, *66*
  mortality and survivorship, 66, *68*
  oscillations and diebacks, 65
  predation and, 66, 67-68 (box)
Population crash, 65, 86
Population growth, human. *See also* Human population
  benefits of, 71
  control of, 82-86
  demographic factors and, 71-77
  demographic transition and, 80-82
  doubling times and, 68 (table)
  effects of, on economic development and technology advances, 97-98
  factors affecting, 77-80
  food production and, in sub-Sahara Africa, *141*
  history of, 66, 68, *69*
  in India, 73-74 (box)
  infant and child survival and, 77-78, 114 (box)
  limits to, 99-101
  Malthusian checks on, 69-70
  Marxist views on, 69, *70*

projected, *71*
  rates of, 72, 73
  technology and, 70-71
Population hurdle, **98**
Population momentum, **75**
Postindustrial economy and level of resource use, 96
Potential energy, **24**
Poverty
  environmental degradation and, 15, 333-34
  ten poorest countries, 14
Power, 24
Precipitation, 35, 228
  acidic, 217-18
  biome type by, *58*
  greenhouse effect and patterns of, 207
  patterns of world, *229*
Predation, 51-52
  population biology and, 66, 67-68 (box)
Predator, **51**
  human-caused extinctions of, 193
  pathogen as, 66
Pressurized water reactor (PWR), **256**
Price elasticity, **97**
Price mechanisms for conserving water, 238
Primary air pollutants, **209**
Primary succession, **55,** *56*
Process-Inherent Ultimate-Safety (PIUS) nuclear reactor, 256, *257*
Producers in ecosystems, 30
Production frontier, **92**
Productivity, **29,** 42-43
  gross primary, in different ecosystems, *44*
Products
  from animals, 192
  green, 323-24
  life cycle analysis, 326, *327*
  plastic vs. paper, 327
  water used in production of, 231 (table)
Promoters (carcinogens), **118**
Pronatalist pressures, 77-78
Proteins, 23. *See also* Enzymes
  nutritional requirements for, 138
Protons, 22, *23*
Public opinion, using media to influence, 330 (table)
Public policy
  establishing environmental standards and, 128-30
  land ownership and, 178-79
  on urbanization, 302-3
Pull factors, immigration, **302**
Purple loosestrife, 196
Push factors, immigration, **302**

R

Radiation
    as health hazard, 117, 119, 259
        (box)
    solar, 205, *206*
    ultraviolet, 25, *26*, 211-12 (box)
Radioactive waste, 260, *261*
Rado, James, 205
Radon gas as indoor air pollutant,
    214-15 (box)
Ragni, Gerome, 205
Rainforest. *See* Tropical forests
Rainwater. *See* Precipitation
Rangelands, 170-73
    buffalo commons on U.S., 172-73
        (box)
    livestock forage from, 171
    management of, 170
    overgrazing and desertification of,
        170-71, 173-74
    in United States, 173-74
    wildlife on, 171-72
Recently industrialized nations (RICs),
    16
Recharge zones, **228,** 245
Recycling, 107, **282-83**
    benefits of, 282-83, 291
    creating incentives for, 283
Reduced tillage systems, **153**
Refuse-derived fuel, **280**
Regenerative agriculture, 151
Rehabilitation of land, **167**
Reilly, William, 14
Renewable resources, **93,** *94*
    soil as, 142-43
Reserves of natural resource, 95
    coal, *251*
Resettlement projects for rural poor,
    179
Resource economics, 101-7
    cost/benefit ratios, 103-4
    discount rates and intergenerational
        justice, 101-3
    distribution of intangible assets, 104
    internal and external costs and, 101
    international economic development
        and, 104-6
    international trade and, 106
    jobs in environmental protection
        and, 106-7
    market-based incentives for
        environmental protection, 101-2
        (box)
Resource partitioning, 49, *51*
Resources. *See* Human resources;
    Natural resources
Respiratory diseases, 115, 116-17
Respiratory fibrotic agents, **116-**17
Restoration of ecosystems, **167**

Rich vs. poor nations, 14-15
Risk
    accepting, 127-28
    assessing, 127
    relative perception of various, *129*
    of select activities, 128 (table)
Rivers
    deforestation and degradation
        of, 165
    soil erosion and, 147 (table)
    water transfer projects with, 235
Roosevelt, Theodore, *11*
RU486 birth control drug, 85-86
Ruffe perch, 195
Run-of-the-river flow, **270**
Rural areas, **299**

S

Saccharin, risk of, 128-29
Salinization, **147,** 240
Salt as water pollutant, 240
Saltwater infiltration, 234
Samuels, Mike, 112
Saudi Arabia, 252
Scarcity of resources, factors
    mitigating, 98
Schistosomiasis, 115, 237
Science and scientific method, 7-8
    (box)
Scientific research, animals used in,
    124-25, 193
Scott, J. Michael, 199
Seafood, 140. *See also* Fish
Sea lamprey, 195
Sea Shepherd, 331, 333
Secondary air pollutants, **209**
Secondary succession, **55-56**
Second law of thermodynamics, **25**
Second World, **15**
Secured landfills, 293, **294**
Sessions, George, 12
Sewage systems, 242
    as urban problem, 303-4
Sexually-transmitted disease, 116
    prevention of, 85
Shallow ecology, **12, 322**
Shanty towns, **306**
Sierra Club, 11, 331
Silent Spring, 11, *12*
Sinkholes, **234**
Slums, **305,** *306*
Smith, Robert Angus, 217
Smog, 209, 213
Smoking, 213, 215
Social ecology, **322**
Socialized countries. *See* Second World
Social justice, **81**
    environmentalism and, 17
    intergenerational, 101-3
    population and, 81

Soil, 134, **142-47**
    conservation of, 152-53
    erosion and degradation of, 144-47
    organisms in, 143
    profiles of, 143, *144*
    as renewable resource, 142-43
    waterlogging and salinization of, 147
Soil Conservation Service, 146
Soil horizons, **143,** *144*
Solar energy, 25, 205, *206,* 249, 263-67
    active solar heat systems, 263, *266*
    high-temperature solar energy
        collection, 264-65
    passive solar heat collectors, 263
    photovoltaic solar energy
        conversion, 265-67
    storing, 267
Solid waste, 276-86
    composting, 284-85 (box)
    disposal methods, 278-80
    energy from, 284-85
    exporting, 280
    incineration of, 280-81
    producing less, 286
    recycling to recover resources from,
        282-83
    shrinking waste stream, 285-86
    waste stream, 277-78
Solubility of toxic chemicals, 122-23
South vs. North division of wealth and
    power, 12-15
Soviet Union
    Aral Sea destruction in, 235-36
        (box)
    Chernobyl nuclear disaster in,
        258-59 (box)
    nuclear reactor design in, 256
Species, **27.** *See also* Biological
    community
    ecological equivalents of, *49*
    extinction of (*see* Extinction of
        species)
    genetic assimilation of, 196-97
    human-caused threats to, 190-97
    indicator, 48, 54
    interaction of, 51-55
    introduced, 57, 194, 195-96 (box)
    keystone, 52
    migration of, 46
    number of living, 184 (table)-85
    pioneer, 55
    spatial distribution of, 45-46, *47*
Species diversity, **43.** *See also*
    Biological resources
    areas important to plant, *200*
    map of, in Hawaii, *199*
    in old-growth forests, *160*
    in tropical forests, 164
Species recovery plan, **198**
Spiny water flea, 195
Squatter towns, **306**

S-shaped population curve, *66*
Stable runoff, **230**
Steady-state economy, **92-93**
Stevenson, Adlai E., II, 2
Stewardship, **9**, 17
  agriculture and, *5*
  as ethic and philosophy, 320
Stoddard, Marion, 329-30 (box)
Stress
  as health hazard, **120**
  as urban problem, 304-5 (box)
Strip-mining, **252**
Strong, Maurice, 13, 334
Structure of biological community, **45-47**
Subatomic particles, 22, *23*
Subsidence, **234**
Subsoil, **144**
Suburbs, redesign of, 313-15
Succession, ecological, **55-57**
Sulfur dioxide, *209*, **210**
  removal techniques for, 220-21
Superfund Act of 1980, 288
Superfund hazardous waste sites, *288-89*
Supply, 96-97
Supply/demand curve, **97**
  at three levels of economic development, *97*
Surface mining, **252**
Surface water, 35, 242. *See also* Water resources
Survivorship, **66**, *68*
Sustainable agriculture, **150-56**
  integrated pest management as, 153-56
  pesticides and, 154-55 (box)
  reduced tillage systems as, 153
  soil conservation as, 152-53
  Thompson family farm in Iowa as example of, 151 (box)
Sustainable development, **12, 107, 334**
  achieving goals of, 334, 335 (table)
  intergenerational justice and, 334
  strategies for, 108 (table)
  of Third World cities, 315-16
Sustainable forestry, 166
Sweden
  city planning in, 312 (box)
  population demography in, *75*
Symbiosis, **53-55**
  lichens as example of, 54 (box)
  nitrogen-fixing bacteria as example of, 36-37, **38**

T

Technological development and market efficiencies, 97
Technological optimists, **16-17**
Technology

benefits of, and optimism for future, 16-17
  environmental future and, 320-22
  population growth and, 70-71
  transfer of, 334, 335 (table)
Technopolis, **313**
Temperate deciduous forests, **59**
Temperature, **24**
  biome types and, *58*
Temperature range of living organisms, 25
Teratogens, **117**, *118*, 125
Terracing, **152**
Terrestrial ecosystems
  food chains in, *29, 30*
  succession in, 55, *56*
Territoriality, **53**
Thermodynamics, laws of, 25
Third World, **15**, 16
  dumping solid and hazardous waste in, 276, 280
  energy sources in, 163, 249
  north/south division between wealthy nations and, 12, 14
  population growth in, *71, 72,* 73-74 (box)
  quality of life indicators for, 15 (table)
  as resource suppliers, 106
  rural and urban population distribution in, *300*
  sustainable development of (*see* Sustainable development)
  sustainable development of cities in, 315-16
  ten economically poorest nations of, 14
  urban problems of, 303-6
Thomas, Lewis, 21
Thompson, Dick and Sharon, 151 (box)
Threatened species, **197**
Tillage systems, 153
Timber industry, 162, 168, 169-70
Time as ecological factor, 50
Tinnitus, 308
Tolerance limits, **48**
Tool-making revolution, **6**
Topsoil, **143,** *144. See also* Erosion
Total fertility rate, **72**
Total growth rate of population, **73**
Tourism, conservation and, 176, 177, 187-88 (box)
Toxic chemicals, 116-18. *See also* Hazardous waste
  air pollution due to, *287*
  bioaccumulation and biomagnification of, 123
  entrance routes of, into body, 122, *123*
  factors in environmental toxicity, 122 (table)

household, 292-93 (box)
  measuring toxicity of, 124-27
  minimizing toxic effects of, 123-24
  natural and synthetic, 119
  persistence of, 123
  solubility of, 122-23
  as water pollutants, 240-42
Toxicity
  acute vs. chronic doses and effects of, 126
  animal testing for measuring, 124-25
  detection limits, 126-27
  mechanisms for minimizing, 123-24
  ratings of, 125
Toxic waste. *See* Hazardous waste
Toxins, **116.** *See also* Toxic chemicals
Trachoma, 115
Trade. *See* International trade
Traffic congestion as urban problem, 303
Transportation. *See also* Automobile
  city growth and, 307-10
  of hazardous waste, 294
Trauma as health hazard, **120**
Treaties and conventions, international, 339-41
  Earth Summit, 13-14 (box), 341 (table)
Trophic levels in ecosystems, 30-31
Tropical Forestry Action Plan, 167
Tropical forests, 60
  destruction of, 164-65
  dry, 60, 167-68 (box)
  murder of Chico Mendes in Amazonian, 165-66 (box)
  protection of, 167-68
  rainforest, 45, 46-47 (box), 60
Tubal ligation, *84*
Tuberculosis, 115
Turtle extruder device (TED), 192

U

Ukraine, Chernobyl nuclear disaster in, 258-59 (box)
Ultraviolet radiation, 25, *26,* 211-12 (box)
Unconventional "noncriteria" air pollutants, 213
Unconventional oil, **253**
Undernourishment, 136, **137**
United Nations Conference on Environment and Development. *See* Earth Summit of 1992
United Nations Environmental Conference of 1972, 13
United States
  acid precipitation over, *218*
  agriculture in, 134, 140
  air pollution in, *222, 287*

birth rates in, between 1910 and
1992, *79*
Buffalo Commons in, 172-73 (box)
cancer in, *118*
cleanup of Nashua River in, 329-30
(box)
composition of domestic waste in,
277
consumption and waste by, 11
(table)
drought in, 230
energy use in, *250*
executive branch of government,
339, *340*
fossil fuel resources in, 253, *254*
Green Seal program in, *326*
hazardous waste in, *288, 289*
hunger in, *135, 136*
judicial branch of government,
337-39
laws protecting biological resources
in, 197-98
laws protecting water resources in,
245
legislation (*see* Legislation, United
States environmental)
legislative branch of government,
336, *337*
life expectancy in, by race and
gender, *75*
materials recovery from solid waste
in, *283*
national parks, forests, and
wilderness in, 174-76
nuclear reactors in, 256
old-growth forests in, 168-70
population demography in, *75*
potential areas of radon pollution
in, *214*
rangelands in, 173-74
response of, to Earth Summit,
13-14 (box)
soil erosion in, 145-46
suburb redesign in, 313-15
sustainable agriculture in, 151 (box)
timber industry in, 162, *163*
types of air pollutants in, *209*
urban crime in, 304 (table)
urbanization in, 299, *300*
urban problems of, 306, 307
water transfer projects in, 235
water use in, 231
Unwin, Raymond, 311
Uranium, 214, 255-56
Urban area, **299.** *See also* Cities
Urbanization, 298, **299**

categories of, 299
worldwide, 299-302

V

Vasectomy, *84*
Vertical city, **313**
Vertical stratification, **50**
Village, **299**
Violence
as health hazard, 120
as urban problem, 304-5 (box)
Vitamins, nutritional requirements for,
139
Volatile organic compounds (VOC),
209, **212**-13
control of, 221-22
Volcanic activity, 207, 217
Vulnerable species, **197**

W

Wallace, Alfred, 49
Ward, Barbara, 12, 13
Warehousing and dumping of
hazardous waste, 289-90
Washington, D.C., city plan for, *311*
Waste. *See* Hazardous waste;
Radioactive waste; Solid waste
Waste lagoons, 289
Waste stream, **277**-78
shrinking, 285-86
Waste-to-energy, **280**
Water, 227-38. *See also* Water
pollution
as agricultural resource, 147
availability of, 229-30
compound of, 22
conservation of, 237-38
fresh, 228-29, 232-34
groundwater (*see* Groundwater)
hydrologic cycle, 32-35, 228
increasing supplies of, 234-37
major compartments of, 228
physical forms of, 22
potential and kinetic energy in, *24*
properties of, and life on planet
Earth, 33-34 (box)
replenishing supplies of, 228-29
uses of, 230-31
Water conservation
individual's role in, 237 (box)
in industry and agriculture, 238
price mechanisms for, 238
Waterlogging, **147**
Water pollution, **238**-42
defined, **238**-39

disease caused by, 239
of drinking water, 148
in oceans, 239
oxygen-demanding waste as, 239,
240
point and nonpoint sources of, 239
salts, acids, and nonmetallic
pollutants as, 240-41
toxic inorganic pollutants as, 240,
241 (box)
toxic organic chemicals as, 241-42
as urban problem, 303-4
Water transfer projects, 233, 234-36
Watson, Paul, 333
Wealthy industrialized nations (WINs),
15-16. *See also* First World
ten richest, 14
Wetlands, *43, 60*
Whales, overharvesting of, 190, *191*
White, Lynn, Jr., 9
Wilderness, 169
in United States, **175**-76
Wilderness Act of 1964, 175
Wilderness Society, 175, 331
Wildlife
Buffalo Commons, 172-73 (box)
as food source, 171-72
overharvesting of, 190, 192
poaching of, *192*
protection of, 11, 169, 197-202
Wind energy, 249, 270-71
Wind farms, **271**
Wise-use movement, 333
Withdrawal of water resources, **230**
Wolves, moose and, on Isle Royale,
67-68 (box)
Woodland, **161**
World Bank, role of, in international
economic development projects,
104-6, 166
World Biosphere Reserves, 176, 201-2
World conservation strategy, **177**

Y

Yellowstone National Park,
biogeographical area of, 199, *200*
Yosemite National Park, *174,* 175
Yunus, Muhammad, 105

Z

Zebra mussel, 195
Zero population growth, **72**
Zero Population Growth, Inc., 64

GEOGRAPHICAL INDEX

A

Africa. *See also* names of countries in Africa
  agricultural fertilizer use in, 148
  disease in, 114, 115, *116*
  famine and hunger in, 135, 139
  forests of, 161, 164
  grazing livestock in, 171-72
  population growth and food production in sub-Sahara, *141*
  population growth in, 72
  poverty in, 14, *15*
  urban growth in, 302
Alaska, 176, 336
Amazon Basin, 46
  deforestation and murder in, 165-66 (box)
Antarctica, 57, 185
  coal in, 251
  international treaty on, 340
Aral Sea, destruction of, 235-36 (box)
Asia, 170, 302. *See also* names of countries in Asia
  agriculture in, 149, 150 (table), 152
  forests of, 164
Australia, 14, 15, 170, 193, 242
  hydropower in, *270*
  skin cancer in, 212
Austria, 218

B

Badlands National Park, U.S., *59*
Bahamas, 305
Bahrain, 229, 234
Bangladesh, 14, 114
  Grammeen Banks in, 105-6 (box)
Belize, 72
Benin, 176
Bermuda, environmental restoration in, 321 (box)
Bhutan, 14
Bolivia, 168, 171
Borneo, 46
Botswana, 105, 171, 176
Brazil, 16, 177, 193, 263
  agriculture in, 156
  biological reserves in, 201
  deforestation of, 164, 165-66 (box)
  Earth Summit in, 13-14 (box), *341*
  environmental cleanup in Cubatao, 223

  land tenure in, 178-79
  urban planning and conservation in Curitiba, 324 (box)
  water resources of, 229
Burma, 164
Butan, 176

C

Cambodia, 164
Cameroon, 164
Canada, 14
  acid rain precipitation in, *218*
  air pollution effects in, *217*
  Environmental Choice program in, *326*
  fishing areas of eastern, *191*
  forests of, 161, 162, 169-70, 218
  fossil fuel resources in, 253, *254*
  laws protecting biological resources in, 197-98
  nuclear reactors in, 256, 257
  population, 72, 73
  radioactive waste in, 260, *261*
  species diversity in, 197-98
  unconventional oil in, 253, *254*
Caribbean Islands, 57
Central Africa Republic, 176
Chad, 14
  famine in, 135
Chernobyl nuclear disaster, 54, 258-59 (box)
China, 14, 178, 316
  air pollution in, 222
  coal in, 251
  food supplies and famine in, 136 (box)
  forests of, 163
  hydropower in, 270
  as importer of wood products, 162
  nuclear power in, 261
  pandas in, *188*
  population growth in, 73
  population stabilization programs in, 76-77 (box)
  soil erosion in, 146, 147 (table)
  water quality in, 244
  water resources of, 229-30
  water transfer projects in, 235
Colombia
  bird species in, 43
  cash crops of, *142*

Colorado River, U.S., 235, 236
Costa Rica, 72, 114, 168, 176
  agriculture in, 156
  deforestation in, *164*
  dry tropical forest restoration in, 167-68 (box)
  ecotourism in, *187*
  medicinal products from forests of, 186
Cuba, 178
Czechoslovakia, air pollution and acid rain in, 218, *219, 223*

D

Denmark, 14
  population of, 72, 73
  toxic waste management in, 294 (box)
  water quality in, 148, 242

E

Earth, *2*
  diversity of life on, 2
  Goode's Homolosine Equal Area Projection map of, *346-47*
  greenhouse effecting climate of, 207-8 (box)
  land-use categories of, *161*
  solar energy affecting, 205, *206*
  as superorganism, 28 (box)
  world precipitation patterns, *229*
Ecuador, 121, 164
Egypt, 115
  Aswan Dam on Nile River in, 236-37
  foreign aid to, 142
  urbanization in, 303
Ethiopia, 14, 81, 105
  famine in, 135
  soil erosion in, 146
Europe, 15, 16
  acid precipitation effects on forests of, 218, *219*
  air pollution in eastern, 218-19, 222-23
  coal in, 251
  famine and plague of 14th century in, 17, 68
  forests of northern, 161
  nutrition in, 139
  species diversity and extinction in, 184, 193, 194

thalidomide tragedy in, 117, *118*
water quality in, 242

F

Finland, 14
    city planning in, 312 (box)
France
    nuclear power in, 256, 261
    water pollution in, 148

G

Gaza, 72, 73, 80
Georgia (nation), longevity in, 121
Georgia (state), Okefenokee Swamp in, *43*
Germany, 218
    air pollution in, 223
    Blue Angel symbols in, 326
    green politics in, *336*
    water quality in, 148, 242
Ghana, 164, 270
Gila National Forest, New Mexico, 175
Gobi Desert, 57
Great Britain
    nuclear power in, 256, 257, 261
    water quality in, 148, 242
Great Lakes, threat of introduced
    species into, 195-96 (box)
Greece, ancient cities of, *298*, 310-11
Greenland, 43
Guanacaste National Park, Costa Rica,
    167-68 (box)
Guatemala
    bird species in, 43
    cash crops in, 142
    overgrazing in, *171*
Guinea, 280
Guinea-Bissau, 14
Guyana, 305

H

Haiti, 81, 178, 276
    deforestation of, 164
    land tenure in, 178
    soil erosion in, 146-47
Hawaii, 6, 57
    endangered species in, *199*
Hong Kong, 16, 79, 261, 305

I

Iceland, 14, 190, 229
India, 14, 81, 114, 164
    agricultural fertilizer use in, 148
    foreign aid to, 142
    hunger and malnutrition in, 137, 139
    land tenure in, 178
    Narmada River dams, 105, 270
    population growth and family size
        in, 73-74 (box)

urban problems in, *303*, 304, 305,
    306
water resources in, *232*, 243
Indonesia, 16, 164, 177, 185, 193, 305,
    315
    agriculture in, *152*
    land tenure in, *179*
    water quality in, *243*
Iraq, 171
Isle Royale National Park, wolves and
    moose populations on, 67-68
    (box)
Israel, foreign aid to, 142

J

Japan, 14, 15, 190, 305
    city planning in, 313
    forests of, 163-64
    ''green'' business in, 107
    as importer of wood products, 162,
        164
    Minamata mercury poisoning in,
        241 (box)
    nuclear power in, 261-62
    nutrition in, 135
    photovoltaic technology in, 267
    population in, 79
    recycling in, 282
    water quality in, 241, 242
Jordan, foreign aid to, 142

K

Kenya
    ecotourism in, 187
    ivory trade and, *192*
Kuwait, 229, 252

L

Latin America, 203. *See also* names of
    countries in Latin America
    forests of, 161, 162, 164
    species diversity of, 184
    urban problems of, 303, 306
    water resources in, 229, 242
Lesotho, 305
Liberia, 164
Los Angeles, California, as urban area,
    309-10
Love Canal, New York, hazardous
    waste disaster of, 290-91 (box)
Luxembourg, 14

M

Madagascar, 6, 164, 168
    medicinal plants from, *186*
Malawi, 14, 72
Malaysia, 16
    forests of, 164, 165
    species diversity in, 184

water resources in, 243
Mali, 14, 72
Mauritania, 306
Mauritius, species extinction in, *190*
Mexico, 168, 178, 315
    deserts of, 57
    population demographics in, 74, *75*,
        76
    Sian Ka'an Reserve in, 201, 202
Mexico City, 278
    air pollution in, 222
    population of, *64*, 300-301, *302*
    urban problems of, *306*, *315*
Middle East. *See also* names of
    countries in Middle East
    oil reserves in, 252-53
    population growth in, 72
    water resources in, 232
Mt. Pinatubo, Philippines, 207, 211
Mozambique, 14, 81
    famine in, 135

N

Nashua River, New England, cleanup
    of, 329-30 (box)
Nepal, ecotourism in, 187
Netherlands, water pollution in, 148
New Zealand, 6, 14, 15, 170
Nicaragua, 178, 339
Niger, 72
Nigeria, 280
Nonsuch Island, Bermuda, 321 (box)
North America, 15. *See also* names of
    countries in North America
    coal in, 251
    forests of, 161, 162
    Great Plains, 58, *59*
    nutrition in, 139
    species diversity and extinction in,
        6, 184, 189
Norway, 14

O

Ogallala Aquifer, U.S., 233 (box)
Oman, 234

P

Pakistan, 121, 171, 315
Papua New Guinea, 46, 185
Peru, 168, 178, 303, 315
Philippines, 193
    agriculture in, 152
    Mt. Pinatubo, 207, 211
    solid waste in, *278*
    water quality in, 244
Poland, 218, *219*, 223, 242
Polynesia, 6

R

Romania, 223
Russia, 161, 242
Rwanda, 72, 176

S

Sahara Desert, 57
Saudi Arabia, 252
Senegal, 176
Sierra Leone, 164
Singapore, 16, 79
Somalia, 14, 135, 171
South Korea, 16, 114, 263, 305
    forests of, 163
    land tenure in, 179
Soviet Union, 178
    air polluton in, 222-23
    Aral Sea destruction in, 235-36
        (box)
    Chernobyl nuclear disaster in, 54,
        258-59 (box)
    coal in, 251
    forest products from, 162
    natural gas reserves in, 253
    nuclear reactor design in, 256, 262
    studies on health effects of
        electromagnetic fields in, 119
Sri Lanka, 114, 315
Sudan, 135, 171
Sumatra, 46
Sweden, 14, 242
    city planning in, 312 (box)
    population demography in, 75, 76
Switzerland, 14, 176, 218
Syria, 80

T

Taiwan, 16, 79, 114, 179
Tanzania, 14, 168, 176, 186, 315
Thailand, 16, 164, 299

U

Uganda, 72
Ukraine, Chernobyl nuclear disaster in,
    54, 258-59 (box)
United States, 14
    acid precipitation over, 218

agriculture in, 134, 140, 150, 151
    (box)
air pollution in, 205, 209, 222, 287
alien species in Great Lakes of,
    195-96 (box)
biomasss energy in, 268
bird species in, 43
birth rates in, between 1910 and
    1992, 79
Buffalo Commons in, 172-73 (box)
cancer in, 118
cleanup of Nashua River in, 329-30
    (box)
consumption and waste by, 10, 11
    (table)
death rate by accidents in, 120
deserts of, 57, 59
drought in, 230
energy conservation in, 262, 263
energy use in, 249, 250
executive branch of government,
    339, 340
forest products from, 162
forests of, 161, 168-70, 218
fossil fuel resources in, 253, 254
Great Plains of, 58, 59
Green Seal program in, 326
hazardous waste in, 288, 289,
    290-91 (box)
hunger in, 135, 136
judicial branch of government,
    337-39
laws protecting biological resources
    in, 197-98
laws protecting water resources in,
    245
legislative branch of government,
    336, 337
life expectancy in, by race and
    gender, 75
Los Angeles, California in, as urban
    area, 309-10
materials recovery from solid waste
    in, 282, 283
migration to, 76
mining in, 251-52
national parks, forests, and
    wilderness in, 174-76, 199, 200
nuclear power in, 255, 256

oil reserves in, 253, 254
pollution charges in, 101-2 (box)
population demographics in, 72, 73,
    74, 75, 79
potential areas of radon pollution
    in, 214
public policy on environmental
    health in, 126, 128-30
radioactive waste in, 260
rangelands in, 171, 173-74
response of, to Earth Summit,
    13-14 (box)
soil erosion in, 145-46
solid waste in, 276-77, 279
solid waste incineration in, 280-81
species diversity and extinction in,
    43, 190-99
studies on health effects of
    electromagnetic fields in, 119-20
suburb redesign in, 313-15
sustainable agriculture in, 151 (box)
timber industry in, 162, 163
urban crime in, 304 (table)
urbanization in, 299, 300
urban problems of, 306-9
water pollution in, 148, 154, 155,
    238, 240, 242
water resources of, 230, 233 (box),
    234, 236
water transfer projects in, 235
water use in, 231
wetlands in, 43
wind power in, 270, 271

V

Vietnam, 164

Y

Yellowstone National Park,
        biogeographical area of, 199, 200
Yosemite National Park, 174, 175

Z

Zambia, 80, 168, 171, 303, 315
Zimbabwe, 176, 280